Modern Concepts of Human Genetics

Modern Concepts of Human Genetics

Edited by Lesley Easton

MURPHY & MOORE
www.murphy-moorepublishing.com

Published by Murphy & Moore Publishing,
1 Rockefeller Plaza,
New York City, NY 10020, USA
www.murphy-moorepublishing.com

Modern Concepts of Human Genetics
Edited by Lesley Easton

International Standard Book Number: 978-1-63987-368-5 (Hardback)

Cataloging-in-Publication Data

Modern concepts of human genetics / edited by Lesley Easton.
p. cm.
Includes bibliographical references and index.
ISBN 978-1-63987-368-5
1. Human genetics. 2. Genetics. 3. Human biology. I. Easton, Lesley.
QH431 .M63 2022
599.935--dc23

TABLE OF CONTENTS

PREFACE

Genes refer to the common factors of qualities of human-inherited traits. The human genome is the complete set of nucleic acid sequences in humans. The study of the inheritance of characteristics in humans is termed as human genetics. This field aims to identify diseases and develop effective treatment for them. It overlaps with a wide range of fields such as developmental genetics, genomics, clinical and classical genetics, biochemical genetics, population genetics, molecular genetics, genomics, etc. The aim of this book is to present researches that have transformed this discipline and aided its advancement. From theories to research to practical applications, case studies related to all contemporary topics of relevance to this field have been included herein. As this field is emerging at a rapid pace, the contents of this book will help the readers understand the modern concepts and applications of the subject.

This book unites the global concepts and researches in an organized manner for a comprehensive understanding of the subject. It is a ripe text for all researchers, students, scientists or anyone else who is interested in acquiring a better knowledge of this dynamic field.

I extend my sincere thanks to the contributors for such eloquent research chapters. Finally, I thank my family for being a source of support and help.

Editor

1

Coupled electrophysiological recording and single cell transcriptome analyses revealed molecular mechanisms underlying neuronal maturation

Xiaoying Chen[1], Kunshan Zhang[1], Liqiang Zhou[1], Xinpei Gao[1], Junbang Wang[1], Yinan Yao[1], Fei He[1], Yuping Luo[1], Yongchun Yu[3,4], Siguang Li[1,4], Liming Cheng[1], Yi E. Sun[1,2,4]✉

[1] Stem Cell Translational Research Center, Tongji Hospital, Tongji University School of Medicine, Shanghai 200065, China
[2] Department of Psychiatry and Biobehavioral Sciences, David Geffen School of Medicine, University of California, Los Angeles, CA 90095, USA
[3] Institute of Neurobiology, Institute of Brain Science and State Key Laboratory of Medical Neurobiology, Fudan University, Shanghai 200032, China
[4] Collaborative Innovation Center for Brain Science, Tongji University, Shanghai 200092, China
✉ Correspondence: yi.eve.sun@gmail.com (Y. E. Sun)

ABSTRACT

The mammalian brain is heterogeneous, containing billions of neurons and trillions of synapses forming various neural circuitries, through which sense, movement, thought, and emotion arise. The cellular heterogeneity of the brain has made it difficult to study the molecular logic of neural circuitry wiring, pruning, activation, and plasticity, until recently, transcriptome analyses with single cell resolution makes decoding of gene regulatory networks underlying aforementioned circuitry properties possible. Here we report success in performing both electrophysiological and whole-genome transcriptome analyses on single human neurons in culture. Using Weighted Gene Coexpression Network Analyses (WGCNA), we identified gene clusters highly correlated with neuronal maturation judged by electrophysiological characteristics. A tight link between neuronal maturation and genes involved in ubiquitination and mitochondrial function was revealed. Moreover, we identified a list of candidate genes, which could potentially serve as biomarkers for neuronal maturation. Coupled electrophysiological recording and single cell transcriptome analysis will serve as powerful tools in the future to unveil molecular logics for neural circuitry functions.

KEYWORDS Patch-Seq, hESC/hiPSC-derived neuron, WGCNA, Biomarkers for neuronal maturation, Ubiquitination and mitochondrial function

INTRODUCTION

The mammalian brain is a fascinating organ, which breeds wisdom, empathy, and creativity, i.e., features unique to the human mind. The brain is also particularly complex, composed of billions of neurons and trillions of glia including both astrocytes and oligodendrocytes. Moreover, billions of neurons form trillions of synapses, which are critical points of communications between and amongst neurons. Through synaptic activities, neurons are integrated and organized in networks and circuitries, where senses, movement, thought, and emotions are generated. Interestingly, it is believed that when an individual is conducting a specific task, only a few neurons or specific neural circuitries are engaged in the activity, which means analysis using brain tissues, as a whole will not provide sufficient resolution for dissection of the molecular gene network underlying the particular brain function. While it is known that cells could be pretty much defined by the genes that they express, and that transcriptome analyses have been shown to be useful to delineate molecular features associated with cell type, cell age, and cell physiological or pathological states, transcriptome analyses of brain tissues had not been particularly useful due to the huge cellular heterogeneity of brain tissues and that cells of interest are often rare populations, such as neurons in particular circuitries involved in specific behaviors. Although there are some bioinformatics tools including WGCNA, which can be used to sort out cell-type specific gene expression signatures from transcriptome of the brain tissue (Miller et al., 2014; Kang et al., 2011; Mirnics 2008), a physical limitation is that transcription signals from rare cells in the tissue will simply not be captured when the whole tissue is subjected to RNA sequencing.

To get around these limitations, single neuron transcriptome analyses were developed, and many papers published on brain cell typing (Junker and van Oudenaarden 2014; Johnson et al., 2015; Pollen et al., 2015). However, most studies using cell sorting including the Fluidigm C1 apparatus completely lost the anatomical positional information of cells by taking them out of their natural habitat. Moreover, the dissociation process is a clear stress, if not devastating to the cells, particularly neurons, which could change the transcription program quite dramatically. It is therefore advantageous to perform transcriptome analyses on neurons that are in electrophysiologically functional state, so that molecular genetic programs underlying neuronal physiological properties could be revealed. In this study, we performed electrophysiological recordings on cultured human neurons differentiated from various sources including human ES cells, iPS cells, and embryonic neural stem cells. After recording, the same neurons were subsequently subjected to RNA sequencing. From single neuron transcriptome analyses, we identified gene programs associated with neuronal maturation judged by electrophysiological parameters. Moreover, by cross-referencing transcriptome data obtained from single human neurons in embryonic and adult human brains (Darmanis et al., 2015), we isolated 39 genes, which could be potentially used as neuronal maturation biomarkers. The methodology established here could also be used to couple single neuron electrophysiological and transcriptome analyses after *in vivo* or brain slices recording to reveal the molecular logic of neural circuitry activities.

RESULTS

In vitro differentiation and maturation of human neurons derived from hESCs and hiPSCs

Ever since Thomson first established human embryonic stem (ES) cell cultures and Yamanaka developed human induced pluripotent stem cell (hiPSC) systems, human neurons could be readily obtained from *in vitro* differentiation and maturation (Wu et al., 2007; Zhang et al., 2013; Hu et al., 2010). Subsequently, human cell-based "disease-in-dish" models became popular approaches for attempting to study human neurological diseases (Mariani et al., 2015; Li et al., 2013; Ma et al., 2012). The step-wise neuronal differentiation protocols with all kinds of variations have been utilized by many laboratories to generate human neurons *in vitro* with high enrichment (Fig. 1A–C). Moreover, these neurons do mature in culture and form synaptic networks, which could be judged anatomically by presynaptic synapsin immunostaining puncta on postsynaptic MAP2-positive dendrites (Fig. 1D), or functionally by the presence of spontaneous excitatory and inhibitory postsynaptic currents (sEPSCs and sIPSCs) (Fig. 1I). As expected, we frequently detected vGlut1 positive glutamatergic excitatory neurons, GABA positive GABAergic inhibitory neurons, and TH positive catecholaminergic neurons (Fig. 1D). Using fluorescent dye injection, neuronal morphology could be precisely revealed and quantifiably measured (Fig. 1E and 1F). Some of these human neurons fire action potentials upon depolarization (Fig. 1G and 1H). However, a great degree of heterogeneity are clearly present in these neuronal cultures, regarding neurotransmitter-based neuronal subtypes, neuronal morphologies, and electrophysiological properties such as action potential firing frequencies, amplitude, and etc.

Coupling of electrophysiological recording and transcriptome (Patch-seq) analyses on the same neurons

In order to delineate the molecular signatures underlying the heterogeneity of the electrophysiological properties of cultured human neurons mentioned above, we carried out "patch-seq" using our own proprietary method (Fig. 2). We measured 9 electrophysiological parameters, among which 6 were related to action potentials, i.e., firing rate, amplitude, halfwidth, threshold, Rin, and rise time (Fig. 2A). We also measured sodium current amplitude, as well as frequencies of sEPSCs and sIPSCs, which are indicative of neural network activities. After recording, neurons were individually extracted by the patch pipet (recording electrode) (Fig. 2B), and subjected to single neuron transcriptome analyses. We do frequently perform technical replications for sequencing to make sure that the sequencing quality is high with low noises, which could be judged by high Pearson Correlation Coefficient (~0.99) of log-transformed whole transcriptome between the replica (Fig. 2C). In this study we sequenced 20 single human neurons with various electrophysiological properties. When single neuronal transcriptome were compared with transcriptome of 21 human peripheral blood samples, we found that these two types of samples are clearly distinguished from each other by dimension 1 of the two-dimensional principle component analyses (PCA) (Fig. 2D). Expression of a list of well-acknowledged neuronal markers could clearly segregate blood samples from single neuron samples (Fig. 2E). Using WGCNA, we identified a gene module (blue, containing 4255 genes) that appeared to be neuronal cell specific (Fig. 2F). Gene Ontology (GO)

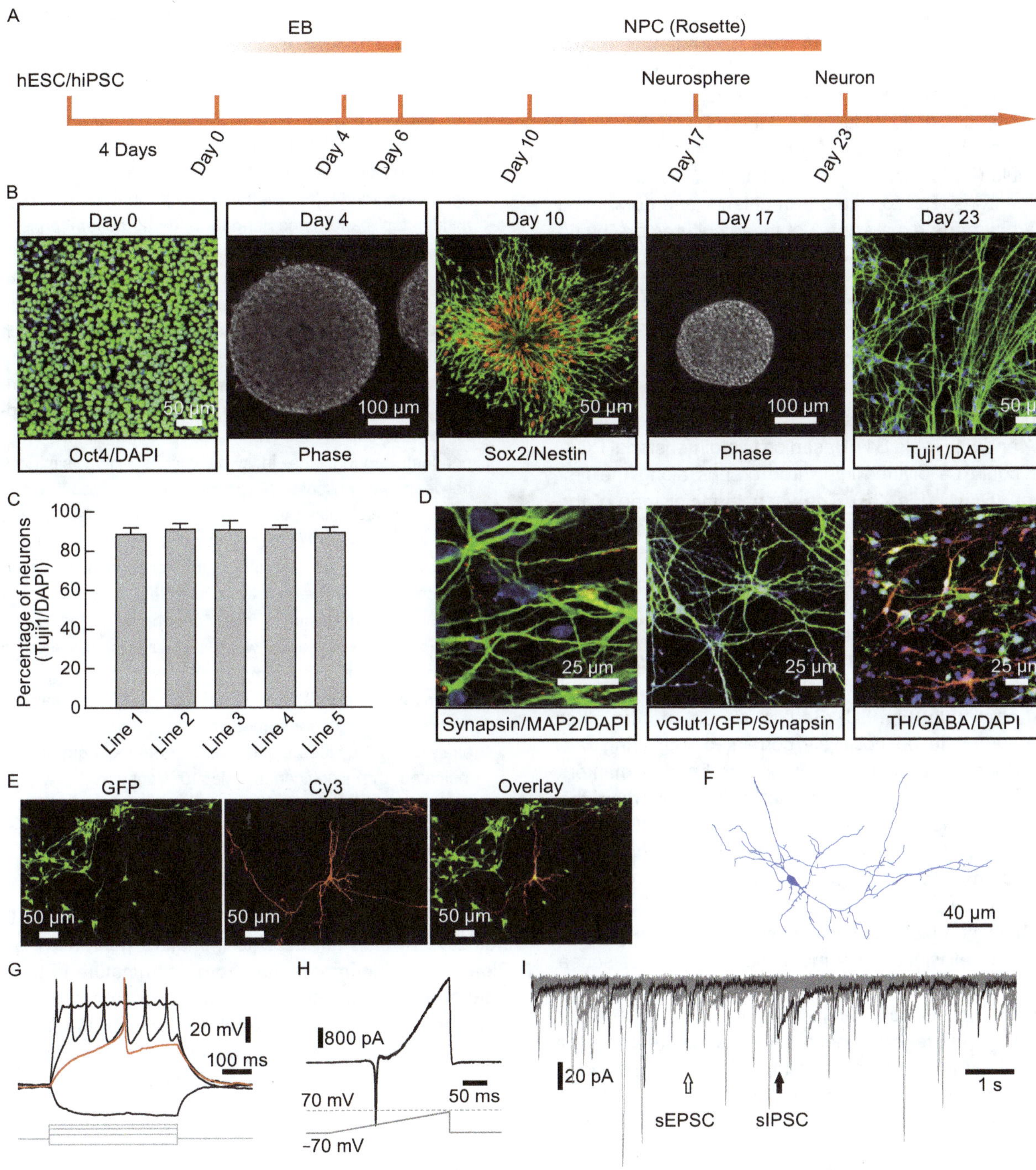

Figure 1. Generation of functional neuron via differentiation of hESC/hiPSC/fetal tissue-derived NSCs. (A) Diagram of the neuronal differentiation protocol (see experimental procedures for details). (B) Representative images showing morphological changes during neural differentiation. hESC/hiPSC are positive for pluripotent markers Oct4. Neural stem/progenitor cell (NSC/NPC) colonies, showing rosette structures, are positive for neural precursor markers Sox2 and Nestin, whereas neurons are positive for Tuj1 (Bar: 100 or 50 μm). (C) Percentage of Tuj1 positive cells showing similar pan neuronal differentiation potentials of different hESC/hiPSC lines. (D) Representative images of mature neurons, co-cultured with mouse astrocytes, showing expression of MAP2, Synapsin, vGlut1, TH, and GABA (Bar: 25 μm). (E) Representative images of a recorded neuron injected with neurobiotin and immunostained (Cy3), surrounded by other GFP expressing lentivirally infected neurons. Bar represents 50 μm. (F) A representative reconstructed image of neurons. Scale bar represents 40 μm. (G) Representative action potential responses of a hiPSC-derived neuron evoked by current injection (Bars: 100 ms and 20 mV). (H) Representative traces from a hiPSC-derived neuron showing inward sodium currents and outward potassium currents in response to a ramp protocol (from −70 mV to +70 mV). Three super imposed traces are shown (Bars: 50 ms and 800 pA). (I) Representative traces from a hiPSC-derived neuron showing spontaneous postsynaptic currents (sPSCs). Ten traces are superimposed. Black arrow, inhibitory sPSC; white arrow, excitatory sPSC (Bars: 1 s and 20 pA).

analysis demonstrated that major GO terms associated with the blue module were indeed related to neurons including "synapse maturation", "neuronal projection extension", "dendrite morphogenesis", "synaptic vesicle endocytosis", and "telencephalon development" (Fig. 2G). Moreover, we also identified the hub-gene network of this neuronal-specific blue module (Fig. 2H). As expected, these hESC and iPSC-derived neurons tend to take on an anterior and dorsal trait as judged by gene expression, yet these cultures are by no means homogenous.

Categorizing immature, maturing, and matured neurons based on electrophysiological properties

The 9 parameters of electrophysiological properties of 20 human neurons were subjected to non-hierarchical clustering (Figs. 3A and S1, Table S1). Based on two-dimensional PCA, neurons could be divided into three groups along the first dimension (Dim1) (Fig. 3B). The variable factor map of the PCA is presented in Fig. 3C, from which it was clear that Dim1 mostly reflected 6 parameters describing different perspectives of action potentials. For example, Firing rate and Amplitude are negatively correlated with Dim1. On the contrary, Halfwidth, Rin, Rise Time, and Threshold are positively correlated with Dim1. This is expected because these 4 parameters are known to be negatively correlated with firing rate and amplitude (Fig. 3C). Sodium current, on the other hand, appeared to be positively correlated with firing rate and amplitude. Since firing rate is one of the best markers for neuronal electrophysiological maturation properties, the –0.911 anti-correlation value between Dim1 and firing rate suggested that either Dim1 or firing rate could be used to bin the 20 neurons into three groups, i.e., the immature group, the maturing group, and the matured group (Fig. 3D and 3E). It is worth noting that frequencies of sEPSCs and sIPSCs do not correlate well with Dim1 or maturation. This makes sense as sEPSCs and sIPSCs only reflect neural network activities imposed on the cell recorded, which does not necessarily mean whether the recorded neuron itself is mature or not, or whether that neuron could transmit the electric signal down to the axon or not (Fig. 3C). Given that Dim1 took all parameters into account, it should be more reliable and less prone to fluctuations as compared to any one parameter including the firing rate. Moreover, since Dim1 correlated with the firing rate very well, it is more sensible to use the Dim1 value to quantify neuronal maturation in order to dissect the underlying gene expression signatures.

Gene programs positively and negatively correlated with neuronal maturation

From transcriptome of 20 single neuronal samples, we identified gene clusters that were negatively correlated with Dim1 of the electrophysiological parameters, which meant that these genes were enriched in matured neurons. We also identified genes that were positively correlated with Dim1, indicative of their affiliation to immature neurons (Fig. 4A). Clearly when neurons transit from an immature to a mature state, they will first shut off signature genes associated with immature neurons (the green cluster, 94 genes) and gradually turn on signature genes affiliated with mature neurons (the magenta cluster, 350 genes) (Fig. 4A). The immature green gene cluster contained genes involved in "DNA conformation changes", "gene expression", and "axon guidance", which were activities associated with early neuronal differentiation. Genes including Sox9, Dlx6, MTA1, LAMA2, ROCK1, and ROCK2 are highly expressed in immature neurons. On the other hand, GO analyses of the magenta cluster revealed genes linked to "neuron development" such as CNTN1, FAIM2, which were also hub-genes in the magenta cluster. Moreover, ATP1B1, a gene encoding a Na^+/K^+ APTase involved in setting up Na^+ and K^+ electrochemical gradient, is enriched in the magenta cluster. DKK3, a head inducer and also a Wnt inhibitor, as well as a pro-apoptotic BEX2 gene are enriched in the magenta cluster, associated with neuronal maturation. Interestingly, genes involved in mitochondrial electron transport such as

Figure 2. Single cell transcriptome analyses reveal neuronal property of collected target cells. (A) A representative patch-clamp recording of a H9 hESC-derived neuron. This neuron was filled with Alexa Fluor 568 hydrazide during whole-cell patch clamping to reveal the morphology (middle and right in top panel). Spontaneous postsynaptic currents were recorded through voltage-clamp mode (middle top panel). Short-term (0.5 s, middle top panel) and long-term (8 s, bottom panels) recordings of action potentials were performed with step-wise current injection through current-clamp mode. (B) Micromanipulator-assisted manual collection of single neurons showing moments of target cell entering the patch pipette, before (left), during (middle), and after (right) the entry. All soma and majority of dendritic parts were collected. (C) Technical replica of two single-cell transcriptome (Cell_2 and Cell_12) were subjected to two independent sequencings in two batches. High similarity between two replicates (Pearson correlation coefficient = 0.99) indicates minimum sequencing batch effect, hence the robustness and accuracy of our sequencing technology. (D) PCA of single-cell transcriptomes including neuron and peripheral blood samples suggested that neurons were distinct from peripheral blood cells. (E) Pan-neuronal genes were exclusively expressed in neurons but not blood samples. (F) Gene clustering of all 20 single cells and 21 blood samples (a1–a21) revealed a neuronal specific gene module (blue module). (G) GO analyses of the blue module (4255 genes). Length of bars indicates the significance (−log10 transferred *P*-value, Fisher's exact test). Genes shown in right are well-known genes with corresponding functions. (H) Hub-gene network of the blue module. ▶

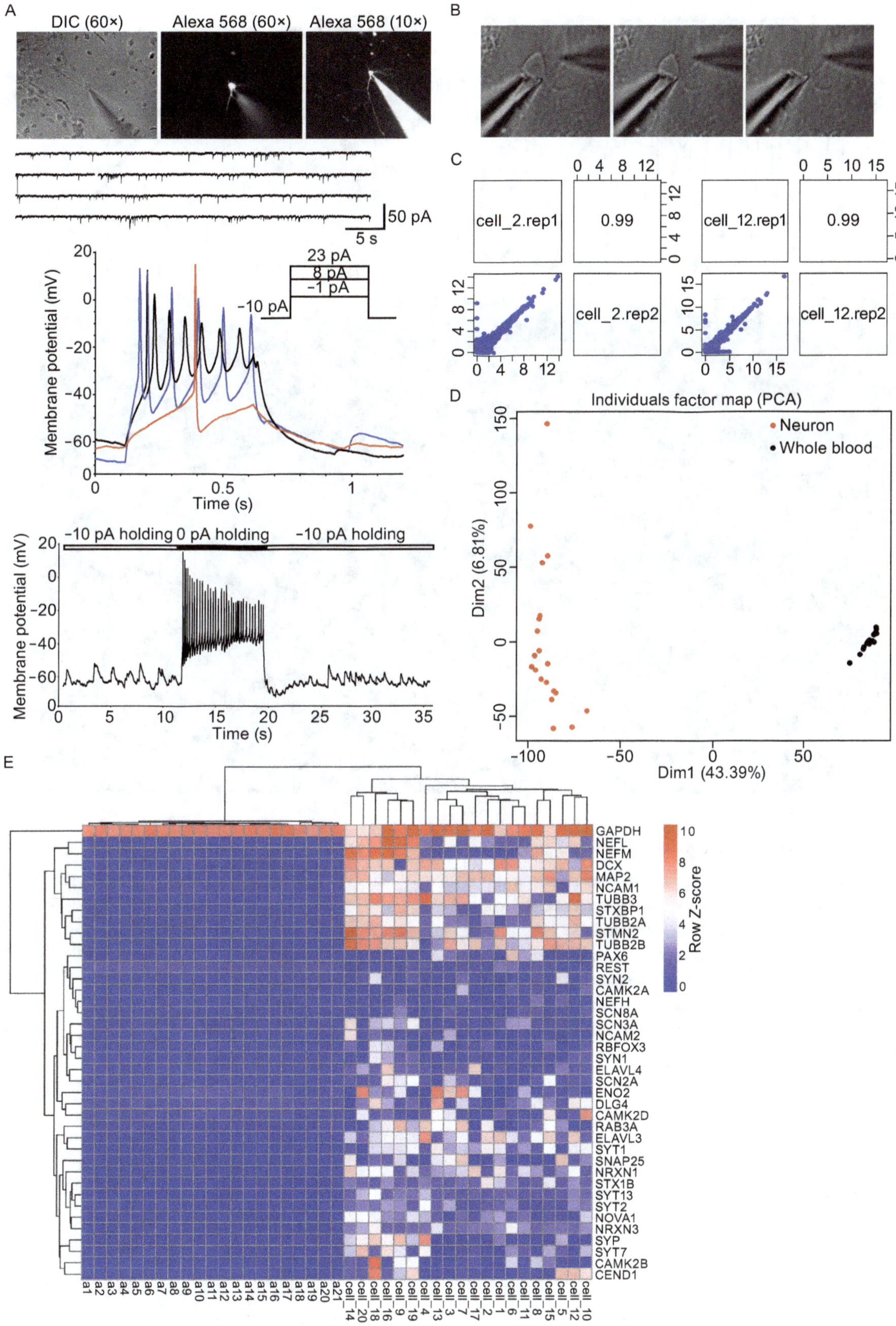
A
DIC (60×)
Alexa 568 (60×)
Alexa 568 (10×)
50 pA
5 s
23 pA
8 pA
−1 pA
−10 pA
Membrane potential (mV)
Time (s)
−10 pA holding
0 pA holding
−10 pA holding
B
C
cell_2.rep1
0.99
cell_2.rep2
cell_12.rep1
0.99
cell_12.rep2
D
Individuals factor map (PCA)
Neuron
Whole blood
Dim2 (6.81%)
Dim1 (43.39%)
E
GAPDH
NEFL
NEFM
DCX
MAP2
NCAM1
TUBB3
STXBP1
TUBB2A
STMN2
TUBB2B
PAX6
REST
SYN2
CAMK2A
NEFH
SCN8A
SCN3A
NCAM2
RBFOX3
SYN1
ELAVL4
SCN2A
ENO2
DLG4
CAMK2D
RAB3A
ELAVL3
SYT1
SNAP25
NRXN1
STX1B
SYT13
SYT2
NOVA1
NRXN3
SYP
SYT7
CAMK2B
CEND1
Row Z-score

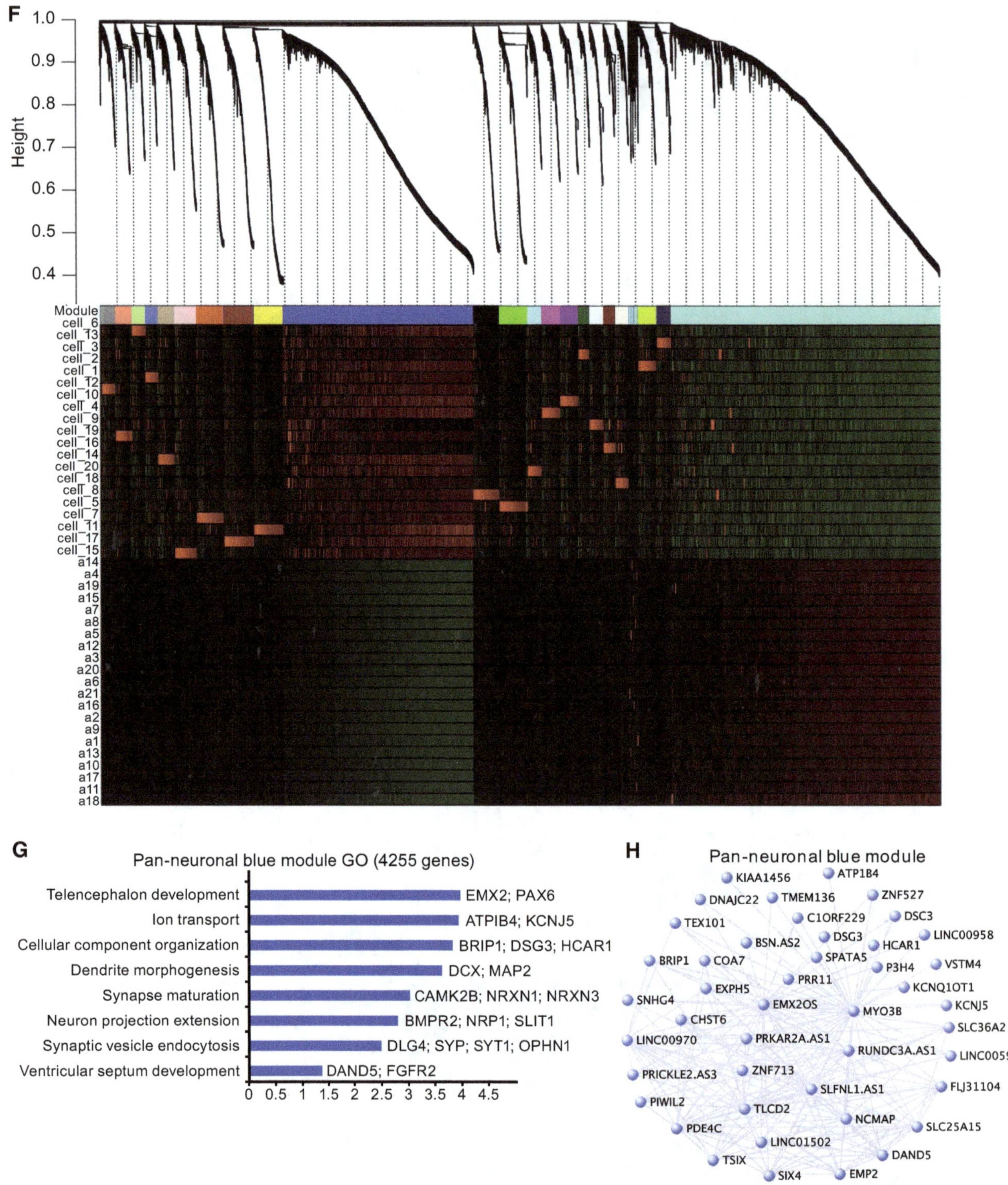

Figure 2. continued.

NDUFA8, NDUFC2, and UQCRB are also enriched in the magenta cluster, suggesting that alterations in mitochondrial electron transport machinery participate in neuronal maturation. As NDUFC2 and UQCRB appeared to be neuronal specific, this could underlie the fact that mature neurons consume large amount of energy, which was supplied by the extraordinary mitochondrial electron transport machinery that neurons specifically poses (Fig. 5).

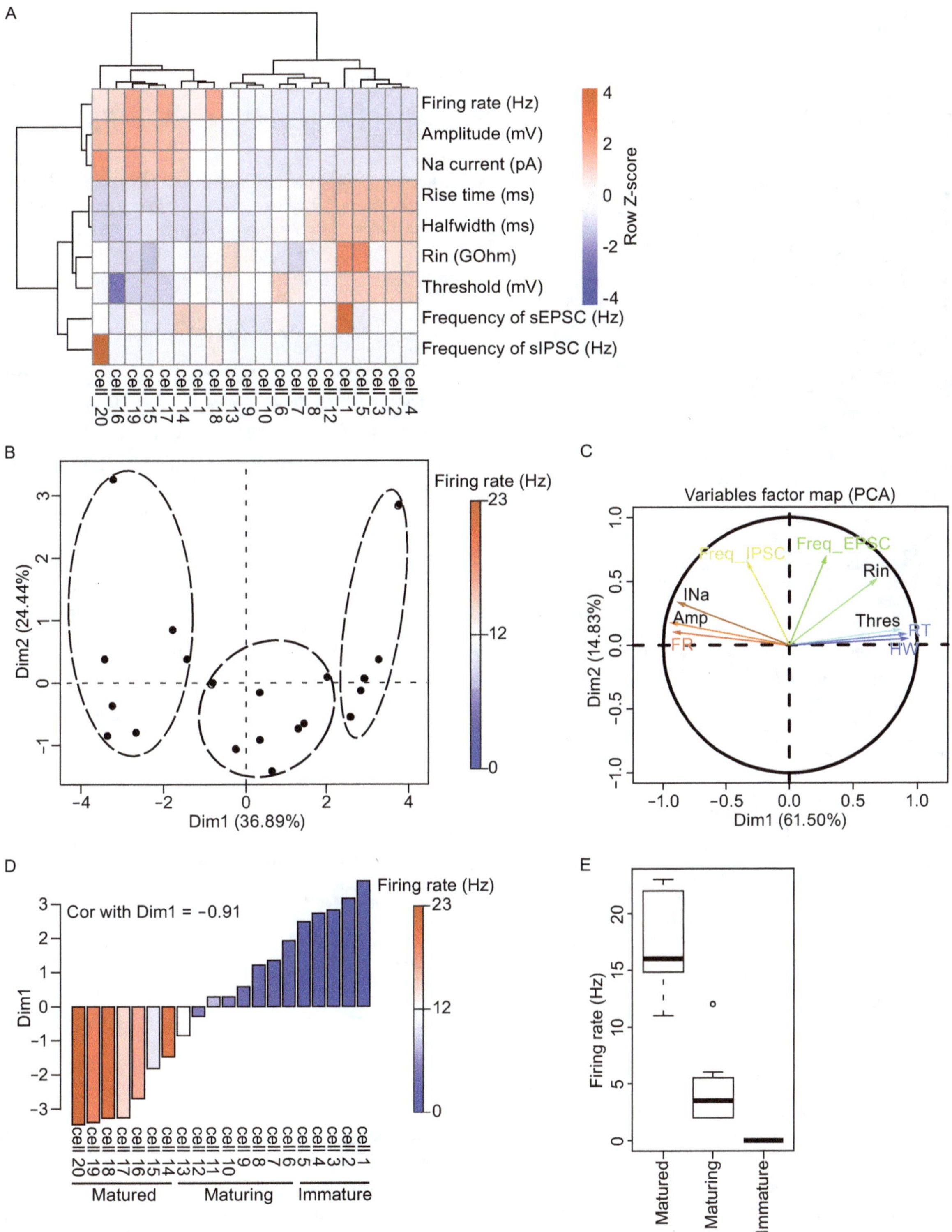

Figure 3. Electrophysiological phenotypes of neurons revealed three groups of neurons based on their maturity. (A) AP and sPSC electrophysiological parameters indicated different maturation stages of 20 single neurons. (B) PCA revealed dimension 1 as an indication of maturity and split 20 neurons into three groups. Colors of each circle correspond to different firing rate of the neurons, which is a highly regarded parameter for neuronal maturity. (C) Variable factor map revealed the correlation vector of each electrophysiological parameter with dimension 1 and dimension 2. Arrow tip denotes correlation coefficient of the respective parameter with each PC. Abbreviation, Thres. (Threshold); Freq_EPSC. (Frequency of sEPSC); RT. (Rise time); HW. (Halfwidth); Freq_IPSC. (Frequency of sIPSC); INa. (Na current); Amp. (Amplitude); FR. (Firing rate). (D) High absolute value of Pearson correlation coefficient (−0.91) between Dim1 and firing rate, indicating Dim1 is highly anti-correlated with neuron maturation. (E) Firing rate of three groups of neuron.

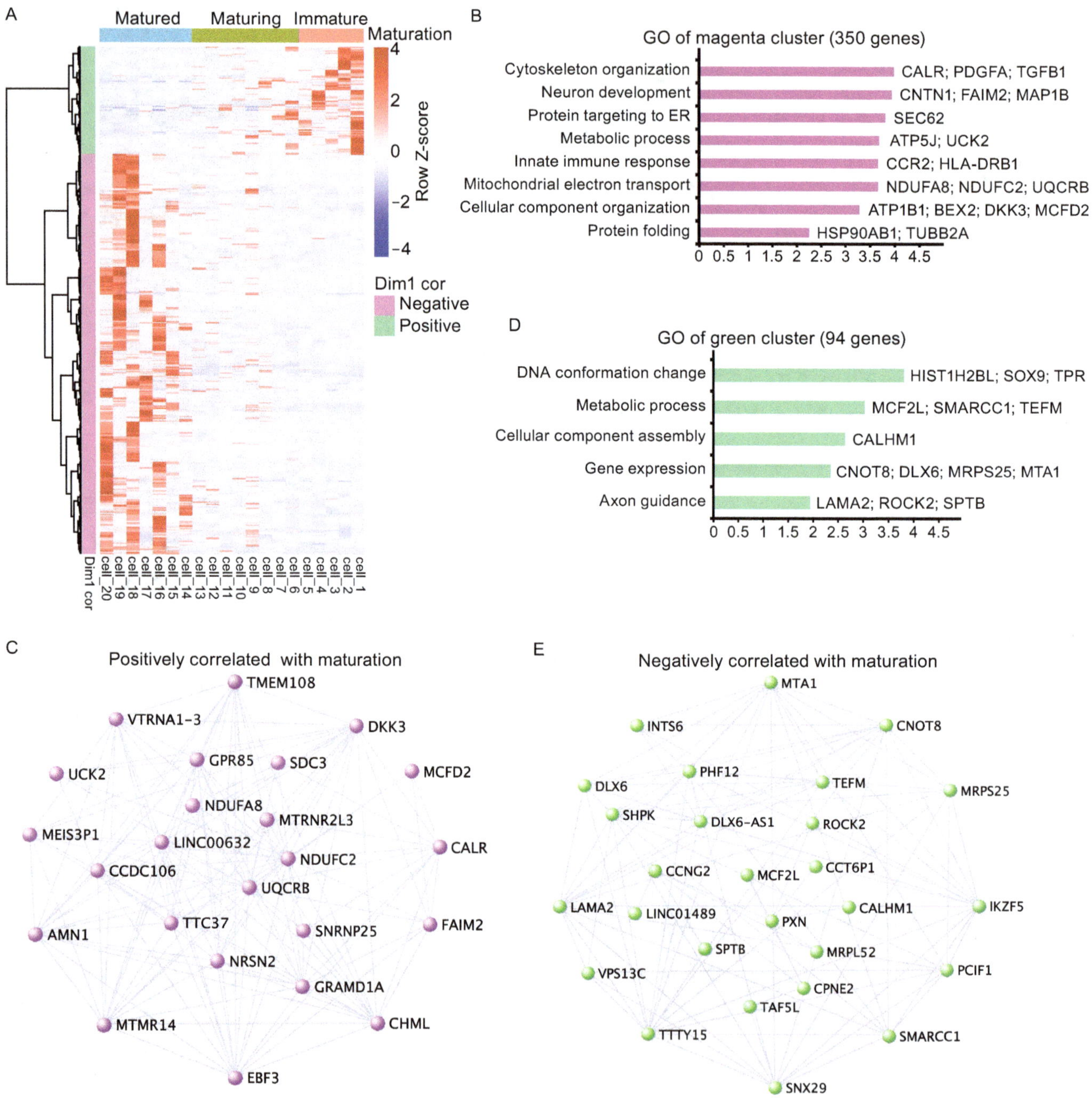

Figure 4. Single cell transcriptome analysis revealed maturation-related genes. (A) Expression of genes positively (green cluster) and negatively (magenta cluster) correlated to dimension 1 of electrophysiological parameters (Correlation > 0.4 and $P < 0.05$, Student correlation *P*-values) in 20 single neurons. (B) GO analyses of genes negatively correlated to dimension 1 (magenta cluster, 350 genes), enriched in mature neurons. Length of bars indicated the significance (−log10 transferred *P*-value, Fisher exact test). (C) Hub-gene network of the magenta gene cluster. (D) GO analyses of genes positively correlated with dimension 1 (green cluster, 94 genes), enriched in immature neurons. Length of bars indicated the significance (−log10 transferred *P*-value, Fisher exact test). (E) Hub-gene network of the green gene cluster.

Identification of candidate genes that potentially mark the maturity of neurons

Through WGCNA, we have already identified a neuronal specific blue module (Fig. 2F–H). We overlapped genes in the magenta cluster and those in the blue module and found 138 neuronal specific genes that were affiliated with neuronal maturation. GO terms of this overlapping cluster are similar to that of the magenta cluster, such as "Neuron development" (CNTN1, FAIM2), "Cellular localization" (PDGFA, SEC62), "Action potential" (ATP1B1), and "Mitochondrial electron transport" (NDUFC2, UQCRB).

Since the 20 neurons we analyzed were all from cultured human neurons differentiated *in vitro* from more primitive stem cells, we wonder whether the neuronal maturation genes (138) we identified could be used generally to mark neuronal maturity. To explore this, we compared recently published data from Ben Barres' laboratory by Darmanis et al. in *PNAS*, where human single neurons from fetal and adult human brains were compared. Darmanis reported 3 dimensional PCA of their data and revealed that dimension 1 could be used to demarcate mature (from adult brain) and immature (from fetal brain) neurons (Darmanis et al., 2015), but without electrophysiological recordings. When we evaluated expression of 1479 genes included in Dim1 by Darmanis et al. in our single neuron samples, we found a subset of the genes had enriched expression in our "matured"

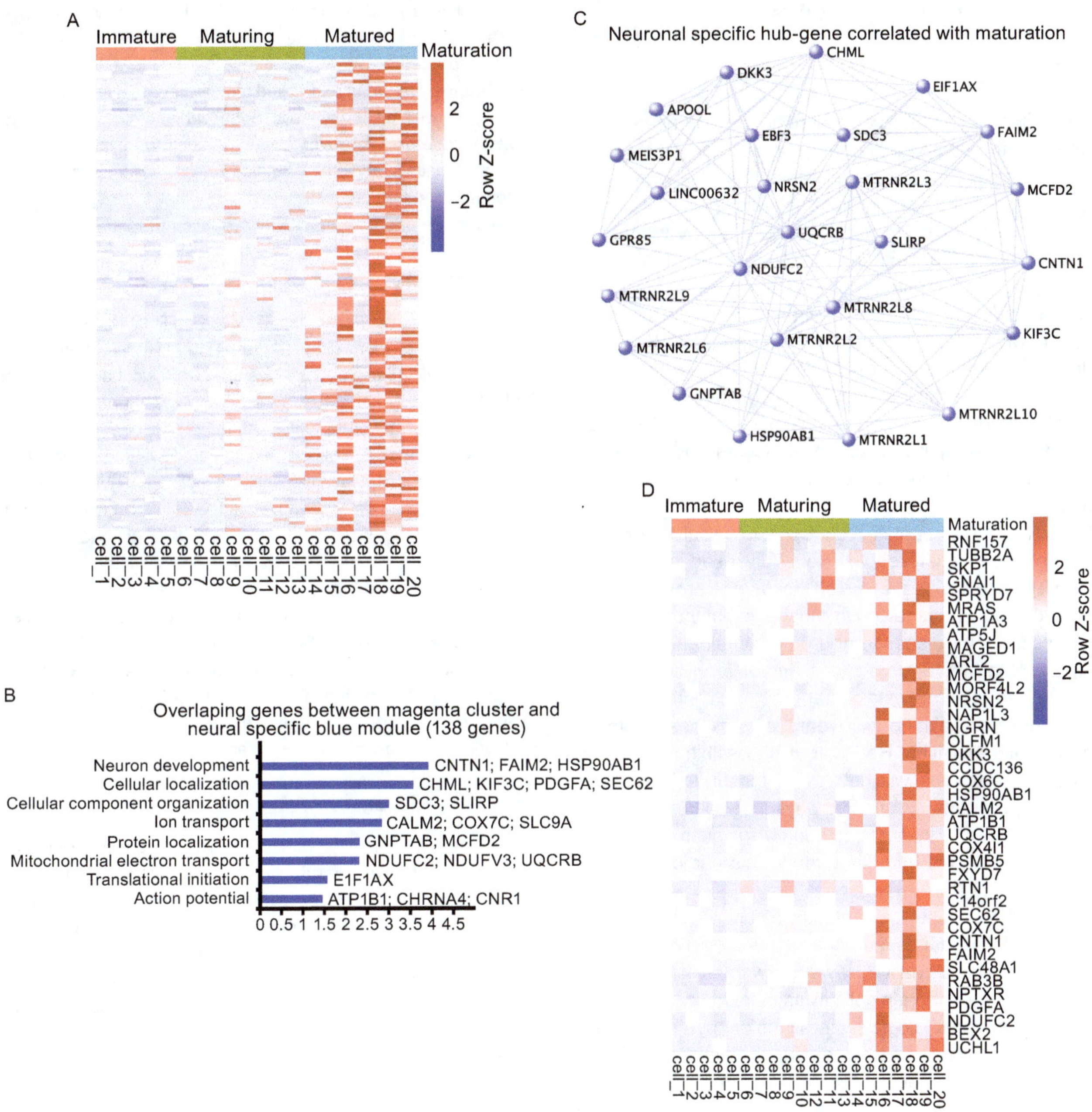

Figure 5. Identification of putative generic neuronal maturation biomarkers. (A) Expression of 138 genes in 20 single cells, which are shared between the magenta cluster and the blue module. (B) GO analyses of 138 genes. Length of bars indicated the significance (−log10 transferred *P*-value, Fisher exact test). (C) Hub-gene network of the 138 genes. (D) 39 putative generic neuronal maturation genes that were also reported in Darmanis et al., (2015).

neuronal samples (Fig. S2). By cross-referencing data from Darmanis et al. and the 138 maturation genes identified in our present work, we isolated 39 genes shared by the two studies, which could potentially serve as generic biomarkers for human neuronal maturation. Among these 39 genes, "mitochondria electron transport" (NDUFC2 and UQCRB) stood out. Moreover, UCHL1, BEX2, PDGFA, FAIM2, CNTN1, SEC62, ATP1B1, and CALM2 are of particular interest, as they not only reappeared in the different rounds of aforementioned analyses but also had relatively high hubness in the network of the gene cluster.

DISCUSSION

Coupled electrophysiological recording and transcriptome analyses on the same single neurons provided much more précised information on gene transcription programs of neurons with very specific electrophysiological functional properties. Such "patch-seq" technology effectively assisted identification of gene programs underlying neuronal maturation as judged by electrophysiological characteristics. It is comforting to discover that our cultured neurons derived from various human stem cell populations revealed a similar maturation program as that revealed by comparing human adult and fetal neurons in the brain, which indirectly reflected neuronal maturity. The 39 genes especially the 10 or so above highlighted genes may be particularly useful in the future to serve as generic biomarkers for evaluation of neuronal maturity.

We predict that coupled electrophysiological recordings and transcription analyses will be wildly used not only in human cell based "disease-in-dish" models to provide highly précised dissection of molecular mechanisms underlying the pathophysiology of diseased neurons with single cell resolution, but also in live brain slices and/or *in vivo* recording studies to pair electrophysiological properties with gene-expression analyses aiming to reveal the "molecular logic" of neural circuitry activities. However, it is possible that simple PCA-type of analyses will not reveal the underlying gene expression programs for specific physiological features one would like to study. In order to search for such programs, modified WCGNA methods as used in this particular study, will be extremely useful.

Along with the fast accumulation of brain mapping and connectome informations, anatomical zip-coding of the brain will soon become a reality. However, without knowing the molecular features of brain cells in each location, one would not easily detect physiological alterations or anything beyond anatomical changes in pathological conditions associated with the diseases, neither would one reveal molecular mechanisms underlying physiological changes related to learning, memory, all kinds of adaptive changes, as well as behavior, emotion, and thoughts. Coupling of brain mapping with single cell resolution transcriptome analyses will pave the way to our ultimate understanding of the brain and the mind.

MATERIALS AND METHODS

Neuron differentiation

hiPSCs and hESCs were differentiated into functional neurons following a modified version of published step-wise differentiation protocol(s). hESC- as well as hiPSC-derived neural stem/progenitor cell (NSC/NPC) colonies formed rosettes and expressed neural precursor markers Nestin and Sox2. Rapid conversion of hESC and hiPSC into functional induced neuronal (iN) cells could be achieved in less than two weeks by overexpression of exogenous Ngn2. The efficiencies of NPC differentiation into Tuj1-positive neurons were similar amongst all lines and were consistently close to 90% from culture to culture. Upon co-culture with mouse astrocytes, hESC/iPSC-derived neurons expressed mature neuronal markers, such as MAP2 as well as presynaptic Synapsin and vGlut1, with typical puncta patterns. Moreover, these neurons were able to produce action potentials, sodium and potassium currents upon stimulation. Spontaneous synaptic activity indicates that the neurons are relatively mature and form functional synaptic networks.

Immunofluorescence and microscope analysis

Cells on coverslips were washed twice with PBS (5 min each), fixed with 4% PFA at room temperature (RT) for 10 min, then rewashed twice with PBS (5 min each). Coverslips were incubated for 1 h at RT in PBS, 0.03% Triton X-100, 3% BSA and 2% normal donkey serum (NDS, Jackson laboratory, #017-000-121) for permeablization and blocking non-specific antibody binding. Following that, incubation of primary antibodies was often carried out overnight at 4°C in the same buffer. Cells were then washed (3 times, 10 min each) in the same buffer and incubated with secondary antibodies (Jackson Laboratory) for 1 h at RT. Cells were finally washed with PBS (3 times, 10 min each) and mounted on glass slide. Staining was viewed and analyzed with Olympus up-right fluorescent microscope (BX51) or with confocal microscope (Leica TCS SP5 II). The following antibodies were used: mouse anti-Oct4 (Millipore), goat anti-Sox2 (Santa Cruz), mouse anti-Nestin (Chemicon), mouse anti-MAP2 (Sigma), rabbit anti-Synapsin (Synaptic Systems), mouse anti-vGlut1 (Synaptic Systems), Rabbit anti-GABA (Sigma).

Morphological analyses

Recorded cells were filled with neurobiotin present in the intracellular solution via positive current injection (400 pA for 250 ms, 1.25 Hz) for at least 6 min. After recording, coverslips were fixed in 4% PFA 0.01 mol/L PBS for 10 min then rinsed with 0.01 mol/L PBS. Neurobiotin labeled cells were visualized via overnight incubation with 1:500 Cy3-conjugated streptavidin (Jackson Immuno Research Laboratories). Whole cell morphological reconstruction and analyses were carried out with Neurolucida software.

Whole-cell recording

The coverslip plated with cultured cells was bathed in an RNase-free recording chamber filled with artificial cerebrospinal fluid (ACSF) containing (in mmol/L) 126 NaCl, 3 KCl, 1.2 NaH_2PO_4, 1.3 $MgCl_2$, 2.4 $CaCl_2$, 26 $NaHCO_3$, and 10 glucose, bubbled with 95% O_2/5% CO_2. An upright infrared-differential interference contrast (IR-DIC) microscope (Olympus) equipped with epi-fluorescence illumination, an ORCA-R2 CCD camera, and two water immersion lenses (10× and 60×) were used to visualize and target recording electrode to cells. Glass recording electrodes (8–12 MΩ resistance) were filled with an intracellular solution consisting of (in mmol/L) 120 K-gluconate, 16 KCl, 2 $MgCl_2$, 0.2 EGTA, 10 HEPES, 2.5 MgATP, 0.5 Na_3GTP, and 10 Na-phosphocreatine (pH = 7.25 and 295 mOsm/kg). Recording was collected using MultiClamp 700B amplifier and pCLAMP10 software (Molecular Device), filtered at 2 KHz and sampled at 5 KHz. After establishing the whole-cell configuration, the spontaneous postsynaptic currents (sPSCs) were recorded for assessing the chemical connection between cells during voltage-clamp recording at −70 mV. To reveal intrinsic electrophysiological properties of the targeted cell, 500 ms current pulses (step at 2 pA) were injected into the cell to induce steady-state action potential (AP). Passive membrane properties were monitored periodically during the course of the experiments to ensure that the cell was healthy. Cells that showed significant rundown were discarded.

To catalog the intrinsic electrophysiological properties of cells, 6 parameters of AP were determined through Clampfit software (Molecular Device): (1) AP threshold determined from the membrane potential at the onset of AP; (2) rise time measured as the time between the threshold and the spike peak; (3) halfwidth determined as the time interval between its half-amplitude that reflects AP duration; (4) AP amplitude measured as the difference between the firing threshold to the peak; (5) firing rate determined as the maximal firing frequency of AP spikes; (6) input resistance (Rin) measured by dividing the initial averaged voltage deflection to the absolute value of first level of injected current pulse. The firing frequency and amplitude of sPSCs were analyzed using mini Analysis Program (Synaptosoft Inc.).

Single Cell Harvest, cDNA and library preparation for next generation sequencing

After recording, the wider tip of another pipette filled with no more than 0.5 μL sterilized cell harvest solution containing (in mmol/L) 144 K-gluconate, 3 $MgCl_2$, 0.5 EGTA and 10 HEPES (pH 7.20 and 295 mOsm/kg) was gently attached to the recorded cell, then a light suction (negative pressure) was applied, through a glass syringe (Thomas Scientific) connected to the pipette. The suction lasted until the entire cell entered the tip of the pipette. The moment the complete cell had been visually located inside the tip, the pipette was quickly removed from the bath. Content of the harvest pipette was expelled into a 0.2 mL PCR tube containing 4.5-μL pre-prepared lysis buffer. Preparation of single cell cDNAs was performed based on previously published protocols with modifications at several steps (Tang et al., 2009; Tang et al., 2010).

After the generation of cDNA from a single cell, 50 ng of cDNA was used for library preparation. Using the Covaris S2 System (Covaris), cDNA was sheared into about 125 bp short fragments according to the manufacturer's instructions. Library preparation was using NEBNext Ultra DNA library Prep Kit for Illumina (E7370). After DNA fragmentation, end repair, adaptor ligation and PCR amplification for 8 cycles, libraries were used for Illumina deep sequencing. Reads were aligning to hg19 with TopHat2 (Kim et al., 2013). Gene expression levels were estimated by Cufflinks (Ghosh and Chan 2016). RNA-Seq data was deposited at GSE 77564.

WGCNA

Weighted gene correlation network analysis (WGCNA) was performed as described before (Luo et al., 2015; Zhang et al., 2014). Briefly, a signed Pearson correlation network was constructed using any gene that was expressed at a FPKM value of 0.1 or higher in at least one of the samples. Soft power parameter was estimated and used to derive a pairwise distance matrix for selected genes using the topological overlap measure, and the dynamic hybrid cut method was used to detect clusters. The node centrality, defined as the sum of within-cluster connectivity measures, was used to rank genes for "hub-ness" within each cluster. For visual analysis of the constructed networks by hard thresholding of edge distances, usually the top 100 edges were represented using VisANT.

Principle component analysis

PCA is performed with R package FactoMineR. Not detected (ND) values were replaced by the maximum value of the parameter when the parameter is negatively correlated with maturation, or replaced by the minimum value of the parameter when the parameter is positively correlated with maturation.

Gene ontology term enrichment analysis

Fisher's exact test was used to score GO terms for significant enrichment.

Graphics

Unless otherwise specified, plots were generated in R.

ACKNOWLEDGEMENTS

This work was supported by the National Basic Research Program (973 Program) (No. 2012CB966303), the National Natural Science Foundation of China (Grant Nos. 81330030, 91319309, and 31271371), the Science and Technology Department of Yunnan Province (Grant 2012HA013), the Yunnan Basic Research Projects (Grant 2014FC004), and the Doctoral Tutor of Education Department of Shanghai, China (Grant 20130072110021), grants from NASAD: BRN & BEHV RES FND 23072, NIH R21:IR21NS095184-01 and finally, a fellowship from China Scholarship Council to X. -Y.C.

ABBREVIATIONS

ACSF, artificial cerebrospinal fluid; AP, action potential; ES, embryonic stem; GO, Gene Ontology; hiPSC, human induced pluripotent stem cell; IR-DIC, infrared-differential interference contrast; NSC/NPC, neural stem/progenitor cell; WGCNA, weighted gene coexpression network analyses.

COMPLIANCE WITH ETHICS GUIDELINES

Xiaoying Chen, Kunshan Zhang, Liqiang Zhou, Xinpei Gao, Junbang Wang, Yinan Yao, Fei He, Yuping Luo, Yongchun Yu, Siguang Li, Liming Cheng, and Yi E. Sun declare that they have no conflict of interest. This article does not contain any studies with human or animal subjects performed by the any of the authors.

REFERENCES

Darmanis S et al (2015) A survey of human brain transcriptome diversity at the single cell level. Proc Natl Acad Sci USA 112:7285–7290. doi:10.1073/pnas.1507125112

Ghosh S, Chan CK (2016) Analysis of RNA-Seq data using TopHat and Cufflinks. Methods Mol Biol 1374:339–361. doi:10.1007/978-1-4939-3167-5_18

Hu BY et al (2010) Neural differentiation of human induced pluripotent stem cells follows developmental principles but with variable potency. Proc Natl Acad Sci USA 107:4335–4340. doi:10.1073/pnas.0910012107

Johnson MB et al (2015) Single-cell analysis reveals transcriptional heterogeneity of neural progenitors in human cortex. Nat Neurosci 18:637–646. doi:10.1038/nn.3980

Junker JP, van Oudenaarden A (2014) Every cell is special: genome-wide studies add a new dimension to single-cell biology. Cell 157:8–11. doi:10.1016/j.cell.2014.02.010

Kang HJ et al (2011) Spatio-temporal transcriptome of the human brain. Nature 478:483–489. doi:10.1038/nature10523

Kim D et al (2013) TopHat2: accurate alignment of transcriptomes in the presence of insertions, deletions and gene fusions. Genome Biol 14:R36. doi:10.1186/gb-2013-14-4-r36

Li Y et al (2013) Global transcriptional and translational repression in human-embryonic-stem-cell-derived Rett syndrome neurons. Cell Stem Cell 13:446–458. doi:10.1016/j.stem.2013.09.001

Luo Y et al (2015) Single-cell transcriptome analyses reveal signals to activate dormant neural stem cells. Cell 161:1175–1186. doi:10.1016/j.cell.2015.04.001

Ma L et al (2012) Human embryonic stem cell-derived GABA neurons correct locomotion deficits in quinolinic acid-lesioned mice. Cell Stem Cell 10:455–464. doi:10.1016/j.stem.2012.01.021

Mariani J et al (2015) FOXG1-dependent dysregulation of GABA/glutamate neuron differentiation in autism spectrum disorders. Cell 162:375–390. doi:10.1016/j.cell.2015.06.034

Miller JA et al (2014) Transcriptional landscape of the prenatal human brain. Nature 508:199–206. doi:10.1038/nature13185

Mirnics K (2008) What is in the brain soup? Nat Neurosci 11:1237–1238. doi:10.1038/nn1108-1237

Pollen AA et al (2015) Molecular identity of human outer radial glia during cortical development. Cell 163:55–67. doi:10.1016/j.cell.2015.09.004

Tang F et al (2009) mRNA-Seq whole-transcriptome analysis of a single cell. Nat Methods 6:377–382. doi:10.1038/nmeth.1315

Tang F et al (2010) RNA-Seq analysis to capture the transcriptome landscape of a single cell. Nat Protoc 5:516–535. doi:10.1038/nprot.2009.236

Wu H et al (2007) Integrative genomic and functional analyses reveal neuronal subtype differentiation bias in human embryonic stem cell lines. Proc Natl Acad Sci USA 104:13821–13826. doi:10.1073/pnas.0706199104

Zhang Y et al (2013) Rapid single-step induction of functional neurons from human pluripotent stem cells. Neuron 78:785–798. doi:10.1016/j.neuron.2013.05.029

Zhang K, Huang K, Luo Y, Li S (2014) Identification and functional analysis of long non-coding RNAs in mouse cleavage stage embryonic development based on single cell transcriptome data. BMC Genom 15:845. doi:10.1186/1471-2164-15-845

2

Global DNA methylation and transcriptional analyses of human ESC-derived cardiomyocytes

Ying Gu[1], Guang-Hui Liu[2,3]✉, Nongluk Plongthongkum[4], Christopher Benner[1], Fei Yi[1], Jing Qu[2], Keiichiro Suzuki[1], Jiping Yang[2], Weiqi Zhang[2], Mo Li[1], Nuria Montserrat[5,6], Isaac Crespo[7], Antonio del Sol[7], Concepcion Rodriguez Esteban[1], Kun Zhang[4]✉, Juan Carlos Izpisua Belmonte[1,5]✉

[1] Gene Expression Laboratory, Salk Institute for Biological Studies, 10010 North Torrey Pines Road, La Jolla, CA 92037, USA
[2] National Laboratory of Biomacromolecules, Institute of Biophysics, Chinese Academy of Sciences, Beijing 100101, China
[3] State key Laboratory of Drug Research, Shanghai Institute of Materia Medica, Chinese Academy of Sciences, Shanghai 201203, China
[4] Department of Bioengineering, University of California at San Diego, La Jolla, CA 92093, USA
[5] Center of Regenerative Medicine in Barcelona, Dr. Aiguader 88, 08003 Barcelona, Spain
[6] Biomedical Research Networking Center in Bioengineering, Biomaterials and Nanomedicine (CIBER-BBN), Poeta Mariano Esquillor s/n, 50018 Sargossa, Spain
[7] Luxembourg Centre for Systems Biomedicine (LCSB), University of Luxembourg, Walferdange, Luxembourg
✉ Correspondence: ghliu@ibp.ac.cn (G.-H. Liu), kzhang@bioeng.ucsd.edu (K. Zhang), belmonte@salk.edu, izpisua@cmrb.eu (J. C. Izpisua Belmonte)

ABSTRACT

With defined culture protocol, human embryonic stem cells (hESCs) are able to generate cardiomyocytes *in vitro*, therefore providing a great model for human heart development, and holding great potential for cardiac disease therapies. In this study, we successfully generated a highly pure population of human cardiomyocytes (hCMs) (>95% cTnT$^+$) from hESC line, which enabled us to identify and characterize an hCM-specific signature, at both the gene expression and DNA methylation levels. Gene functional association network and gene-disease network analyses of these hCM-enriched genes provide new insights into the mechanisms of hCM transcriptional regulation, and stand as an informative and rich resource for investigating cardiac gene functions and disease mechanisms. Moreover, we show that cardiac-structural genes and cardiac-transcription factors have distinct epigenetic mechanisms to regulate their gene expression, providing a better understanding of how the epigenetic machinery coordinates to regulate gene expression in different cell types.

KEYWORDS human cardiomyocyte, DNA methylation, microarray, heart development

INTRODUCTION

Heart failure, caused by massive loss or dysfunction of hCMs, is the main cause of death and morbidity in the developed world. Treatments for this devastating disorder are inefficient and usually focused around symptomatic alleviation, while the main cause of the disease, that is, loss of hCMs and associated contractile function, remains unchallenged. Thus, identification of factors reducing fibrotic scarring or promoting hCM proliferation is of the utmost importance for public health. Achieving this goal is complicated by the difficulty of obtaining large numbers of pure, fully differentiated hCMs and their intermediate stages.

hESCs have the potential to differentiate into cells of all lineages, therefore providing an ideal *in vitro* model to study organ development and disease mechanisms. Recent efforts have successfully developed several protocols, by which hESCs can be differentiated into hCMs (Yang et al., 2008; Kattman et al., 2011; Lian et al., 2012; Willems et al.,

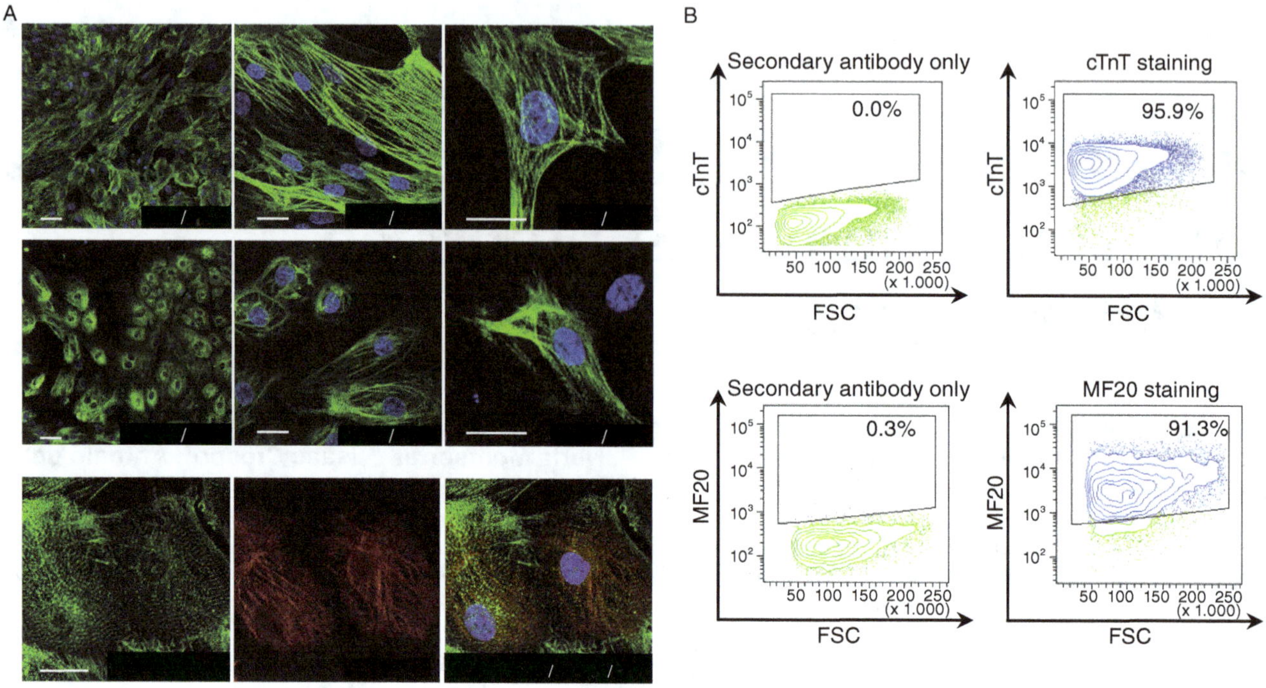

Figure 1. Characterization of hESC-derived hCMs. (A) Immunofluorescence analyses showing the expression of key cardiac markers in d25 hCMs derived from H9 hESCs. Top panel: cTnT (green). Middle panel: MF20 (green). Bottom Panel: alpha-Actinin (green) and MLC-2v (red). Scale bar: 20 μm. (B) Flow cytometry analysis of cells expressing cTnT (top panel) or MF20 (bottom panel). Cells were collected on day 25 of hCM differentiation.

2012; Zhang et al., 2012). Furthermore, hESC-derived hCMs have been analyzed by microarray and a group of cardio-specific genes were revealed (Beqqali et al., 2006; Cao et al., 2008; Synnergren et al., 2008). However, one inevitable concern is that by these published differentiation protocols, it is difficult to obtain pure hCMs. Hence, any following study using the mixed cell population may potentially give some misleading results. To overcome this, an improved protocol is needed in order to achieve a satisfactory purity of hESC-derived hCMs. Moreover, while the gene expression profile for hCMs has been provided, a more comprehensive investigation on gene expression together with epigenetic regulation, such as DNA methylation, during the hCM differentiation is still lacking.

In an effort to overcome these limitations, we have developed a new cardiac differentiation protocol starting from hESCs that yields a highly pure population of hCMs (>95%) suitable for genomic studies. As proof-of-concept, and in an effort to uncover new cardiac-specific targets relevant for therapeutic applications, we performed global epigenetic and transcriptional analyses during cardiac specification using this protocol. We performed transcriptional profiling and genome-wide DNA methylation analyses of hCMs and compared them to undifferentiated hCMs and hESC-derived neural stem cells (hNSCs). Our results provide a step forward towards the characterization of hCMs at both the transcriptional and epigenetic levels, and offer a powerful tool towards better understanding heart physiology and disease.

RESULTS

Derivation of highly enriched cardiomyocytes from hESCs

Following Palecek's previous protocol (Lian et al., 2012), hESCs were seeded as single cells on Matrigel and maintained in mTeSR. The GSK3 specific inhibitor CHIR99021 was added on the first day of differentiation, followed by the Wnt inhibitor IWP4 on day 3. After 15 days, a relatively pure and contracting cardiomyocyte population was obtained (Movie S1). We enriched this fraction by collecting and washing the contracting hCM sheets and re-plating them on fresh Matrigel plates (Movies S2 and S3). These subcultured hCMs expressed the CM-specific markers cardiac troponin T (cTnT) and sarcomeric myosin (MF20), and exhibited normal cardiac sarcomere organization, as indicated by alpha-Actinin and MLC2v co-staining (Fig. 1A). Flow cytometry analysis indicated a majority of definitive cardiac cells were present at day 25 (96% $cTnT^+$ cells and 91% $MF20^+$, Fig. 1B).

Global gene expression profiling in hCMs

We obtained RNA from undifferentiated hESCs and hCMs and used it for microarray analysis. hESC-derived hNSCs

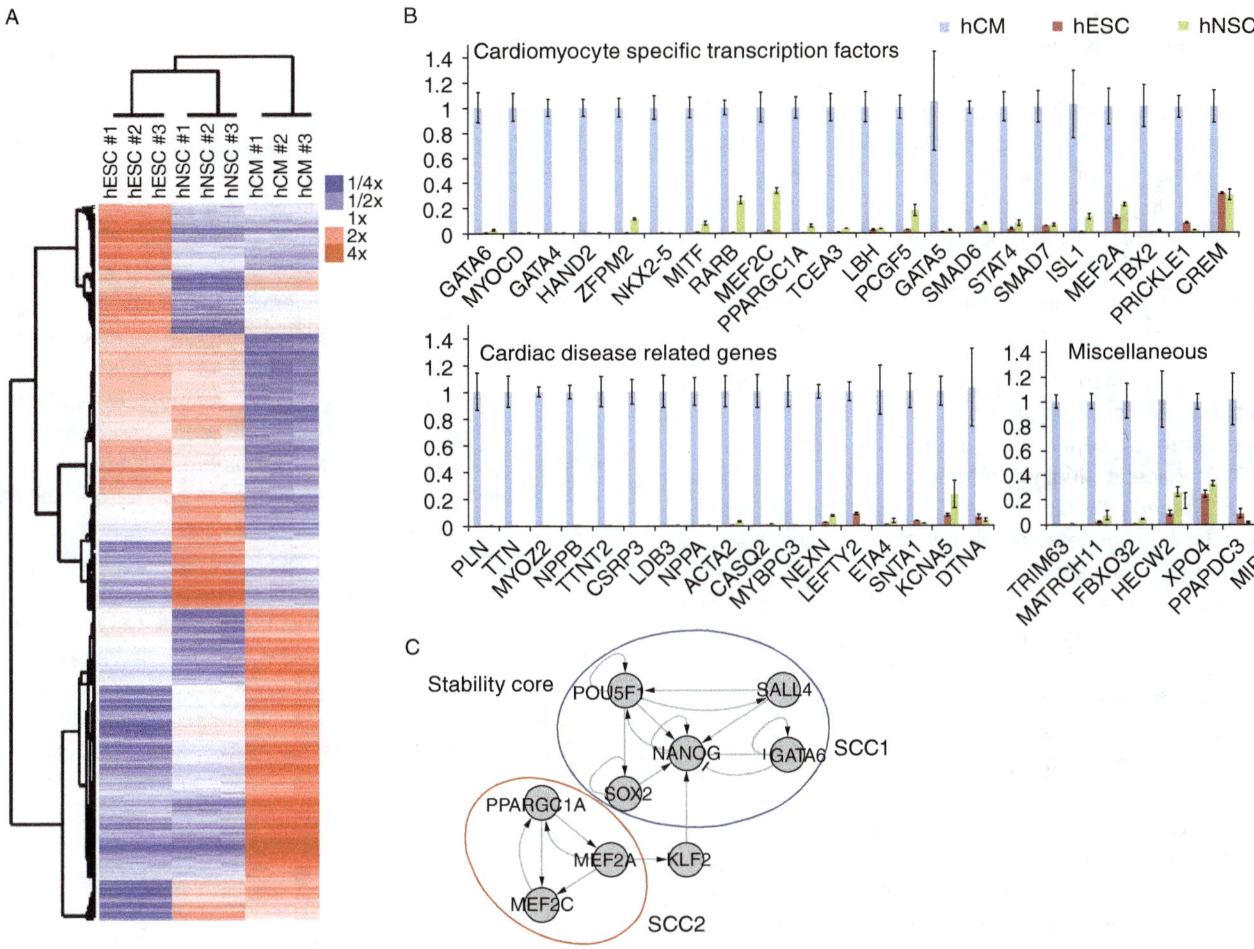

Figure 2. Global gene expression profiling of hCMs. (A) Heatmap and hierarchical clustering analysis of gene expression profiles of hESC, hNSCs, and hCMs performed in triplicate. Color represents the expression level relative to mean. (B) RT-qPCR analysis of transcript expression in hESCs, hCMs, and hNSCs. The expression levels of genes in hCMs were set to one. Data are shown as mean ± s.d., n = 3. (C) The gene regulation network that includes a stability core of 9 genes distributed in two strongly connected components (SCCs). These two SCCs or clusters of circuits, named as SCC1 and SCC2 could be broadly linked to pluripotency cellular features and the hCM-specific properties, respectively. Nodes represent genes and edges regulatory interactions positive ("→") and negative ("—|").

(Liu et al., 2012) were used as a control population. All the cells shared the same genetic background (H9), allowing for an unbiased side-by-side comparison of their gene expression profile. Three biological replicates from each cell type were measured with PrimeView Human Gene Expression Arrays, covering more than 36,000 transcripts and variants. All of the replicates were highly reproducible, supporting the purity and reliability of the method. Clustering data indicated that hESCs and hNSCs were closer to each other in terms of expression, while hCMs showed a more distinct expression pattern (Fig. 2A). Among represented transcripts, we identified 695 genes that showed at least a two-fold up-regulation and 401 genes that showed at least a two-fold down-regulation in hCMs compared to both hESCs and hNSCs (Tables S1 and S2). A group of the cardiac-enriched genes were validated by qRT-PCR (Fig. 2B).

To gain further insight into the functions of these hCM differentially expressed genes, we performed Gene Ontology (GO) analyses using the BiNGO (Maere et al., 2005) Cytoscape (Shannon et al., 2003) plugin. Interestingly, hCM up-regulated genes were significantly over-represented in cardiac function-related GO terms (complete lists of GO terms are shown in Tables S3–5), including muscle contraction, heart development, and sarcomeric structures. In contrast, hCM differentially down-regulated genes were significantly clustered into GO terms such as M phase, nuclear division, and mitosis (complete lists of GO terms are shown in Tables S6–8), suggesting that mitosis in hCMs is strongly repressed, as has been consistently observed in hCMs during maturation (Zhang et al., 2012).

Next, we analyzed differentially regulated targets in the context of gene regulatory networks. We could identify the minimal combinations of reprogramming determinants responsible for the transition of hESCs towards hCMs. Specifically, our computational model defined a gene regulatory network stability core with two major components associated with both pluripotency and hCMs. Perturbation of these genes (up- or down-regulation, depending on the original state) triggered a regulatory chain reaction resulting on the transition of hESCs to hCMs (Fig. 2C).

Interaction network and gene-disease network analyses of hCM-enriched genes

To better understand the functional interaction between the identified hCM-specific genes, all of the 695 hCM-enriched genes were screened into GeneMANIA (Montojo et al., 2010) Cytoscape plugin to produce a functional association network based on their relationships, such as co-expression, co-localization, genetic interaction, and physical interaction (a complete information of the interactions is shown in Table S9). Subnetworks of functional associations between genes involved in muscle contraction, heart development and cardiac transcriptional regulation were also generated (Fig. 3, complete information of interactions are shown in Tables S10–12). These networks may help in describing new relationships and provide a systematic resource for cardiac gene function prediction.

Phenotype-genotype relationship studies facilitate the understanding of the normal and diseased physiology. We used the DisGeNET (Bauer-Mehren et al., 2010) plugin of Cytoscape to generate a gene-disease network of highly expressed hCM-specific genes. Of the top 50 up-regulated hCM-specific genes we analyzed, 27 of them (54%) showed associations with at least one known cardiovascular diseases and 5 of them (NPPB, TNNT2, NPPA, RYR2, and PLN) were linked to more than 10 different types of cardiovascular diseases (Fig. 4A, Table S13). In addition, a total number of 142 disease-associated mutations were found in those top hCM-enriched genes (mutations listed in Fig. 4B, references listed in Table S14), suggesting a preferred enrichment of disease causing mutations in hCM-specific genes responsible for critical human heart functions.

Genome-wide DNA methylation analysis of hCMs

Up to date, relatively few studies have analyzed global DNA methylation status in human cardiac lineages. hESCs, hCMs, and hNSCs were collected for genome-wide DNA methylation profiling using methylation sequencing with bisulfite padlock probes as previously described (Diep et al., 2012). hCMs displayed the highest global DNA methylation level (Fig. 5A) and 985 genes showed an increase in their promoter mCpG levels (5% or more), whereas 195 genes exhibited a 5% or more decrease of promoter DNA methylation compared to hESCs and hNSCs (Tables S15 and S16). Interestingly, these hCM-specific demethylated genes were significantly over-represented in our cardiac-related GO terms search, suggesting that promoter DNA demethylation may contribute to the regulation of cardiac functions (Tables S17–19). Since DNA methylation represents an essential epigenetic mechanism for gene repression, we speculated whether decreased DNA methylation correlated with the increased expression in cardiac-related genes. We combined DNA methylation and microarray data together and identified a group of 29 genes that exhibited hCM-specific promoter demethylation, as well as at least a two-fold up-regulation in gene expression (Table S20). These genes were highly enriched in cardiac-structural related GO terms such as muscle contraction, structural muscle constituents, and sarcomere-related transcripts (Tables S21–23). Again, this data supported the notion that genes encoding cardiac-structural proteins are largely regulated by DNA methylation in hCMs. However, and since the majority of demethylated genes did not show significant changes in gene expression, other layers of regulation, such as histone modifications, might also play a significant role (Xie et al., 2013).

Among those cardiac-structural genes that were specifically demethylated in hCMs, a majority of them exhibited hypermethylated mCpG (>70%) at the promoter regions in hESC and hNSCs (Fig. 5B), indicating that DNA methylation in the promoter region of these genes may function as a major repressive signal in non-cardiac lineages. In contrast, cardiac-specific transcription factors (TFs), such as NKX2.5, GATA6, GATA4, MYOCD, HAND2, TBX5, and TBX18, were more expressed in hCMs than in hESCs or hNSCs, while displaying a similar pattern of low promoter mCpG levels (<10%) in all three lineages, suggesting that the regulation of the expression of cardiac-specific TFs is largely independent of DNA methylation (Fig. 5B).

DISCUSSION

In this study, we report on an improved protocol for the generation of hCMs from hESCs, which enabled us to identify and characterize a hCM-specific signature, at both the gene expression and DNA methylation levels. The interaction networks we present in this study describe the precise mutual relationships between every hCM-enriched genes, as well as the critical genes involved in the fundamental heart functions, therefore providing an informative and systematic resource for gene function prediction in cardiac research field. In regard to understanding disease mechanisms, we identified a tight correlation between various heart diseases and hCM highly up-regulated genes. These findings not only highlight the importance of these cardiac-specific genes in maintaining normal heart development and functions, but may also provide novel gene targets for uncovering heart disease mechanisms.

By combining the promoter DNA methylation and gene expression profile, we got an overall glance at the gene regulation by DNA methylation in hCMs. Interestingly, we found that a group of cardiac-structural genes exhibited

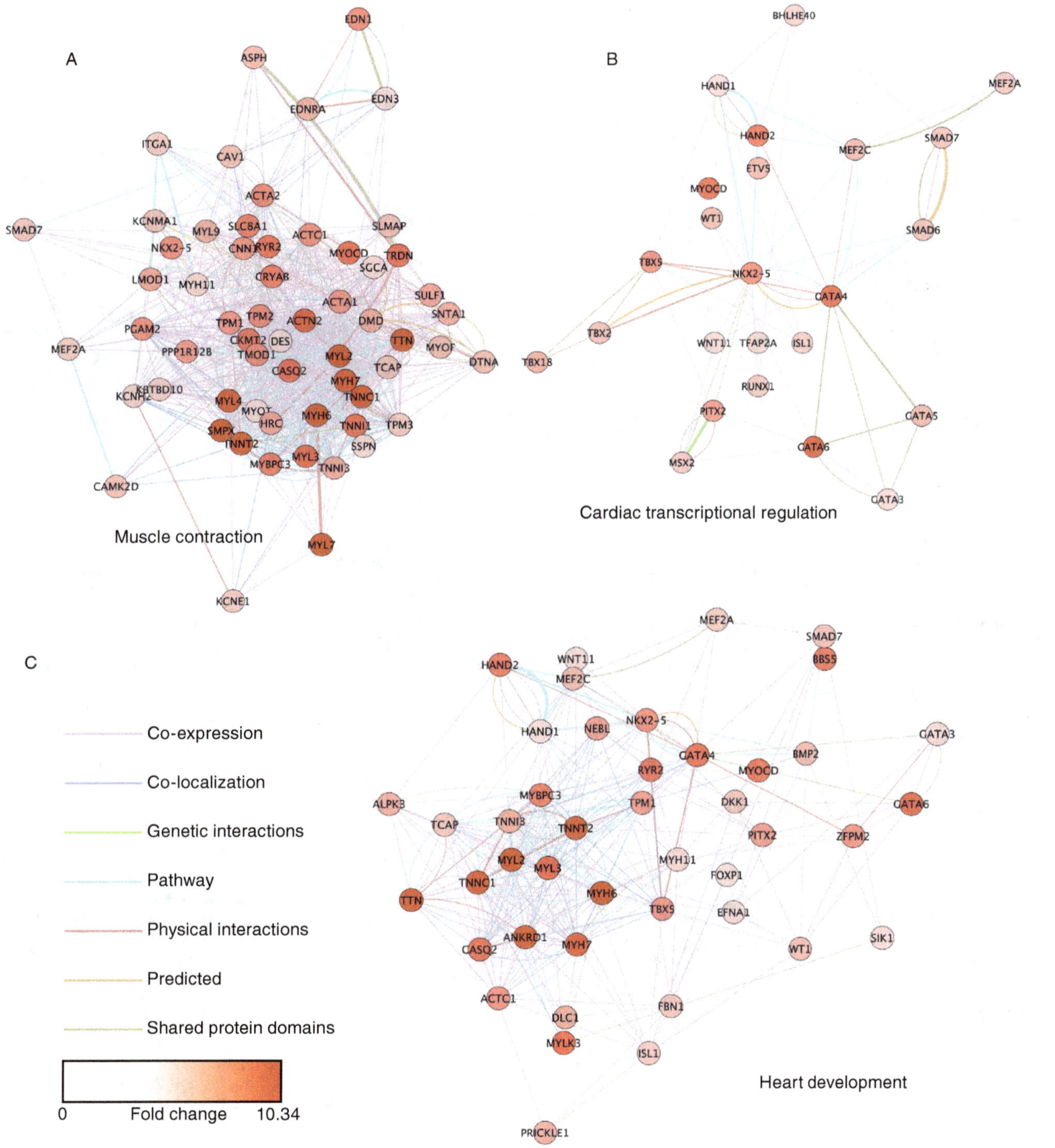

Figure 3. Association networks of hCM-enriched genes involved in function groups of muscle contraction, heart development, and cardiac transcriptional regulation. Networks were generated by GeneMANIA Cytoscape plugin. Nodes represent genes involved in the selected function group, and edges represent the relationships between linked genes. Node colors represent the average expression fold changes in hCMs compared to both hESCs and hNSCs.

hypermethylated mCpG at the promoter regions in hESCs and hNSCs, but were demethylated in hCMs, suggesting that promoter DNA methylation may be employed as a major transcription repression mechanism in these genes. Supporting this notion, cardiac-structural genes were minimally expressed and often lacked another essential repressive mark, H3K27me3, in human ESCs and early stage cells of multiple non-cardiac lineages (Paige et al., 2012; Xie

A

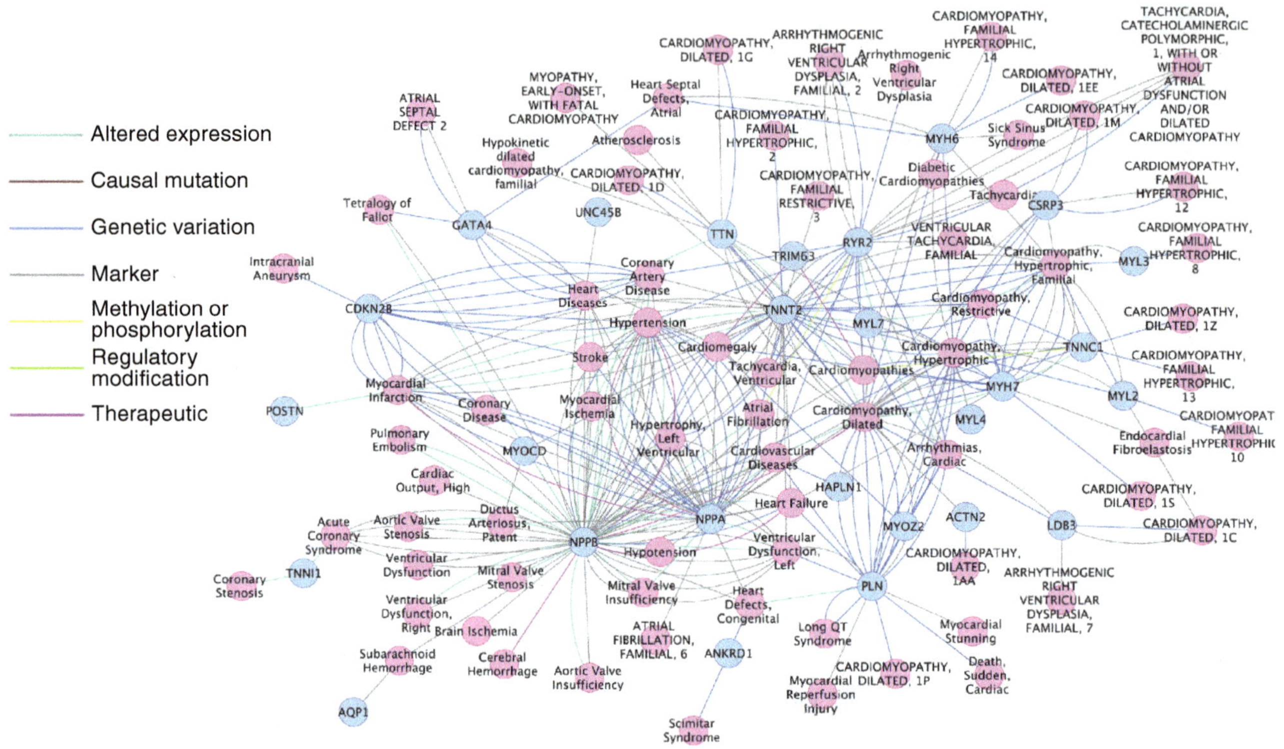

B

Gene	Diseases	Mutations in DNA or protein sequence
PL	HCM	-42C>G, -77A>G
	DCM	R9C, R14del
MYH6	HCM or DCM	R795Q, P830L, A1004S, Q1065H, E1457K
	ASD	I820N
MYOZ2	HCM	S48P, I246M
MYL2	HCM	A13T, F18L, E22K, N47K, R58Q, K103E, IVS6-1G>C
CSRP3	HCM	L44P, S54R/E55G, C58G, 172T>G
	DCM	W4R
TTN	HCM	R740L, S3799Y
	DCM	V54M, A743V, W930R, R25618Q, Q4053ter, S4465D, 2-bp insertion in exon 326
LDB3	DCM	D117N, K136M, S196L, T213I, I352M, D626N
	LVNC	D117N, K136M, S196L,
MYH7	DCM	P211L, L390V, R403W, R403Q, Y501C, H576R, R663C, Q734P, I736T, G741W, A797T, R870H, A901G, D928N, E930K, R1053Q, E1356K, A1379T, A1454T, R1500W, S1776G
	DCM	I201T, T412N, A550V, T1019N, R1193S, E1426K, R1634S
	LVNC	M531R
ANKTD1	TAPVR	10;21 translocation

Gene	Diseases	Mutations in DNA or protein sequence
NPPB	HCM	D175N
	EH	Variable number of tandem repeat of 5'-flanking region
TNNT2	DCM	I79N, R141W, K210del
	HCM	I79N, I90M, R92C, R92W, F110I, K124N, R130C, R278C, G287A, Δexon 14, a five-base (CTTCT) insertion/deletion in intron 3
	RCM	I79N, Q96del
NPPA	Isehemic Stroke	G66A
	AF	c.456–457delAA
	EH	V7M, −664G
	LVH	-2843A>G
TNNC1	HCM	A8V, L29Q, A31S, C84Y, E134D, D145E
	DCM	Y5H, M103I, D145E, I148V
ACTN2	DCM	Q9R
MYL3	HCM	D778E
GATA4	CHD	A6V, S52F, P163S, E216D, E359K, S377G, P407Q, S429T, A442V, IVS4 + 55C>A c.1074delC
RYR2	CPVT	R414C, E1724K, A2254V, R2267H, A2394G, V2475F, F4020L, E4076K, N4104I, N4104K, H4108N, H4108Q, R4497C, S4565R, G4662S, H4762P, V4771I, P4902S, genomic deletion involved RYR2 exon-3
MYOCD	CHD	K259R
TRIM63	HCM	A48V, I130M, Q247del

DCM	Dilated Cardiomyopathy
HCM	Hypertrophic Cardiomyopathy
RCM	Restrictive Cardiomyopathy
ASD	Atrial Septal Defect
LVNC	Left Ventricular Non-Compaction
TAPVR	Total Anomalous Pulmonary Venous Return
EH	Essential Hypertension
AF	Atrial Fibrillation
LVH	Left Ventricular Hypertrophy
CPVT	Catecholaminergic Polymorphic Ventricular Tachycardia
CHD	Congenital Heart Disease
ARVC	Arrhythmogenic Right Ventricular Cardiomyopathy

◀ **Figure 4. Disease associations of hCM-enriched genes.** (A) Gene-disease network of hCM-enriched genes and cardiovascular diseases. Networks were generated by DisGeNET Cytoscape plugin. Nodes represent genes that showed association with cardiovascular diseases, and edges represent the relationship between linked genes and diseases. (B) List of cardiovascular disease-associated mutations in the top hCM-enriched genes. Among the top 50 expressed genes, the ones with known disease-related mutations were listed.

et al., 2013). In contrast, a group of cardiac TFs display similar low promoter mCpG levels in all three linages we tested. These observations are also in line with the idea that these developmental-related genes are more likely to locate in large genomic domains devoid of DNA methylation in most lineages (Xie et al., 2013). Interestingly, these cardiac-specific TFs were previously found to have high levels of the repressive histone mark H3K27me3 in hESCs and many other non-cardiac lineages. However, during hESC differentiation towards hCM, the levels of H3K27me3 were decreased in hCM specific TFs, highlighting this mark as a major repressing factor in non-cardiac cell types (Paige et al., 2012; Xie et al., 2013). Whether DNA methylation offers a more stable regulation for cardiac-structural genes expression and whether the highly flexible nature of histone modifications is required for the dynamic cardiac-TF expression in a time-sensitive manner will be interesting questions to ask in future studies.

Overall, our analysis provides new insights into the mechanisms of hCM transcriptional regulation, and stands as an informative and rich resource for investigating cardiac gene functions, as well as providing a better understanding of how the epigenetic machinery coordinates to regulate gene expression in different cell types. Furthermore, the cardiac-specific molecular hallmarks and epigenetic signatures presented in this study might be of relevance for clinical applications as biomarkers for diagnosis and treatment of heart-related diseases.

MATERIALS AND METHODS

Cell culture and differentiation

H9 hESCs were purchased from WiCell Research, and cultured under standard hESC culture conditions as previously described (Liu et al., 2011). Differentiation of H9 hESC into hNSC was followed by the protocol described previously (Liu et al., 2012).

The hCM derivation from H9 hESCs was performed as previously described (Lian et al., 2012), with important modification to further improve the purity. Briefly, single cell suspension of H9 hESCs were seeded onto Matrigel (BD Biosciences) pre-coated cell culture dishes at a density of 100,000 cells per cm^2 in mTeSR (StemCell Technologies) in the presence of ROCK inhibitor Y-27632 (Sigma-Aldrich) for 24 h. Cells were then cultured in mTeSR for another 2 days. Differentiation was initiated by treatment with 12 μmol/L CHIR99021 (Selleck) in RPMI/B27-insulin (Life Technologies) for 24 h (day 0 to day 1), and medium was then changed to RPMI/B27-insulin. On day 3, 5 μmol/L Wnt inhibitor IWP4 (Stemgent) was added into RPMI/B27-insulin and cells were cultured without medium change for 48 h. From day 5, cells were maintained in RPMI/B27 with medium change every 2–3 days. On day 15, contracting hCM sheets or clusters were collected by intensive mechanic washing or manually picking, re-plated onto fresh Matrigel pre-coated dishes and maintained in RPMI/B27.

Immunofluorescence

Cells were fixed with 4% paraformaldehyde, permeabilized in 0.3% Triton X-100/PBS and incubated with primary antibodies overnight at 4°C at following dilutions: mouse anti-cTnT (1:200, Lab Vision), mouse anti-MF-20 (1:20 Developmental Studies Hybridoma Bank), rabbit anti-MLC2v (1:200, ProteinTech Group), mouse anti-alpha-Actinin (1:200, Sigma-Aldrich). Cells were then washed and incubated with Alex Fluor 488 goat anti-mouse IgG (1:500, Life Technologies) and/or Alex Fluor 568 goat anti-rabbit IgG (1:500, Life Technologies) for 1 h at RT. Cell nuclei were counter stained with DAPI (1 μg/mL, Sigma-Aldrich). Images were captured using a Zeiss LSM780 confocal microscope, and were analyzed in ZEN 2011 software.

Flow cytometry

Flow cytometry analyses were performed as described previously (Zhang et al., 2012), with antibody concentrations as following: mouse anti-cTnT (1:200, Lab Vision), mouse anti-MF-20 (1:20 Developmental Studies Hybridoma Bank), and Alex Fluor 488 goat anti-mouse IgG (1:500, Life Technologies). Data were collected and analyzed on a LSRII flow cytometer (Becton-Dickinson).

Microarray analysis

hESC, hNSC, and hCM samples were prepared in biological triplicates. Total RNA was extracted using Trizol Reagent (Life Technologies) and further purified by RNeasy Mini Kit (Qiagen). Microarray experiments were performed using Affymetrix GeneChip PrimeView Human Gene Expression Arrays in the Functional Genomics Core Facility at the Salk Institute for Biological Studies according to the manufacturer's protocol (Affymetrix, Santa Clara, CA). Expression signals were scanned on Affymetrix GeneChip Scanner 3000 7G. Gene expression values were calculated and normalized with RMA using R/Bioconductor. Expression levels were clustered using Cluster 3.0 and visualized using Java TreeView. All data have been deposited in NCBI's Gene Expression Omnibus and are accessible through GEO Series accession number GSE48257. http://www.ncbi.nlm.nih.gov/geo/query/acc.cgi?token=fxqpxgigmoaggzg&acc=GSE48257

Quantitative RT-PCR

RT-qPCR was performed as described previously (Liu et al., 2011). The primer sequences were listed in Table S24.

Genome-wide DNA methylation analysis

Bisulfite padlock probe sequencing was performed as previously described (Diep et al., 2012). Briefly, genomic DNA was extracted

A

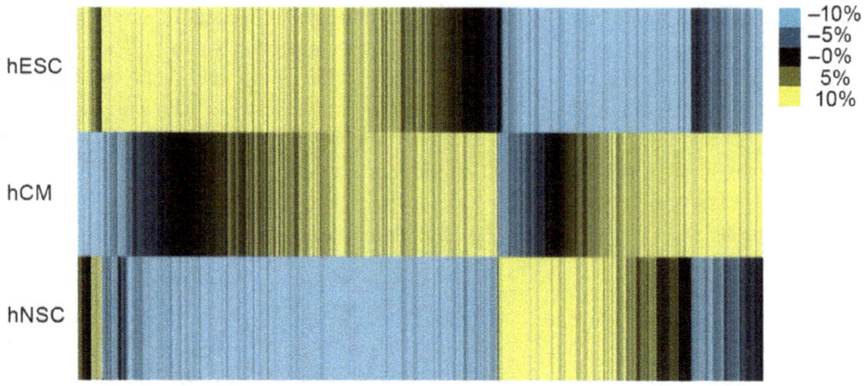

B

Gene Name	Cardiac-structural factors: DNA methylation hESC-mCpG	hNSC-mCpG	hCM-mCpG	Gene expression hCM vs hESC	hCM vs hNSC
TRIM55	0.89	0.67	0.34	7.37	7.27
ASPH	0.83	0.93	0.53	4.28	3.12
KBTBD10	0.90	0.91	0.72	2.91	2.67
MYL4	0.92	0.81	0.64	9.22	9.00
CRYAB	0.94	0.85	0.68	5.94	6.31
MYOZ2	0.97	0.96	0.83	9.49	9.51
OBSCN	0.90	0.86	0.73	2.60	2.40
ABLIM1	0.83	0.83	0.73	2.26	2.65
SYNPO2L	0.90	0.93	0.82	9.26	9.42
ANKRD1	0.25	0.18	0.10	9.71	8.64
TNNI1	0.79	0.70	0.62	6.71	6.79
MYH6	0.87	0.76	0.69	9.52	9.79
ABRA	0.97	0.92	0.85	5.46	5.34
TPM1	0.77	0.88	0.70	3.64	6.14
MYOM1	0.92	0.90	0.84	5.93	5.94
MYBPC3	0.89	0.83	0.77	6.02	6.12
TNS1	0.74	0.62	0.57	2.64	2.55
FLNC	0.23	0.32	0.18	5.20	4.37

Gene Name	Cardiac-transcription factors: DNA methylation hESC-mCpG	hNSC-mCpG	hCM-mCpG	Gene expression hCM vs hESC	hCM vs hNSC
GATA6	0.01	0.01	0.01	8.01	8.05
GATA4	0.01	0.01	0.01	7.04	7.09
MYOCD	0.05	0.03	0.04	6.70	6.91
HAND2	0.09	0.07	0.14	6.43	6.41
NKX2.5	0.03	0.04	0.03	5.28	5.32
TBX5	0.03	0.02	0.18	5.11	5.15
TBX18	0.01	0.01	0.03	4.45	3.54

0 Fold change 10 0 mCpG 1

Figure 5. Genome-wide DNA methylation analysis of hCMs. (A) Heatmap and hierarchical clustering of CpG methylation levels measured within 1 kb of promoter regions. Color represents the mCpG/CpG level relative to mean. (B) List of cardiac-structural factors and cardiac-transcription factors with their mCpG levels and relative fold changes in their gene expression.

from H9-hESC, H9-hNSC, and H9-hCM using QIAamp DNA Micro Kit (QIAGEN), and approximately 1000 ng of genomic DNA was bisulfite converted with EZ-96 DNA Methylation-Lightning MagPrep kit (Zymo Research). Approximately 250 ng of bisulfite converted genomic DNAs were mixed with normalized amount of the genome-wide scale padlock probe set. The annealed padlock probes were polymerized, gap-filled, and ligated to generate circularized DNA. The circularized captured targets were amplified and barcoded by PCR using the library-free BSPP protocol as previously described (Diep et al., 2012). The resulting bisulfite sequencing libraries were pooled in the same molar ratio, size-selected at the fragment size approximately 375 bp in 6% TBE PAGE gel (Life Technologies), and sequenced by Illumina HiSeq2000 sequencer (110 bp, paired-end reads). The bisulfite reads were mapped to the *in silico* bisulfite-converted human genome sequences (hg19) by bisREADMapper (Diep et al., 2012). DNA methylation frequency (at level from 0–1) at each CpG site with minimum 10× depth coverage was calculated. Only CpG sites with 10× coverage in each cell type were used for differential analysis. Promoter CpG methylation levels were calculated by averaging the mCpG/CpG ratio for all CpG dinucleotides with 10× coverage within 1 kb of the TSS. Clustering of promoter methylation ratios was carried out using Cluster 3.0 and visualized using Java TreeView.

Gene Ontology (GO) analysis

Functional properties of the differentially expressed or methylated genes were categorized using the BiNGO Cytoscape plugin as described previously (Maere et al., 2005). A *P*-value of less than 0.05 was considered statistically significant.

Gene regulatory networks reconstruction

Differentially expressed genes obtained after the performance of a *t*-test with a *P*-value < 0.05 were connected to expression regulatory interactions from literature. For this specific purpose we use the information contained in the ResNet mammalian database from Ariadne Genomics (http://www.ariadnegenomics.com/) (Novichkova et al., 2003; Daraselia et al., 2004). We selected only the interactions included in the category of Promoter Binding and Direct Regulation. In order to contextualize the network to the biological conditions under which the expression data was obtained we applied an algorithm that exploits the consistency between predicted and known stable states from experimental data to guide an iterative network pruning. The algorithm predicted missing expression values in gene regulatory networks, and could be applied to contextualize the network when all the expression values in two attractors are known. The method assumes a Boolean model to compute attractors of networks that are iteratively pruned by means of an evolutionary algorithm. The evolutionary algorithm samples the probability distribution of positive circuits and individual interactions within the subpopulation of the best-pruned networks at each iteration. The resulting contextualized network is based not only on previous knowledge about local connectivity but also on a global network property (stability).

Attractor computation was performed assuming a Boolean model applying a synchronous updating scheme (Garg et al., 2008) that updates all gene states simultaneously at each step until the system reaches an attractor. For this purpose we used our own implementation (Crespo et al., 2013) written in Perl of the algorithm described by Garg and co-workers (Garg et al., 2007). We implemented the Johnsons algorithm (Johnson, 1975) to detect all elementary circuits in the network. Both elementary circuit detection and positive circuits sorting scripts were implemented in Perl.

Gene functional association network analysis

Networks of gene associations (co-expression, co-localization, genetic interaction, pathway, and more) were generated using the GeneMANIA Cytoscape plugin as described previously (Montojo et al., 2010). The "query gene based" weighting method was used in this study. Sub-networks were generated by selecting genes that were categorized in the same functional GO group.

Gene-disease association network analysis

Network of gene-disease associations (altered expression, casual mutation, genetic variation, marker, and more) were generated using DisGeNET Cytoscape plugin as described previously (Bauer-Mehren et al., 2010). The top 50 hCM up-regulated genes were inquired individually for disease association in the cardiovascular disease class. All the individual gene-disease networks were then combined based on the shared disease type.

ACKNOWLEDGEMENTS

We thank Dr. Yue Ma for helpful discussion. J.Q. is supported by the National Basic Research Program (973 Program) (No. 2014CB910500). G.H.L. is supported by the Strategic Priority Research Program of the Chinese Academy of Sciences (XDA01020312), the National Basic Research Program of China (973 Program) (No. 2014CB964600), the National Natural Science Foundation (Grant Nos. 81271266 and 31222039), the National Laboratory of Biomacromolecules, and the State Key Laboratory of Drug Research (SIMM1302KF-17). K.S. and M.L. are supported by a CIRM training grant. K.Z is supported by California Institute for Regenerative Medicine (RB3-05083) and NIH (R01 GM097253). J.C.I.B is supported by TERCEL-ISCIII-MINECO, CIBER-BBN, Cardiocel, Fundacion Cellex, G. Harold and Leila Y. Mathers Charitable Foundation, The Leona M. and Harry B. Helmsley Charitable Trust and The Ellison Medical Foundation.

ABBREVIATIONS

cTnT, cardiac troponin T; GO, Gene Ontology; hESCs, human embryonic stem cells; hCM, human cardiomyocyte; hNSCs, hESC-derived neural stem cells; TFs, transcription factors.

COMPLIANCE WITH ETHICS GUIDELINES

Ying Gu, Guang-Hui Liu, Nongluk Plongthongkum, Christopher Benner, Fei Yi, Jing Qu, Keiichiro Suzuki, Jiping Yang, Weiqi Zhang, Mo Li, Nuria Montserrat, Isaac Crespo, Antonio del Sol, Concepcion Rodriguez Esteban, Kun Zhang and Juan Carlos Izpisua Belmonte declare that they have no conflict of interest.

This article does not contain any studies with human or animal subjects performed by the any of the authors.

REFERENCES

Bauer-Mehren A, Rautschka M, Sanz F, Furlong LI (2010) DisGeNET: a Cytoscape plugin to visualize, integrate, search and analyze gene-disease networks. Bioinformatics 26:2924–2926

Beqqali A, Kloots J, Ward-van Oostwaard D, Mummery C, Passier R (2006) Genome-wide transcriptional profiling of human embryonic stem cells differentiating to cardiomyocytes. Stem Cells 24:1956–1967

Cao F, Wagner RA, Wilson KD, Xie X, Fu JD, Drukker M, Lee A, Li RA, Gambhir SS, Weissman IL et al (2008) Transcriptional and functional profiling of human embryonic stem cell-derived cardiomyocytes. PLoS One 3:e3474

Crespo I, Krishna A, Le Bechec A, del Sol A (2013) Predicting missing expression values in gene regulatory networks using a discrete logic modeling optimization guided by network stable states. Nucleic Acids Res 41:e8

Daraselia N, Yuryev A, Egorov S, Novichkova S, Nikitin A, Mazo I (2004) Extracting human protein interactions from MEDLINE using a full-sentence parser. Bioinformatics 20:604–611

Diep D, Plongthongkum N, Gore A, Fung HL, Shoemaker R, Zhang K (2012) Library-free methylation sequencing with bisulfite padlock probes. Nat Methods 9:270–272

Garg A, Xenarios I, Mendoza L, DeMicheli G (2007) Lecture notes in computer science vol. 4453. In: Speed T, Huang H (eds). Springer, Berlin, p 62–76

Garg A, Di Cara A, Xenarios I, Mendoza L, De Micheli G (2008) Synchronous versus asynchronous modeling of gene regulatory networks. Bioinformatics 24:1917–1925

Johnson DB (1975) Finding all the elementary circuits of a directed graph. SIAM J Comput 4:77–84

Kattman SJ, Witty AD, Gagliardi M, Dubois NC, Niapour M, Hotta A, Ellis J, Keller G (2011) Stage-specific optimization of activin/nodal and BMP signaling promotes cardiac differentiation of mouse and human pluripotent stem cell lines. Cell Stem Cell 8:228–240

Lian X, Hsiao C, Wilson G, Zhu K, Hazeltine LB, Azarin SM, Raval KK, Zhang J, Kamp TJ, Palecek SP (2012) Robust cardiomyocyte differentiation from human pluripotent stem cells via temporal modulation of canonical Wnt signaling. Proc Natl Acad Sci USA 109:E1848–E1857

Liu GH, Barkho BZ, Ruiz S, Diep D, Qu J, Yang SL, Panopoulos AD, Suzuki K, Kurian L, Walsh C et al (2011) Recapitulation of premature ageing with iPSCs from Hutchinson–Gilford progeria syndrome. Nature 472:221–225

Liu GH, Qu J, Suzuki K, Nivet E, Li M, Montserrat N, Yi F, Xu X, Ruiz S, Zhang W et al (2012) Progressive degeneration of human neural stem cells caused by pathogenic LRRK2. Nature 491:603–607

Maere S, Heymans K, Kuiper M (2005) BiNGO: a Cytoscape plugin to assess overrepresentation of gene ontology categories in biological networks. Bioinformatics 21:3448–3449

Montojo J, Zuberi K, Rodriguez H, Kazi F, Wright G, Donaldson SL, Morris Q, Bader GD (2010) GeneMANIA Cytoscape plugin: fast gene function predictions on the desktop. Bioinformatics 26: 2927–2928

Novichkova S, Egorov S, Daraselia N (2003) MedScan, a natural language processing engine for MEDLINE abstracts. Bioinformatics 19:1699–1706

Paige SL, Thomas S, Stoick-Cooper CL, Wang H, Maves L, Sandstrom R, Pabon L, Reinecke H, Pratt G, Keller G et al (2012) A temporal chromatin signature in human embryonic stem cells identifies regulators of cardiac development. Cell 151:221–232

Shannon P, Markiel A, Ozier O, Baliga NS, Wang JT, Ramage D, Amin N, Schwikowski B, Ideker T (2003) Cytoscape: a software environment for integrated models of biomolecular interaction networks. Genome Res 13:2498–2504

Synnergren J, Akesson K, Dahlenborg K, Vidarsson H, Ameen C, Steel D, Lindahl A, Olsson B, Sartipy P (2008) Molecular signature of cardiomyocyte clusters derived from human embryonic stem cells. Stem Cells 26:1831–1840

Willems E, Cabral-Teixeira J, Schade D, Cai W, Reeves P, Bushway PJ, Lanier M, Walsh C, Kirchhausen T, Izpisua Belmonte JC et al (2012) Small molecule-mediated TGF-beta type II receptor degradation promotes cardiomyogenesis in embryonic stem cells. Cell Stem Cell 11:242–252

Xie W, Schultz MD, Lister R, Hou Z, Rajagopal N, Ray P, Whitaker JW, Tian S, Hawkins RD, Leung D et al (2013) Epigenomic analysis of multilineage differentiation of human embryonic stem cells. Cell 153:1134–1148

Yang L, Soonpaa MH, Adler ED, Roepke TK, Kattman SJ, Kennedy M, Henckaerts E, Bonham K, Abbott GW, Linden RM et al (2008) Human cardiovascular progenitor cells develop from a KDR+ embryonic-stem-cell-derived population. Nature 453:524–528

Zhang J, Klos M, Wilson GF, Herman AM, Lian X, Raval KK, Barron MR, Hou L, Soerens AG, Yu J et al (2012) Extracellular matrix promotes highly efficient cardiac differentiation of human pluripotent stem cells: the matrix sandwich method. Circ Res 111:1125–1136

3

Enrichment analysis of *Alu* elements with different spatial chromatin proximity in the human genome

Zhuoya Gu[1], Ke Jin[2], M. James C. Crabbe[3,4], Yang Zhang[5], Xiaolin Liu[6], Yanyan Huang[1], Mengyi Hua[1], Peng Nan[1✉], Zhaolei Zhang[2,7✉], Yang Zhong[1,8✉]

[1] School of Life Sciences, Fudan University, Shanghai 200433, China
[2] Banting and Best Department of Medical Research, Donnelly Centre, University of Toronto, Toronto, ON M5S 1A1, Canada
[3] Department of Zoology, University of Oxford, Tinbergen Building, South Parks Road, Oxford OX1 3PS, UK
[4] Institute of Biomedical and Environmental Science & Technology, Department of Life Sciences, University of Bedfordshire, Park Square, Luton LU1 3JU, UK
[5] Department of Bioengineering, University of Illinois at Urbana-Champaign, Champaign, IL 61801, USA
[6] School of Public Health, University of Michigan, Ann Arbor, MI 48109, USA
[7] Department of Molecular Genetics, University of Toronto, Toronto, ON M5S 1A1, Canada
[8] Institute of Biodiversity Science and Institute of High Altitude Medicine, Tibet University, Lhasa 850012, China
✉ Correspondence: nanpeng@fudan.edu.cn (
Peng Nan), zhaolei.zhang@utoronto.ca (Z. Zhang), yangzhong@fudan.edu.cn (Y. Zhong)

ABSTRACT

Transposable elements (TEs) have no longer been totally considered as "junk DNA" for quite a time since the continual discoveries of their multifunctional roles in eukaryote genomes. As one of the most important and abundant TEs that still active in human genome, *Alu*, a SINE family, has demonstrated its indispensable regulatory functions at sequence level, but its spatial roles are still unclear. Technologies based on 3C (chromosome conformation capture) have revealed the mysterious three-dimensional structure of chromatin, and make it possible to study the distal chromatin interaction in the genome. To find the role TE playing in distal regulation in human genome, we compiled the new released Hi-C data, TE annotation, histone marker annotations, and the genome-wide methylation data to operate correlation analysis, and found that the density of *Alu* elements showed a strong positive correlation with the level of chromatin interactions (hESC: $r = 0.9$, $P < 2.2 \times 10^{16}$; IMR90 fibroblasts: $r = 0.94$, $P < 2.2 \times 10^{16}$) and also have a significant positive correlation with some remote functional DNA elements like enhancers and promoters (Enhancer: hESC: $r = 0.997$, $P = 2.3 \times 10^{-4}$; IMR90: $r = 0.934$, $P = 2 \times 10^{-2}$; Promoter: hESC: $r = 0.995$, $P = 3.8 \times 10^{-4}$; IMR90: $r = 0.996$, $P = 3.2 \times 10^{-4}$). Further investigation involving GC content and methylation status showed the GC content of *Alu* covered sequences shared a similar pattern with that of the overall sequence, suggesting that *Alu* elements also function as the GC nucleotide and CpG site provider. In all, our results suggest that the *Alu* elements may act as an alternative parameter to evaluate the Hi-C data, which is confirmed by the correlation analysis of *Alu* elements and histone markers. Moreover, the GC-rich *Alu* sequence can bring high GC content and methylation flexibility to the regions with more distal chromatin contact, regulating the transcription of tissue-specific genes.

KEYWORDS chromatin interaction, alternative parameter of Hi-C data, open chromatin, methylation potential

INTRODUCTION

Transposable elements (TEs) are DNA sequences able to move and replicate within the genome of a single cell and

Zhuoya Gu and Ke Jin have contributed equally to this work.

have been found in virtually all eukaryotic genomes so far sequenced (Feschotte 2008; Kidwell and Lisch 2001). Large-scale genome sequencing has revealed that TEs comprise more than 45% of the human genome, and also represent an abundant part of the genomes in fungi and metazoans (3%–20% in fungi and 3%–45% in metazoans) (Wicker et al. 2007; Lander et al. 2001). According to the mechanism of transposition, TEs are usually divided into two classes (Class I or retrotransposons, and Class II or DNA transposons). Retrotransposons can be further grouped into three main orders (long terminal repeats (LTRs), long interspersed nuclear elements (LINE) and short interspersed nuclear elements (SINE)). In humans, *Alu* elements are the most abundant SINEs with over one million inserted copies, resulting from their continuous proliferation activity over the past ~65 million years (Myr) (Cordaux and Batzer 2009), and comprise about 10% of the DNA sequences of the human genome. According to the current model, modern *Alu*s are ~300 bp in length and emerged from a head to tail fusion of two distinct fossil *Alu* monomers (FAMs) that originated from the 7SL RNA gene 55 Myr ago with the expansion of primates (Quentin 1992a, b). During the course of primate evolution, *Alu* elements have spread and formed several distinct subfamilies on the basis of different rates of amplification. The three major subfamilies are: the oldest *Alu*J, intermediately aged *Alu*S, and the youngest *Alu*Y. Previous studies have indicated that *Alu*J and *Alu*S appeared in ancient genome 65–25 Myr, and *Alu*Y 25 Myr (Kapitonov and Jurka 1996). Due to the short evolutionary history of the *Alu*Y class, some *Alu*Y repeats have not went through the purify selection and are also unable to reach the fixation, resulting in the fact that, currently, most of *Alu* elements in the human genome belong to the *Alu*S class.

Alu elements, together with other repetitive elements, were originally thought of as genomic parasites and had been long dismissed as selfish or 'junk' DNA (Kidwell and Lisch 2001; Brookfield 2005). Accumulating evidence has demonstrated that widespread *Alu* elements are distributed throughout the human genome in a non-random manner (Grover et al. 2003) and play an important role in genome (Kazazian 2004) and gene evolution (Nekrutenko and Li 2001), epigenetic and gene regulation (Feschotte 2008; Lynch et al. 2011; Teng et al. 2011; Polak and Domany 2006). For example, recent studies have indicated that *Alu* elements are predominant in isochores (Hackenberg et al. 2005) and segmental duplications (Jurka et al. 2004). Likewise, the distribution of *Alu* elements was found to have a highly positive correlation with local GC contents and the density of genes or introns (Grover et al. 2004). Moreover Polak et al. demonstrated that the upstream regions of the transcription start site (TSS) are enriched with *Alu* elements that contain many putative binding sites for transcription factors (Polak and Domany 2006). A subsequent study further highlighted that primate-specific *Alu* elements had derived many promoters and were thus considered to contribute a lot to lineage-specific patterns of gene expression in humans (Huda et al. 2011). In addition, except for the involvement of transcriptional regulation, several lines of evidence have shown that *Alu* elements also play a role in the post-transcriptional regulation by deriving target sites for miRNAs in the 3′ untranslated region (UTR). For example, Jordan and colleagues illustrated that nearly 20% of the human genes contain TEs in the 3′ UTR (Jordan et al. 2003), and it was also observed that some *Alu* elements within the 3′ UTR of human mRNAs are highly conserved and provide perfect complementary target sites for miRNAs (Smalheiser and Torvik 2006).

Transcription of eukaryotic genes is a highly complicated process requiring the ultra-precise cooperation of a batch of interactions among functionally diversified proteins and DNA sequences (Maston et al. 2006). Regulation of transcription is performed by a myriad of *cis*-regulatory elements, including enhancers, promoters, silencers and insulators, and fulfilled mainly by enhancers, which are portions of DNA that can activate transcription regardless of their location, distance or orientation relative to the promoters of genes by binding a variety of transcription factors (Ong and Corces 2011). This functional flexibility has indeed diversified the transcriptional regulation in different tissues or cell types. For example, a recent study examined the genome-wide histone H3 lysine 4 (H3K4) methylation patterns in two cancer cell lines—K562, a human erythroleukemia, and HeLa, a human cervical carcinoma, and predicted about 24,000 to 36,000 enhancers in each cell line, and found that only 5000 (14%–21%) are present in both, indicating the high cell-type-specificity of enhancer activity (Heintzman et al. 2009). Nevertheless, this flexibility impedes efforts to comprehensively investigate and record the full list of enhancers and promoters they regulated within the genome.

The systematic identification of a high-resolution *cis*-regulatory map between enhancers and their target promoters in mammalian systems has so far been limited. However, recently, 3C (chromosome conformation capture) based technologies have enabled large-scale spatiotemporal interactions, between distal sequence elements (mainly enhancers and promoters) within genomic loci located throughout the mammalian genome, to be routinely investigated by high-throughput microarray or deep sequencing approaches (Lieberman-Aiden et al. 2009; Wang et al. 2011; Eskeland et al. 2010; Kagey et al. 2010). 3C-based protocols were originally used to evaluate long-range chromatin interactions (CIs) between a pair of pre-specified genomic sites (Dekker et al. 2002). The extensions of the 3C approach include 4C (chromosome conformation capture-on-chip) (Simonis et al. 2006), 5C (chromosome conformation capture carbon copy) (Dostie et al. 2006), ChIA-PET (Chromatin Interaction Analysis with Paired-End Tag) (Fullwood et al. 2009) and Hi-C (Lieberman-Aiden et al. 2009) (reviewed in de Wit and de Laat 2012). Among these methods, Hi-C allows the detection of chromatin interactions between any pair of loci across an entire genome at a higher resolution (Lieberman-Aiden et al. 2009). In Hi-C experiments, CIs can be determined by the number of Hi-C reads and the read count is negatively correlated with the three-dimensional (3D) distance between two loci, suggesting that Hi-C read count can serve as a powerful proxy to measure the CI level (Lieberman-Aiden et al. 2009). Recently, Dixon et al. reported an investigation of the 3D organization of the human and mouse genome in four different cell types at unprecedented resolution using Hi-C method (Dixon et al. 2012) and identified large, mega-base-sized local chromatin interaction domains, termed "topological domains", as a pervasive structural feature of the genome organization in human and mouse. Such a pervasive structural characteristic was subsequently confirmed by an independent study (Nora et al. 2012). Further investigation reveals that the overall domain structure is generally conserved during mammalian evolution, and largely stable between different cell types, whereas intra-domain contributes most (>96%) of dynamic interacting regions, and thus potentially participate in cell-type-specific regulatory events (Dixon et al. 2012). Furthermore, they also observed that *Alu*/B1 and B2 elements in mice and *Alu* elements in humans are enriched at boundary regions of domains, indicating a role for TEs in the spatial organization of the genome. In contrast to the intense research on the spatial characteristics of TEs at boundary regions, TEs themselves, as the most abundant part in mammalian genomes, many unanswered questions associated with their function and involvement in intra-domains have largely remained uncharacterized. For example, whether TEs actively take part in the cell-type-specific regulation in intra-domains, and if so, how, is still mysterious.

To address these issues, in this study, we used newly released, genome-wide chromatin interaction data provided by Dixon et al. (Dixon et al. 2012) to investigate the possible roles for TEs, especially *Alu* elements, in intra-domain regions, and further link the TEs with EP interactions in human. First, we explored the association between TEs and CIs. Our results showed that SINE coverage is positively correlated with CI frequency in both pluripotent and differentiated cell lines, while other TE families are not. Further analyses indicated that such positive correlation is solely contributed by *Alu* elements. Second, we investigated whether the positive correlation between *Alu* coverage and CI frequency is due to the involvement of *Alu* elements in enhancer-promoter interactions. We found that enhancers or promoters are increased during the growing of CI frequency, resulting in the increasing of their possible interaction pairs. We then studied the relationship between *Alu* coverage and enhancer-promoter interactions, and found they indeed had a significantly positive correlation in different cell types, which indicated that the amount of *Alu* elements may be an alternative parameter to evaluate the distal interaction between enhancers and promoters. To test whether this pattern holds true in the human genome, we validated these results using a data set coming from an independent study. The examination confirmed our result to be a general feature in the human genome independent from tissue type.

Besides deriving enhancers and being involved in enhancer-promoter interaction, the *Alu* elements also play an important role in providing potential methylation sites. Previous studies showed that G or C nucleotides will accumulate in regions with more chromatin interactions (Dostie et al. 2006). And the *Alu* elements, especially the old *Alu*J and *Alu*S, have been known to be enriched in GC-rich regions ever since the sequencing of the human genome (Lander et al. 2001). Our results not only prove the phenomenon found in old studies again, but also for the first time showed that the *Alu* elements may be derived from increasing of GC content in high-CI regions, and CpG sites also, which may raise the methylation potentiality as a result. The high-CI regions are reported to be associated with open chromatin (Smit 1999) and are highly related to tissue-specific transcription by the regulation of distal DNA elements, nucleosome modification, and so on. The *Alu* elements can be therefore involved in tissue-specific regulation by raising methylation flexibility and deriving binding sites for transcriptional factors.

Taken together, our results show for the first time that *Alu* elements play a role in chromatin interactions, and its coverage can be served as a surrogate of CI frequency and be used to detect potential enhancer-promoter interaction. Additionally, the more *Alu* elements in regions with more chromatin contacts will endow the regions with methylation capacities and flexibilities, and may be important for tissue-specific transcription. This means that the spatial characteristics of TEs may play a more important role in conformation of chromatin structure and gene activity.

RESULTS

SINE elements are over-presented in the genomic loci of intra-domains with high frequencies of chromatin interactions

The recently released whole-genome CI profiling data in mammals include human embryonic stem cells (hESCs), human IMR90 fibroblasts, mouse embryonic stem cells and mouse cortex (Dixon et al. 2012), among which here we focused our analysis on two human cell lines. Several lines of evidence have previously demonstrated that the majority of functional *cis*-regulatory interactions occur within the same topological domain (Dixon et al. 2012; Nora et al. 2012). For example, although the distance between the Shh enhancer and the target shh gene exceeds nearly 1 Mb, they still locate within the same topological domain (Smallwood and Ren 2013). The frequency of intra-domain CIs is higher than inter-domain CIs and a higher frequency means a higher probability to be a real interaction (Li et al. 2010), suggesting that CIs between two genomic loci, located at different domains, may be largely non-informative or background noise. In other words, frequency of inter-domain CIs could serve as an effective threshold to remove false-positive CIs. Based on this assumption, we filtered the frequency of intra-domain CIs using mean frequency of inter-domain ones as a threshold, and removed the bin-pairs whose frequencies are lower than the threshold in each of the two human cell lines (see MATERIALS AND METHODS). Furthermore, it was also observed that bin-pairs with extremely high frequencies are scattered, indicating them as possible outliers. We subsequently discarded the bin-pairs whose frequencies are larger than the thresholds of 153 and 133 in hESCs and IMR90 fibroblasts, respectively. The upper thresholds were chosen according to the following reasons: (1) they are the first window without interacting bin-pairs in each cell lines, when we binned the frequency of intra-domain CIs by the step of 1; (2) after applying the upper threshold in each cell line, the removed bin-pairs only occupied 0.0085% and 0.0076% of valid ones in human ESCs and IMR90 fibroblasts, respectively (Fig. S1). Finally, we ambiguously retained 897,867 and 850,305 intra-domain CI bin-pairs (hereafter referred to as CI-bin-pairs, unless otherwise mentioned) in hESCs and IMR90 fibroblasts, respectively.

We first sought to examine whether a similar distribution pattern of intra-domain CI frequency (hereafter referred to as CI frequency, unless otherwise mentioned) exists in four different TE families. To do this, we divided CI frequencies into 50 bins (referred to as frequency bin), and computed density for each of the four TE families in each frequency bin. The density is the fraction of TEs belonging to the same family in one CI-bin-pair, divided by the maximum possible same-family TEs, which is the ratio of the length of each CI-bin-pair (80 kilobases) to the mean length of that TE family (Fig. 1). The density was calculated for each of the four TE families and plotted across different bins in hESCs and IMR90 fibroblasts, respectively (see Table S1 for TE families with calculated density in each cell line). Higher density implies that more same-family TEs are in a frequency bin, whereas a lower one means less same-family TEs. We noticed that two TE families, LTR and DNA transposons, have comparable low densities, yet SINE and LINE have higher densities across different bins in both cell lines (Fig. 1). This phenomenon may be due to the reason that DNA transposons and LTRs either are currently not mobile or have a very limited activity in the human genome compared to SINEs and LINEs, where their subfamilies, such as *Alu*s and L1s, are still active in humans and have proliferated during primate evolution (Cordaux and Batzer 2009). Then, the Pearson correlation coefficient (PCC) score was computed for each of four different TE families in each cell type, and we found that SINE density is positively correlated with CI frequency in hESC ($r = 0.89$, $P < 2.2 \times 10^{-16}$, Fig. 1A), and this trend for SINEs is also observed in IMR90 fibroblasts ($r = 0.91$, $P < 2.2 \times 10^{-16}$, Fig. 1B). But the densities of other three TE families (LINE, LTR, and DNA) are negatively correlated in both cell lines (Fig. 1C–H). To evaluate whether the above results will be affected by bin lengths, we further separated CI frequency into 20 and 100 bins, respectively. The results illustrated that the trends of different TE families, previously observed in the bin size of 50, were still hold true for the bin sizes of 20 and 100 (Figs. S2 and S3).

Since the above results are not affected by the different bin sizes, we then tested whether these results are significantly different from those of non-intra-domains. The correlation between density and CI frequency in non-intra-domains were subsequently calculated for each of the four TE families, respectively. We found that no TE families are correlated with interaction frequency in non-intra-domains, except for DNA transposons in IMR90 fibroblasts with a weakly negative correlation ($r = -0.377$, $P = 0.013$, Table S1). We further compared the correlations of DNA transposon between intra- and non-intra-domains using a z-test, and found that they are significantly different ($P = 0.0016$). Thus, these results clearly demonstrate that the density of four TE families in intra-domains are significantly different from those in non-intra-domains, and SINE is the only TE family whose distribution shows strong correlations with the frequency of intra-domain CIs in both investigated cell lines. Such correlation is also supported by a recent study, in which they indicated that multigene chromatin interacted regions are enriched with SINE density.

Alu is the major player of TEs in intra-domain chromatin interactions

As *Alu*s are the most successful active TEs in the human genome in terms of copy number, and occupy two-thirds of the SINE elements, this prompted us to contemplate whether *Alu*s contributed the most to the correlation between SINE density and CI frequency in intra-domains. To prove this hypothesis, we divided SINE elements into two subgroups, *Alu*s and SINE/non-*Alu*s, and investigated them separately. Compared with SINE/non-*Alu*s, we found that *Alu* density showed very strong correlations with CI frequencies in both hESC ($r = 0.90$, $P < 2.2 \times 10^{-16}$) and IMR90 fibroblasts ($r = 0.94$, $P < 2.2 \times 10^{-16}$), in sharp contrast with PCC scores of −0.082 ($P = 0.57$) and −0.4 ($P = 0.0044$) for SINE/non-*Alu*s in hESC and IMR90 fibroblasts, respectively (Fig. 2A and 2B). Likewise, we noticed that SINE/non-*Alu*s are even not associated with CI frequencies due to the insignificant *P*-value ($P > 0.05$). The difference between *Alu*s and SINE/non-*Alu*s may be due to the reason that, except for *Alu*s, most of SINEs have lost their mobilization activity in mammals, and such a strong correlation between *Alu* density and CI frequency in both investigated cell lines is consistent with our assumption that *Alu*s are the major contributors of SINE elements in intra-domain interactions.

As mentioned previously, *Alu*s can be grouped into three major groups (old *Alu*J, middle *Alu*S, and the young *Alu*Y) that emerged at different time points during the course of primate evolution. We further ask whether a similar tendency can be observed in different *Alu* subfamilies. To do this, we investigated the characterizations of each *Alu* subfamilies by calculating their density across different frequency bins. The distribution of density of three subfamilies in two cell types

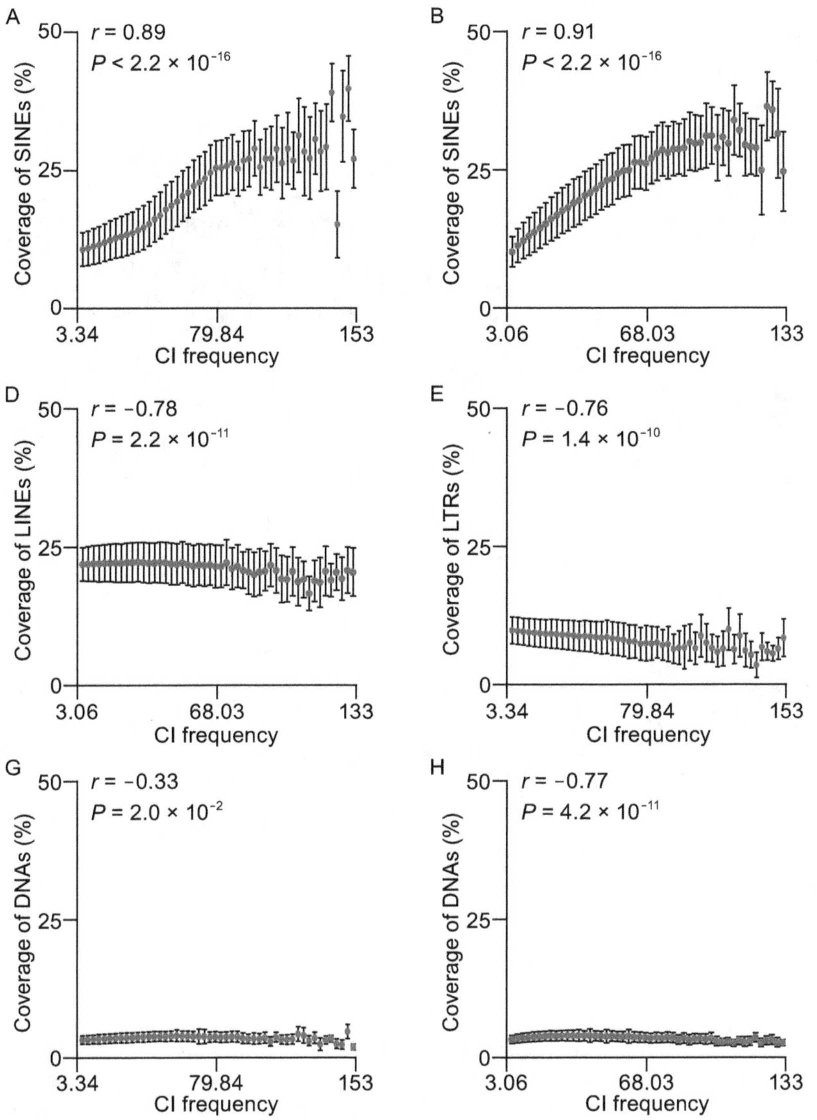

Figure 1. Mean coverage (MC) of four TE types in each 50 frequency bin. (A) The MC of SINE in 50 frequency bins in hESC. (B) The MC of SINE in 50 frequency bins in IMR90. (C) The MC of LINE in 50 frequency bins in hESC. (D) The MC of SINE in 50 frequency bins in IMR90. (E) The MC of LTR in 50 frequency bins in hESC. (F) The MC of SINE in 50 frequency bins in IMR90. (G) The MC of DNA in 50 frequency bins in hESC. (H) The MC of SINE in 50 frequency bins in IMR90.

was plotted in Fig. S4. The densities of three major subfamilies are all positively correlated with CI frequencies in both hESC and IMR90 fibroblasts, suggesting that they have similar distribution tendency in intra-domain interactions.

Considering that *Alu*s are active TEs in the human genome, we doubted whether other active TE families in the human genome shared a similar pattern as *Alu*s. Besides *Alu*s, there are two other TE subfamilies, L1 and SVA, which can actively transpose in humans (Chen et al. 2005). Since SVA is very sparse in the human genome with only 1750–3500 copies, we only focused on the L1 family by exploring the relationship between L1 density and CI frequency. Fig. 2C and 2D clearly demonstrated that L1 densities are not correlated with CI frequencies in hESC ($r = 0.11$, $P = 0.47$), and negatively correlated in IMR90 fibroblasts ($r = -0.42$, $P = 0.0021$). Although L1s and *Alu*s are both active in the human genome, they show no similar distribution patterns in each of the two cell lines. Such a huge difference between them may be due to the reason that *Alu*s have a greater tendency to be retained in the genomic loci with higher frequency of intra-domain interactions, whereas L1 elements do not or even tend to deplete in the higher frequency loci, although both of them maybe integrate into genome randomly. Our finding strongly indicated that more than 81% (r^2) of the variation in *Alu* density could be explained by CI frequency, suggesting that *Alu*s may play an

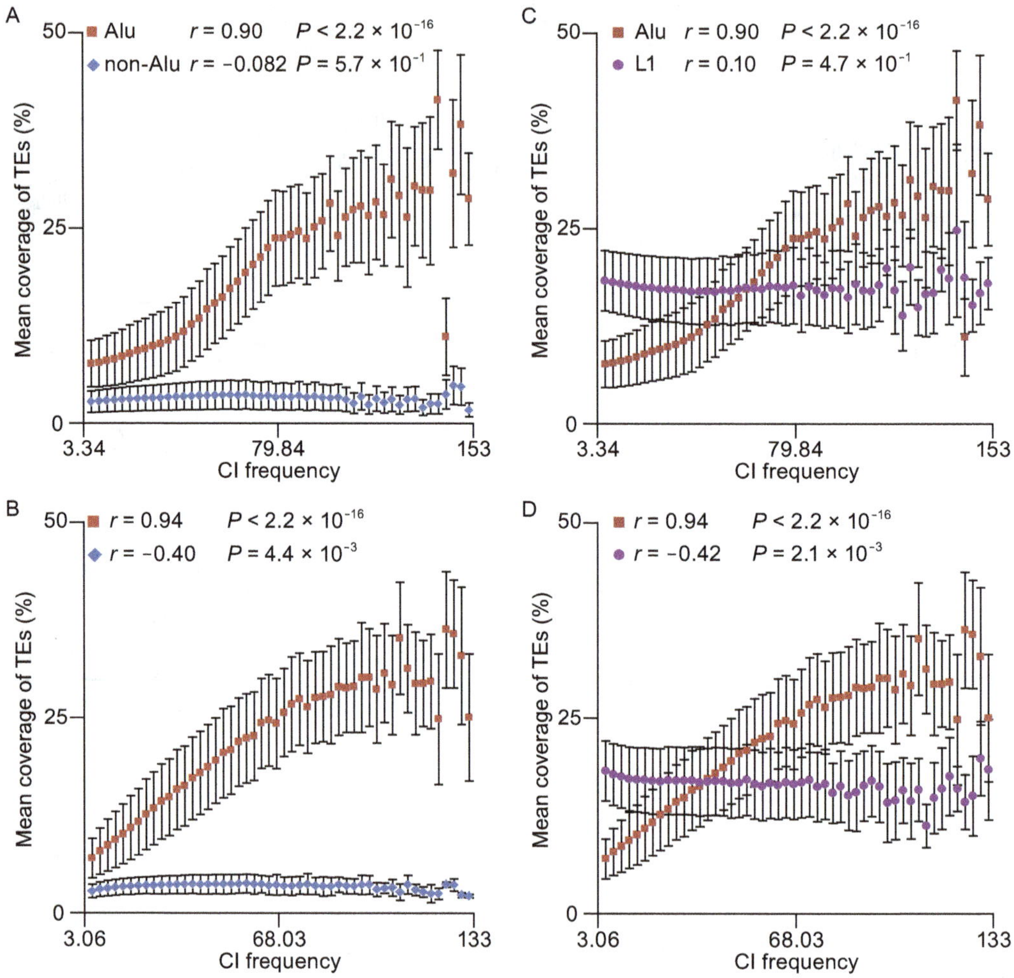

Figure 2. MC of SINE/*Alu*, SINE/non-*Alu*, and LINE/L1 in each frequency bin. (A) In hESC, the MCs of *Alu* were positively correlated with CI frequencies ($r = 0.90$, $P < 2.2 \times 10^{16}$), but the MCs of SINE/non-*Alu* were not correlated with CI frequencies ($r = -0.082$, $P = 0.57$). (B) Although the MCs of *Alu* were also positively correlated with CI frequencies in IMR90 fibroblasts ($r = 0.94$, $P < 2.2 \times 10^{16}$), the MCs of SINE/non-*Alu* in IMR90 fibroblasts were negatively correlated with CI frequencies ($r = -0.40$, $P = 0.0044$). (C) Different patterns of the correlation of the MCs of *Alu* and L1 with CI frequencies were observed in hESC, with a positive correlation of *Alu* and no correlation of L1 ($r = 0.11$, $P = 0.47$). (D) In IMR90 fibroblasts, on the contrary of the positive correlation of the MCs of *Alu*, the MCs of L1 were negatively correlated with CI frequencies ($r = -0.42$, $P = 0.002$).

important functional role in intra-domain interactions. Taken together, our results strongly emphasized that *Alu*s are over-presented in genomic loci of intra-domains with high CI frequencies.

Abundant *Alu*s in highly interacted CI regions contribute more GC content and CpG sites, and probably facilitate to maintain an open chromatin status

The SINE elements are long known to be appeared in GC-rich regions that are associated to open chromatin (Smit 1999). Previous studies have indicated that chromatin interactions are enriched in higher GC-rich regions (Li et al. 2012), so as the SINE elements (Korenberg and Rykowski 1988). However, it is still a mystery what relationships there are among CI frequency, GC content and SINE density.

To answer this question, we first divided total GC content (background) in each CI-pair into three major groups (*Alu*, SINE/non-*Alu* and non-SINE) according to their origins, and then calculated the arithmetic mean of GC content values for each CI frequency bin. We found that high CI frequencies are significantly enriched in GC-rich regions in both hESC ($r = 0.92$, $P < 2.2 \times 10^{-16}$, green triangle) and IMR90 ($r = 0.9$, $P < 2.2 \times 10^{-16}$, green triangle), which is consistent with the previous study (Li et al. 2012) (Fig. 3A and 3B). Interestingly, further investigation indicates that the GC content contributed by *Alu*s are significantly correlated with CI

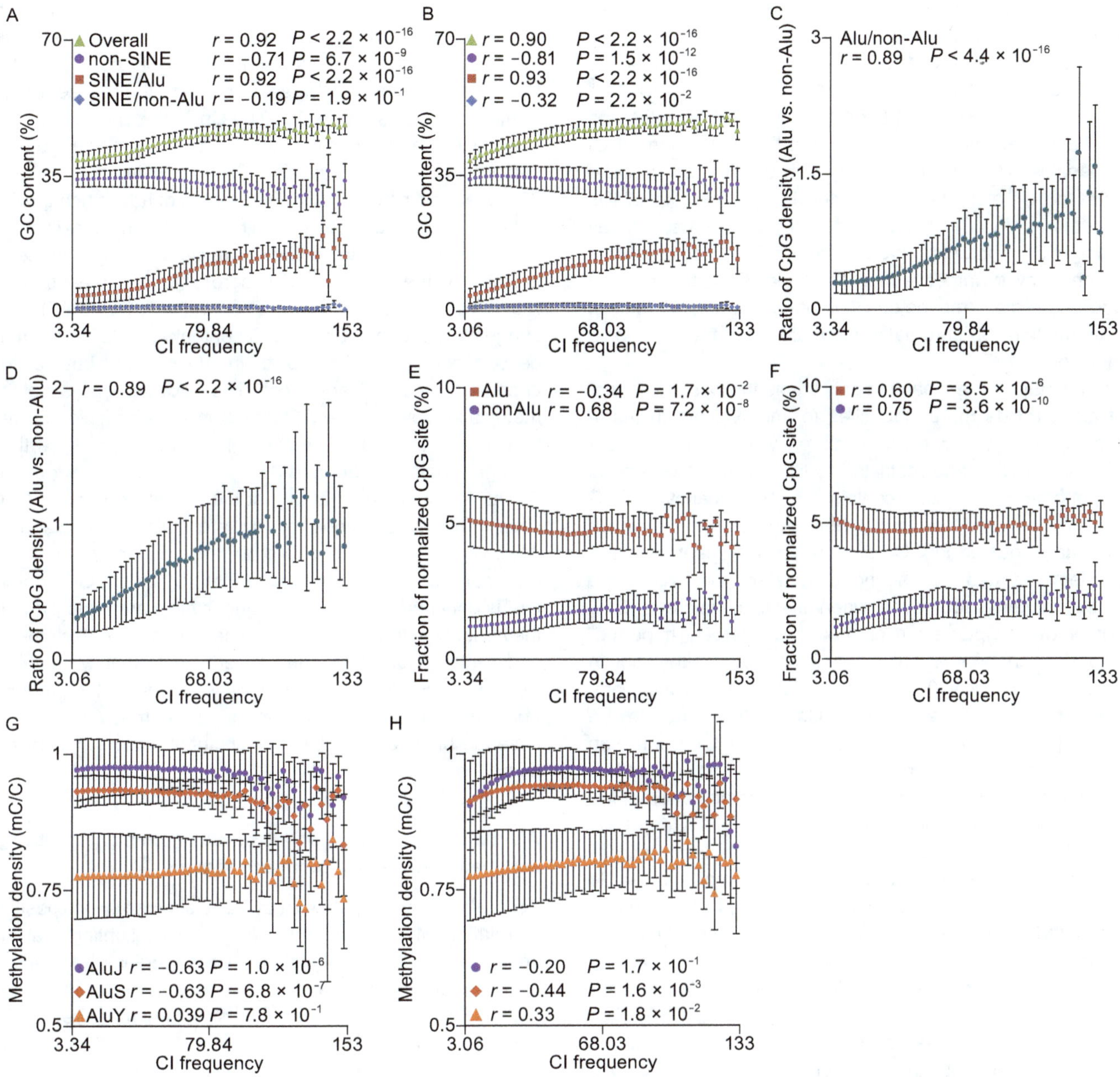

Figure 3. The GC content, CpG content, and methylation rate of bin-pairs and SINEs. (A and B) GC content of genome background, *Alu* elements regions, SINE/non-*Alu* regions (other transposable elements in SINE type) and non-SINE regions (regions that aren't covered by SINEs) in each frequency bin (A: hESC; B: IMR90 fibroblasts). (C and D) The ratio of CpG density of *Alu* regions and genome non-*Alu* regions (*Alu* regions + SINE/non-*Alu* regions + non-SINE regions) increased as the increasing of CI frequencies. (C: hESC; D: IMR90 fibroblasts). (E and F) The fraction of normalized CpG sites in *Alu* regions and non-*Alu* regions (*Alu* regions + SINE/non-*Alu* regions + non-SINE regions) (E: hESC; F: IMR90 fibroblasts). (G and H) Methylation density (the ratio of the methylated cytosine in CpG site) of three *Alu* subfamilies, *Alu*J, *Alu*S, and *Alu*Y (G: hESC; H: IMR90 fibroblasts).

frequencies in hESC ($r = 0.92$, $P < 2.2 \times 10^{-16}$, red square) and IMR90 ($r = 0.93$, $P < 2.2 \times 10^{-16}$, red square), whereas those contributed by SINE/non-*Alu*s are not or negatively correlated with CI frequencies in hESC ($r = -0.19$, $P = 0.19$, blue diamond) and IMR90 ($r = -0.32$, $P = 0.022$, blue diamond), respectively (Fig. 3A and 3B). On the other hand, GC contents derived from *Alu*s, on average, are generally higher than those from SINE/non-*Alu*s, suggesting that large proportion of GC contents in SINE elements are contributed by *Alu*s. We also observed that there are strong negative correlations between GC contents contributed by non-SINEs and CI frequencies in both cell types ($r = -0.71$, $P = 6.7 \times 10^{-9}$ for hESC; $r = -0.81$, $P = 1.5 \times 10^{-12}$ for IMR90, Fig. 3A and 3B). These results demonstrated that the enrichment of CI frequencies in GC-rich regions is solely due to the increased copies of *Alu* elements, suggesting that enrichment of *Alu* elements in high CI frequency loci may be not just a result of higher GC contents but a cause of it. We concluded that the tendency toward overrepresentation of *Alu*s in high CI frequency regions maybe due to important functional roles in the formation of chromatin structure and transcriptional regulation.

Higher GC content can not only stable the DNA structure, but also provide more CpG sites for methylation. In mammals, cytosine methylation is, currently, the only covalent DNA modification and restricted to CpG sites. CpG islands are genomic regions that contain a high frequency of CpG dinucleotides, and commonly represent promoters, which are usually located in GC dense regions. Recent research has demonstrated that strong CpG island promoters are mostly unmethylated, even when inactive, whereas CpG poor or weak CpG island promoters are largely hypermethylated, or are preferential targets for *de novo* methylation in human, respectively (Weber et al. 2007). This result indicates that CpG islands with CpG sites tend to be hypomethylated to allow an open chromatin organization and facilitate neighboring gene expression. As *Alu*s encompass ~25% of all CpG sites in the human genome, Xie et al. investigated the epigenetic status of CpG sites of *Alu* elements in human ependymomas and found that methylation status of the majority of CpG sites within or adjacent to *Alu* elements remained unchanged, and differentially methylated CpG sites, in normal control and ependymomas, are enriched in the loci with low CpG density in the flanking regions of *Alu* elements, rather than within the *Alu* sequences themselves (Xie et al. 2010), suggesting CpG sites within *Alu* elements may be more resistant to alteration in methylation than those in the flanking regions. Based on this evidence, we then asked whether the increased GC contents contributed by *Alu*s in high CI frequency loci result in elevated CpG density and further function in maintaining an open chromatin structure for highly active CIs. To tackle this question, we analyzed CpG density contributed by three above-mentioned categories (*Alu*, SINE/non-*Alu* and non-SINE) and compare them with background (see MATERIALS AND METHODS). We found that background CpG density is correlated positively with CI frequency ($r = 0.90$, $P < 2.2 \times 10^{-16}$ for hESC, Fig. S5A; $r = 0.95$, $P < 2.2 \times 10^{-16}$ for IMR90, Fig. S5B). Similar to GC content, strong correlations between CpG density contributed by *Alu*s and CI frequency were observed ($r = 0.88$, $P < 2.2 \times 10^{-16}$ for hESC, Fig. S5A; $r = 0.95$, $P < 2.2 \times 10^{-16}$ for IMR90, Fig. S5B). In contrast, CpG density contributed by SINE/non-*Alus* is not correlated with CI frequency and, on average, is much lower than that from *Alus*, suggesting that the majority of CpG sites contributed by SINEs come from *Alu* elements. We then performed the correlation analysis comparing the CpG sites provided by *Alu* sequence and the rest sequence termed "non-*Alu* sequences". The CpG sites provided by *Alu* sequences increase significantly faster than the non-*Alu* sequences as the increase of CI frequencies (Fig. 3C and 3D). We then looked into the proportion of how much the GC content becoming CpG sites; the results showed that the proportion of CpG sites in the GC nucleotide in *Alu* elements is stable as the increase of CI scores, but is increased in the region not covering by *Alu* elements (Fig. 3E and 3F). This suggests to the increase of GC content and CpG sites may be connected with the growth in *Alu* number, but not the changing of the *Alu* sequences. The following analysis of the methylation level of the CpG sites in *Alu* elements showed that older *Alu* elements have higher methylation level than younger ones (Fig. 3G and 3H).

Our result demonstrated that the increased parts of CpG sites in higher CI frequency loci are mainly contributed by *Alu*s, which is consistent with our assumption. In summary, our analysis strongly suggested that, in human, GC content that is enriched in higher CI frequency loci are mainly due to the tendency towards overrepresentation of *Alu* elements and further significantly increased CpG density in these regions. Such elevated CpG density may protect the regions from methylation and allows an open chromatin structure to increase binding probability of ubiquitous transcription factors (Deaton and Bird 2011).

Alu-derived Enhancers and Promoters are significantly enriched in higher CI frequency regions

According to our results, *Alu* coverage showed strong positive correlation with CI frequency (Fig. 2), highlighting that the enrichment of *Alu*s in high CI frequency regions will leads to elevated GC contents and CpG sites that might function in maintaining an open chromatin structure to increase the binding probability of transcriptional factors (Fig. 3A–D). However, such enrichment will also evolve in distal DNA elements regulation directly, like deriving enhancers and promoters that forming distal interactions. Previous studies indeed supported this hypothesis that enhancers are enriched in TE-derived genomic sequences. For example, an *Alu*-containing enhancer regulates the *CD8a* gene in human T cells (Hambor et al. 1993). Likewise, promoter regions are also hotspots of TEs that provided alternative promoters or binding sites for transcription factor (TF). For instance, the LTR-derived promoter significantly increases the expression of human EBR gene in placenta (Medstrand et al. 2001), and *Alu* elements provide binding motifs for three zinc-finger TFs (Sp1, estrogen receptor alpha, and YY1) (Oei et al. 2004). Because chromatin interaction is usually used to exam enhancer-promoter interactions, we determined to investigate the relationship between enhancer-promoter interactions and CI frequency using histone modification marks. Since active enhancers are marked by mono-methylation of H3 lysine 4 (H3K4me1) together with acetylation of H3 lysine 27 (H3K27ac), and active promoters are associated with tri-methylation of H3 lysine 4 (H3K4me3), we analyzed the data downloaded from the UCSD Human Reference Epigenome Mapping Project (see MATERIALS AND METHODS). We first computed the expression levels of H3K4me3-marked genes, and found that these genes indeed expressed higher than non-active ones (Fig. S6), which is consistent with the previous evidence (Santos-Rosa et al. 2002; Santos-Rosa et al. 2003; Schneider et al. 2004).

After mapping active enhancers and TSSs to the chromatin interaction bin-pairs, we found that the number of enhancers and TSSs are positively correlated with CI frequencies in both hESC and IMR90 cell lines, indicating that the enhancers and promoters are enriched in the interacted genomic loci with high CI frequencies (Fig. 4A and 4B). Since *Alu* coverage increased with growing CI frequency, we asked whether *Alu* elements contribute a lot to the increasing trends of enhancers and promoters during the increase of CI frequencies. To answer this question, we divided the genome into 40 kb bins and assigned each bin to 5 groups based on the Alu coverage of the sequence. We then compared the distribution of regulatory elements in the bins of the 5 groups. By correlation analysis of the coverage of *Alu* elements and the number of enhancers and TSSs, we found that both the enhancers and TSSs showed striking positive correlation with *Alu* coverage (Enhancer: hESC: $r = 0.997$, $P = 2.3 \times 10^{-4}$; IMR90: $r = 0.934$, $P = 2 \times 10^{-2}$; Promoter: hESC: $r = 0.995$, $P = 3.8 \times 10^{-4}$; IMR90: $r = 0.996$, $P = 3.2 \times 10^{-4}$, Fig. 4C and 4D), suggesting the *Alu* coverage can partially indicate the enrichment of active enhancers and TSSs. This result was confirmed by further examination by using two other independent data sources, the experimental reviewed enhancers in VISTA enhancer browser, and predicted enhancers of K562 and HeLa cell lines predicted by Heintzman ND et al. (Heintzman et al. 2009). We grouped the enhancers based on whether they are covered by *Alu* elements, and analyzed the 40 kb sequence centered on the enhancers. The results demonstrated that the regions around *Alu*-derived enhancers have significantly high *Alu* coverage compared with the randomly selected regions in both two cell lines and VISTA dataset ($P < 0.05$). The result was verified in flanking regions of the enhancers of different length (Fig. S8). The enrichment of active enhancers and TSSs in regions with more distal chromatin interaction and higher *Alu* coverage showed that *Alu* elements may be able to predicted distal enhancer-promoter interaction as a complement of Hi-C data, and may have some biological function in distal regulation of gene transcription.

DISCUSSION

Alu elements are highly evolved in genome-wide chromatin interaction

Alu elements, the active SINE in human genome, are highly enriched in regions with high level of chromatin contact and can be an additional criterion to conduct Hi-C data quality control. Transposable elements have been regarded as noncoding sequences with potential regulatory functions in genomes instead of the orthodoxy "junk DNA" for quite a time. Previous studies of mammalian genomes revealed important functions of TEs, including promoting the evolution of mammalian pregnancy pathway (Lynch et al. 2011; Xie et al. 2010) and tissue-specific transcriptions (Bourque et al. 2008; Faulkner et al. 2009; Lowe et al. 2007; Jin et al. 2012), rewiring the regulatory network in embryo development (Kunarso et al. 2010; Gifford et al. 2013), the microRNA mediated post-transcriptional regulation (Ahn et al. 2013; Berezikov 2011), and the arrangement of CTCF (CCCTC-binding factor) binding sites (Schmidt et al. 2012). The newly studied distal regulatory roles of SINE, the wide-spread retrotransposons in the human genome, are enlightened by chromatin construction capture methods, but only find SINE enriched in regions that separate interaction-rich domains (Lieberman-Aiden et al. 2009; Wang et al. 2011; Eskeland et al. 2010; Kagey et al. 2010). By testing the correlation between different TE types and chromatin interaction levels, we found only SINE but not other TEs enriched in region with high CI scores. Besides, further analysis ruled out non-*Alu* elements in SINE, elucidating that *Alu* element tend to enriched in regions with high CI score for the first time, which suggestion that the *Alu* elements may be significant for distal regulations. High level of chromatin interactions indicates more adjustable transcriptional regulatory events like enhancer-promoter interaction, and has strong nexus with tissue-specific expression (Dixon et al. 2012). There are more than 1 million *Alu* copies in human genome. Besides causing diseases in most time they transpose (Batzer and Deininger 2002; Ule 2013; Winkler et al. 2013; Gallus et al. 2010), the fixed ones often went through selection, and some of them have obtained various important functional roles including inducing enhancers and transcriptional factor binding sites in promoters (Huda et al. 2011; Pastor and Pagani 2011; Antonaki et al. 2011; Cui et al. 2011), deriving alternative spliced exons (Sorek et al. 2002; Shen et al. 2011), and forming alternative promoters, which lead to tissue-specific expression (Faulkner et al. 2009; Lin et al. 2008). Although L1 of LINE, the other active TE in human genome, also can provide source for expression change though mechanisms like generating pseudogenes (Esnault et al. 2000), exonization (Kaer et al. 2011), our results showed that it is *Alu* elements that play the role in distal controlling of genome but not the other TEs, even the other activate ones.

We further suggest *Alu* elements be a new complementary parameter to CI scores on the basis of the significantly positive correlation between the density of *Alu* elements and the score of chromatin interaction (Fig. 2). The Hi-C approach for knowing the genome-wide chromatin contact is undoubtedly remarkable, which makes the search for good ways to eliminate the bias and screen out high-quality pairs important. More than one model was raised to ensure the

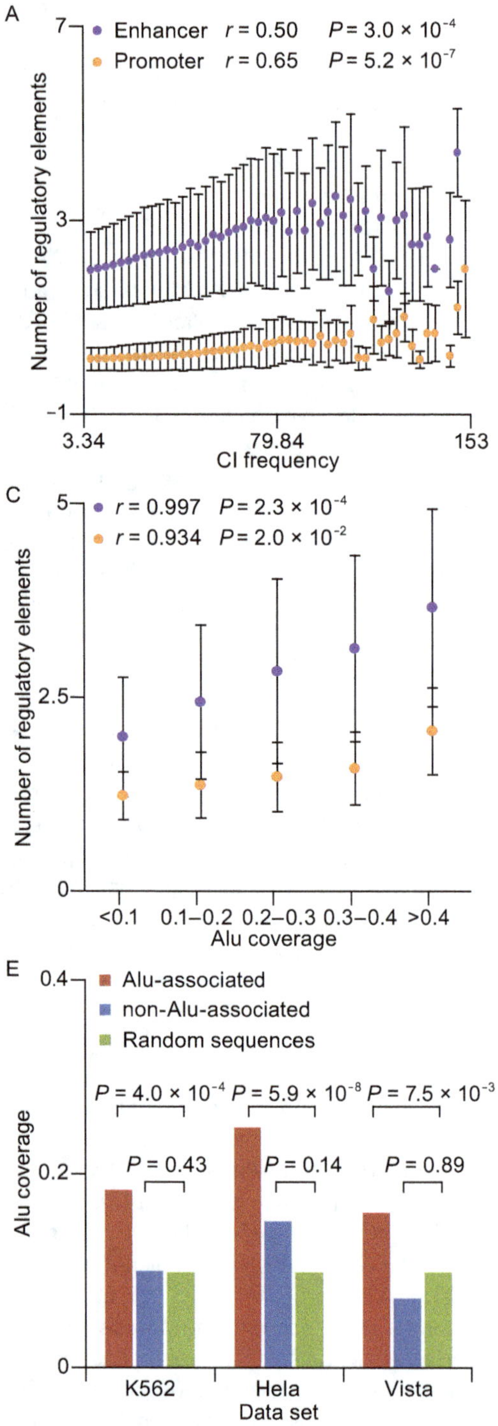

Figure 4. Mean numbers of enhancers and promoters demonstrated correlation with CI frequencies. (A and B) The mean numbers of active enhancer regions and active TSS regions were both positively correlated with CI frequencies in two cell lines (A: hESC; B: IMR90 fibroblasts). (C and D) The mean numbers of active enhancer regions and active TSS regions were both positively correlated with *Alu* coverage of the bin-pairs they belonged to in two cell lines (C: hESC; D: IMR90 fibroblasts). (E) The *Alu* coverage in the 40 kb regions centered on the predicted enhancer peaks of K562 cell line, HeLa cell line, and in VISTA enhancer browser. The *Alu*-overlapped enhancers in the two cell lines and VISTA enhancer browser were all significantly different from the background, which is the mean *Alu* coverage of a random 40 kb region in human genome (Binomial test, $P < 0.01$), and the non-*Alu*-overlapped enhancers in HeLa cell line were also significantly different from the random distribution (Binomial test, $P < 0.05$).

reliability of Hi-C data, which included procedures like reads quality and length controlling, epigenetic features testing, and controlling GC effects (Yaffe and Tanay 2011; Lu et al. 2013). Considering the significantly high correlation between *Alu* elements and Hi-C data, the *Alu* may also be involved in the Hi-C-normalization models as an important parameter. Association analysis between the amounts of *Alu* and distal regulatory DNA elements like enhancers demonstrated that enhancers are more abundant in regions not only with higher CI scores, but also higher *Alu* coverage (Fig. 4), which indicates that the density of *Alu* can act as an ev*Alu*ator of CI analysis results.

The enhancer sometimes interacts with the promoter distantly and bidirectionally, which is the typical case of distal chromatin interaction that leads to tissue-specific regulation (Banerji et al. 1983; Gillies et al. 1983). Further tests showed that active enhancers regions marked by histone markers H3K4me1 and H3K27ac increased remarkably as the increase of CI frequencies, so as the active TSSs marked by H3K4me3 histone marker (Fig. 4A and 4B). When using the sequence coverage of *Alu* elements as the independent variable, the number of enhancers and promoters also follow the increasing pattern with a correlation *P* value less than 0.05 in both hESC and IMR90 fibroblasts (Fig. 4C and 4D), suggesting that the number of *Alu* elements could act as the one of the measures or filter criteria when estimating genome-wide active enhancer-promoter interaction (Fig. 5). The results above were then confirmed by enhancers of K562 leukemia cells and HeLa cells predicted by ChIP-chip approach and experimental-verified enhancers in a VISTA enhancer browser (Heintzman et al. 2009; Visel et al. 2007). In the investigation, the 40 kb regions with *Alu*-derived enhancers as the center were covered by significantly more *Alu* elements than random sequences, revealing the spatial consistency of *Alu* elements and enhancers. In short, we concluded, for the first time, the indicator potentiality that *Alu* may play in distal chromatin interactions, which is corroborated by the data of two other cell lines from an independent study and a widely-accepted experimental-reviewed database.

The high GC content in chromatin interaction hotspot is provided by *Alu* elements

Alu elements are GC-rich sequences, of which the presence will raise the GC content of certain sequences, provide more CpG sites for methylation, resulting in the remaining of open chromatin status and higher regulatory flexibility. The sequencing step in the Hi-C method for the data we used suffered from the interference of nucleotide composition, also regarded as GC content, which needed to be converted and normalized to limit the effects (Dixon et al. 2012; Yaffe and Tanay 2011). However, although the Hi-C data we investigated have been normalized to rule out GC effect, it still showed a significant correlation with CI scores (Fig. 3). This phenomenon should be caused by functional DNA elements enrichment, which is consistent with other studies. Investigation of the contribution *Alu* elements made to GC content demonstrated that the correlation of *Alu*-provided GC is also positively correlated with CI score, and the pattern is share consistency with the genome background, while the regions not covered by *Alu* elements contribute less GC content to the regions with higher CI scores. The increasing total GC content consisting of increased share of *Alu* regions and decreased share of non-*Alu* regions illustrate clearly that the presence of *Alu* elements raise the GC content in the regions with more chromatin contacts.

The arguments for the intriguing and complicated question why the GC content of *Alu*-surrounded DNA in human genomes is high have lasted for more than a decade. Despite the *Alu* elements making use of the insertion pathway of LINEs, which have a preference for AT-rich integration sites and lead to the accumulation of LINEs in AT-rich region, the human genome project strongly supports that the *Alu* elements tend to be distributed in GC-rich regions (Lander et al. 2001; Jurka 1997). Additionally, further analysis showed that the distribution of "young" *Alu* elements were still in accordance with the pattern of LINE, while the "old" *Alu* elements tended to accumulated in GC-rich regions, indicating the functional role *Alu* may play (Lander et al. 2001). Other studies soon followed up trying to determine the reason why the *Alu* distribution is relative to the GC content (Kunarso et al. 2010; Brookfield 2001), but none of them denied the phenomenon that most *Alu* elements, especially the old ones, have a surrounded area with high GC content. As expected, our results also proved the old finding that there are more *Alu* elements displayed in regions with higher GC content. However, it seemed that it is the GC-rich *Alu* elements themselves raise the GC content of the whole region, but not the sequences not belonging to *Alu*s. That is to say, the regions with high CI frequencies own more *Alu*s that have more guanine and cytosine, and the whole GC content of these regions are higher than the regions with less chromatin contact as a result (Fig. 5). To summarise, the finding that the *Alu* contributes more GCs in regions with higher CI scores combined and answered the two long controversial questions together, pointed out that the *Alu* elements are the reason why active interacted region have a higher GC content, and it is the *Alu* elements gathered together that raise the GC content of the region, not only the surrounding DNA but also itself.

The amount of GC can affect the chromatin structure directly, while a large part that forms its biological function is that DNA can be methylated at CpG sites, which will accumulate when there is more GC as our results show (Fig. S5). By comparing the genome-wide CpG density in different part of the human genome, we found that the although the CpG density of each region showed positive correlation with CI score regardless of whether the sequences are covered by *Alu* elements, the higher the CI score is, the more CpG comes from *Alu* elements than non-*Alu* regions (Fig. 3B), which indicates that *Alu* elements not merely generate GC

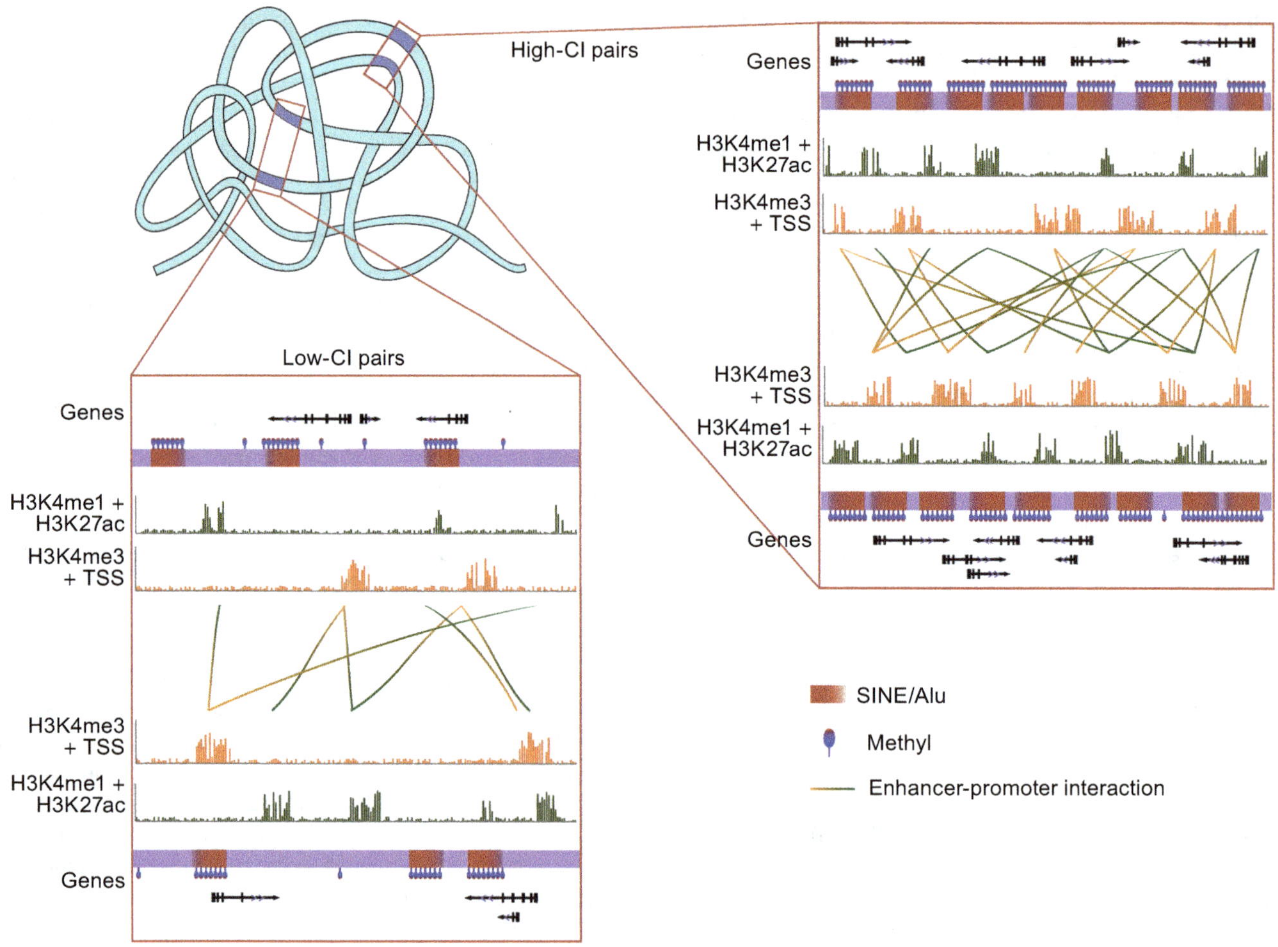

Figure 5. Different sequence component of regions with high or low CI frequencies. The right top square represents the high CI region, and the left bottom square represents the low CI region. In each square, the two long purple strips displayed vertically symmetrical demonstrate the two regions interacting with each other. The gradually-changed red squares on the strips show the *Alu* elements with certain direction. The methyl group is represented by a blue ellipse with a red tip on the sequence, which is denser in *Alu* elements than other sequence regions. The active enhancer regions are marked by histone markers H3K4me1 and H3K27ac, and the active TSS are marked by H3K4me3, which are shown by the green and yellow histogram separately. The interaction between enhancer regions and promoters is represented by lines with one end yellow (promoter) and the other arrows.

content rise in regions with high CI score, they also provide more CpG sites that can be methylated. When we looked into the proportion of how much the GC content becomes CpG sites, the results showed that the increasing GC content and CpG sites are connected with the growth in *Alu* number, but not the changing of the *Alu* sequences (Fig. 3C). The results shown above highlight again that the more *Alu* elements in regions with more chromatin contact can provide additional GC content as well as CpG sites.

The CpG sites in vertebrates are known for typical DNA methylation. The methylation process at CpG sites is by adding a methyl group to cytosine and turning it to 5-methylcytosine, which may change the structure of chromatin and result in various biological consequences. For example, the methylation surrounding TSS will block the initiation of transcription, while the methylation in the gene body will not block the transcription but even sometimes stimulate it (Suzuki and Bird 2008). The TEs are often highly methylated since most transposable events are deleterious, and the host genome must try to suppress the activation of TEs (Yoder et al. 1997; Bestor 1998). In our results, the *Alu*s showed a very high level of methylation in the whole genome with more than 90% of the CpG sites in *Alu* regions methylated regardless the cell types (Fig. S5C and S5D). This is consistent with the previous studies, and elucidated clearly that the transposition of *Alu* is not welcomed by genome and should be under control. The methylation rate showed a slightly negative correlation in hESC cell line, which maybe result from the several abnormal values at the high-CI regions, and it is not correlated with CI correlation in IMR90 fibroblasts. The high methylation in *Alu* elements also raises the methylation level of promoters and gene bodies when *Alu* is inserted into genes (Fig. S7C–F), which is quite reasonable because the transposition of *Alu* elements in actively transcribing genes should be strictly restricted. Nevertheless, the promoter region without *Alu* elements possessed more percentage of methylated CpG sites in higher-CI regions. Obviously, the promoters in regions with more chromatin contacts are actively regulated by DNA elements, so they have loose chromatin structures with less methylation. But the significant decreasing pattern of methylation did not happen in the gene body, which only had a weak decreasing trend as the CI score increased. As mentioned above, the methylation in CpG sites will not affect the transcription in the gene body, sometimes even promote it, which make our results rational no matter where the gene is actively regulated of trancribing, the methylation density in the gene body will not change much. Moreover, the methylation density in the gene body is always higher than the promoter as our results demonstrate (Fig. S7A and S7B), which confirmed again that the CpG islands are barely methylated when located around transcription start sites (TSSs), but will be methylated when in gene bodies in a tissue-specific manner.

Different *Alu* subfamilies follow the chronological methylation

Alu elements were noticed to have the GC-rich region preference after the human genome was sequenced, with the "old" ones tending to accumulate in GC-rich regions and "young" ones AT-rich regions, (Lander et al. 2001). Here we tested the methylation rate of the CpG sites in *Alu* elements with different ages, and found that besides the difference in preference for GC content, the "old" *Alu* elements were more highly methylated than the "young" ones (Fig. 3D). Those young *Alu* elements are generally more transcriptionally active than old ones results from various reasons. One of the reasons is the accumulation of mutations in *Alu* elements disables the recognition of transposase; others include methylation and RNA interferance by the host genome, which are more important. The mutations will accumulate as time passes, so the older ones must contain more mutated sites; however the methylation status will not always become increasingly intensive. The host genome cannot recognize new *Alu* repeats immediately after insertion, and forbid its future transposition with methylation, so the young *Alu*Y may have less methylated CpG sites and have less mutations in key regions related to transposition as our results indicate. As the *Alu* elements stay in the genome for more time, the host genome will discover this sequence and methylate it more often, and the sequence will change at a certain rate itself. Under the pressure of evolution, the methylation of CpG will eventually reach saturation or just enough to suppress the move and remain stable, and the mutation will not stop. When the sequence is converted so much that it could not be identified by transposase and will be silent forever, the deleterious nature of the *Alu* is fully removed, and the host genome may release it from its high methylation status as time goes by. According to our investigation of old *Alu*J, median *Alu*S and young *Alu*Y subfamilies, the *Alu*J have a slightly higher methylation density than *Alu*S, and both of them are almost fully methylated with more than 90% CpG sites methylated, while only less than 80% CpG sites in *Alu*Y were methylated. The average age of *Alu*J and *Alu*S is 81 million years and 36 million years (Kapitonov and Jurka 1996), while *Alu*Y sequences are very young and still fully active in human genome. If the methylation rate of the host genome is stable, we can estimate that the methylation rate of CpG sites in *Alu* elements is about 0.004% per million-year using the methylation density in our result, which means in another 10 million years, all the *Alu*Y will be fully methylated and silent as the presents *Alu*J. However, the new insertions will still be active and may form another subfamily that can still affect the host genome in various ways.

CONCLUSIONS

The *Alu* elements are highly enriched in the genome region with high levels of chromatin interaction. Correlation analysis reveals that the sequence character of *Alu* elements and its enrichment have endowed the actively regulating regions with a higher GC content, together with more CpG sites, and also a higher methylation density. The increase in methylation potentiality will provide flexibility to the sequences and transcriptional regulation as shown in Fig. 5. That is why the high CI frequencies always relate to tissue-specific expression. In contrast, other TEs including the active L1 in human genome are not enriched in high-CI regions and show a random distribution when evaluated by CI frequency.

MATERIALS AND METHODS

Processing chromatin interaction data

The recently released genome-wide chromatin interaction profiling of two cell lines in human, includes human embryonic stem (ES) cells and human IMR90 fibroblasts corresponding to pluripotent and differentiated cells. In these two cell lines, the datasets with the bin size of 40 kilobases (kb) were adopted, since chromatin interaction (CI) frequencies were normalized for biases in the data. The coordinates of domain-boundary annotations were converted from Hg18 to Hg19 using the UCSC Genome Browser liftover utility (http://genome.ucsc.edu/cgi-bin/hgLiftOver). We discarded the domains and bins, of which coordinates failed to remap, which is about 6.68% of the bins of Hg18. We then calculated the mean inter-domain CI frequencies in each cell line and used these values as the lower threshold for each cell line (3.34 for hESC and 3.06 for IMR90) and discarded all the bin-pairs whose CI frequencies are lower than the threshold in the corresponding cell lines (assuming they are non-informative or background noises in hESC and IMR90 cell lines). We then calculated the number of bin-pairs to be filtered out by setting upstream thresholds increased from the downstream threshold by 1 as the step each time, and finally removed the bin-pairs whose CI frequencies are larger than 153 in hESC and 133 in IMR90, respectively (assuming they are outlier in each cell line). Finally, we were able to retain 897,867 bin-pairs for hESC and 850,305 bin-pairs for IMR90.

Compiling TEs

The coordinates of TEs in the human genome were downloaded from RepeatMasker (http://www.repeatmasker.org, release version 3.3.0, Repbase library version 20110920) (Hubley and Green 1996). Based on the classification of RepeatMasker program, we divided all the TEs to four major types (LINE, SINE, LTR, and DNA) and discarded the TEs with uncertain categories. The coverage of certain TE type or TE family was defined as the number of nucleotide in the chromatin interaction bin-pairs by using the length and annotation of TEs. For the *Alu* subfamilies, *Alu*J, *Alu*S, and *Alu*Y, we counted the number of the TE copies of certain *Alu* subfamilies in chromatin interaction bin-pairs regarded as the enrichment score. The reason why we used coverage for four TE types, but used enrichment score for *Alu* subfamilies is that the average lengths of the four TE types are different, while those of the *Alu* subfamilies are nearly the same. The coverage of TE copies and the enrichment score of *Alu* subfamilies in bin-pairs with different CI frequencies were then used in the Pearson correlation coefficient analysis with other parameters.

Determination of the methylation density

The coordinates of methylated CpG sites of both hESC cell line and IMR90 fibroblasts were based on the results of recent work (http://neomorph.salk.edu/human_methylome) (Lister et al. 2009) and were converted from Hg18 to Hg19 by using UCSC Genome Browser liftover utility (http://genome.ucsc.edu/cgi-bin/hgLiftOver). To evaluate the methylation level, we calculated the percentage of methylated sites of all the CpG sites in each bin-pair, termed methylation density. Similarly, the CpG density was defined as the percentage of CpG sites in all the C/G nucleotide in each bin-pair, while the GC content is the percentage of C/G nucleotide in all the nucleotide in each bin-pair.

Identification of active genes and cis-elements

The active genes were identified in both hESC cell lines and IMR90 fibroblasts by combining the gene annotation and histone markers. We used annotated mRNAs in RefSeq genes and UCSC genes, and de-redundant by keeping the longest gene loci in gene loci with the same gene symbols. The ChIP-Seq data of H3K4me1, H3K27ac, and H3K4me3 histone markers were downloaded from GEO database (GSE16256), and processed by Model-based Analysis for ChIP-Seq (MACS) to rule out the false positive peaks with the false discovery rate (FDR) value over 5% (Heintzman et al. 2009; Zhang et al. 2008; Hawkins et al. 2010). We then combined the processed transcription start sites (TSS) annotation and H3K4me3 peaks to obtain the active markers. The active enhancer regions were defined by the overlapping regions of H3K4me1 and H3K27ac markers. Active TSSs were defined by H3K4me3 peaks on TSS. If a gene locus has multiple TSS, we choose the TSS with most H3K4me3 peak as the promoter of the whole gene locus. To evaluate the difference between enhancer regions and TSS regions in two different cell lines, we regarded the enhancers or promoters from two datasets a same one if the distance between two peaks is less than 250 bp. The hESC and IMR90 groups share 2126 enhancers, which are 8.2% in hESC and 7.2% in IMR90. The hESC and IMR90 groups share 5292 TSSs, which are 80.7% in hESC and 83.8% in IMR90 (Fig. S9A).

Correlation analysis between TEs and other chromatin characters

The PCC between the coverage of different TE families and the CI frequencies in each bin-pairs were calculated. The number of TE copies of *Alu* subfamilies and the CI frequencies were also evaluated by PCC. The number of GC content, CpG content and methylation density were investigated by PCC with CI frequencies in different regions in each bin-pairs including regions covered by *Alu* elements, regions covered by SINE elements but not *Alu*, and regions not covered by SINE. The PCC between the number of promoters and enhancers annotation by histone markers and the CI frequencies were calculated.

Processing enhancer data of K562 and HeLa cell lines, and VISTA database

The predicted enhancers of K562 and HeLa cell lines were adopted from Bing Ren's study (Heintzman et al. 2009). We also downloaded the reviewed data in the VISTA enhancer browser (http://enhancer.lbl.gov/frnt_page_n.shtml). We then grouped all the enhancers to *Alu*-derived class and non-*Alu*-derived class depending on whether their peaks are in *Alu*s, and then calculated the mean *Alu* coverage in 20 kb upstream to 20 kb downstream in the two classes. We evaluated the enhancers in three datasets by comparing the distance between their peaks. If the distance between two peaks is less than 250 bp, we regarded them as a same enhancer. The three groups had very different patterns that K562 and HeLa groups share 1200 enhancers, which are 3.3% in HeLa and 4.9% in K562. The K562 and Vista groups share 22 enhancers, which are 0.09% in K562 and 2.7% in Vista data. The HeLa and Vista groups share 28 enhancers, which are 0.08% in HeLa and 3.5% in Vista data. And only 4 enhancers are shared by all three groups, which are 0.01% in HeLa, 0.02% in K562, and 0.5% in Vista data (Fig. S9B).

AUTHOR CONTRIBUTIONS

Y. Zhong and P. Nan conceived and initiated the project. Z. Gu and K. Jin planed and performed the experiments and statistic analysis. Z. Zhang, Y. Zhang, and M. J. C. Crabbe contributed in data analysis. Z. Gu, K. Jin, Yanyan Huang, Mengyi Hua, M. J. C. Crabbe, and Y. Zhong wrote the manuscript. All authors reviewed and contributed to the final manuscript.

ACKNOWLEDGEMENTS

The authors thank the National Natural Science Foundation of China (Grant No. 91131901), Fudan Graduate Students Innovative Grant (EZH1322383/001/002) and PSCIRT for financial support.

ABBREVIATIONS

3C, chromosome conformation capture; 3D, three-dimensional; CIs, chromatin interactions; CTCF, CCCTC-binding factor; FAMs, fossil *Alu* monomers; FDR, false discovery rate; hESCs, human embryonic stem cells; LINE, long interspersed nuclear elements; LTRs, long terminal repeats; Myr, million years; SINE, short interspersed nuclear elements; TEs, transposable elements; TSS, transcription start site; UTR, untranslated region.

COMPLIANCE WITH ETHICS GUIDELINES

Zhuoya Gu, Ke Jin, M. James C. Crabbe, Yang Zhang, Xiaolin Liu, Yanyan Huang, Mengyi Hua, Peng Nan, Zhaolei Zhang, and Yang Zhong declare that they have no conflict of interest. This article does not contain any studies with human or animal subjects performed by the any of the authors.

REFERENCES

Ahn K, Gim JA, Ha HS, Han K, Kim HS (2013) The novel MER transposon-derived miRNAs in human genome. Gene 512 (2):422–428

Antonaki A, Demetriades C, Polyzos A, Banos A, Vatsellas G, Lavigne MD, Apostolou E, Mantouvalou E, Papadopoulou D, Mosialos G et al (2011) Genomic analysis reveals a novel nuclear factor-kappaB (NF-kappaB)-binding site in *Alu*-repetitive elements. J Biol Chem 286(44):38768–38782

Banerji J, Olson L, Schaffner W (1983) A lymphocyte-specific cellular enhancer is located downstream of the joining region in immunoglobulin heavy chain genes. Cell 33(3):729–740

Batzer MA, Deininger PL (2002) *Alu* repeats and human genomic diversity. Nat Rev Genet 3(5):370–379

Berezikov E (2011) Evolution of microRNA diversity and regulation in animals. Nat Rev Genet 12(12):846–860

Bestor TH (1998) The host defence function of genomic methylation patterns. Novartis Foundation Symposium 214:187–195; discussion 195–189, 228–132

Bourque G, Leong B, Vega VB, Chen X, Lee YL, Srinivasan KG, Chew JL, Ruan Y, Wei CL, Ng HH et al (2008) Evolution of the mammalian transcription factor binding repertoire via transposable elements. Genome Res 18(11):1752–1762

Brookfield JF (2001) Selection on *Alu* sequences? Curr Biol 11(22): R900–901

Brookfield JF (2005) The ecology of the genome—mobile DNA elements and their hosts. Nat Rev Genet 6(2):128–136

Chen JM, Stenson PD, Cooper DN, Ferec C (2005) A systematic analysis of LINE-1 endonuclease-dependent retrotranspositional events causing human genetic disease. Human Genet 117(5):411–427

Cordaux R, Batzer MA (2009) The impact of retrotransposons on human genome evolution. Nat Rev Genet 10(10):691–703

Cui F, Sirotin MV, Zhurkin VB (2011) Impact of *Alu* repeats on the evolution of human p53 binding sites. Biol Direct 6:2

de Wit E, de Laat W (2012) A decade of 3C technologies: insights into nuclear organization. Genes Dev 26(1):11–24

Deaton AM, Bird A (2011) CpG islands and the regulation of transcription. Genes Dev 25(10):1010–1022

Dekker J, Rippe K, Dekker M, Kleckner N (2002) Capturing chromosome conformation. Science 295(5558):1306–1311

Dixon JR, Selvaraj S, Yue F, Kim A, Li Y, Shen Y, Hu M, Liu JS, Ren B (2012) Topological domains in mammalian genomes identified by analysis of chromatin interactions. Nature 485(7398):376–380

Dostie J, Richmond TA, Arnaout RA, Selzer RR, Lee WL, Honan TA, Rubio ED, Krumm A, Lamb J, Nusbaum C et al (2006) Chromosome Conformation Capture Carbon Copy (5C): a massively parallel solution for mapping interactions between genomic elements. Genome Res 16(10):1299–1309

Eskeland R, Leeb M, Grimes GR, Kress C, Boyle S, Sproul D, Gilbert N, Fan Y, Skoultchi AI, Wutz A et al (2010) Ring1B compacts chromatin structure and represses gene expression independent of histone ubiquitination. Mol Cell 38(3):452–464

Esnault C, Maestre J, Heidmann T (2000) Human LINE retrotransposons generate processed pseudogenes. Nat Genet 24(4):363–367

Faulkner GJ, Kimura Y, Daub CO, Wani S, Plessy C, Irvine KM, Schroder K, Cloonan N, Steptoe AL, Lassmann T et al (2009) The regulated retrotransposon transcriptome of mammalian cells. Nat Genet 41(5):563–571

Feschotte C (2008) Transposable elements and the evolution of regulatory networks. Nat Rev Genet 9(5):397–405

Fullwood MJ, Liu MH, Pan YF, Liu J, Xu H, Mohamed YB, Orlov YL, Velkov S, Ho A, Mei PH et al (2009) An oestrogen-receptor-alpha-bound human chromatin interactome. Nature 462(7269):58–64

Gallus GN, Cardaioli E, Rufa A, Da Pozzo P, Bianchi S, D'Eramo C, Collura M, Tumino M, Pavone L, Federico A (2010) *Alu*-element insertion in an OPA1 intron sequence associated with autosomal dominant optic atrophy. Mol Vis 16:178–183

Gifford WD, Pfaff SL, Macfarlan TS (2013) Transposable elements as genetic regulatory substrates in early development. Trends Cell Biol 23(5):218–226

Gillies SD, Morrison SL, Oi VT, Tonegawa S (1983) A tissue-specific transcription enhancer element is located in the major intron of a rearranged immunoglobulin heavy chain gene. Cell 33(3):717–728

Grover D, Majumder PP, Rao CB, Brahmachari SK, Mukerji M (2003) Nonrandom distribution of *Alu* elements in genes of various functional categories: insight from analysis of human chromosomes 21 and 22. Mol Biol Evol 20(9):1420–1424

Grover D, Mukerji M, Bhatnagar P, Kannan K, Brahmachari SK (2004) *Alu* repeat analysis in the complete human genome: trends and variations with respect to genomic composition. Bioinformatics 20(6):813–817

Hackenberg M, Bernaola-Galvan P, Carpena P, Oliver JL (2005) The biased distribution of *Alu*s in human isochores might be driven by recombination. J Mol Evol 60(3):365–377

Hambor JE, Mennone J, Coon ME, Hanke JH, Kavathas P (1993) Identification and characterization of an *Alu*-containing, T-cell-specific enhancer located in the last intron of the human CD8 alpha gene. Mol Cell Biol 13(11):7056–7070

Hawkins RD, Hon GC, Lee LK, Ngo Q, Lister R, Pelizzola M, Edsall LE, Kuan S, Luu Y, Klugman S et al (2010) Distinct epigenomic landscapes of pluripotent and lineage-committed human cells. Cell Stem Cell 6(5):479–491

Heintzman ND, Hon GC, Hawkins RD, Kheradpour P, Stark A, Harp LF, Ye Z, Lee LK, Stuart RK, Ching CW et al (2009) Histone modifications at human enhancers reflect global cell-type-specific gene expression. Nature 459(7243):108–112

Huda A, Bowen NJ, Conley AB, Jordan IK (2011) Epigenetic regulation of transposable element derived human gene promoters. Gene 475(1):39–48

Jin P, Qin S, Chen X, Song Y, Li-Ling J, Xu X, Ma F (2012) Evolutionary rate of human tissue-specific genes are related with transposable element insertions. Genetica 140(10–12):513–523

Jordan IK, Rogozin IB, Glazko GV, Koonin EV (2003) Origin of a substantial fraction of human regulatory sequences from transposable elements. Trends Genet 19(2):68–72

Jurka J (1997) Sequence patterns indicate an enzymatic involvement in integration of mammalian retroposons. Proc Nat Acad Sci USA 94(5):1872–1877

Jurka J, Kohany O, Pavlicek A, Kapitonov VV, Jurka MV (2004) Duplication, coclustering, and selection of human *Alu* retrotransposons. Proc Nat Acad Sci USA 101(5):1268–1272

Kaer K, Branovets J, Hallikma A, Nigumann P, Speek M (2011) Intronic L1 retrotransposons and nested genes cause transcriptional interference by inducing intron retention, exonization and cryptic polyadenylation. PLoS One 6(10):e26099

Kagey MH, Newman JJ, Bilodeau S, Zhan Y, Orlando DA, van Berkum NL, Ebmeier CC, Goossens J, Rahl PB, Levine SS et al (2010) Mediator and cohesin connect gene expression and chromatin architecture. Nature 467(7314):430–435

Kapitonov V, Jurka J (1996) The age of *Alu* subfamilies. J Mol Evol 42(1):59–65

Kazazian HH Jr (2004) Mobile elements: drivers of genome evolution. Science 303(5664):1626–1632

Kidwell MG, Lisch DR (2001) Perspective: transposable elements, parasitic DNA, and genome evolution. Evolution 55(1):1–24

Korenberg JR, Rykowski MC (1988) Human genome organization: *Alu*, lines, and the molecular structure of metaphase chromosome bands. Cell 53(3):391–400

Kunarso G, Chia NY, Jeyakani J, Hwang C, Lu X, Chan YS, Ng HH, Bourque G (2010) Transposable elements have rewired the core regulatory network of human embryonic stem cells. Nat Genet 42 (7):631–634

Lander ES, Linton LM, Birren B, Nusbaum C, Zody MC, Baldwin J, Devon K, Dewar K, Doyle M, FitzHugh W et al (2001a) Initial sequencing and analysis of the human genome. Nature 409 (6822):860–921

Lander ES, Linton LM, Birren B, Nusbaum C, Zody MC, Baldwin J, Devon K, Dewar K, Doyle M, FitzHugh W et al (2001b) Initial sequencing and analysis of the human genome. Nature 409 (6822):860–921

Li G, Fullwood MJ, Xu H, Mulawadi FH, Velkov S, Vega V, Ariyaratne PN, Mohamed YB, Ooi HS, Tennakoon C et al (2010) ChIA-PET tool for comprehensive chromatin interaction analysis with paired-end tag sequencing. Genome Biol 11(2):R22

Li G, Ruan X, Auerbach RK, Sandhu KS, Zheng M, Wang P, Poh HM, Goh Y, Lim J, Zhang J et al (2012) Extensive promoter-centered chromatin interactions provide a topological basis for transcription regulation. Cell 148(1–2):84–98

Lieberman-Aiden E, van Berkum NL, Williams L, Imakaev M, Ragoczy T, Telling A, Amit I, Lajoie BR, Sabo PJ, Dorschner MO et al (2009) Comprehensive mapping of long-range interactions reveals folding principles of the human genome. Science 326(5950):289–293

Lin L, Shen S, Tye A, Cai JJ, Jiang P, Davidson BL, Xing Y (2008) Diverse splicing patterns of exonized *Alu* elements in human tissues. PLoS Genet 4(10):e1000225

Lister R, Pelizzola M, Dowen RH, Hawkins RD, Hon G, Tonti-Filippini J, Nery JR, Lee L, Ye Z, Ngo QM et al (2009) Human DNA methylomes at base resolution show widespread epigenomic differences. Nature 462(7271):315–322

Lowe CB, Bejerano G, Haussler D (2007) Thousands of human mobile element fragments undergo strong purifying selection near developmental genes. Proc Nat Acad Sci USA 104 (19):8005–8010

Lu Y, Zhou Y, Tian W (2013) Combining Hi-C data with phylogenetic correlation to predict the target genes of distal regulatory elements in human genome. Nucleic Acids Res 41(22):10391–10402

Lynch VJ, Leclerc RD, May G, Wagner GP (2011) Transposon-mediated rewiring of gene regulatory networks contributed to the evolution of pregnancy in mammals. Nat Genet 43(11):1154–1159

Maston GA, Evans SK, Green MR (2006) Transcriptional regulatory elements in the human genome. Annu Rev Genom Human Genet 7:29–59

Medstrand P, Landry JR, Mager DL (2001) Long terminal repeats are used as alternative promoters for the endothelin B receptor and apolipoprotein C-I genes in humans. J Biol Chem 276(3):1896–1903

Nekrutenko A, Li WH (2001) Transposable elements are found in a large number of human protein-coding genes. Trends Genet 17 (11):619–621

Nora EP, Lajoie BR, Schulz EG, Giorgetti L, Okamoto I, Servant N, Piolot T, van Berkum NL, Meisig J, Sedat J et al (2012) Spatial partitioning of the regulatory landscape of the X-inactivation centre. Nature 485(7398):381–385

Oei SL, Babich VS, Kazakov VI, Usmanova NM, Kropotov AV, Tomilin NV (2004) Clusters of regulatory signals for RNA polymerase II transcription associated with *Alu* family repeats and CpG islands in human promoters. Genomics 83(5):873–882

Ong CT, Corces VG (2011) Enhancer function: new insights into the regulation of tissue-specific gene expression. Nat Rev Genet 12 (4):283–293

Pastor T, Pagani F (2011) Interaction of hnRNPA1/A2 and DAZAP1 with an *Alu*-derived intronic splicing enhancer regulates ATM aberrant splicing. PLoS One 6(8):e23349

Polak P, Domany E (2006) *Alu* elements contain many binding sites for transcription factors and may play a role in regulation of developmental processes. BMC Genom 7:133

Quentin Y (1992a) Origin of the *Alu* family: a family of *Alu*-like monomers gave birth to the left and the right arms of the *Alu* elements. Nucleic Acids Res 20(13):3397–3401

Quentin Y (1992b) Fusion of a free left *Alu* monomer and a free right *Alu* monomer at the origin of the *Alu* family in the primate genomes. Nucleic Acids Res 20(3):487–493

Santos-Rosa H, Schneider R, Bannister AJ, Sherriff J, Bernstein BE, Emre NC, Schreiber SL, Mellor J, Kouzarides T (2002) Active genes are tri-methylated at K4 of histone H3. Nature 419 (6905):407–411

Santos-Rosa H, Schneider R, Bernstein BE, Karabetsou N, Morillon A, Weise C, Schreiber SL, Mellor J, Kouzarides T (2003) Methylation of histone H3 K4 mediates association of the Isw1p ATPase with chromatin. Mol Cell 12(5):1325–1332

Schmidt D, Schwalie PC, Wilson MD, Ballester B, Goncalves A, Kutter C, Brown GD, Marshall A, Flicek P, Odom DT (2012) Waves of retrotransposon expansion remodel genome organization and CTCF binding in multiple mammalian lineages. Cell 148 (1–2):335–348

Schneider R, Bannister AJ, Myers FA, Thorne AW, Crane-Robinson C, Kouzarides T (2004) Histone H3 lysine 4 methylation patterns in higher eukaryotic genes. Nat Cell Biol 6(1):73–77

Shen S, Lin L, Cai JJ, Jiang P, Kenkel EJ, Stroik MR, Sato S, Davidson BL, Xing Y (2011) Widespread establishment and regulatory impact of *Alu* exons in human genes. Proc Nat Acad Sci USA 108(7):2837–2842

Simonis M, Klous P, Splinter E, Moshkin Y, Willemsen R, de Wit E, van Steensel B, de Laat W (2006) Nuclear organization of active and inactive chromatin domains uncovered by chromosome conformation capture-on-chip (4C). Nat Genet 38(11):1348–1354

Smalheiser NR, Torvik VI (2006) *Alu* elements within human mRNAs are probable microRNA targets. Trends Genet 22(10):532–536

Smallwood A, Ren B (2013) Genome organization and long-range regulation of gene expression by enhancers. Curr Opin Cell Biol 25(3):387–394

Smit AF (1999) Interspersed repeats and other mementos of transposable elements in mammalian genomes. Curr Opin Genet Dev 9(6):657–663

Smit A, Hubley R, Green P (1996–2010) RepeatMasker Open-3.0. http://www.repeatmaskerorg

Sorek R, Ast G, Graur D (2002) *Alu*-containing exons are alternatively spliced. Genome Res 12(7):1060–1067

Suzuki MM, Bird A (2008) DNA methylation landscapes: provocative insights from epigenomics. Nat Rev Genet 9(6):465–476

Teng L, Firpi HA, Tan K (2011) Enhancers in embryonic stem cells are enriched for transposable elements and genetic variations associated with cancers. Nucleic Acids Res 39(17):7371–7379

Ule J (2013) *Alu* elements: at the crossroads between disease and evolution. Biochem Soc Trans 41(6):1532–1535

Visel A, Minovitsky S, Dubchak I, Pennacchio LA (2007) VISTA Enhancer Browser–a database of tissue-specific human enhancers. Nucleic acids research 35(Database issue):D88–92

Wang KC, Yang YW, Liu B, Sanyal A, Corces-Zimmerman R, Chen Y, Lajoie BR, Protacio A, Flynn RA, Gupta RA et al (2011) A long noncoding RNA maintains active chromatin to coordinate homeotic gene expression. Nature 472(7341):120–124

Weber M, Hellmann I, Stadler MB, Ramos L, Paabo S, Rebhan M, Schubeler D (2007) Distribution, silencing potential and evolutionary impact of promoter DNA methylation in the human genome. Nat Genet 39(4):457–466

Wicker T, Sabot F, Hua-Van A, Bennetzen JL, Capy P, Chalhoub B, Flavell A, Leroy P, Morgante M, Panaud O et al (2007) A unified classification system for eukaryotic transposable elements. Nat Rev Genet 8(12):973–982

Winkler S, van Leeuwen K, Deboer M, Rosen-Wolff A, Roos D, Roesler J (2013) *Alu* repeat-induced deletions in chronic granulomatous disease: a cause not only for p67-phox, but also for p47-phox deficiency. Ann Hematol 92(7):1003–1004

Xie H, Wang M, Bonaldo MDF, Rajaram V, Stellpflug W, Smith C, Arndt K, Goldman S, Tomita T, Soares MB (2010a) Epigenomic analysis of *Alu* repeats in human ependymomas. Proc Nat Acad Sci USA 107(15):6952–6957

Xie D, Chen CC, Ptaszek LM, Xiao S, Cao X, Fang F, Ng HH, Lewin HA, Cowan C, Zhong S (2010b) Rewirable gene regulatory networks in the preimplantation embryonic development of three mammalian species. Genome Res 20(6):804–815

Yaffe E, Tanay A (2011) Probabilistic modeling of Hi-C contact maps eliminates systematic biases to characterize global chromosomal architecture. Nat Genet 43(11):1059–1065

Yoder JA, Walsh CP, Bestor TH (1997) Cytosine methylation and the ecology of intragenomic parasites. Trends Genet 13(8):335–340

Zhang Y, Liu T, Meyer CA, Eeckhoute J, Johnson DS, Bernstein BE, Nusbaum C, Myers RM, Brown M, Li W et al (2008) Model-based analysis of ChIP-Seq (MACS). Genome Biol 9(9):R137

4

The genetic and epigenetic alterations in human hepatocellular carcinoma: a recent update

Ming Liu, Lingxi Jiang, Xin-Yuan Guan✉

Department of Clinical Oncology, University of Hong Kong, Hong Kong, China
✉ Correspondence: xyguan@hkucc.hku.hk (X.-Y. Guan)

ABSTRACT

Hepatocellular carcinoma (HCC) is one of the most frequent human malignancies worldwide with very poor prognosis. It is generally accepted that the progression of HCC is a long-term process with accumulation of multiple genetic and epigenetic alterations, which further lead to the activation of critical oncogenes or inactivation of tumor suppressor genes. HCC is characterized with multiple cancer hallmarks including their ability to proliferate, anti-apoptosis, invade, metastasis, as well as the emerging features such as stem cell properties and energy metabolic switch. The irreversible alterations at genetic level could be detected as early as in the pre-neoplastic stages and accumulate during cancer progression. Thus, they might account for the cancer initiating steps and further malignant transformation. In addition to genetic alterations, epigenetic alterations can affect the cancer transcriptome more extensively. Alterations in DNA methylation, histone modification, miRNAs, RNA editing, and lncRNAs might result in disrupted gene regulation networks and substantially contribute to HCC progression. In this review, the genetic and epigenetic alterations which significantly contribute to the malignant capabilities of HCC will be updated and summarized in detail. Further characterization of those critical molecular events might better elucidate the pathogenesis of HCC and provide novel therapeutic targets for treatment of this deadly disease.

KEYWORDS hepatocellular carcinoma (HCC), cancer hallmarks, genetic regulation, epigenetic regulation, therapeutic targets, HCC progression

INTRODUCTION

Hepatocellular carcinoma (HCC) is one of the most frequent human malignancies worldwide. It is the sixth most prevalent cancer in the world and the third leading cause of cancer-related mortality (Parkin et al., 2005). The prevalence of HCC varies markedly in different regions. The highest incidence of HCC was found in Asia-pacific area (>20/100,000), while low incidence was found in Northern Europe and Northern America (<5/100,000) (Venook et al., 2010). China is a typical high-risk region which may account for more than 50% of HCC cases in the world (Yuen et al., 2009). The uneven distribution of HCC incidence among different geographic regions suggests that multiple genetic and environmental factors may interplay in the progression of this disease. Almost 70%–90% of HCC patients accompany with liver cirrhosis, which is believed to be the most important risk factor for HCC (Fattovich et al., 1997). Thus, all levels of viral infection, liver cytotoxicity, chronic inflammation which can lead to liver cirrhosis are important risk factors in the development of HCC (El-Serag and Rudolph, 2007).

It is widely accepted that carcinogenesis is a multi-step process with accumulation of genetic alterations in critical genes which regulate cell proliferation, growth, survival, apoptosis, adhesion, and metabolism (Vogelstein and Kinzler, 1993). The stepwise accumulation of genetic alterations in oncogenes and tumor suppressor genes will transform a normal cell and finally leads to carcinogenesis (Farber, 1984). The pathogenesis of HCC is also believed to be a long-term process which begins from the pre-malignant stage to the dysplastic stage and finally proceeds to the malignant stage (Thorgeirsson and Grisham, 2002). Better understanding of the genetic and epigenetic changes and their interactions at all the stages during HCC progression

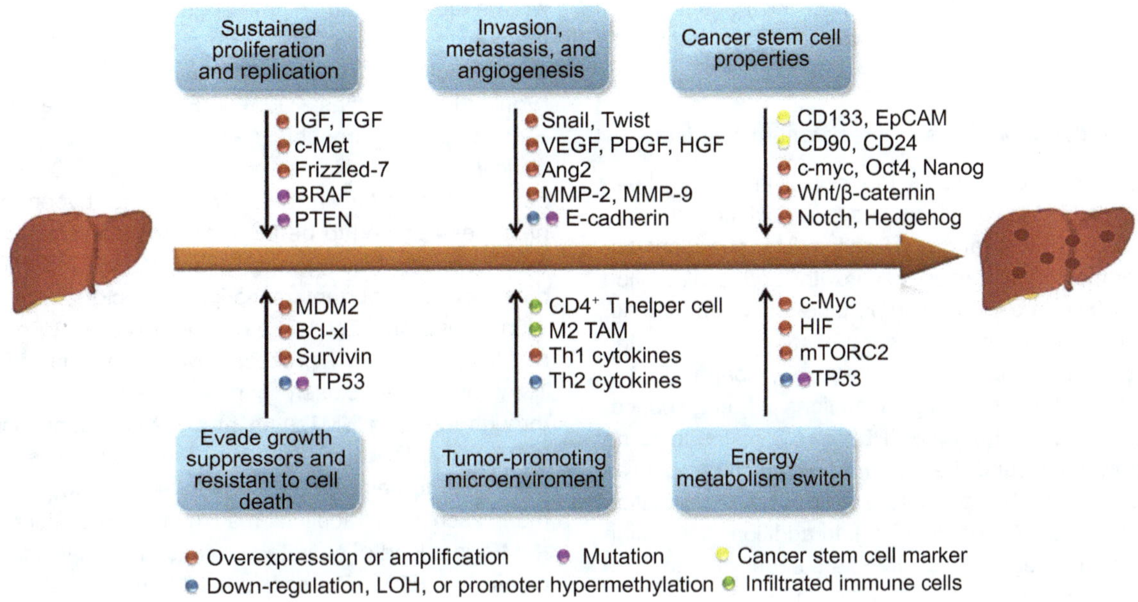

Figure 1. Multiple cancer hallmarks and the underlying molecular alterations in the progression of HCC. The progression of HCC is a multi-step process, which is characterized by several cancer hallmarks including sustained cell proliferation, evading growth suppressors, resistant to cell death, invasion, metastasis, angiogenesis, and deregulated energy metabolism. Multiple cellular and molecular alterations such as amplification or overexpression of oncogenes, hypermethylation or mutation of tumor suppressor genes, activation of cancer stem cells, and infiltration of immune cells, contribute to the malignant transformation of HCC.

will greatly facilitate to elucidate the pathogenesis of HCC. In this review, we summarize the current knowledge of genetic and epigenetic alterations in the progression of HCC.

HALLMARKS OF HUMAN HEPATOCELLULAR CARCINOMA

For decades, scientists are trying to unveil the underlied molecular mechanisms of cancer initiation and progression. However, the diverse characteristics and heterogeneity of cancer usually make confusion. Now, it is widely accepted that cancer evolves progressively from normal cells to malignant stages. During the multistep process, cancer cells acquired several hallmark capabilities which enable them to become tumorigenic and show all kinds of malignant phenotypes (Hanahan and Weinberg, 2000, 2011). Like other solid tumors, HCC is also characterized with those cancer hallmarks such as sustained cell proliferation, evading growth suppressors, resistant to cell death, invasion, metastasis, angiogenesis, and deregulated energy metabolism. The diverse malignant phenotypes of cancer cells usually associate with several genetic or epigenetic alterations of critical oncogenes or tumor suppressor genes. Thus, linking the hallmarks of HCC with genetic or epigenetic alterations will help to identify potential molecular mechanisms and find out novel targets for HCC treatment. Multiple cancer hallmarks and the underlying molecular alterations in the progression of HCC are summarized in Fig. 1.

Sustained proliferation and replication

The growth and proliferation of normal cells are strictly regulated to maintain a homeostasis of cell number and tissue architecture. However, in tumors, cell growth and proliferation are usually deregulated, and sustained cell proliferation is one of the most common traits of cancer cells. Cancer cells have several ways to obtain the ability to proliferate and replicate rapidly. One mechanism is that they can autocrine or paracrine growth factors which will further activate the mitogenic signaling pathways (Lemmon and Schlessinger, 2010). In HCC, cell growth factors such as IGF, FGF were reported to be overexpressed in a way of autocrine or paracrine and further promote cell growth and proliferation (Kim et al., 1998; Yoshiji et al., 2002). Growth factor receptors are usually overexpressed or mutated in tumor cells, which results in persistent activation of downstream mitogenic signals. In HCC, the overexpression of HGF receptor c-Met can activate the downstream Ras/Raf/MEK signaling pathway (Ueki et al., 1997). Overexpression of the Frizzled-7 receptor leads to the activation of Wnt/beta-caternin signaling pathway in HCC (Merle et al., 2004). In addition to direct overexpressing growth factors and receptors, cancer cells can also activate the mitogenic pathways through affecting the upstream and downstream signal transducers or disruption of the negative feed back loop. For example, the oncogenic signal transducer BRAF is frequently mutated and consistently activated in HCC (Colombino et al., 2012); loss of function mutation of PTEN, which is the negative feed back regulator of PI3K/AKT pathway, was also frequently observed in HCC patients (Yao et al., 1999).

Evading growth suppressors and resistant to cell death

In contrast to the proliferation stimulating signals, cells have also developed a growth inhibition system, which acts as a guardian of cell growth and proliferation. The guardian system is composed of series of growth and proliferation inhibitors, which are usually important tumor suppressor genes. The well-known tumor suppressor TP53 is at the center of the guardian system. Somatic mutation of TP53 is one of the most frequent genetic alterations in human cancer (Olivier et al., 2010). In HCC, TP53 was also found to be frequently mutated, and the common risk factors such as AFB1, HBV, and HCV are reported to cause TP53 mutation (Hussain et al., 2007; Ozturk, 1991). In addition to mutation of TP53 itself, the regulators of TP53 are usually found to be altered in cancer, such as MDM2. MDM2 ubiquitinates TP53 and promotes the proteasome-mediated degradation of TP53 (Kubbutat et al., 1997). Amplification and overexpression of MDM2 was frequently observed in HCC, and this might also account for the deregulated TP53 signaling pathway in HCC (Jablkowski et al., 2005). In addition to cell growth inhibition, cells can undergo apoptosis upon receiving extrinsic or intrinsic signals (Hengartner, 2000). The apoptotic signal is controlled by a group of counteracting pro- and anti-apoptotic proteins. The pro-apoptotic proteins such as Bax and Bak can enhance the permeability of mitochondria membranes and promote the release of cytochrome c, which further activates the caspase cascade. The anti-apoptotic proteins such as Bcl-2, Bcl-xl will counteract the pro-apoptotic proteins (Adams and Cory, 2007). Other regulatory proteins which interfere with the apoptotic signaling cascade, such as survivin, are also important anti-apoptotic components (Adams and Cory, 2007). The overexpression of anti-apoptotic proteins such as Bcl-xl and survivin are frequently observed in HCC patients (Shiraki et al., 2000; Takehara et al., 2001).

Invasion, metastasis, and angiogenesis

Invasion and metastasis is one of the most common hallmarks of cancer, especially those with high grade malignancy. To gain invasive abilities, cancer cells usually undergo morphological changes termed "Epithelial-to-Mesechymal transition" (EMT). In epithelial tissues, cells are usually attached to each other or to the extracellular matrix (ECM) through adhesion molecules. However, in metastatic tumors, the adhesion molecules such as E-cadherin are usually down-regulated or mutated, and the loosened cell contact enables tumor cells to invade out from the primary niche (Cavallaro and Christofori, 2004). Conversely, adhesion molecules associated with cell migration such as N-cadherin, which are usually expressed in migrating cells, will be up-regulated in the aggressive cancer cells. The morphology of the cell will also change from the epithelial-like phenotype to fibroblastic mesenchymal-like phenotype (Voulgari and Pintzas, 2009). In HCC, altered expression of E-cadherin was frequently observed and correlated with clinical pathological features (Wei et al., 2002). Loss of heterozygosity (LOH) and CpG island hypermethylation have been proved to be the major mechanisms accounting for E-cadherin inactivation in HCC (Kanai et al., 1997). To date, several important transcriptional factors, such as Snail, Slug, Twist, and Zeb1/2, have been proved to be the key regulator of the EMT process. Overexpression of Snail and Twist has been closely correlated with HCC metastasis through inducing EMT (Lee et al., 2006; Sugimachi et al., 2003). In order to migrate from the original tissue, cancer cells need to degrade the barriers which hinder their movement, such as extracellular matrix. Matrix metalloproteinases (MMP) is a kind of secreted protease family, which can help digest the ECM (Stamenkovic, 2000). MMPs are synthesized in an inactive form, which could be activated after removing the pro-peptide domain (Pei et al., 2000). Overexpression and activation of MMPs are frequently observed in cancer cells, especially those with high metastatic ability (Rundhaug, 2003). Overexpression of MMPs, such as MMP-2 and MMP-9, are frequently observed in HCC patients and has been associated with cancer invasive potential (Arii et al., 1996; Giannelli et al., 2002).

Angiogenesis is another important feature of cancer. When tumor mass grows, the tumor cells need blood vesicles to provide enough nutrient and oxygen. Formation of tumor vessels can accelerate the proliferation, growth, and metastasis of cancer cells (Carmeliet and Jain, 2000). In the process of angiogenesis, tumor cells will secret several critical growth factors such as FGF, VEGF, which will activate the proliferation of endothelial cells or fibroblasts (Yancopoulos et al., 2000). VEGF has been proved to play a critical role in tumor angiogenesis including HCC. Overexpression of VEGF was correlated with HCC angiogenesis and vascular formation (Mise et al., 1996). Monoclonal antibodies targeting VEGF or small molecules inhibiting VEGF receptors have already being used in HCC treatment (Finn and Zhu, 2009). In addition to VEGF, other proangiogenic factors including platelet-derived growth factor (PDGF), hepatoctye growth factor (HGF), basic fibroblast growth factor (bFGF), and angiopoietin-2 (Ang2) are also elevated in the HCC plasma and make a substantial contribution to HCC angiogenesis (Semela and Dufour, 2004; Zhu et al., 2011).

Tumor-promoting microenvironment

The initiation, growth, and metastasis of tumor not only depend on the malignant characteristics of cancer cells themselves, but also the tumor-promoting microenvironment (Joyce, 2005). Tumor grows in a complicated microenvironment, which is composed of stromal fibroblasts, endothelial cells, and infiltrating immune cells. These cells in the microenvironment can either secret growth factors to support tumor cell growth and angiogenesis, or produce pro-inflammatory cytokines and chemokines, which favors malignant transformation (Whiteside, 2008). For HCC, the tumor microenvironment might play a critical role in tumor initiation and progression. Etiological studies indicated that HCC mainly developed from liver cirrhosis, which is caused by chronic hepatitis virus infection, fatty liver disease, and alcohol abuse. The common trait during hepatocarcinogenesis is the sustained liver damage and regeneration, which leads to an inflammatory microenvironment in the liver. The inflammatory microenvironment supports the recruitment and activation of hepatic stellate cells and macrophages, which further produce components of the ECM, growth factors, and chemokines for angiogenesis and fibrosis (Hernandez-Gea et al., 2013).

There are several kinds of cells closely associated with HCC tumor microenvironment. Like other solid tumors, the most common cells observed in the tumor microenvironment are immune cells. In response to inflammatory signal, immune cells including T cells, B cells, macrophages, and dendritic cells will infiltrate into the tumor mass, and produce several kinds of cytokines, which either inhibit or promote tumor growth (Hernandez-Gea et al., 2013). The most common tumor-infiltrating lymphocyte is CD^{4+} T helper cells. The cytokines secreted by Th cells could further be divided into two groups including Th1-like cytokines and Th2-like cytokines. A unique signature of increased Th1 cytokines (IL-1, IL-2, TNFα, etc.) but decreased Th2 cytokines (IL-4, IL-8, IL-10, etc.) was frequently observed in HCC tumor microenvironment and associated with poor prognosis of HCC patients (Ye et al., 2003). Tumor associated macrophage (TAM) is another important subset of infiltrated immune cells in the tumor microenvironment. High density of infiltrated TAMs usually associated with poor prognosis of HCC patients (Ding et al., 2009). TAMs can either secret tumor-promoting growth factor, cytokines, chemokines, etc. to facilitate tumor growth, or suppress the anti-tumor immunity in HCC tissues. The macrophages can also be divided into two subgroups like T helper cells. TAMs resemble the M2 macrophages, which provide the formation of Th2 tumor microenvironment (Bingle et al., 2002).

Cancer stem cell properties

According to the cancer stem cell (CSC) model, cancer originates from a subset of stem like cells that have self-renewal properties. Malignant cancer cells usually have similar properties as embryonic cells characterized with elevated stemness markers and maintained in a dedifferentiated status (Reya et al., 2001). Assessing the differentiation level of tumor is often conducted in the clinic, and the poorly-differentiated tumors are closely associated with patient prognosis. The histologically poorly-differentiated tumors usually show an embryonic-like gene expression signature (Ben-Porath et al., 2008). In HCC, certain cell populations have been identified as potential cancer stem cells. The "Oval cells" which give rise to hepatoblast cells and primitive bile duct cells during liver development are considered to be origins of liver cancer stem cells. Oval cells express cellular markers of both hepatocytes and bile duct, and have the potential to differentiate into both lineages. Therefore, oval cells are considered to be liver progenitor cells, which might be initiating cells in hepatocarcinogenesis (Mishra et al., 2009). A small group of cells known as "side population" (SP), which are able to pump out nucleus dye via ABCG2-transporters, are also considered to be potential liver cancer stem cells. SP cells have enhanced self-renewal ability *in vitro* and tumorigenic ability *in vivo*. Molecular characterization of SP cells indicated that several oncogenic signaling pathways associated with cancer self-renewal and differentiation, are activated in SP cells (Marquardt et al., 2011). In addition to oval cells and SP cells, several cell surface markers have been identified as cancer stem cell markers, including CD133, EpCam, CD90, CD24, etc. The isolated cancer cells using those markers all shown strong self-renewal properties and tumorigenic ability with only few cells injected in xenograph mouse model (Lee et al., 2011; Ma et al. 2007; Yamashita et al., 2009; Yang et al., 2008)

Several oncogenic signaling pathways have been proved to play important roles in regulating cancer stemness and differentiation. The well-known oncogene c-myc was reported to account for the embryonic stem cell like phenotype of cancer cells (Kim et al., 2010). Inactivation of the myc network can induce the differentiation of HCC cells (Shachaf et al., 2004). Wnt/β-caternin is another important signaling pathway in regulating cancer stemness and differentiation. The secreted wnt will inhibit the cytoplasmic degradation of β-caternin, which further activates the β-caternin/TCF transcriptional machinery and promotes the transcription of several stemness markers such as Epcam, Ck19, and CD44, etc. (Fodde and Brabletz, 2007; Yamashita et al., 2007). In addition, other critical signaling pathways involved in regulating stem cell self-renewal and differentiation, such as Oct4, Nanog, Sox2, and Notch/Hedgehog have been reported to be important in maintaining the pluripotency of liver cancer progenitor cells and are frequently activated in HCC (Patil et al., 2006; Yuan et al., 2010).

Energy metabolism switch

Altered energy metabolism switch from oxidative phosphorylation to glycolysis, known as Warburg effect, has been widely accepted as an emerging hallmark of cancer. Early in the 1930s, oncologists and scientists have noticed the alteration of energy metabolism in malignant tumors. Glycolysis was preferentially used as the main program for energy metabolism in tumor cells even in the presence of oxygen (Cairns et al., 2011). In normal cells, ATP is mainly generated from the tricarboxylic acid (TCA) cycle, followed by oxidative phosphorylation in the mitochondria. Oxidative phosphorylation generates 36 molecules of ATP from one molecule of glucose. In contrast, the glycolysis only generates two molecules of ATP from one molecule of glucose (Kroemer and Pouyssegur, 2008). Although glycolysis is not efficient in generating energy, it can provide a large amount of nucleotides, fatty acids, membrane lipids to support the synthesis of macromolecules, which are required for rapid tumor growth. In compensation, cancer cells increased the glucose intake by up-regulating glucose transporters and enhancing the usage of glutamine. The Warburg-like metabolic switch found to be present in many rapidly dividing embryonic tissues further supported the hypothesis that glycolysis could generate diverse intermediates for biosynthetic programs that are important for active cell proliferation (Hsu and Sabatini, 2008).

The reduced dependence of cancer cells on oxidative phosphorylation is not only due to defects in the cellular components of TCA cycle, but also strictly regulated by series of oncogenes and tumor suppressor genes. Oncogenic activation of c-myc, Ras, Akt, and HIF, or inactivation of tumor suppressors such as TP53 can drive metabolism changes in cancer cells (Levine and Puzio-Kuter, 2010). Like other solid tumors, metabolic remodeling from mitochondrial oxidation to aerobic glycolysis is common in human HCC (Beyoglu et al., 2013). c-Myc was reported to induce mouse liver tumors with elevated glucose and glutamine catabolism (Yuneva et al., 2012). Hepatic mTORC2 can activate glycolysis and lipogenesis through phosphorylating AKT (Hagiwara et al., 2012). Multi-kinase inhibitor such as sorafenib, which targets those oncogenic signaling pathways, was reported to be able to reverse the metabolic reprogramming in HCC (Fiume et al., 2011).

GENETIC ALTERATIONS IN HCC

Genetic alteration is one of the most important mechanisms associated with HCC initiation and progression. Genetic changes could be observed as early as in the pre-neoplastic lesions of cirrhotic liver, and it is thought to be the initiating events in hepatocarcinogenesis. The irreversible genetic abnormalities accumulate in hepatocytes, which further cause disrupted gene expression, and finally lead to malignant transformation. Genetic alterations could be divided into several types, including large chromosomal amplification, translocation, deletion, small fraction loss, and single nucleotide variation. The genetic changes at all levels usually result in the activation or loss-of-function of certain important oncogenes or tumor suppressor genes, which govern cell growth and proliferation.

Chromosomal instability

Chromosomal instability is the most common genetic changes in HCC. It could be induced by either error during mitosis or disruption in DNA replication and repair. The chromosome abnormalities could be observed as the gain and loss of whole chromosome arms, or just amplification and deletion of small chromosomal fragments. According to the comparative genomic hybridization (CGH) data, chromosome 1q and 8q are frequently amplified, while chromosome 1p, 4q, 6q, 9p, 16p, 16q, and 17p are frequently lost in HCC (Guan et al., 2000). The observation of chromosomal alterations in preneoplastic liver tissues indicated that chromosome instability may occur in the early stage of HCC, and accumulates during tumor progression (Kondo et al., 2000). Thus chromosome instability may activate certain cancer driver genes during hepatocarcinogenesis.

Amplification of chromosome 1q is one of the most frequently observed chromosome abnormalities in HCC. The minimal region of 1q21 was found to be amplified in more than 50% of HCC patients. A well characterized oncogene CHD1L is localized in that region (Ma et al., 2008). CHD1L was found to have several oncogenic roles such as inhibiting cell apoptosis, regulating cell mitosis, and promoting cell epithelial-to-mesenchymal (EMT) transition during hepatocarcinogenesis (Chan et al., 2012; Chen et al., 2010; Chen et al., 2009a). In a transgenic mice model, CHD1L could induce spontaneous liver tumors formation (Chen et al., 2009b). In addition, CHD1L could regulate p53 stability, potentially via interacting with SCYLIBP1, which modulates the pirh2-mediated ubiquitin degradation of p53 (Hu et al., 2012). Adenovirus-mediated silencing of CHD1L could inhibit HCC tumorigenesis in xenograft mouse model further suggested CHD1L as a potential therapeutic target in HCC treatment (Chen et al., 2011). In addition to chromosome 1q21, a recent study indicated that a novel potential oncogene Maelstrom (MAEL) at 1q24, could induce EMT and enhance stemness properties of HCC cells (Liu et al., 2013). Chromosome 8q is another highly amplified chromosome arm in HCC, especially at the 8q24 region (Wang et al., 2002). Well-known oncogenes including *c-Myc* and *PTK2* are located at this region, and have been characterized for their oncogenic effects on HCC development (Okamoto et al., 2003; Santoni-Rugiu et al., 1998). In addition to 8q24, the chromosomal region proximal to the centromere is also frequently amplified in HCC (Parada et al., 1998). A serine/threonine kinase SGK3, which shares great similarity with AKT, was found to be frequently amplified and confer AKT-independent oncogenic roles in HCC (Liu et al., 2012).

Chromosome segmental loss is also frequently observed in HCC. The minimal region of 1p35-36 was found to be deleted in more than 50% HCC patients. Several tumor suppressors such as 14-3-3 σ and Rb-interacting zinc finger 1 (RIZ1) were located in this region (Iwata et al., 2000). Loss of the short arm of chromosome 8 has been recurrently observed in HCC. A minimal region of 8p21-22 was found to be frequently deleted in HCC. Deleted in liver cancer 1 (DLC-1), which is a homolog of the rat RhoGAP gene, is located in that region (Yuan et al., 1998). DLC-1 is frequently deleted in HCC tissues due to allele

Table 1. Aberration of chromosome and candidate target genes reported in HCC.

Chromosome	Type of aberration	Frequent aberration region and candidate target genes (Location)	References
1q	Gain	*CKS1B* (1q21.2), *CHD1L* (1q21.1), *JTB* (1q21), *MDM4* (1q32.1)	Chen et al., 2010; Kim et al., 2008
1p	Loss	*p18* (1p32), *14-3-3σ* (1p35), *p73* (1p36.3), *RIZ* (1p36.13-p36.23)	Nishimura et al., 2005; Iwata et al., 2000; Fang et al., 2000
3q	Gain	*Gankyrin* (3q28)	Higashitsuji et al., 2000
3p	LOH, CpG methylation	*RASSF1A* (3p21.3), *CTNNB1* (3p21), *TGF-1βR11* (3q22)	Zhang et al., 2002; Miyoshi et al., 1998
4q	LOH		Zondervan et al., 2000
6p	Gain		Chochi et al., 2009
6q	LOH	*M6P/IGF2R* (6q26-q27)	Oka et al., 2002
8q	Gain	*c-Myc* (8q24.21), *PTK2* (8q24.3), *EIF3S3* (8q23.3), *SGK3* (8q13.1)	Santoni-Rugiu et al., 1998; Okamoto et al., 2003; Liu et al., 2012
8p	LOH, CpG methylation	*DLC-1* (8p21.3-22), *LPTS* (8p23), *CSMD1* (8p23.2)	Yuan et al., 1998
9p	LOH, CpG methylation	*CDKN2A* (9p21), *CDKN2B* (9q21),	Wang et al., 2000
10q	LOH	*PTEN/MMAC1* (10q23.3)	Fujiwara et al., 2000
11q	Gain	*cyclinD1* (11q13)	Nishida et al., 1994
11p	LOH, CpG methylation	*KAI1* (11p11.2), *IGF-2* (11p15), *TSLC1* (11q23.2)	Tsujiuchi et al., 2007
13q	LOH	*Rb1* (13q14.2), *BRCA2* (13q12.3), *Tg737* (13q12.1), *TFDP1* (13q34), *CUL4A* (13q34), *CDC1* (13q34)	Kuroki et al., 1995; Yasui et al., 2002
16q	LOH, CpG methylation	*CDH1* (16q22.1)	Wang et al., 2000
16p	CpG methylation	*Axin1* (16p13.3), *SOCS*-1 (16p13.3)	Li et al., 2013c; Ko et al., 2008
17p	LOH	*p53* (17p13.1), *HIC-1* (17p13.3), *HCCS1* (17p13.3)	Nishida et al., 1993; Kanai et al., 1999; Zhao et al., 2001

loss and promoter hypermethylation (Wong et al., 2003). Restore DLC-1 expression in hepatoma cells could induce cell apoptosis, and inhibit tumor growth (Zhou et al., 2004). Chromosome 16q is another region with frequent deletion in HCC. Cell adhesion molecule E-cadeherin (CDH1), which inhibits cell proliferation and metastasis, is located on 16q22 (Kanai et al., 1997). Another tumor suppressor gene Tyrosine aminotransferase (TAT), which might contribute to the pathogenesis of HCC, is also located on 16q22 (Fu et al., 2010). Recently, a significant allele-specific imbalance was identified in the 16q23 region in a cohort of HCC patient due to LOH. The affected gene *Oxidative Stress-Induced Growth Inhibitor 1 (OSGIN1)* can directly induce cell apoptosis in HCC cells and contributes significantly to the progression of HCC (Liu et al., 2014). The well-known tumor suppressor TP53 is mapped to 17p13.1, which is also a recurrently lost region in HCC. The 17p13 region was characterized with DNA hypermethylation, and loss of 17p13.1 was closely associated with TP53 mutation (Nishida et al., 1993). Summary of chromosome alterations and candidate target genes reported in HCC was listed in Table 1.

Genomic mutations

In addition to large chromosomal alterations, genomic mutation is another important genetic alteration which contributes to tumor initiation and progression. Genomic mutations could be divided into germline mutations and somatic mutations. A germline mutation is usually inherited, and exists in all cell types of a body. Germline mutations are rare, and usually account for cancer risk in certain families. In contrast, somatic mutations usually exist in tumor tissues or preneuplastic tissues, and accumulate during cancer progression. Somatic mutations are more common, and might account for malignant transformation of sporadic tumors. Missense genomic mutations at the open reading frame can either lead to loss-of-function of tumor suppressors or gain-of-function of oncogenes. In addition, mutations at the non-coding region of the genome can also affect cancer risk and progression, for they may change the transcription, translation, or stability of the gene product.

With the development of the next-generation high throughput deep sequencing technology, scientists now are able to identify somatic mutation patterns in a certain tumor tissue, like HCC. Recently, two groups have sequenced the whole genome of several HCC tumor tissues and their paired non-tumor tissues (Fujimoto et al., 2012; Guichard et al., 2012). Recurrent somatic mutations were enriched in several signaling pathways including wnt/β-catenin, p53/cell cycle control, chromatin remodeling, PI3K/Ras signaling, and oxidative and endoplasmic reticulum stress. The wnt signaling pathway was found to be the most frequently altered in HCC. Activating mutation in CTNNB1 was found in 32.8% of HCC patients. While the inactivating mutations of AXIN1 and APC was found in 15.2% and 1.6% of HCC patients, respectively. The second most altered pathway in HCC is the p53 signaling pathway. Inactivation mutation of p53 was present in 20.8% of HCC patients and mutation in CDKN2A was identified in 8% of HCC patients. In addition to other traditional signaling pathways, which are frequently mutated in cancer, several recent studies reported that the components of the chromosome remodeling complex are frequently mutated in many cancer types including HCC. The mutation of the SWI/SNF chromatin remodeling complex component ARID1A was detected in more than 20% of HCC patients. These indicated that the chromatin remodeling complex might play important roles in cancer initiation and progression.

Cancer susceptibility genes

It is widely accepted that genetic polymorphisms at cancer susceptibility genes can affect the cancer risk of certain population. Unlike genetic mutations, which directly cause loss-of-function or gain-of-function of gene products, and usually affect important oncogenes or tumor suppressor genes involved in critical signaling pathways, nucleotide changes in cancer susceptibility alleles might not directly cause dramatic functional changes of a protein. Instead, cancer susceptible genetic variations might slightly affect the function of a protein, for example the efficiency of an enzyme, thus confer an increased cancer risk for certain population. A wide range of genes are associated with cancer risk, including carcinogen metabolism genes, antitumor immune response genes, and genes associated with cellular response to stress (Antoniou et al., 2010).

Genome-wide association study (GWAS) is emerging as a powerful tool to identify cancer susceptibility alleles in tumorigenesis. GWAS examines common genetic variants in different individuals and identifies variants associated with certain disease. In contrast to mendelian linkage analysis, which aims to identify highly penetrant tumorigenic mutations, GWAS is powerful in identifying less penetrant tumor susceptibility alleles, which are more common and might be important in cancer initiation and progression. Several GWASs have been performed to identify susceptibility alleles associated with HCC. Intronic SNP (rs17401966) in KIF1B on chromosome 1p36.22 has been linked to HBV-associated HCC (Zhang et al., 2010). Chromosome loci 6p21.32 and 21q21.3 have also been associated with HCC in chronic HBV carriers (Li et al., 2012). A recent study indicated that genetic variations in STAT4 and HLA-DQ genes may confer risk of HBV-related HCC (Jiang et al., 2013). SNP (rs2596542) in the 5′ flanking region of MICA on 6p21.33 has been linked to HCV-associated HCC (Kumar et al., 2011).

MicroRNAs (miRNAs)

MicroRNA, a class of non-coding RNAs, has been identified as important regulators of gene expression at post transcriptional levels. Emerging evidences indicated that miRNAs are associated with the development and progression of HCC. In recent years, intensive investigations have been conducted to find out the abnormally expressed miRNAs and their roles in HCC development and progression. Some miRNAs can regulate the proliferation pathways via modulating cyclins or cyclin-dependent kinases, such as miR-122a and miR-221 (Gramantieri et al., 2007). Some miRNAs can help HCC cells to escape from apoptosis by targeting pro-apoptotic protein. For example, Bmf, a proapoptotic protein, is a target of miR-221 (Gramantieri et al., 2009). On the contrary, other miRNAs can promote HCC apoptosis. For example, the anti-apoptotic proteins Bcl-2 and Mcl-1 are two direct targets of miR-29 (Xiong et al., 2010). As two of the most critical hallmarks of HCC, invasion and metastasis are also regulated by miRNAs. On the one hand, the pro-metastatic miRNAs can promote cell migration and spreading in HCC. For example, miR-106b can promote HCC cell migration and invasion by activating epithelial-mesenchymal transition (EMT) process (Yau et al., 2013). On the other hand, several miRNAs such as let-7g, miR-139, and miR-195 can suppress metastasis and progression of HCC (Ji et al., 2010; Wang et al., 2013b). Additionally, some miRNAs have been reported to enhance the ability of self-renewal and tumorigenicity of HCC. MiR-130b can regulate CD133(+) liver cancer stem cells via silencing TP53INP1 (Ma et al., 2010). Inhibition of miR-181 can result in a reduction in EpCAM(+) HCC cell quantity. Exogenous miR-181 expression in HCC cells led to an enrichment of EpCAM (+) HCC cells and promote tumor initiating ability (Ji et al., 2009). Summary of the abnormally expressed miRNAs and their functions are listed in Table 2.

RNA editing

The RNA transcripts are usually faithfully transcribed from the genome without sequential changes after RNA processing. However RNA editing is a molecular process which could result in nucleotide changes at specific sites of the RNA transcripts. Thus, RNA editing could add great diversity to the posttranscriptional regulation of gene expression (Gott and Emeson, 2000). RNA editing can modify the transcribed RNA sequences via nucleotide insertion, deletion, and substitution. The most common type of RNA editing in human is

Table 2. MiRNAs aberrantly expressed and validated target genes in hepatocellular carcinoma.

miRNA	Expression	Gene targets	Function	References
miR-17-5p	Up	p38, MAPK pathway, E2F-1, c-MYC	Promote tumor growth and metastasis.	Yang et al., 2010; El Tayebi et al., 2013
miR-18a	Up	ERα	Promote proliferation.	Liu et al., 2009
miR-18b	Up	TNRC6B	Promote cell proliferation and loss of cell adhesion.	Murakami et al., 2013
miR-21	Up	PTEN, RECK, PDCD4	Inhibit apoptosis, promote cell migration and invasion.	Meng et al., 2007; Zhou et al., 2013
miR-106b	Up	E2F1, RhoGTPases, RhoA, RhoC	Promote cell migration and stress fiber formation.	Yau et al., 2013
miR-130b	Up	TP53INP1	Promote CD133(+) liver cancer stem cell growth and self-renewal.	Ma et al., 2010
miR-143	Up	FNDC3B	Promote tumor metastasis.	Zhang et al., 2009
miR-151	Up	RhoGDIA, FAK,	Stimulate tumor invasion and metastasis.	Ding et al., 2010; Luedde, 2010
miR-181b	Up	TIMP3	Promote tumor metastasis.	Wang et al., 2010a
miR-181	Up	CDX2, GATA6,NLK	Promote EpCAM(+) liver cancer stem cell growth and self-renewal.	Ji et al., 2009
miR-185	Up	KCNN3	Association with HCC venous metastasis.	Budhu et al., 2008
miR-210	Up	VMP1	Promote hypoxia-induced HCC cell metastasis.	Ying et al., 2011
miR-221/222	Up	CDKN1B/p27,CDKN1C/p57, DDIT4, PTEN, Bmf, TIMP3, PPP2R2A	Inhibit apoptosis, promote tumor growth and metastasis.	Fornari et al., 2008; Gramantieri et al., 2009
miR-224	Up	API-5, CDC42, CDH1, PAK2, BCL-2, MAPK1, PPP2R1B.	Promote cell proliferation, migration, invasion, and inhibit cell apoptosis.	Wang et al., 2008; Zhang et al., 2013
miR-1	Down	FoxP1, MET, HDAC4.	Inhibition of cell growth and reduced replication potential.	Datta et al., 2008
let-7	Down	c-Myc, p16, Bcl-xl, COLIA2.	Inhibition of cell growth and proliferation.	Wang et al., 2010b; Ji et al., 2010
miR-26a	Down	Cyclin D2, Cyclin E2,Cyclin E1, CDK6, IL-6	Inhibit tumor growth, metastasis, and invasion.	Yang et al., 2013
miR-29	Down	MEG3, Bcl-2, Mcl-1	Promotion of apoptosis and inhibition of tumor growth	Xiong et al., 2010
miR-34a	Down	c-Met	Inhibition of cell growth, migration, and invasion.	Li et al., 2009
miR-122	Down	CyclinG1, ADAM10, SRF, IGF1R, PTTG1, PBF,CUTL1, NDRG3, MDR-1	Inhibit viral replication and cell proliferation.	Song et al., 2012; Li et al., 2013a; Xu et al., 2010; Gramantieri et al., 2007
miR-124	Down	ROCK2, EZH2, PIK3CA	Inhibit tumor growth, invasion, and metastatic potential of HCC.	Zheng et al., 2012; Lang and Ling, 2012
miR-126	Down	ROCK2, c-Fos	Inhibit cell invasion and migration.	Wong et al., 2011

Table 2 continued

miRNA	Expression	Gene targets	Function	References
miR-145	Down	OCT4, IRS1, IRS2, IGF signaling, HDAC2.	Inhibit cell proliferation, migration, and invasion.	Wang et al., 2013a; Law et al., 2012; Noh et al., 2013
miR-148a	Down	HPIP, AKT/ERK/FOXO4/ATF5 pathway	Inhibit tumorigenesis.	Xu et al., 2013
miR-195	Down	cyclin D1, CDK6, E2F3, LATS2, VEGF, VAV2, CDC42, IKKα and TAB3, TNF-α/NF-κB pathway	Inhibit G1/S transition, angiogenesis, and metastasis, promote apoptosis.	Xu et al., 2009; Wang et al., 2013b; Ding et al., 2013
miR-199a-3p	Down	mTOR, c-Met, CD44	Inhibit cell growth and metastasis	Fornari et al., 2010; Henry et al., 2010
miR-214	Down	XBP-1, HDGF, EZH2, CTNNB1, β-catenin signaling pathway	Inhibit cell proliferation, promote cell apoptosis, and suppress tumor vascularity.	Shih et al., 2012; Xia et al., 2012
miR-223	Down	stathmin1	Inhibit cell proliferation	Wong et al., 2008
miR-375	Down	YAP, AEG-1, ATG7	Inhibit tumorigenesis	Liu et al., 2010; He et al., 2012; Chang et al., 2012

the adenosine to inosine (A-to-I editing), which is mediated by theadenosine deaminase acting on dsRNA (ADAR) family of enzymes. The A-to-I editing can affect various types of RNA molecules including mRNAs, microRNAs, viral RNAs, etc. (Athanasiadis et al., 2004). RNA editing at the coding regions of the transcripts may lead to non-synonymous amino acid changes in the gene products, which might affect the biological functions of the proteins. Recent studies have linked A-to-I RNA editing to hepatocarcinogenesis. Through next-generation RNA sequencing technology, an A-to-I editing event within the AZIN1 transcript was identified in the tumor tissues from HCC patients. Hyper-editing of AZIN1 transcripts in the tumor cells resulted in a recording of AZIN1 protein from Serine to Glysine at coden 367. The edited AZIN1 showed strong oncogenic phenotypes on HCC cell lines and mouse models, compared with the wild type form. The frequency of RNA editing in the tumor tissues also significantly associated with the prognosis of HCC patients (Chen et al., 2013). The disrupted RNA editing was found to be mediated by differential expression of ADARs in HCC (Chan et al., 2013). Further characterization of the RNA editing events in HCC might help elucidate the pathogenesis of this disease (Li et al., 2013b).

EPIGENETIC ALTERATIONS IN HCC

Genetic alterations are irreversible changes that affect the DNA sequence of the genome. In contrast, epigenetic regulations do not change the sequence of the genome but affect the chromatin structure and gene transcription. Epigenetic regulations affect gene products at multiple levels, including both transcriptional level and post-transcriptional regulation, which added great diversity to the gene regulation network. DNA methylation, histone modification, and recently emerging lncRNA, are major forms of epigenetic regulations. Alterations at cellular machineries governing those processes are frequently observed in cancer cells including HCC. The epigenetic alterations usually result in the activation of oncogenes or inactivation of tumor suppressor genes, which further contribute to malignant cancer hallmarks. Increasing evidences suggested that epigenetic alterations are evolving as an important mechanism in cancer initiation and progression (Momparler, 2003).

DNA methylation

In a normal cell, DNA methylation and demethylation is an important mechanism in regulating gene expression and chromatin structure. DNA methylase (DNMT) catalyze the methylation of cytosine at CpG islands at the promoter region of a gene. However, in tumor cells, the promoter methylation pattern is usually changed. Aberrant DNA methylation at the promoter region is an important mechanism of tumor suppressor gene inactivation. The hypermethylated CpG islands at the promoter region will prevent the binding of RNA polymerase and transcriptional factors, thus inhibit the transcription of the target genes. In addition, the hypermethylated protein will recruit m^{5}CpG-binding domain (MBD) containing proteins, which will be an obstacle for the binding of transcriptional factors to the promoters, thus inhibit gene transcription (Hendrich and Bird, 1998).

In HCC, CpG island hypermethylation is frequently observed at the promoter region of important tumor suppressor genes. Suppressor of cytokine signaling (SOCS-1), which regulates the JAK/STAT signaling pathway, was found to be silenced in more than 60% of HCC patients due to promoter hypermethylation (Yoshikawa et al., 2001). The well-known tumor suppressor APC and E-cadeherin were also hypermethylated in 53% and 49% of HCC patients, respectively (Yang et al., 2003). Methylation profiling of multistep HCC tumors revealed that the number of genes methylated showed stepwise increase with the progression of cancer stage. The observation of tumor suppressor gene hypermethylation in the para-tumor liver tissues and cirrhotic livers indicated that aberrant promoter methylation occurs in the early stage of hepatocarcinogenesis and increased progressively during cancer progression (Lee et al., 2003). In addition, genome-wide DNA methylation analysis revealed that epigenetic silencing of multiple tumor suppresors in HCC could result in the activation of several oncogenic signaling pathways including Ras, JAK/STAT, and Wnt/β-catenin (Calvisi et al., 2007).

There are several proposed hypotheses for the aberrant DNA methylation in cancer. One possible mechanism is the aberrant expression of DNMT1. As part of the DNA replication complex, DNMT1 transfer the methyl to the DNA immediately after DNA replication. In cancer cells, DNMT1 is usually abnormally expressed, and this will commit methylation errors during DNA replication (Vertino et al., 2002). Significant increase of DNMT1 was observed in HCC patients (Saito et al., 2003). In addition to DNMT1, which mainly accounts for the maintenance of methylation pattern of the genome, other DNMT family members such as DNMT3A and DNMT3B can directly add methyl groups to unmethylated DNA. DNMT3A and DNMT3B are responsible for novel methylation pattern formation in the genome (Okano et al., 1999). DNMT3A and DNMT3B were reported to be associated with hypermethylation of several important tumor suppressor genes, such as CDKN2A, CDKN2B, CDH1, and Rb1 (Mizuno et al., 2001). The expression of DNMT3A and DNMT3B are both significantly overexpressed in HCC compared with the non-cancerous liver tissues (Oh et al., 2007).

Histone modification and chromatin remodeling

Chromatin is the fundamental structure of the genome, which is constituted by nucleosome particles. The chromatin structure is important to gene transcription. In active transcription sites, the chromatin will be loosened, so that the DNA can be exposed to transcriptional factors for transcription initiation. This open chromatin structure is termed "euchromatin". In contrary, some of the chromatin structure is heavily condensed and the transcriptions of those genes within those regions are inhibited. The condensed chromatin structure is termed "heterochromatin". Thus, chromatin structure is of critical importance in regulating gene expression in a temporal and spatial dependent manner (Wang et al., 2007a).

Histone modification is playing a central role in chromatin structure regulation. Covalent modification of histones with methylation or acetylation will result in the chromatin structural change and could be used as markers for chromatin structure. There are two histone modification markers which represent an active transcription. Trimethylation of H3 lysine 4 (H3K4Me3) is often observed at the promoter region of actively transcribed genes. Trimethylation of H3 lysine 36 (H3K36Me3) is also closely associated with active transcription. In contrary, trimethylation of H3 lysine 27 (H3K27Me3) and trimethylation of H3 lysine 9 (H3K9Me3) are associated with repressed transcription (Kouzarides, 2007). It is recognized that histone modification is catalyzed by several enzymes which modulate the histone markers. The histone modifiers include histone methyltransferases (HMT), histone acetylatransferase (HAT), and histone deacertylase (HDAC), etc. Abnormal expression of those histone modifiers which further drives epigenetic alterations are frequently observed in cancer cells. In HCC, overexpression of EZH2, which is the histone methyltransferase for H3K27Me3, has been proven to contribute to the malignant transformation and poor prognosis of HCC (Chen et al., 2007; Sudo et al., 2005). The P300/CBP-associated factor (PCAF), which is a well-known HAT, was expressed at low level in HCC, and has been proven to inhibit HCC tumorigenesis both *in vitro* and *in vivo* (Zheng et al., 2013). HDAC inhibitors have been suggested to specifically induce apoptosis in hepatoma cells but not in primary hepatocytes. And these results greatly supported the potential application of HDAC inhibitors in clinical treatment of HCC patients (Armeanu et al., 2005; Pathil et al., 2006).

In addition to histone modifiers, the ATP-dependent chromatin remodeling complex, which utilize ATP to mobilize nucleosomes along DNA, are also closely involved in tumorigenesis. The ATP-dependent chromatin remodeling family could be further divided into four subfamilies including: the SWI/SNF (Switching defective/, sucrose non-fermenting) family, the ISWI family (imitation SWI), the NuRD/CHD (Nucleosome remodeling and deacetylation/Chromodomain helicase, DNA binding) family, and the INO80 (inositol requiring 80) family (Wang et al., 2007). Whole-genome sequencing has identified recurrent somatic mutations in genes associated chromatin remodeling complex, including ARID1A, ARID2, and SMARCA4 (Guichard et al., 2012; Li et al., 2011). The frequently observed inactivating mutations indicated the important roles of chromatin remodeling complex in HCC development. The ATPase and putative DNA helicase RuvB-like 2 (RUVBL2) was found to be overexpressed in HCC and has contributed to the malignant transformation (Rousseau et al., 2007). Copy number loss or down-regulation of SWI/SNF chromatin remodelling subunit-BRG1 and BRM were also frequently observed in HCC

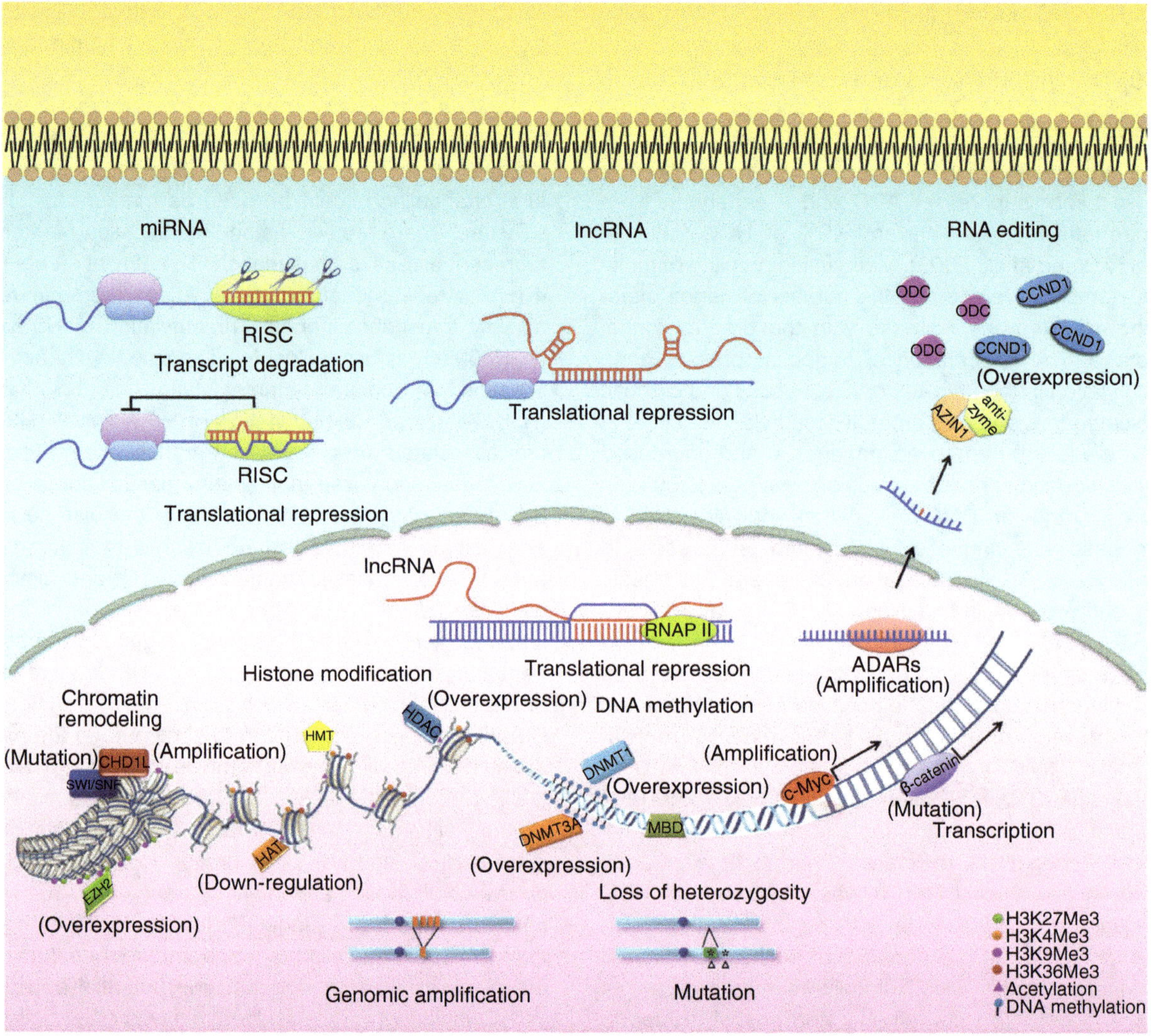

Figure 2. Genetic and epigenetic alterations in HCC. Genetic and epigenetic alterations interplay during cancer initiation and progression. The alterations exist at multiple levels including large chromosomal gain or loss, single nucleotide variations or mutations, overexpression or down-regulation of miRNAs and lncRNAs, disrupted RNA editing events, hyperactivation or inactivation of chromatin remodeling components, and aberrant DNA methylation at the promoter region of critical tumor suppressor genes et al.

patients (Endo et al., 2013). In addition, the CHD family member Chromodomain helicase DNA binding 1 like (CHD1L) has been proven to have diverse oncogenic roles in hepatocarcinogenesis (Chen et al., 2010).

Long non-coding RNAs (lncRNAs)

A large number of non-protein coding transcripts exist in the genome. In the past, those long non-coding RNAs were considered as "rubbish" of the genome for their unknown functions. Recently, emerging evidences suggested that lncRNAs might play important roles in regulating gene expression at post-transcriptional level. LncRNAs can regulate gene transcription either through directly binding to the RNA polymerase II, or modifying the activity of the transcriptional co-regulators (Mercer et al., 2009). In addition to transcriptional regulation, lncRNAs can also control the post-transcriptional mRNA processing such as mRNA splicing and translation. Furthermore, lncRNAs were also reported to be involved in regulating histone methylation and chromatin remodeling, which are the most important epigenetic regulatory machinery in regulating gene expression (Guttman and Rinn 2012). Altered expression of lncRNAs has been observed in tumors including HCC and they are suggested to play critical roles during tumorigenesis. High expression of lncRNA-HEIH is significantly associated with HCC recurrence and poor prognosis. *In vitro* and *in vivo* functional studies revealed that the overexpression of lncRNA-HEIH can promote HCC tumorigenesis and might function through EZH2 (Yang et al., 2011a). In addition, overexpression of Long Non-coding RNA HOTAIR and MALAT-1 could help predict tumor recurrence and prognosis of HCC patients (Lai et al., 2012; Yang et al., 2011b). All these evidences indicated that lncRNAs might be important in HCC initiation and progression.

SUMMARY AND PERSPECTIVES

Like other solid tumors, HCC is characterized with multiple hallmarks including sustained proliferation, evading growth suppressive signals, metastasis to other organs, promoting angiogenesis, tumor-promoting microenvironment, cancer stem cell properties, and energy metabolism switch, etc. Genetic and epigenetic alterations interplay during cancer initiation and progression. The genomic changes vary from large chromosomal gain or loss to single nucleotide variations or mutations. Genetic alterations are irreversible alterations, which could be observed as early as in the preneoplastic stages. The early onset of genetic alterations indicated that they might be the tumor initiating steps in the development of cancer. Chromosome instability is the most common type of genetic alteration. Chromosome 1q and 8q are frequently amplified, while chromosome 1p, 4q, 6q, 9p, 16p, 16q, and 17p are frequently lost in HCC. Those hot regions usually harbor important oncogenes or tumor suppressor genes, which might significantly contribute to hepatocarcinogenesis. In addition to large chromosomal segmental changes, single nucleotide changes in the genome also make a substantial contribution to cancer progression. Nucleotide changes known as mutations or variations can lead to either gain-of-function or loss-of-function of oncogenes and tumor suppressor genes. Non-coding nucleotide changes can also affect gene transcription, and post-transcriptional regulations of critical tumor related genes, which may directly trigger oncogenesis or enhance cancer risk. Epigenetic alteration is another important mechanism for oncogenesis. Epigenetic regulation includes a wide range of regulations at transcriptional or post-transcriptional levels, such as DNA methylation, histone modification, chromatin remodeling, and lncRNAs. Alterations at the epigenetic regulation machinery may lead to disrupted gene expression, which can also cause the activation of oncogenes or inactivation of tumor suppressor genes. The genetic and epigenetic alterations in HCC are summarized in Fig. 2. Small molecules or monoclonal antibodies, which specifically target the altered onco-proteins, have already been proven to be efficient in treating several types of cancer. For example, imatinib, which specifically target the BCR-ABL fusion kinase, is used in treating chronic myeloid leukemia; transtuzumab, a monoclonal antibody targeting the amplified tyrosine kinase receptor HER2, is used to treat advanced-stage breast cancer. However, the targeted therapies which are effective in treating HCC are still limited. Better understanding and characterization of novel genetic and epigenetic alterations, which are important to hepatocarcinogenesis, may help understand the molecular pathogenesis of HCC, as well as providing novel therapeutic targets for HCC treatment.

ACKNOWLEDGEMENTS

This work was supported by Hong Kong Research Grant Council (RGC) General Research Fund (HKU/7668/11M), RGC Collaborative Research Funds (HKU7/CRG/09, HKBU5/CRG/10 and HKU3/CRF/11R), Hong Kong Theme-based Research Scheme fund (T12-403/11), and grants from National Basic Research Program (973 Program) (No. 2012CB967001) and the National Natural Science Foundation of China (Grant Nos. 81272416 and 81172338).

ABBREVIATIONS

Ang2, angiopoietin-2; bFGF, basic fibroblast growth factor; CGH, comparative genomic hybridization; CSC, cancer stem cell; ECM, extracellular matrix; GWAS, genome-wide association study; HCC, hepatocellular carcinoma; HGF, hepatoctye growth factor; lncRNAs, long non-coding RNAs; LOH, loss of heterozygosity; MMP, matrix metalloproteinases; PDGF, platelet-derived growth factor; RIZ1, Rb-interacting zinc finger 1; TAM, tumor associated macrophage; TCA, tricarboxylic acid.

COMPLIANCE WITH ETHICS GUIDELINES

Ming Liu, Lingxi Jiang, and Xin-Yuan Guan declare that they have no conflict of interest. This article does not contain any studies with human or animal subjects performed by the any of the authors.

REFERENCES

Adams JM, Cory S (2007) The Bcl-2 apoptotic switch in cancer development and therapy. Oncogene 26:1324–1337

Antoniou AC, Beesley J, McGuffog L, Sinilnikova OM, Healey S, Neuhausen SL, Ding YC, Rebbeck TR, Weitzel JN, Lynch HT et al (2010) Common breast cancer susceptibility alleles and the risk of breast cancer for BRCA1 and BRCA2 mutation carriers: implications for risk prediction. Cancer Res 70:9742–9754

Arii S, Mise M, Harada M, Furutani M, Ishigami S, Niwano M, Mizumoto M, Fukumoto M, Imamura M (1996) Overexpression of matrix metalloproteinase 9 gene in hepatocellular carcinoma with invasive potential. Hepatology 24:316–322

Armeanu S, Pathil A, Venturelli S, Mascagni P, Weiss TS, Gottlicher M, Gregor M, Lauer UM, Bitzer M (2005) Apoptosis on hepatoma cells but not on primary hepatocytes by histone deacetylase inhibitors valproate and ITF2357. J Hepatol 42:210–217

Athanasiadis A, Rich A, Maas S (2004) Widespread A-to-I RNA editing of Alu-containing mRNAs in the human transcriptome. PLoS Biol 2:e391

Ben-Porath I, Thomson MW, Carey VJ, Ge R, Bell GW, Regev A, Weinberg RA (2008) An embryonic stem cell-like gene expression signature in poorly differentiated aggressive human tumors. Nat Genet 40:499–507

Beyoglu D, Imbeaud S, Maurhofer O, Bioulac-Sage P, Zucman-Rossi J, Dufour JF, Idle JR (2013) Tissue metabolomics of hepatocellular carcinoma: tumor energy metabolism and the role of transcriptomic classification. Hepatology 58:229–238

Bingle L, Brown NJ, Lewis CE (2002) The role of tumour-associated macrophages in tumour progression: implications for new anticancer therapies. J Pathol 196:254–265

Budhu A, Jia HL, Forgues M, Liu CG, Goldstein D, Lam A, Zanetti KA, Ye QH, Qin LX, Croce CM et al (2008) Identification of metastasis-related microRNAs in hepatocellular carcinoma. Hepatology 47:897–907

Cairns RA, Harris IS, Mak TW (2011) Regulation of cancer cell metabolism. Nat Rev Cancer 11:85–95

Calvisi DF, Ladu S, Gorden A, Farina M, Lee JS, Conner EA, Schroeder I, Factor VM, Thorgeirsson SS (2007) Mechanistic and prognostic significance of aberrant methylation in the molecular pathogenesis of human hepatocellular carcinoma. J Clin Invest 117:2713–2722

Carmeliet P, Jain RK (2000) Angiogenesis in cancer and other diseases. Nature 407:249–257

Cavallaro U, Christofori G (2004) Cell adhesion and signalling by cadherins and Ig-CAMs in cancer. Nat Rev Cancer 4:118–132

Chan THM, Chen LL, Liu M, Hu L, Zheng BJ, Poon VKM, Huang PZ, Yuan YF, Huang JD, Yang J et al (2012) Translationally controlled tumor protein induces mitotic defects and chromosome missegregation in hepatocellular carcinoma development. Hepatology 55:491–505

Chan TH, Lin CH, Qi L, Fei J, Li Y, Yong KJ, Liu M, Song Y, Chow RK, Ng VH et al (2013) A disrupted RNA editing balance mediated by ADARs (adenosine deaminases that act on RNA) in human hepatocellular carcinoma. Gut 63(5):832–843

Chang Y, Yan W, He XX, Zhang LM, Li CJ, Huang H, Nace G, Geller DA, Lin JS, Tsung A (2012) miR-375 inhibits autophagy and reduces viability of hepatocellular carcinoma cells under hypoxic conditions. Gastroenterology 143:177.e8–187.e8

Chen YC, Lin MC, Yao H, Wang H, Zhang AQ, Yu J, Hui CK, Lau GK, He ML, Sung J et al (2007) Lentivirus-mediated RNA interference targeting enhancer of zeste homolog 2 inhibits hepatocellular carcinoma growth through down-regulation of stathmin (vol. 46, p. 200, 2007). Hepatology 46:1314

Chen LL, Hu L, Chan THM, Tsao GSW, Xie D, Huo KK, Fu L, Ma S, Zheng BJ, Guan XY (2009a) Chromodomain helicase/adenosine triphosphatase DNA binding protein 1-like (CHD1L) gene suppresses the nucleus-to-mitochondria translocation of Nur77 to sustain hepatocellular carcinoma cell survival. Hepatology 50:122–129

Chen MH, Huang JD, Hu L, Zheng BJ, Chen LL, Tsang SL, Guan XY (2009b) Transgenic CHD1L expression in mouse induces spontaneous tumors. PLoS One 4(8):e6727

Chen LL, Chan THM, Yuan YF, Hu L, Huang J, Ma S, Wang J, Dong SS, Tang KH, Xie D et al (2010) CHD1L promotes hepatocellular carcinoma progression and metastasis in mice and is associated with these processes in human patients. J Clin Invest 120:1178–1191

Chen LL, Yuan YF, Li Y, Chan THM, Zheng BJ, Huang J, Guan XY (2011) Clinical significance of CHD1L in hepatocellular carcinoma and therapeutic potentials of virus-mediated CHD1L depletion. Gut 60:534–543

Chen LL, Li Y, Lin CH, Chan THM, Chow RKK, Song YY, Liu M, Yuan YF, Fu L, Kong KL et al (2013) Recoding RNA editing of AZIN1 predisposes to hepatocellular carcinoma. Nat Med 19:209–216

Chochi Y, Kawauchi S, Nakao M, Furuya T, Hashimoto K, Oga A, Oka M, Sasaki K (2009) A copy number gain of the 6p arm is linked with advanced hepatocellular carcinoma: an array-based comparative genomic hybridization study. J Pathol 217:677–684

Colombino M, Sperlongano P, Izzo F, Tatangelo F, Botti G, Lombardi A, Accardo M, Tarantino L, Sordelli I, Agresti M et al (2012) BRAF and PIK3CA genes are somatically mutated in hepatocellular carcinoma among patients from South Italy. Cell Death Dis 3: e259

Datta J, Kutay H, Nasser MW, Nuovo GJ, Wang B, Majumder S, Liu CG, Volinia S, Croce CM, Schmittgen TD et al (2008) Methylation mediated silencing of microRNA-1 gene and its role in hepatocellular carcinogenesis. Cancer Res 68:5049–5058

Ding T, Xu J, Wang F, Shi M, Zhang Y, Li SP, Zheng LM (2009) High tumor-infiltrating macrophage density predicts poor prognosis in patients with primary hepatocellular carcinoma after resection. Hum Pathol 40:381–389

Ding J, Huang SL, Wu SQ, Zhao YJ, Liang LH, Yan MX, Ge C, Yao J, Chen TY, Wan DF et al (2010) Gain of miR-151 on chromosome 8q24.3 facilitates tumour cell migration and spreading through downregulating RhoGDIA. Nat Cell Biol 12:390–399

Ding J, Huang SL, Wang Y, Tian Q, Zha RP, Shi HB, Wang QF, Ge C, Chen TY, Zhao YJ et al (2013) Genome-wide screening reveals that miR-195 targets the TNF-/NF-B pathway by downregulating IB kinase alpha and TAB3 in hepatocellular carcinoma. Hepatology 58:654–666

El Tayebi HM, Omar K, Hegy S, El Maghrabi M, El Brolosy M, Hosny KA, Esmat G, Abdelaziz AI (2013) Repression of miR-17-5p with elevated expression of E2F-1 and c-MYC in non-metastatic hepatocellular carcinoma and enhancement of cell growth upon reversing this expression pattern. Biochem Biophys Res Commun 434:421–427

El-Serag HB, Rudolph KL (2007) Hepatocellular carcinoma: epidemiology and molecular carcinogenesis. Gastroenterology 132:2557–2576

Endo M, Yasui K, Zen Y, Gen Y, Zen K, Tsuji K, Dohi O, Mitsuyoshi H, Tanaka S, Taniwaki M et al (2013) Alterations of the SWI/SNF chromatin remodelling subunit-BRG1 and BRM in hepatocellular carcinoma. Liver Int 33:105–117

Fang W, Piao Z, Simon D, Sheu JC, Huang S (2000) Mapping of a minimal deleted region in human hepatocellular carcinoma to 1p36.13–p36.23 and mutational analysis of the RIZ (PRDM2) gene localized to the region. Genes chromosomes cancer 28:269–275

Farber E (1984) The multistep nature of cancer development. Cancer Res 44:4217–4223

Fattovich G, Giustina G, Degos F, Tremolada F, Diodati G, Almasio P, Nevens F, Solinas A, Mura D, Brouwer JT et al (1997) Morbidity and mortality in compensated cirrhosis type C: a retrospective follow-up study of 384 patients. Gastroenterology 112:463–472

Finn RS, Zhu AX (2009) Targeting angiogenesis in hepatocellular carcinoma: focus on VEGF and bevacizumab. Expert Rev Anticancer Ther 9:503–509

Fiume L, Manerba M, Vettraino M, Di Stefano G (2011) Effect of sorafenib on the energy metabolism of hepatocellular carcinoma cells. Eur J Pharmacol 670:39–43

Fodde R, Brabletz T (2007) Wnt/beta-catenin signaling in cancer stemness and malignant behavior. Curr Opin Cell Biol 19:150–158

Fornari F, Gramantieri L, Ferracin M, Veronese A, Sabbioni S, Calin GA, Grazi GL, Giovannini C, Croce CM, Bolondi L et al (2008) MiR-221 controls CDKN1C/p57 and CDKN1B/p27 expression in human hepatocellular carcinoma. Oncogene 27:5651–5661

Fornari F, Milazzo M, Chieco P, Negrini M, Calin GA, Grazi GL, Pollutri D, Croce CM, Bolondi L, Gramantieri L (2010) MiR-199a-3p Regulates mTOR and c-Met to influence the doxorubicin sensitivity of human hepatocarcinoma cells. Cancer Res 70:5184–5193

Fu L, Dong SS, Xie YW, Tai LS, Chen L, Kong KL, Man K, Xie D, Li Y, Cheng Y et al (2010) Down-regulation of tyrosine aminotransferase at a frequently deleted region 16q22 contributes to the pathogenesis of hepatocellular carcinoma. Hepatology 51:1624–1634

Fujimoto A, Totoki Y, Abe T, Boroevich KA, Hosoda F, Nguyen HH, Aoki M, Hosono N, Kubo M, Miya F et al (2012) Whole-genome sequencing of liver cancers identifies etiological influences on mutation patterns and recurrent mutations in chromatin regulators. Nat Genet 44:760–764

Fujiwara Y, Hoon DS, Yamada T, Umeshita K, Gotoh M, Sakon M, Nishisho I, Monden M (2000) PTEN / MMAC1 mutation and frequent loss of heterozygosity identified in chromosome 10q in a subset of hepatocellular carcinomas. JJpn J Cancer Res 91:287–292

Giannelli G, Bergamini C, Marinosci F, Fransvea E, Quaranta M, Lupo L, Schiraldi O, Antonaci S (2002) Clinical role of MMP-2/TIMP-2 imbalance in hepatocellular carcinoma. Int J Cancer 97:425–431

Gott JM, Emeson RB (2000) Functions and mechanisms of RNA editing. Annu Rev Genet 34:499–531

Gramantieri L, Ferracin M, Fornari F, Veronese A, Sabbioni S, Liu CG, Calin GA, Giovannini C, Ferrazzi E, Grazi GL et al (2007) Cyclin g1 is a target of miR-122a, a microRNA frequently down-regulated in human hepatocellular carcinoma. Cancer Res 67:6092–6099

Gramantieri L, Fornari F, Ferracin M, Veronese A, Sabbioni S, Calin GA, Grazi GL, Croce CM, Bolondi L, Negrini M (2009) MicroRNA-221 targets Bmf in hepatocellular carcinoma and correlates with tumor multifocality. Clin Cancer Res 15:5073–5081

Guan XY, Fang Y, Sham JST, Kwong DLW, Zhang YQ, Liang QW, Li HM, Zhou H, Trent JM (2000) Recurrent chromosome alterations in hepatocellular carcinoma detected by comparative genomic hybridization. Genes Chromosomes Cancer 29:110–116

Guichard C, Amaddeo G, Imbeaud S, Ladeiro Y, Pelletier L, Ben Maad I, Calderaro J, Bioulac-Sage P, Letexier M, Degos F et al (2012) Integrated analysis of somatic mutations and focal copy-number changes identifies key genes and pathways in hepatocellular carcinoma. Nat Genet 44:694–698

Guttman M, Rinn JL (2012) Modular regulatory principles of large non-coding RNAs. Nature 482(7385):339–346

Hagiwara A, Cornu M, Cybulski N, Polak P, Betz C, Trapani F, Terracciano L, Heim MH, Ruegg MA, Hall MN (2012) Hepatic mTORC2 activates glycolysis and lipogenesis through Akt, glucokinase, and SREBP1c. Cell Metab 15:725–738

Hanahan D, Weinberg RA (2000) The hallmarks of cancer. Cell 100:57–70

Hanahan D, Weinberg RA (2011) Hallmarks of cancer: the next generation. Cell 144:646–674

He XX, Chang Y, Meng FY, Wang MY, Xie QH, Tang F, Li PY, Song YH, Lin JS (2012) MicroRNA-375 targets AEG-1 in hepatocellular carcinoma and suppresses liver cancer cell growth in vitro and in vivo. Oncogene 31:3357–3369

Hendrich B, Bird A (1998) Identification and characterization of a family of mammalian methyl-CpG binding proteins. Mol Cell Biol 18:6538–6547

Hengartner MO (2000) The biochemistry of apoptosis. Nature 407:770–776

Henry JC, Park JK, Jiang JM, Kim JH, Nagorney DM, Roberts LR, Banerjee S, Schmittgen TD (2010) miR-199a-3p targets CD44 and reduces proliferation of CD44 positive hepatocellular carcinoma cell lines. Biochem Biophys Res Commun 403:120–125

Hernandez-Gea V, Toffanin S, Friedman SL, Llovet JM (2013) Role of the microenvironment in the pathogenesis and treatment of hepatocellular carcinoma. Gastroenterology 144:512–527

Higashitsuji H, Itoh K, Nagao T, Dawson S, Nonoguchi K, Kido T, Mayer RJ, Arii S, Fujita J. (2000). Reduced stability of retinoblastoma protein by gankyrin, an oncogenic ankyrinrepeat protein overexpressed in hepatomas. Nat Med 6(1):96–99

Hsu PP, Sabatini DM (2008) Cancer cell metabolism: Warburg and beyond. Cell 134:703–707

Hu L, Liu M, Chen LL, Chan THM, Wang J, Huo KK, Zheng BJ, Xie D, Guan XY (2012) SCYL1 binding protein 1 promotes the ubiquitin-dependent degradation of Pirh2 and has tumor-suppressive function in the development of hepatocellular carcinoma. Carcinogenesis 33:1581–1588

Hussain SP, Schwank J, Staib F, Wang XW, Harris CC (2007) TP53 mutations and hepatocellular carcinoma: insights into the etiology and pathogenesis of liver cancer. Oncogene 26:2166–2176

Iwata N, Yamamoto H, Sasaki S, Itoh F, Suzuki H, Kikuchi T, Kaneto H, Iku S, Ozeki I, Karino Y et al (2000) Frequent hypermethylation of CpG islands and loss of expression of the 14-3-3 sigma gene in human hepatocellular carcinoma. Oncogene 19:5298–5302

Jablkowski M, Bocian A, Bialkowska J, Bartkowiak J (2005) A comparative study of P53/MDM2 genes alterations and P53/MDM2 proteins immunoreactivity in liver cirrhosis and hepatocellular carcinoma. J Exp Clin Cancer Res 24:117–125

Ji JF, Yamashita T, Budhu A, Forgues M, Jia HL, Li CL, Deng CX, Wauthier E, Reid LM, Ye QH et al (2009) Identification of microRNA-181 by genome-wide screening as a critical player in

EpCAM-positive hepatic cancer stem cells. Hepatology 50:472–480

Ji J, Zhao L, Budhu A, Forgues M, Jia HL, Qin LX, Ye QH, Yu J, Shi X, Tang ZY et al (2010) Let-7g targets collagen type I alpha2 and inhibits cell migration in hepatocellular carcinoma. J Hepatol 52:690–697

Jiang DK, Sun JL, Cao GW, Liu Y, Lin DX, Gao YZ, Ren WH, Long XD, Zhang HX, Ma XP et al (2013) Genetic variants in STAT4 and HLA-DQ genes confer risk of hepatitis B virus-related hepatocellular carcinoma. Nat Genet 45(1):72–75

Joyce JA (2005) Therapeutic targeting of the tumor microenvironment. Cancer Cell 7:513–520

Kanai Y, Ushijima S, Hui AM, Ochiai A, Tsuda H, Sakamoto M, Hirohashi S (1997) The E-cadherin gene is silenced by CpG methylation in human hepatocellular carcinomas. Int J Cancer 71:355–359

Kanai Y, Hui AM, Sun L, Ushijima S, Sakamoto M, Tsuda H, Hirohashi S (1999) DNA hypermethylation at the D17S5 locus and reduced HIC-1 mRNA expression are associated with hepatocarcinogenesis. Hepatology 29:703–709

Kim KW, Bae SK, Lee OH, Bae MH, Lee MJ, Park BC (1998) Insulin-like growth factor II induced by hypoxia may contribute to angiogenesis of human hepatocellular carcinoma. Cancer Res 58:348–351

Kim TM, Yim SH, Shin SH, Xu HD, Jung YC, Park CK, Choi JY, Park WS, Kwon MS, Fiegler H et al (2008) Clinical implication of recurrent copy number alterations in hepatocellular carcinoma and putative oncogenes in recurrent gains on 1q. Int J Cancer 123:2808–2815.

Kim J, Woo AJ, Chu JL, Snow JW, Fujiwara Y, Kim CG, Cantor AB, Orkin SH (2010) A Myc network accounts for similarities between embryonic stem and cancer cell transcription programs. Cell 143:313–324

Ko E, Kim SJ, Joh JW, Park CK, Park J, Kim DH (2008) CpG island hypermethylation of SOCS-1 gene is inversely associated with HBV infection in hepatocellular carcinoma. Cancer Lett 271:240–250

Kondo Y, Kanai Y, Sakamoto M, Mizokami M, Ueda R, Hirohashi S (2000) Genetic instability and aberrant DNA methylation in chronic hepatitis and cirrhosis - A comprehensive study of loss of heterozygosity and microsatellite instability at 39 loci and DNA hypermethylation on 8 CpG islands in microdissected specimens from patients with hepatocellular carcinoma. Hepatology 32:970–979

Kouzarides T (2007) Chromatin modifications and their function. Cell 128:693–705

Kroemer G, Pouyssegur J (2008) Tumor cell metabolism: cancer's Achilles' heel. Cancer Cell 13:472–482

Kubbutat MHG, Jones SN, Vousden KH (1997) Regulation of p53 stability by Mdm2. Nature 387:299–303

Kumar V, Kato N, Urabe Y, Takahashi A, Muroyama R, Hosono N, Otsuka M, Tateishi R, Omata M, Nakagawa H et al (2011) Genome-wide association study identifies a susceptibility locus for HCV-induced hepatocellular carcinoma. Nat Genet 43:455–458

Kuroki T, Fujiwara Y, Nakamori S, Imaoka S, Kanematsu T, Nakamura Y (1995) Evidence for the presence of two tumour-suppressor genes for hepatocellular carcinoma on chromosome 13q. Br J Cancer 72:383–385

Lai MC, Yang Z, Zhou L, Zhu QQ, Xie HY, Zhang F, Wu LM, Chen LM, Zheng SS (2012) Long non-coding RNA MALAT-1 overexpression predicts tumor recurrence of hepatocellular carcinoma after liver transplantation. Med Oncol 29:1810–1816

Lang QB, Ling CQ (2012) MiR-124 suppresses cell proliferation in hepatocellular carcinoma by targeting PIK3CA. Biochem Biophys Res Commun 426:247–252

Law PTY, Ching AKK, Chan AWH, Wong QWL, Wong CK, To KF, Wong N (2012) MiR-145 modulates multiple components of the insulin-like growth factor pathway in hepatocellular carcinoma. Carcinogenesis 33:1134–1141

Lee S, Lee HJ, Kim JH, Lee HS, Jang JJ, Kang GH (2003) Aberrant CpG island hypermethylation along multistep hepatocarcinogenesis. Am J Pathol 163:1371–1378

Lee TK, Poon RTP, Yuen AP, Ling MT, Kwok WK, Wang XH, Wong YC, Guan XY, Man K, Chau KL et al (2006) Twist overexpression correlates with hepatocellular carcinoma metastasis through induction of epithelial-mesenchymal transition. Clin Cancer Res 12:5369–5376

Lee TK, Castilho A, Cheung VC, Tang KH, Ma S, Ng IO (2011) CD24 (+) liver tumor-initiating cells drive self-renewal and tumor initiation through STAT3-mediated NANOG regulation. Cell Stem Cell 9:50–63

Lemmon MA, Schlessinger J (2010) Cell signaling by receptor tyrosine kinases. Cell 141:1117–1134

Levine AJ, Puzio-Kuter AM (2010) The control of the metabolic switch in cancers by oncogenes and tumor suppressor genes. Science 330:1340–1344

Li N, Fu H, Tie Y, Hu Z, Kong W, Wu Y, Zheng X (2009) miR-34a inhibits migration and invasion by down-regulation of c-Met expression in human hepatocellular carcinoma cells. Cancer Lett 275:44–53

Li J, Fu H, Xu C, Tie Y, Xing R, Zhu J, Qin Y, Sun Z, Zheng X (2010) miR-183 inhibits TGF-beta1-induced apoptosis by downregulation of PDCD4 expression in human hepatocellular carcinoma cells. BMC Cancer 10:354

Li M, Zhao H, Zhang X, Wood LD, Anders RA, Choti MA, Pawlik TM, Daniel HD, Kannangai R, Offerhaus GJ et al (2011) Inactivating mutations of the chromatin remodeling gene ARID2 in hepatocellular carcinoma. Nat Genet 43:828–829

Li SP, Qian J, Yang Y, Zhao WT, Dai JC, Bei JX, Foo JN, McLaren PJ, Li ZQ, Yang JM et al (2012) GWAS identifies novel susceptibility loci on 6p21.32 and 21q21.3 for hepatocellular carcinoma in chronic hepatitis B virus carriers. PLoS Genet 8(7): e1002791

Li CF, Wang YZ, Wang SF, Wu B, Hao JL, Fan HX, Ju Y, Ding YP, Chen LZ, Chu XY et al (2013a) Hepatitis B virus mRNA-mediated miR-122 inhibition upregulates PTTG1-binding protein, which promotes hepatocellular carcinoma tumor growth and cell invasion. J Virol 87:2193–2205

Li Y, Chen LL, Chan THM, Guan XY (2013b) Hepatocellular carcinoma: transcriptome diversity regulated by RNA editing. Int J Biochem Cell B 45:1843–1848

Li J, Quan H, Liu Q, Si Z, He Z, Qi H (2013c) Alterations of axis inhibition protein 1 (AXIN1) in hepatitis B virus-related hepatocellular carcinoma and overexpression of AXIN1 induces apoptosis in hepatocellular cancer cells. Oncol Res 20:281–288

Liu WH, Yeh SOH, Lu CC, Yu SL, Chen HY, Lin CY, Chen DS, Chen PJ (2009) MicroRNA-18a prevents estrogen receptor-alpha expression, promoting proliferation of hepatocellular carcinoma cells. Gastroenterology 136:683–693

Liu AM, Poon RTP, Luk JM (2010) MicroRNA-375 targets Hippo-signaling effector YAP in liver cancer and inhibits tumor properties. Biochem Biophys Res Commun 394:623–627

Liu M, Chen LL, Chan THM, Wang J, Li Y, Li Y, Zeng TT, Yuan YF, Guan XY (2012) Serum and glucocorticoid kinase 3 at 8q13.1 promotes cell proliferation and survival in hepatocellular carcinoma. Hepatology 55:1754–1765

Liu L, Dai Y, Chen J, Zeng T, Li Y, Chen L, Zhu YH, Li J, Xie D, Yuan YF et al (2013) Maelstrom promotes hepatocellular carcinoma metastasis by inducing epithelial–mesenchymal transition via Akt/GSK-3beta/snail signaling. Hepatology 59(2):531–543

Liu M, Li Y, Chen L, Chan THM, Song Y, Fu L, Zeng TT, Dai YD, Zhu YH, Li Y et al (2014) Allele-specific imbalance of oxidative stress-induced growth inhibitor 1 associates with progression of hepatocellular carcinoma. Gastroenterology 146(4):1084–1096

Luedde T (2010) MicroRNA-151 and its hosting gene FAK (focal adhesion kinase) regulate tumor cell migration and spreading of hepatocellular carcinoma. Hepatology 52:1164–1166

Ma S, Chan KW, Hu L, Lee TK, Wo JY, Ng IO, Zheng BJ, Guan XY (2007) Identification and characterization of tumorigenic liver cancer stem/progenitor cells. Gastroenterology 132:2542–2556

Ma NF, Hu L, Fung JM, Xie D, Zheng BJ, Chen LL, Tang DJ, Fu L, Wu Z, Chen M et al (2008) Isolation and characterization of a novel oncogene, amplified in liver cancer 1, within a commonly amplified region at 1q21 in hepatocellular carcinoma. Hepatology 47:503–510

Ma S, Tang KH, Chan YP, Lee TK, Kwan PS, Castilho A, Ng I, Man K, Wong N, To KF et al (2010) miR-130b promotes CD133(+) liver tumor-initiating cell growth and self-renewal via tumor protein 53-induced nuclear protein 1. Cell Stem Cell 7:694–707

Marquardt JU, Raggi C, Andersen JB, Seo D, Avital I, Geller D, Lee YH, Kitade M, Holczbauer A, Gillen MC et al (2011) Human hepatic cancer stem cells are characterized by common stemness traits and diverse oncogenic pathways. Hepatology 54:1031–1042

Meng FY, Henson R, Wehbe-Janek H, Ghoshal K, Jacob ST, Patel T (2007) MicroRNA-21 regulates expression of the PTEN tumor suppressor gene in human hepatocellular cancer. Gastroenterology 133:647–658

Mercer TR, Dinger ME, Mattick JS (2009) Long non-coding RNAs: insights into functions. Nat Rev Genet 10:155–159

Merle P, De La Monte S, Kim M, Herrmann M, Tanaka S, Von dem Bussche A, Kew MC, Trepo C, Wands JR (2004) Functional consequences of frizzled-7 receptor overexpression in human hepatocellular carcinoma. Gastroenterology 127:1110–1122

Mise M, Arii S, Higashituji H, Furutani M, Niwano M, Harada T, Ishigami SI, Toda Y, Nakayama H, Fukumoto M et al (1996) Clinical significance of vascular endothelial growth factor and basic fibroblast growth factor gene expression in liver turner. Hepatology 23:455–464

Mishra L, Banker T, Murray J, Byers S, Thenappan A, He AR, Shetty K, Johnson L, Reddy EP (2009) Liver stem cells and hepatocellular carcinoma. Hepatology 49:318–329

Miyoshi Y, Iwao K, Nagasawa Y, Aihara T, Sasaki Y, Imaoka S, Murata M, Shimano T, Nakamura Y (1998) Activation of the beta-catenin gene in primary hepatocellular carcinomas by somatic alterations involving exon 3. Cancer Res 58:2524–2527

Mizuno S, Chijiwa T, Okamura T, Akashi K, Fukumaki Y, Niho Y, Sasaki H (2001) Expression of DNA methyltransferases DNMT1, 3A, and 3B in normal hematopoiesis and in acute and chronic myelogenous leukemia. Blood 97:1172–1179

Momparler RL (2003) Cancer epigenetics. Oncogene 22:6479–6483

Murakami Y, Tamori A, Itami S, Tanahashi T, Toyoda H, Tanaka M, Wu WH, Brojigin N, Kaneoka Y, Maeda A et al (2013) The expression level of miR-18b in hepatocellular carcinoma is associated with the grade of malignancy and prognosis. BMC Cancer 13:99

Nishida N, Fukuda Y, Kokuryu H, Toguchida J, Yandell DW, Ikenega M, Imura H, Ishizaki K (1993) Role and mutational heterogeneity of the P53-gene in hepatocellular-carcinoma. Cancer Res 53:368–372

Nishida N, Fukuda Y, Komeda T, Kita R, Sando T, Furukawa M, Amenomori M, Shibagaki I, Nakao K, Ikenaga M et al (1994) Amplification and overexpression of the cyclin D1 gene in aggressive human hepatocellular carcinoma. Cancer Res 54:3107–3110

Nishimura T, Nishida N, Itoh T, Komeda T, Fukuda Y, Ikai I, Yamaoka Y, Nakao K (2005) Discrete breakpoint mapping and shortest region of overlap of chromosome arm 1q gain and 1p loss in human hepatocellular carcinoma detected by semiquantitative microsatellite analysis. Genes Chromosomes Cancer 42:34–43

Noh JH, Chang YG, Kim MG, Jung KH, Kim JK, Bae HJ, Eun JW, Shen Q, Kim SJ, Kwon SH et al (2013) MiR-145 functions as a tumor suppressor by directly targeting histone deacetylase 2 in liver cancer. Cancer Lett 335:455–462

Oh BK, Kim H, Park HJ, Shim YH, Choi J, Park C, Park YN (2007) DNA methyltransferase expression and DNA methylation in human hepatocellular carcinoma and their clinicopathological correlation. Int J Mol Med 20:65–73

Oka Y, Waterland RA, Killian JK, Nolan CM, Jang HS, Tohara K, Sakaguchi S, Yao T, Iwashita A, Yata Y et al (2002) M6P/IGF2R tumor suppressor gene mutated in hepatocellular carcinomas in Japan. Hepatology 35:1153–1163

Okamoto H, Yasui K, Zhao C, Arii S, Inazawa A (2003) PTK2 and EIF3S3 genes may be amplification targets at 8q23-q24 and are associated with large hepatocellular carcinomas. Hepatology 38:1242–1249

Okano M, Bell DW, Haber DA, Li E (1999) DNA methyltransferases Dnmt3a and Dnmt3b are essential for de novo methylation and mammalian development. Cell 99:247–257

Olivier M, Hollstein M, Hainaut P (2010) TP53 mutations in human cancers: origins, consequences, and clinical use. Cold Spring Harb Perspect Biol 2(1):a001008

Ozturk M (1991) P53 mutation in hepatocellular-carcinoma after aflatoxin exposure. Lancet 338:1356–1359

Parada LA, Hallen M, Tranberg KG, Hagerstrand I, Bondeson L, Mitelman F, Johansson B (1998) Frequent rearrangements of chromosomes 1, 7, and 8 in primary liver cancer. Genes Chromosomes Cancer 23:26–35

Parkin DM, Bray F, Ferlay J, Pisani P (2005) Global cancer statistics, 2002. CA Cancer J Clin 55:74–108

Pathil A, Armeanu S, Venturelli S, Mascagni P, Weiss TS, Gregor M, Lauer UM, Bitzer M (2006) HDAC inhibitor treatment of hepatoma cells induces both TRAIL-independent apoptosis and restoration of sensitivity to TRAIL. Hepatology 43:425–434

Patil MA, Zhang J, Ho C, Cheung ST, Fan ST, Chen X (2006) Hedgehog signaling in human hepatocellular carcinoma. Cancer Biol Ther 5:111–117

Pei DQ, Kang TB, Qi HX (2000) Cysteine array matrix metalloproteinase (CA-MMP)/MMP-23 is a type II transmembrane matrix metalloproteinase regulated by a single cleavage for both secretion and activation. J Biol Chem 275:33988–33997

Reya T, Morrison SJ, Clarke MF, Weissman IL (2001) Stem cells, cancer, and cancer stem cells. Nature 414:105–111

Rousseau B, Menard L, Haurie V, Taras D, Blanc JF, Moreau-Gaudry F, Metzler P, Hugues M, Boyault S, Lemiere S et al (2007) Overexpression and role of the ATPase and putative DNA helicase RuvB-like 2 in human hepatocellular carcinoma. Hepatology 46:1108–1118

Rundhaug JE (2003) Matrix metalloproteinases, angiogenesis, and cancer—Commentary re: A. C. Lockhart et al., reduction of wound angiogenesis in patients treated with BMS-275291, a broad spectrum matrix metalloproteinase inhibitor. Clin Cancer Res 9:551–554

Saito Y, Kanai Y, Nakagawa T, Sakamoto M, Saito H, Ishii H, Hirohashi S (2003) DNA methyltransferase (DNMT) 1 protein expression is significantly increased in human hepatocellular carcinomas with malignant potential and may be a biological predictor of prognosis in hepatocellular carcinoma patients. Hepatology 38:593a–593a

Santoni-Rugiu E, Jensen MR, Thorgeirsson SS (1998) Disruption of the pRb/E2F pathway and inhibition of apoptosis are major oncogenic events in liver constitutively expressing c-myc and transforming growth factor alpha. Cancer Res 58:123–134

Semela D, Dufour JF (2004) Angiogenesis and hepatocellular carcinoma. J Hepatol 41:864–880

Shachaf CM, Kopelman AM, Arvanitis C, Karlsson A, Beer S, Mandl S, Bachmann MH, Borowsky AD, Ruebner B, Cardiff RD et al (2004) MYC inactivation uncovers pluripotent differentiation and tumour dormancy in hepatocellular cancer. Nature 431:1112–1117

Shih TC, Tien YJ, Wen CJ, Yeh TS, Yu MC, Huang CH, Lee YS, Yen TC, Hsieh SY (2012) MicroRNA-214 downregulation contributes to tumor angiogenesis by inducing secretion of the hepatoma-derived growth factor in human hepatoma. J Hepatol 57:584–591

Shiraki K, Sugimoto K, Fujikawa K, Yamanaka T, Takase K, Nakano T (2000) Survivin promotes cell proliferation in human hepatocellular carcinoma. Gastroenterology 118:A906–A906

Song K, Han C, Wu T (2012) Epigenetic regulation of miR-122 expression by PPAR gamma/RXR alpha complex and HBx in hepatocellular carcinoma. Hepatology 56:609a–610a

Stamenkovic I (2000) Matrix metalloproteinases in tumor invasion and metastasis. Semin Cancer Biol 10:415–433

Sudo T, Utsunomiya T, Mimori K, Nagahara H, Ogawa K, Inoue H, Wakiyama S, Fujita H, Shirouzu K, Mori M (2005) Clinicopathological significance of EZH2 mRNA expression in patients with hepatocellular carcinoma. Br J Cancer 92:1754–1758

Sugimachi K, Tanaka S, Kameyama T, Taguchi KI, Aishima SI, Shimada M, Sugimachi K, Tsuneyoshi M (2003) Transcriptional repressor snail and progression of human hepatocellular carcinoma. Clin Cancer Res 9:2657–2664

Takehara T, Liu X, Fujimoto J, Friedman SL, Takahashi H (2001) Expression and role of Bcl-xL in human hepatocellular carcinomas. Hepatology 34:55–61

Thorgeirsson SS, Grisham JW (2002) Molecular pathogenesis of human hepatocellular carcinoma. Nat Genet 31:339–346

Tsujiuchi T, Sugata E, Masaoka T, Onishi M, Fujii H, Shimizu K, Honoki K (2007) Expression and DNA methylation patterns of Tslc1 and Dal-1 genes in hepatocellular carcinomas induced by N-nitrosodiethylamine in rats. Cancer Sci 98:943–948

Ueki T, Fujimoto J, Suzuki T, Yamamoto H, Okamoto E (1997) Expression of hepatocyte growth factor and its receptor, the c-met proto-oncogene, in hepatocellular carcinoma. Hepatology 25:619–623

Venook AP, Papandreou C, Furuse J, de Guevara LL (2010) The incidence and epidemiology of hepatocellular carcinoma: a global and regional perspective. Oncologist 15(Suppl 4):5–13

Vertino PM, Sekowski JA, Coll JM, Applegren N, Han S, Hickey RJ, Malkas LH (2002) DNMT1 is a component of a multiprotein DNA replication complex. Cell Cycle 1:416–423

Vogelstein B, Kinzler KW (1993) The multistep nature of cancer. Trends Genet 9:138–141

Voulgari A, Pintzas A (2009) Epithelial–mesenchymal transition in cancer metastasis: mechanisms, markers and strategies to overcome drug resistance in the clinic. BBA-Rev Cancer 1796:75–90

Wang G, Huang CH, Zhao Y, Cai L, Wang Y, Xiu SJ, Jiang ZW, Yang S, Zhao T, Huang W et al (2000) Genetic aberration in primary hepatocellular carcinoma: correlation between p53 gene mutation and loss-of-heterozygosity on chromosome 16q21–q23 and 9p21–p23. Cell Res 10:311–323

Wang Y, Wu MC, Sham JST, Zhang WG, Wu WQ, Guan XY (2002) Prognostic significance of c-myc and AIB1 amplification in hepatocellular carcinoma—a broad survey using high-throughput tissue microarray. Cancer-Am Cancer Soc 95:2346–2352

Wang GG, Allis CD, Chi P (2007a) Chromatin remodeling and cancer, part I: covalent histone modifications. Trends Mol Med 13:363–372

Wang GG, Allis CD, Chi P (2007b) Chromatin remodeling and cancer, Part II: ATP-dependent chromatin remodeling. Trends Mol Med 13:373–380

Wang Y, Lee ATC, Ma JZI, Wang JB, Ren JW, Yang YC, Tantoso E, Li KB, Ooi LLPJ, Tan P et al (2008) Profiling microRNA expression in hepatocellular carcinoma reveals microRNA-224 up-regulation and apoptosis inhibitor-5 as a microRNA-224-specific target. J Biol Chem 283:13205–13215

Wang B et al (2010a) TGFbeta-mediated upregulation of hepatic miR-181b promotes hepatocarcinogenesis by targeting TIMP3. Oncogene 29:1787–1797

Wang Y, Lu YW, Toh ST, Sung WK, Tan P, Chow P, Chung AYF, Jooi LLP, Lee CGL (2010b) Lethal-7 is down-regulated by the hepatitis B virus x protein and targets signal transducer and activator of transcription 3. J Hepatol 53:57–66

Wang L, Guo ZY, Zhang R, Xin B, Chen R, Zhao J, Wang T, Wen WH, Jia LT, Yao LB et al (2013a) Pseudogene OCT4-pg4 functions as a natural micro RNA sponge to regulate OCT4 expression by competing for miR-145 in hepatocellular carcinoma. Carcinogenesis 34:1773–1781

Wang R, Zhao N, Li S, Fang JH, Chen MX, Yang J, Jia WH, Yuan Y, Zhuang SM (2013b) MicroRNA-195 suppresses angiogenesis and metastasis of hepatocellular carcinoma by inhibiting the expression of VEGF, VAV2, and CDC42. Hepatology 58:642–653

Wei Y, Van Nhieu JT, Prigent S, Srivatanakul P, Tiollais P, Buendia MA (2002) Altered expression of E-cadherin in hepatocellular carcinoma: correlations with genetic alterations, beta-catenin expression, and clinical features. Hepatology 36:692–701

Whiteside TL (2008) The tumor microenvironment and its role in promoting tumor growth. Oncogene 27:5904–5912

Wong CM, Lee JMF, Ching YP, Jin DY, Ng IOL (2003) Genetic and epigenetic alterations of DLC-1 gene in hepatocellular carcinoma. Cancer Res 63:7646–7651

Wong QWL, Lung RWM, Law PTY, Lai PBS, Chan KYY, To KF, Wong N (2008) MicroRNA-223 is commonly repressed in hepatocellular carcinoma and potentiates expression of Stathmin1. Gastroenterology 135:257–269

Wong CCL, Wong CM, Tung EKK, Au SLK, Lee JMF, Poon RTP, Man K, Ng IOL (2011) The MICRORNA miR-139 suppresses metastasis and progression of hepatocellular carcinoma by down-regulating rho-kinase 2. Gastroenterology 140:322–331

Xia HP, Ooi LLPJ, Hui KM (2012) MiR-214 Targets beta-catenin pathway to suppress invasion, stem-like traits and recurrence of human hepatocellular carcinoma. PLoS One 7(9):e44206

Xiong YJ et al (2010) Effects of microRNA-29 on apoptosis, tumorigenicity, and prognosis of hepatocellular carcinoma. Hepatology 51:836–845

Xu T, Zhu Y, Xiong Y, Ge YY, Yun JP, Zhuang SM (2009) MicroRNA-195 suppresses tumorigenicity and regulates G1/S transition of human hepatocellular carcinoma cells. Hepatology 50:113–121

Xu H, He JH, Xiao ZD, Zhang QQ, Chen YQ, Zhou H, Qu LH (2010) Liver-enriched transcription factors regulate microRNA-122 that targets CUTL1 during liver development. Hepatology 52:1431–1442

Xu X, Fan Z, Kang L, Han J, Jiang C, Zheng X, Zhu Z, Jiao H, Lin J, Jiang K et al (2013) Hepatitis B virus X protein represses miRNA-148a to enhance tumorigenesis. J Clin Invest 123:630–645

Yamashita T, Budhu A, Forgues M, Wang XW (2007) Activation of hepatic stem cell marker EpCAM by Wnt-beta-catenin signaling in hepatocellular carcinoma. Cancer Res 67:10831–10839

Yamashita T, Ji J, Budhu A, Forgues M, Yang W, Wang HY, Jia H, Ye Q, Qin LX, Wauthier E et al (2009) EpCAM-positive hepatocellular carcinoma cells are tumor-initiating cells with stem/progenitor cell features. Gastroenterology 136:1012–1024

Yancopoulos GD, Davis S, Gale NW, Rudge JS, Wiegand SJ, Holash J (2000) Vascular-specific growth factors and blood vessel formation. Nature 407:242–248

Yang B, Guo MZ, Herman JG, Clark DP (2003) Aberrant promoter methylation profiles of tumor suppressor genes in hepatocellular carcinoma. Am J Pathol 163:1101–1107

Yang ZF, Ho DW, Ng MN, Lau CK, Yu WC, Ngai P, Chu PW, Lam CT, Poon RT, Fan ST (2008) Significance of CD90+ cancer stem cells in human liver cancer. Cancer Cell 13:153–166

Yang F, Yin YX, Wang F, Wang YQ, Zhang L, Tang Y, Sun SH (2010) miR-17-5p promotes migration of human hepatocellular carcinoma cells through the P38 mitogen-activated protein kinase-heat shock protein 27 pathway. Hepatology 51:1614–1623

Yang F, Zhang L, Huo XS, Yuan JH, Xu D, Yuan SX, Zhu N, Zhou WP, Yang GS, Wang YZ et al (2011a) Long noncoding RNA high expression in hepatocellular carcinoma facilitates tumor growth through enhancer of zeste homolog 2 in humans. Hepatology 54:1679–1689

Yang Z, Zhou L, Wu LM, Lai MC, Xie HY, Zhang F, Zheng SS (2011b) Overexpression of long non-coding RNA HOTAIR predicts tumor recurrence in hepatocellular carcinoma patients following liver transplantation. Ann Surg Oncol 18:1243–1250

Yang X, Liang L, Zhang XF, Jia HL, Qin Y, Zhu XC, Gao XM, Qiao P, Zheng Y, Sheng YY et al (2013) MicroRNA-26a suppresses tumor growth and metastasis of human hepatocellular carcinoma by targeting interleukin-6-Stat3 pathway. Hepatology 58:158–170

Yao YJ, Ping XL, Zhang H, Chen FF, Lee PK, Ahsan H, Chen CJ, Lee PH, Pleacocke M, Santella RM et al (1999) PTEN/MMAC1 mutations in hepatocellular carcinomas. Oncogene 18:3181–3185

Yasui K, Arii S, Zhao C, Imoto I, Ueda M, Nagai H, Emi M, Inazawa J (2002) TFDP1, CUL4A, and CDC16 identified as targets for amplification at 13q34 in hepatocellular carcinomas. Hepatology 35:1476–1484

Yau WL, Lam CSC, Ng L, Chow AKM, Chan STC, Chan JYK, Wo JYH, Ng KTP, Man K, Poon RTP et al (2013) Over-expression of miR-106b promotes cell migration and metastasis in hepatocellular carcinoma by activating epithelial–mesenchymal transition process. PLoS One 8(3):e57882

Ye QH, Qin LX, Forgues M, He P, Kim JW, Peng AC, Simon R, Li Y, Robles AI, Chen YD et al (2003) Predicting hepatitis B virus-positive metastatic hepatocellular carcinomas using gene expression profiling and supervised machine learning. Nat Med 9:416–423

Ying Q, Liang L, Guo W, Zha R, Tian Q, Huang S, Yao J, Ding J, Bao M, Ge C et al (2011) Hypoxia-inducible microRNA-210 augments the metastatic potential of tumor cells by targeting vacuole membrane protein 1 in hepatocellular carcinoma. Hepatology 54:2064–2075

Yoshiji H, Kuriyama S, Yoshii J, Ikenaka Y, Noguchi R, Hicklin DJ, Huber J, Nakatani T, Tsujinoue H, Yanase K et al (2002) Synergistic effect of basic fibroblast growth factor and vascular endothelial growth factor in murine hepatocellular carcinoma. Hepatology 35:834–842

Yoshikawa H, Matsubara K, Qian GS, Jackson P, Groopman JD, Manning JE, Harris CC, Herman JG (2001) SOCS-1, a negative regulator of the JAK/STAT pathway, is silenced by methylation in human hepatocellular carcinoma and shows growth-suppression activity. Nat Genet 28:29–35

Yuan BZ, Miller MJ, Keck CL, Zimonjic DB, Thorgeirsson SS, Popescu NC (1998) Cloning, characterization, and chromosomal localization of a gene frequently deleted in human liver cancer (DLC-1) homologous to rat RhoGAP. Cancer Res 58:2196–2199

Yuan F, Zhou W, Zou C, Zhang Z, Hu H, Dai Z, Zhang Y (2010) Expression of Oct4 in HCC and modulation to wnt/beta-catenin and TGF-beta signal pathways. Mol Cell Biochem 343:155–162

Yuen MF, Hou JL, Chutaputti A (2009) Hepatocellular carcinoma in the Asia pacific region. J Gastroenterol Hepatol 24:346–353

Yuneva MO, Fan TWM, Allen TD, Higashi RM, Ferraris DV, Tsukamoto T, Mates JM, Alonso FJ, Wang CM, Seo Y et al (2012) The metabolic profile of tumors depends on both the responsible genetic lesion and tissue type. Cell Metab 15:157–170

Zhang XY, Liu SR, Hu TS, Liu SP, He Y, Sun SH (2009) Up-regulated MicroRNA-143 transcribed by nuclear factor kappa B enhances hepatocarcinoma metastasis by repressing fibronectin expression. Hepatology 50:490–499

Zhang YJ, Ahsan H, Chen Y, Lunn RM, Wang LY, Chen SY, Lee PH, Chen CJ, Santella RM (2002) High frequency of promoter hypermethylation of RASSF1A and p16 and its relationship to aflatoxin B1-DNA adduct levels in human hepatocellular carcinoma. Mol Carcinog 35:85–92

Zhang HX, Zhai Y, Hu ZB, Wu C, Qian J, Jia WH, Ma FC, Huang WF, Yu LX, Yue W et al (2010) Genome-wide association study identifies 1p36.22 as a new susceptibility locus for hepatocellular carcinoma in chronic hepatitis B virus carriers. Nat Genet 42 (9):755–758

Zhang YZ, Takahashi S, Tasaka A, Yoshima T, Ochi H, Chayama K (2013) Involvement of microRNA-224 in cell proliferation, migration, invasion, and anti-apoptosis in hepatocellular carcinoma. J Gastroenterol Hepatol 28:565–575

Zhao XT, Li JJ, He YH, Lan F, Fu LL, Guo JY, Zhao RJ, Ye Y, He M, Chong WM et al (2001) A novel growth suppressor gene on chromosome 17p13.3 with a high frequency of mutation in human hepatocellular carcinoma. Cancer Res 61:7383–7387

Zheng F, Liao YJ, Cai MY, Liu YH, Liu TH, Chen SP, Bian XW, Guan XY, Lin MC, Zeng YX et al (2012) The putative tumour suppressor microRNA-124 modulates hepatocellular carcinoma cell aggressiveness by repressing ROCK2 and EZH2. Gut 61:278–289

Zheng X, Gai XH, Ding FH, Lu ZT, Tu KS, Yao YM, Liu QG (2013) Histone acetyltransferase PCAF up-regulated cell apoptosis in hepatocellular carcinoma via acetylating histone H4 and inactivating AKT signaling. Mol Cancer 12:96

Zhou XL, Thorgeirsson SS, Popescu NC (2004) Restoration of DLC-1 gene expression induces apoptosis and inhibits both cell growth and tumorigenicity in human hepatocellular carcinoma cells. Oncogene 23:1308–1313

Zhou L, Yang ZX, Song WJ, Li QJ, Yang F, Wang DS, Zhang N, Dou KF (2013) MicroRNA-21 regulates the migration and invasion of a stem-like population in hepatocellular carcinoma. Int J Oncol 43:661–669

Zhu AX, Duda DG, Sahani DV, Jain RK (2011) HCC and angiogenesis: possible targets and future directions. Nat Rev Clin Oncol 8:292–301

Zhu Y, Lu Y, Zhang Q, Liu JJ, Li TJ, Yang JR, Zeng CX, Zhuang SM (2012) MicroRNA-26a/b and their host genes cooperate to inhibit the G1/S transition by activating the pRb protein. Nucleic Acids Res 40:4615–4625

Zondervan PE, Wink J, Alers JC, JN IJ, Schalm SW, de Man RA, van Dekken H (2000) Molecular cytogenetic evaluation of virus-associated and non-viral hepatocellular carcinoma: analysis of 26 carcinomas and 12 concurrent dysplasias. J Pathol 192:207–215

5

Mammalian mitochondrial RNAs are degraded in the mitochondrial intermembrane space by RNASET2

Peipei Liu, Jinliang Huang, Qian Zheng, Leiming Xie, Xinping Lu, Jie Jin, Geng Wang✉

MOE Key laboratory of Bioinformatics, Cell Biology and Development Center, School of Life Sciences, Tsinghua University, Beijing 100084, China

✉ Correspondence: wangeng@biomed.tsinghua.edu.cn (G. Wang)

ABSTRACT

Mammalian mitochondrial genome encodes a small set of tRNAs, rRNAs, and mRNAs. The RNA synthesis process has been well characterized. How the RNAs are degraded, however, is poorly understood. It was long assumed that the degradation happens in the matrix where transcription and translation machineries reside. Here we show that contrary to the assumption, mammalian mitochondrial RNA degradation occurs in the mitochondrial intermembrane space (IMS) and the IMS-localized RNASET2 is the enzyme that degrades the RNAs. This provides a new paradigm for understanding mitochondrial RNA metabolism and transport.

KEYWORDS mitochondria, intermembrane space, ribonuclease, mtRNA, RNA degradation, decay, RNASET2, RNase T2, inner membrane, transport, RNA trafficking

INTRODUCTION

Mammalian mitochondria contain a small circular genome, which encodes 2 ribosomal RNAs, 22 tRNAs, and 13 essential protein subunits in the OXPHOS pathway (Anderson et al., 1981). The transcription and translation machineries in the mitochondrial matrix require both mitochondrion-encoded RNAs and nucleus encoded protein and RNA factors (Bonawitz et al., 2006; Rubio et al., 2008; Wang et al., 2010). Mitochondrial RNAs (mtRNAs) are first transcribed as polycistronic transcripts from both H-strand and L-strand, and the transcripts are dissected by RNA processing enzymes into individual rRNAs, tRNAs, and mRNAs (Hallberg and Larsson, 2014; Mercer et al., 2011). These rRNAs and tRNAs together with translational factors imported from cytosol then synthesize the OXPHOS proteins from the mRNAs (Hallberg and Larsson, 2014).

Much is understood about mtRNA synthesis and processing (Sanchez et al., 2011; Schafer et al., 2005). By contrast, less attention has been given to understanding how mtRNAs are degraded. Yet, mitochondrial RNA homeostasis is one of the key elements in regulating mitochondrial functions. Mitochondria have to quickly respond to external stimuli and are constantly going through fusion and fission, which is directly linked to their biosynthesis and requires fast changes of their gene expression (Mishra and Chan, 2016). RNA decay adds a new dimension to the regulation network by changing the abundance of transcripts and removing aberrant RNA molecules and intermediates of transcription and processing. The process must be tightly regulated and often is performed by multi-protein complexes (Szczesny et al., 2012).

In yeast, mtRNAs are degraded by a degradosome complex consisting of a helicase Suv3 and an RNase II like protein Dss1 (Dziembowski et al., 1998; Margossian et al., 1996; Szczesny et al., 2013). Suv3 unwinds and feeds RNA substrates to Dss1, which acts as a 3′ to 5′ exoribonuclease that yields nucleoside monophosphates and a four-nucleotide residual core (Malecki et al., 2010).

In comparison, the degradation machinery of mammalian mtRNAs is much less clear. Even though the role of hSuv3p helicase has been demonstrated, the ribonuclease partner has not been identified (Khidr et al., 2008). Human polynucleotide phosphorylase PNPASE has been shown to be involved in mtRNA degradation and appears to colocalize with Suv3 in distinct mitochondrial foci, but the protein is mainly localized in the mitochondrial intermembrane space (IMS) and functions in mitochondrial RNA import from cytosol (Borowski et al., 2013; Chen et al., 2006; Chujo et al., 2012; Sato et al., 2017; Vedrenne et al., 2012; von Ameln et al., 2012; Wang et al., 2010). Whether its function in mitochondrial RNA degradation is that of a ribonuclease has never been proven. Other ribonucleases such as LACTB2 and Endo G have also been identified in mitochondria (Cote and Ruiz-Carrillo, 1993; Levy et al., 2016; Ohsato et al., 2002; Zhou et al., 2016). However, the exact mechanisms of their involvement in mtRNA degradation have never been resolved. It has long been assumed that mitochondrial RNA decay would occur in the mitochondrial matrix as both yeast and human Suv3 proteins were identified in the matrix. Here we show that contrary to the assumption, mtRNA degradation happens in the mitochondrial intermembrane space (IMS) and IMS-localized RNASET2 is the ribonuclease that carries out the degradation.

RESULTS

Characterization of in organello mtRNA degradation

The identity of the ribonuclease responsible for mammalian mitochondrial RNA degradation had been elusive. It has been shown that human polynucleotide phosphorylase PNPASE and the helicase Suv3 are involved in mtRNA degradation (Borowski et al., 2013; Chujo et al., 2012). However, mammalian PNPASE is localized mainly in the IMS and functions in import of cytosolic RNA into mitochondria (Chen et al., 2006; Sato et al., 2017; Vedrenne et al., 2012; von Ameln et al., 2012; Wang et al., 2010), so it is hard to deduce whether its role in mtRNA degradation is that of a ribonuclease. The ribonuclease activity of mammalian PNPASE has been examined before with the protein expressed in bacteria (Portnoy et al., 2008; Slomovic et al., 2008). We have purified the human PNPASE from both human mitochondria and bacteria as previously described (Wang et al., 2010) and compared the ribonuclease activities. Surprisingly, PNPASE showed no ribonuclease activity in human mitochondria but strong activity when expressed in bacteria (Fig. 1A–C), consistent with a previous report showing mammalian PNPASE has little to no activity in a mammalian system (Sarkar et al., 2005). Total mitochondrial membrane where PNPASE localizes also showed no ribonuclease activity (Fig. 1D). In comparison, strong ribonuclease activity was observed in soluble fraction of the mitochondria that contains no PNPASE (Fig. 1D). Taken together, these results suggest that the function of PNPASE or PNPASE-containing complex in mammalian mtRNA degradation is not that of a ribonuclease. In this report, we will focus on identification of the ribonuclease instead of addressing how mammalian PNPASE ribonuclease activity is inactivated in mitochondria and how PNPASE is involved in mtRNA degradation.

To identify the ribonuclease for mtRNA degradation, an *in vitro* assay is not sufficient, as multiple ribonucleases have been identified in mitochondria (Bruni et al., 2013; Cote and Ruiz-Carrillo, 1993; Levy et al., 2016; Zhou et al., 2016). The ribonuclease that functions directly in mtRNA degradation should have characteristics similar to those of mtRNA degradation. In order to find the ribonuclease, we designed a system that allows in organello characterization of mtRNA degradation and examined the dependency of mtRNA degradation on temperature, pH, ATP, and metal ions. Newly synthesized mtRNAs in isolated mitochondria were labeled with P32 or Biotin. Controls have been performed to show that the isolated mitochondria had no nuclear DNA contamination and the labeled RNAs were mtRNAs in the intact mitochondria as no labeling was observed in the *rho0* mitochondria (Fig. S1A–E). Degradation of the newly synthesized mtRNAs was examined under various chasing conditions. Slower degradation was observed at 25°C than at 37°C (Fig. 1E), and lower pH had an inhibitory effect on the decay (pH 6.5 vs. pH 7.4) (Fig. 1F). The effects of ATP and di-valence metal ions were more complicated: Cu^{2+} had a strong inhibitory effect even at a low concentration (0.5 mmol/L) (Fig. 1G), while Mg^{2+} had little effect on the degradation at a low concentration (0.5 mmol/L) but an inhibitory effect at a higher concentration (20 mmol/L) (Fig. 1H). ATP had a similar effect as that of Mg^{2+}, little effect at a low concentration (0.5 mmol/L) but inhibitory at a higher concentration (8 mmol/L) (Fig. 1I).

mtRNAs are degraded in the mitochondrial intermembrane space (IMS)

Equipped with the information about mtRNA degradation, we next investigated the possible sub-mitochondrial localization of the mtRNA degradation activity. Using biochemical approaches, we separated mitochondria into four fractions: total soluble, total membranes, IMS, and matrix (Fig. 2A). Biotinylated RNA was used as a substrate to examine the ribonuclease activity of each fraction. Majority of the activity appeared to be in the IMS and the total soluble that includes both IMS and the matrix (Fig. 2B). Since the matrix had little activity, IMS should be where majority of the activity resides (Fig. 2B). To rule out the possibility of a cytosolic contamination, ribonuclease activity was examined using the intact mitochondria. No activity was observed when the mitochondria were incubated with an RNA substrate in the isotonic buffer, but the RNA substrate was quickly degraded when the outer membrane was ruptured using hypotonic buffer, suggesting the outer surface of the purified mitochondria harbors no ribonuclease activity and the IMS activity has no cytosolic contamination (Fig. 2C).

The IMS activity was further characterized and was shown to be inhibited by Cu^{2+} as the in organello mtRNA degradation was, but not by EDTA (Fig. 2D). The activity also degraded total mtRNAs with similar characteristics, and like

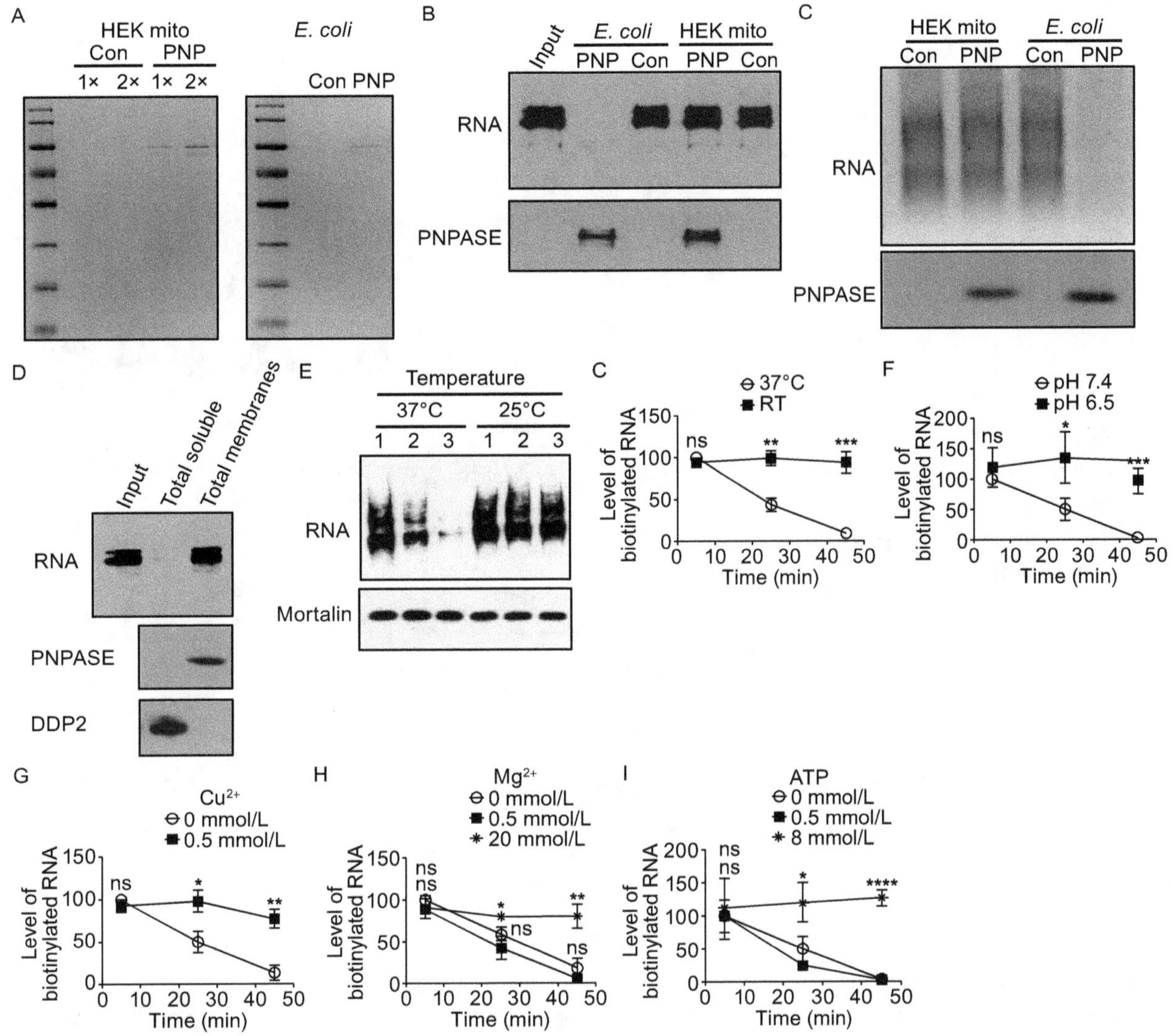

Figure 1. Characterization of mtRNA degradation using an in organello system. (A) Coomassie staining of human PNPASE samples purified from HEK293 mitochondria (HEK mito) or *E. coli*. For PNPASE purified from HEK mitochondria, two concentrations of samples were loaded. Con: same volume of eluate from cells harboring the empty vector; PNP: eluate from PNPASE-HisPC expressing cells. (B) PNPASE purified from *E. coli* or HEK mitochondria was incubated with Biotin-labeled *UCP2* RNA. The lower panel shows immunoblotting of PNPASE. (C) PNPASE purified from *E. coli* or HEK mitochondria was incubated with total cytosolic RNA and the samples were resolved on an agarose gel. The lower panel shows immunoblotting of PNPASE. (D) Mitochondria were separated into total soluble and total membrane and the two fractions were examined for ribonuclease activity using biotinylated *UCP2* mRNA as a substrate. Lower panels show Immunoblotting of PNPASE (Membrane) and DDP2 (Soluble). (E) Effect of temperature on in organello mtRNA degradation. Degradation was performed at 37°C (the temperature used for the other experiments if not specified) or 25°C. The three numbers (1, 2, and 3) represent three time points (5 min, 25 min, and 45 min). Top panel on the left shows the remaining labeled mtRNAs. Bottom panel is an immunoblot of mitochondrial protein Mortalin showing the amount of mitochondria taken out at each time point. Right panel shows the quantification of labeled mtRNAs (n = 3). (F) Effect of pH on in organello mtRNA degradation. Degradation was performed at pH 7.4 (the pH used for the other experiments if not specified) or pH 6.5. (G) Effect of Cu^{2+} on in organello mtRNA degradation. Two concentrations (0 mmol/L and 0.5 mmol/L) of Cu^{2+} were used. (H) Effect of Mg^{2+} on in organello mtRNA degradation. Three concentrations (0 mmol/L, 0.5 mmol/L, and 20 mmol/L) of Mg^{2+} were used. (I) Effect of ATP on in organello mtRNA degradation. Three concentrations (0 mmol/L, 0.5 mmol/L, and 8 mmol/L) of ATP were used. Statistical comparisons are performed using unpaired t-tests (n = 3 if not specified); $^{*}P < 0.05$, $^{**}P < 0.01$, $^{***}P < 0.001$, $^{****}P < 0.0001$. Data are presented as mean ± standard error of the mean (s.e.m.).

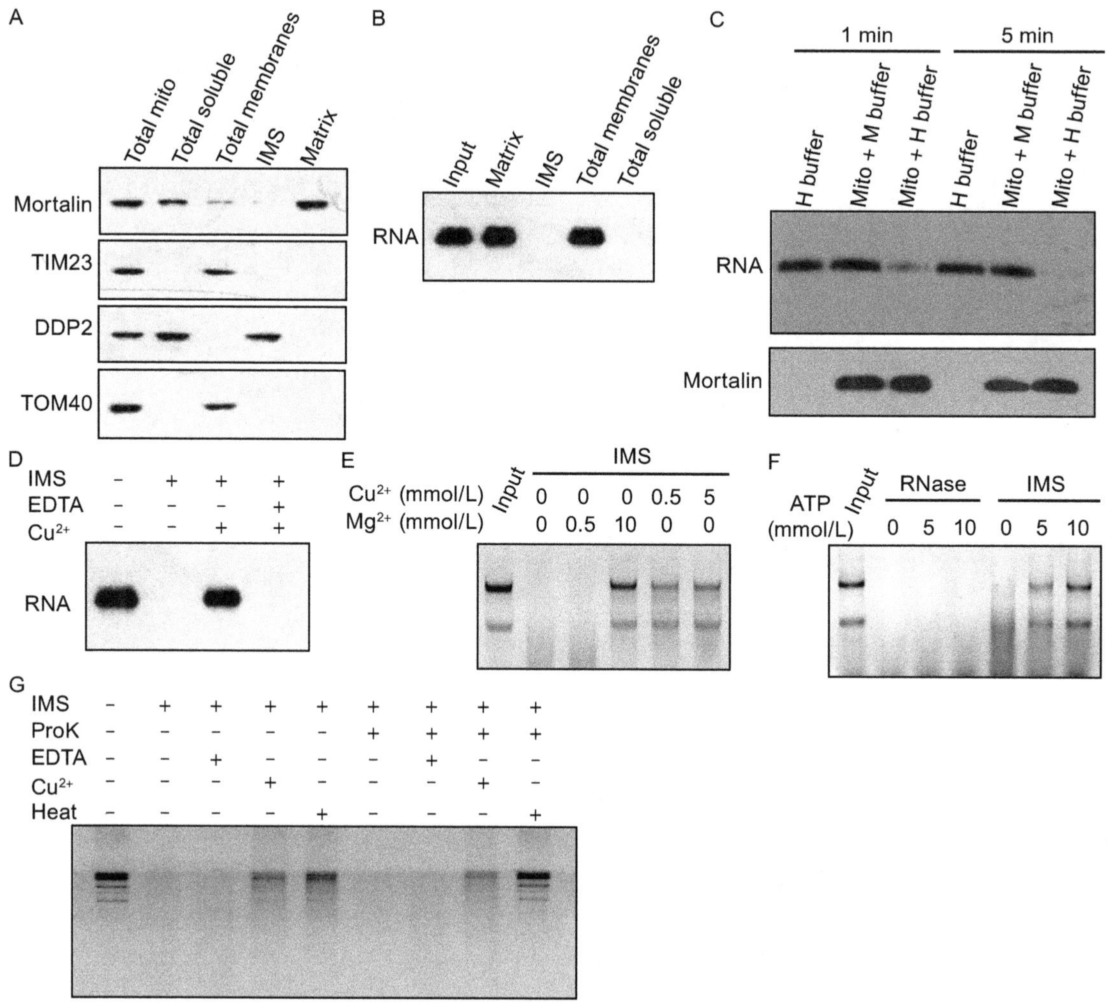

Figure 2. Characterization of a ribonuclease activity in the mitochondrial IMS. (A) Immunoblots of total mitochondria and fractions. Mitochondria were separated into four fractions: total soluble (including IMS and Matrix), total membrane, IMS and matrix. Immunoblotting was performed using antibodies for Mortalin (Matrix), TIM23 (Inner membrane), DDP2 (IMS), and TOM40 (Outer membrane). (B) Four mitochondrial fractions were examined for ribonuclease activity using biotinylated *UCP2* mRNA as a substrate. (C) No ribonuclease activity localizes at the outer surface of mitochondrial outer membrane. Isolated mitochondria were resuspended in mitoprep buffer (M buffer) or hypotonic buffer (H buffer) that ruptures the mitochondrial outer membrane. Biotinylated *UCP2* RNA was added to the mixture and incubated at 37°C for 1 min or 5 min before the reaction was terminated. (D) IMS ribonuclease activity was tested for its sensitivity to EDTA (2 mmol/L) and Cu^{2+} (0.5 mmol/L) using biotinylated *UCP2* mRNA as a substrate. (E) IMS ribonuclease activity was tested for its sensitivity to different concentrations of Cu^{2+} (0.5 mmol/L and 5 mmol/L), and Mg^{2+} (0.5 mmol/L and 10 mmol/L) using RNAs purified from isolated mitochondria as substrates. (F) IMS ribonuclease activity and RNaseI were tested for sensitivity to different concentrations of ATP (0 mmol/L, 5 mmol/L, and 10 mmol/L). (G) IMS ribonuclease activity was tested for sensitivity to Proteinase K (ProK), EDTA (2 mmol/L), Cu^{2+} (0.5 mmol/L), and heat (90°C 10 min).

the in organello mtRNA degradation, its inhibition by Mg^{2+} or ATP was also concentration dependent (Fig. 2E and 2F). In addition, it was inactivated by heating at 90°C, but not by proteinase K treatment (Fig. 2G).

The localization of a ribonuclease activity in the IMS with characteristics similar to those of in organello mtRNA degradation suggests that mtRNA degradation happens in the IMS. The possibility was investigated using the in organello mtRNA degradation assay. Mitoplasting was performed after the in organello RNA synthesis to rupture the mitochondrial outer membrane, wash away the IMS and rid of some inner membrane associated proteins (Fig. 3A). The inner membrane of the mitoplast did appear more fragile as shown by immunoblotting of the matrix protein Mortalin

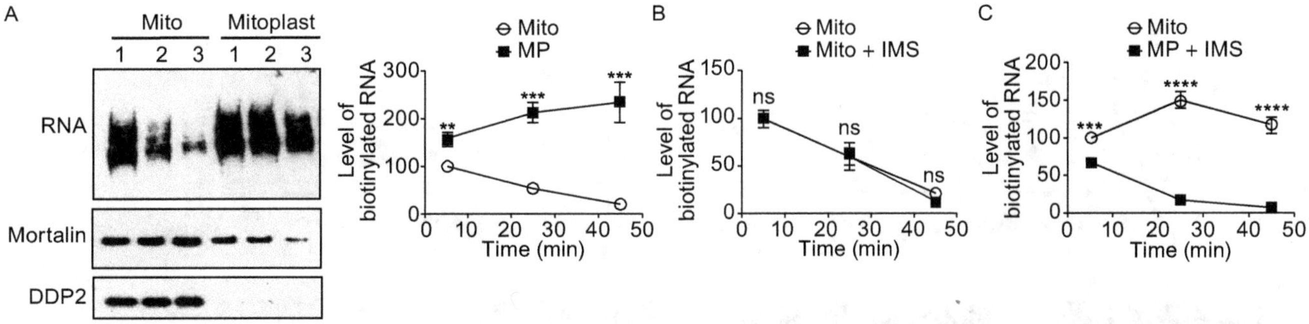

Figure 3. mtRNAs are degraded in the mitochondrial IMS. (A) Effect of mitoplasting on in organello mtRNA degradation. The three numbers (1, 2, and 3) represent three time points (5 min, 25 min, and 45 min). Top panel on the left shows the labeled mtRNAs. Middle panel is an immunoblot of mitochondrial matrix protein Mortalin showing the amount of mitochondria taken out at each time point with intact inner membrane. Bottom panel is an immunoblot of mitochondrial IMS protein DDP2 showing the efficiency of mitoplasting. Right panel shows the quantification of the remaining labeled mtRNAs normalized to Mortalin signal ($n = 3$). (B) Effect of adding isolated IMS (0.15 mg/mL) on in organello mtRNA degradation in intact mitochondria. (C) Effect of adding isolated IMS (0.15 mg/mL) on in organello mtRNA degradation in mitoplasts. Statistical comparisons are performed using unpaired t-tests ($n = 3$ if not specified); $^{*}P < 0.05$, $^{**}P < 0.01$, $^{***}P < 0.001$, $^{****}P < 0.0001$. Data are presented as mean ± standard error of the mean (s.e.m.).

(Fig. 3A). However, the remaining mitoplast with matrix still protected by the inner membrane appeared to have its mtRNA degradation stopped (Fig. 3A). The newly synthesized mtRNA level actually increased to about 1.5 fold of its original level, suggesting that even under the chase condition there was active mtRNA synthesis and the degradation observed with the intact mitochondria was a result of the balance being tipped towards degradation (Fig. 3A). More interestingly, the stopped mtRNA degradation in the mitoplast was rescued when IMS was added into the degradation buffer and the addition of IMS had no effect on the intact mitochondria (Fig. 3B and 3C). Taken together, these results provide compelling evidence that mtRNA degradation happens in the mitochondrial IMS.

Identification of an IMS-localized ribonuclease RNASET2

With the localization of the ribonuclease for mammalian mtRNA degradation confirmed, we next tried to identify the activity (Fig. S2A and S2B). Based on the mass-spectrometry results of the purified samples (Tables S1 and S2), we singled out RNASET2 for further testing. Mammalian RNASET2s are endoribonucleases that are targeted for secretion and also to lysosomes (Acquati et al., 2011; Luhtala and Parker, 2010). Optiprep gradient fractionation showed HeLa cells have a distinct mitochondrial pool of endogenous RNASET2 (Fig. 4A). Overexpression of RNASET2 did not cause a dramatic change of the gradient centrifugation profile (Fig. 4B), indicating a similar localization of the tagged protein to that of the endogenous RNASET2. Colocalization of GFP-tagged RNASET2 and mitoTracker Red was also observed using confocal microscopy (Fig. 4C). A mutation (C184R) that disrupts lysosomal localization led to a better colocalization of RNASET2 with mitochondria (Fig. 4D), suggesting a dynamic balance among different subcellular pools.

Since the IMS ribonuclease activity is insensitive to proteinase K treatment, we tested whether RNASET2 could be digested by proteinase K. Consistent with the IMS result, both the endogenous RNASET2 and the tagged version were resistant to proteinase K digestion, while TIM23 was readily degraded by the protease (Fig. 4E). The tagged version ran lower on SDS-PAGE after proteinase K treatment simply because the tag was digested by the protease as shown by a lack of immunoblotting signal for the tag after the treatment (Fig. 4E). Because RNASET2 was proteinase K insensitive, Mitoplasting and NaCl wash instead of the traditional Mitoplasting and protease treatment was performed to examine the sub-mitochondrial localization of the endogenous RNASET2. Endogenous RNASET2 showed a fractionation profile similar to that of an IMS protein DDP2 but different from those of a membrane protein BAP37 and a matrix protein Mortalin, suggesting an IMS localization of RNASET2 within mitochondria (Fig. 4F).

To rule out the possibility that the localization of RNASET2 in mitochondria we observed was due to contamination from other cellular compartments, we took advantage of a new technique called proximity-based labeling, which has been used to overcome the problem of fractionation contamination (Han et al., 2017; Kim and Roux, 2016; Williams et al., 2014). Ascorbate peroxidase (APEX) biotinylates proteins in close proximity when biotin-phenol and H_2O_2 are added to the medium of live cells (Jan et al., 2014). APEX was used to tag RNASET2 and a mitochondrial IMS protein MIA40. Subcellular fractionation and mitoplasting with or without proteinase K treatment was performed to ensure that

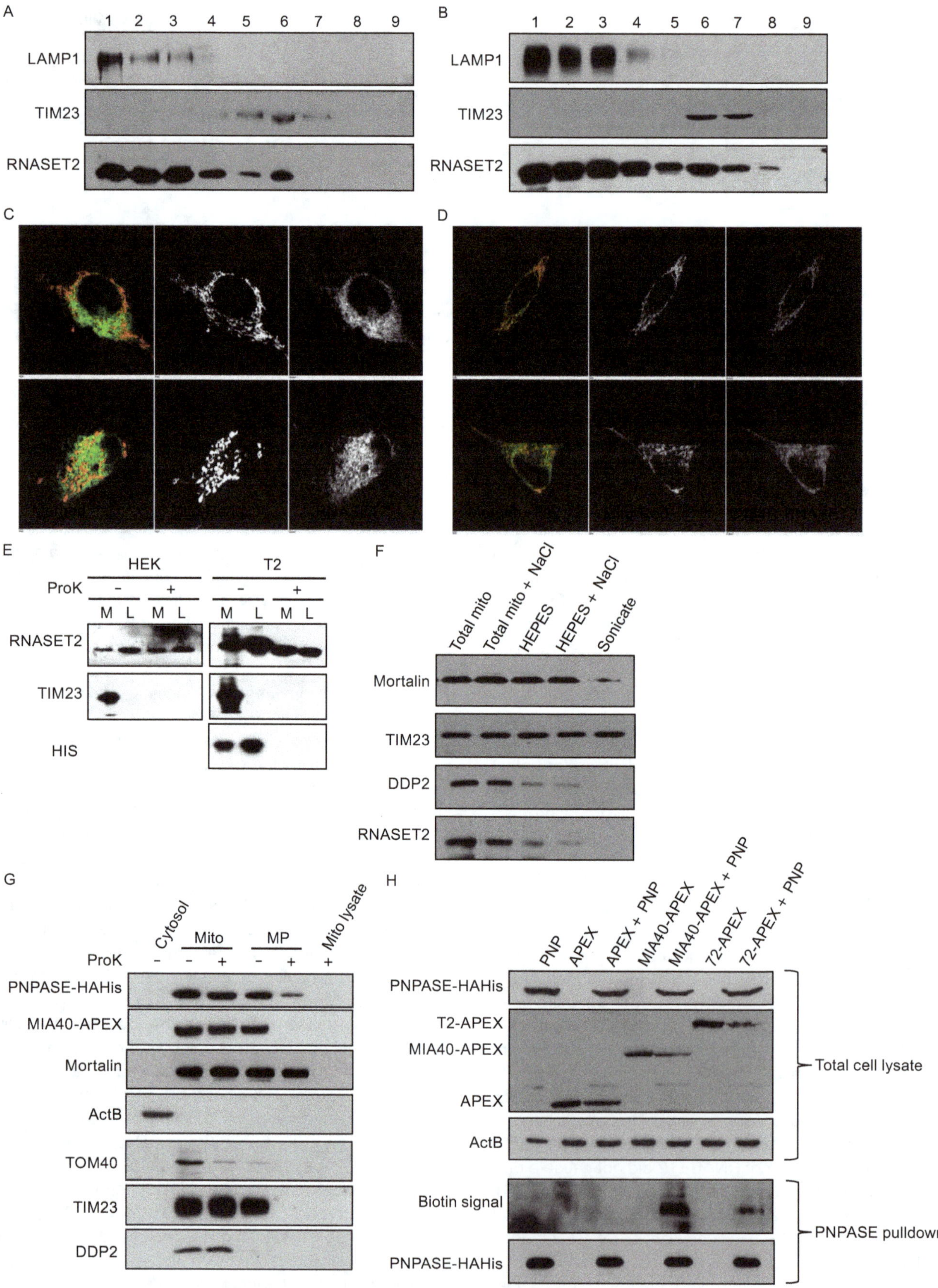
A
1 2 3 4 5 6 7 8 9
LAMP1
TIM23
RNASET2
B
1 2 3 4 5 6 7 8 9
LAMP1
TIM23
RNASET2
C
D
E
HEK
T2
ProK
- + - +
M L M L M L M L
RNASET2
TIM23
HIS
F
Total mito
Total mito + NaCl
HEPES
HEPES + NaCl
Sonicate
Mortalin
TIM23
DDP2
RNASET2
G
Cytosol
Mito
MP
Mito lysate
ProK
- - + - + +
PNPASE-HAHis
MIA40-APEX
Mortalin
ActB
TOM40
TIM23
DDP2
H
PNP
APEX
APEX + PNP
MIA40-APEX
MIA40-APEX + PNP
72-APEX
72-APEX + PNP
PNPASE-HAHis
T2-APEX
MIA40-APEX
APEX
ActB
Total cell lysate
Biotin signal
PNPASE-HAHis
PNPASE pulldown

◄ **Figure 4. RNASET2 localizes in mitochondrial IMS.** (A) Immunoblots of different fractions from Optiprep gradient centrifugation of crude HeLa mitochondria with antibodies for lysosomal marker LAMP1, mitochondrial marker TIM23 and RNASET2. (B) Immunoblots of different fractions from Optiprep gradient centrifugation of crude HEK mitochondria overexpressing tagged RNASET2. (C) Confocal microscopy of HeLa cells expressing GFP tagged RNASET2. MitoTracker red (Mito-Red) was used for mitochondrion staining. (D) Confocal microscopy of HeLa cells expressing GFP tagged RNASET2 mutant (C184R). (E) Immunoblots of mitochondrial (M) lysates and lysosomal (L) lysates from control HEK cells or RNASET2-overexpressing cells (T2) with or without proteinase K (ProK) treatment. TIM23 was used as a positive control for Proteinase K treatment. HIS antibody was used to detect the integrity of the tag on overexpressed RNASET2. (F) Immunoblots of mitochondria after undergoing different treatments that exposed different sub-mitochondrial compartments: NaCl (0.3 mol/L) wash to rid the membrane of electrostatically associated proteins, hypotonic treatment to rupture the outer membrane (HEPES), or sonication to rupture both the outer and the inner membrane. Antibodies for Mortalin (matrix), TIM23 (membrane), and DDP2 (IMS) were used. (G) Immunoblots of the cytosol, mitochondria (Mito), mitoplasts (MP, mitochondria with outer membrane ruptured), or mitochondrial lysate with (+) or without (−) proteinase K treatment. PNPASE-HisPC and MIA40-APEX-HAHis (Mia40-Apex) were detected with anti-His antibody. ActB: cytosol; Mortalin: mitochondrial matrix; TOM40: mitochondrial outer membrane; TIM23: mitochondrial inner membrane; DDP2: mitochondrial intermembrane space. (H) Biotinylation of PNPASE by RNASET2-APEX. Top three panels are immunoblots of total lysates from cells expressing PNPASE-HAHis (PNP), APEX-HisPC (APEX), Mia40-APEX-HisPC (MIA40-APEX), RNASET2-APEX-HisPC (T2-APEX) or PNPASE-HAHis together with one of the APEX constructs. Anti-His antibody was used to compare the relative expression levels of the APEX and the fusion proteins (all containing a His tag). Bottom two panels are biotin detection blot and an immunoblot of the PNPASE-HAHis pulldown samples from mitochondria of the six strains listed. (I) Immunoblots of the cytosol, mitochondria (Mito), mitoplasts (MP), or mitochondrial lysate with (+) or without (−) proteinase K treatment. BirA-HAHis (BirA), TOM22-BirA-HAHis (TOM22-BirA), and MIA40-Avi-His-Flag (Mia40-Avi) were detected with anti-His antibody. (J) Biotinylation of RNASET2-Avi-HisFlag (RNASET2-Avi) by TOM22-BirA-HAHis (TOM22-BirA). Upper four panels are immunoblots of total lysates from cells expressing BirA-HAHis with RNASET2-Avi-HisFlag, TOM22-BirA-HAHis with MIA40-Avi-HisFlag, or TOM22-BirA-HAHis with RNASET2-Avi-His-Flag. Anti-His antibody was used to detect BirA and the fusion proteins (all containing a His tag), and anti-Flag antibody was used to detect Avi fusion proteins. Bottom panel is a biotin detection blot.

APEX and other tags had no effect on the localization of the proteins (Fig. 4G). Biotinylation of mitochondrial IMS protein PNPASE was observed only when MIA40-APEX or RNASET2-APEX was expressed but not when cytosolic APEX was expressed at a similar level, further proving a mitochondrial localization of RNASET2 (Fig. 4H). This technique, however, lacks specificity; so another proximity labeling approach was adopted. Biotin ligase (BirA) biotinylates Avi-tagged proteins when they are in close proximity (Roux et al., 2012). BirA was used to tag a mitochondrial outer membrane protein TOM22 with its C-terminus facing the IMS and Avi was used to tag RNASET2 and MIA40. Subcellular fractionation and mitoplasting with or without proteinase K treatment was again performed to ensure that BirA, Avi, and other tags had no effect on the localization of the proteins and that BirA on the C-terminus of TOM22 is in the IMS (Fig. 4I). TOM22-BirA but not cytosolic BirA quickly biotinylated both MIA40-Avi and RNASET2-Avi, when Biotin was added to the medium, proving once again RNASET2 is localized in the mitochondrial IMS (Fig. 4J).

RNASET2 functions in mtRNA degradation and indirectly affects mtRNA synthesis

To verify the involvement of RNASET2 in mtRNA degradation, it was overexpressed in the HEK293 cells (Fig. 5A). IMS was isolated from the RNASET2-overexpressing mitochondria and the control mitochondria, and the ribonuclease activity examined. Compared to the control IMS, RNASET2-overexpressing IMS was less sensitive to Cu^{2+} or ATP inhibition (Fig. 5B and 5C). The in organello mtRNA synthesis was faster for the RNASET2-overexpressing mitochondria (Fig. 5D) and the in organello mtRNA degradation rate was also higher (Fig. 5E). Under *in vivo* condition that was partially inhibitory to mitochondrial RNA synthesis (H_2O_2 treatment), RNASET2 overexpression led to a significant decrease of the mtRNA levels (Fig. 5F). RNASET2 knockdown with shRNAs showed opposite effects both on in organello mtRNA synthesis and degradation with mtRNA synthesis and degradation rates decreased in the knockdown mitochondria (Fig. 5G and 5H). The effects were not restricted to human cells, as overexpression of mouse RNASET2 in a mouse cell line TM6 also led to an increase of both in organello mtRNA synthesis and degradation rates (Fig. S3A–C).

Since RNASET2 is a protein with multiple subcellular localizations, the effects of overexpression and knockdown on mtRNA degradation could be indirect and be due to changes of its functions in other cellular compartments. To rule out the possibility, we fused an enzymatically dead mutant (C65R, C118R) without the N-terminal 24 amino acids to the targeting sequence of cytochrome C1 that targeted the protein to the IMS (C1ΔNT2M). Unlike the wild-type RNASET2, C1ΔNT2M localizes predominately in the

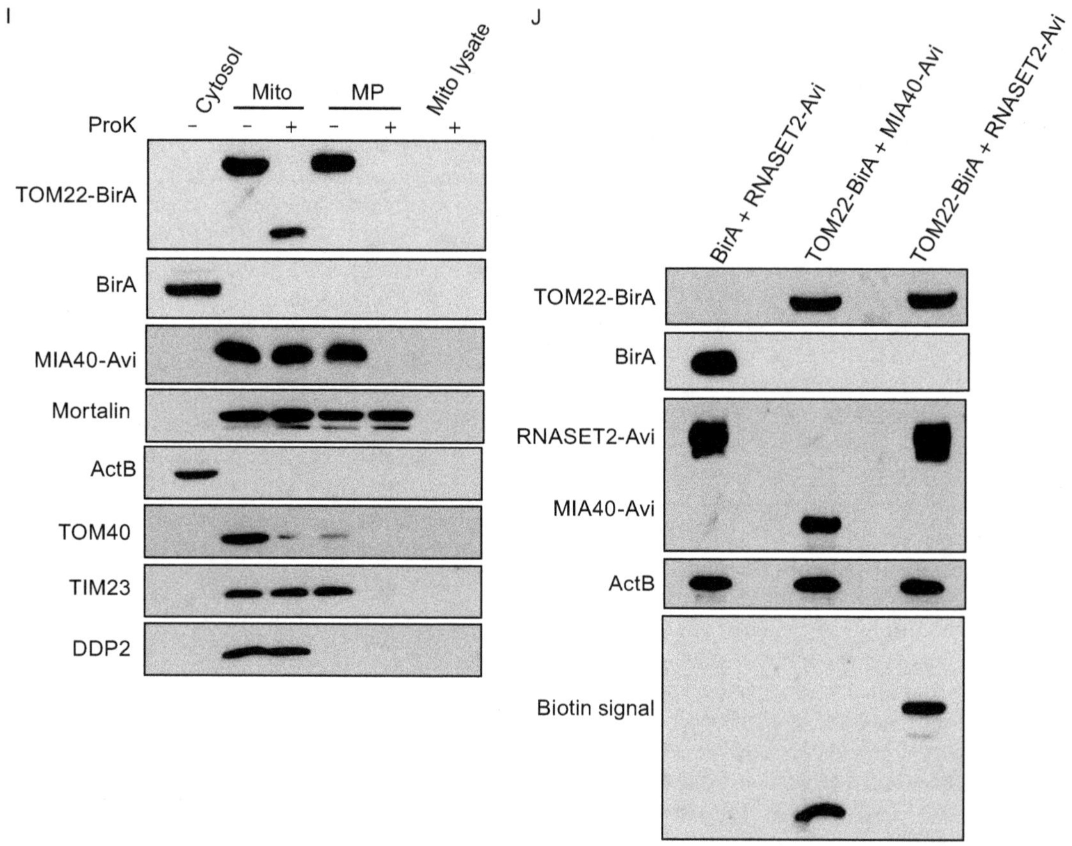

Figure 4. continued.

mitochondria (Fig. 5J). This enzymatically dead mutant, when targeted to the IMS, acting like a dominant negative mutant, had similar effects on the in organello mtRNA synthesis and degradation as those of RNASET2 knockdown (Fig. 5K and 5L). Under *in vivo* condition that was inhibitory to mitochondrial RNA synthesis, C1ΔNT2M expressing mitochondria contained significantly higher levels of mitochondrial rRNAs and showed less structural defects (Figs. 5M and S4). The mild effect of C1ΔNT2M on mitochondrial mRNA levels is possibly due to the feedback down-regulation of mtRNA synthesis, as there is no efficient approach to quickly turn off mtRNA synthesis without dramatically affecting other cellular processes (Fig. 5K). Taken together, these data suggest the mitochondrial pool of RNASET2 indeed functions in mtRNA degradation and its ribonuclease activity is required.

Purified RNASET2 has characteristics similar to those of in organello mtRNA degradation

A dual tagged human RNASET2 was expressed in HEK293 cells and purified under native conditions from the IMS (Fig. 6A). The purified proteins contained a ribonuclease activity that was also insensitive to proteinase K treatment and was inhibited by Cu^{2+} (Fig. 6B). RNASET2 purified under denaturing conditions showed same characteristics as that purified under native conditions, suggesting the activity came from RNASET2 itself instead of interacting proteins (Fig. 6C and 6D). Conditions tested on in organello mtRNA degradation assay were thoroughly tested on the purified protein. The responses of the protein to temperature, pH, ATP, Cu^{2+} and Mg^{2+}, and those of the in organello system showed a striking match, while other ribonucleases showed vastly different responses: RNaseI only responded to Cu^{2+} treatment with its activity inhibited, and RNaseA was insensitive to all the treatments (Fig. 6E–H).

The response of purified RNASET2 to pH is of great interest. Protozoan RNASET2s are mostly acidic ribonucleases (Irie, 1999), but the human RNASET2 appears to favor more of a neutral pH (Fig. 6F). This change could be due to changes of the amino acid sequence or post-translational modification in mammalian cells. To find the reason behind the change, human RNASET2 was expressed in bacteria and the protein was purified under denaturing conditions (Fig. S5A). Again, the purified protein was insensitive to proteinase K treatment (Fig. S5B). The responses to temperature, pH, ATP, Cu^{2+} and Mg^{2+} were also identical to the protein purified from mammalian cells (Fig. S5C–F), suggesting the activity of human RNASET2 is intrinsically different from its protozoan counterparts.

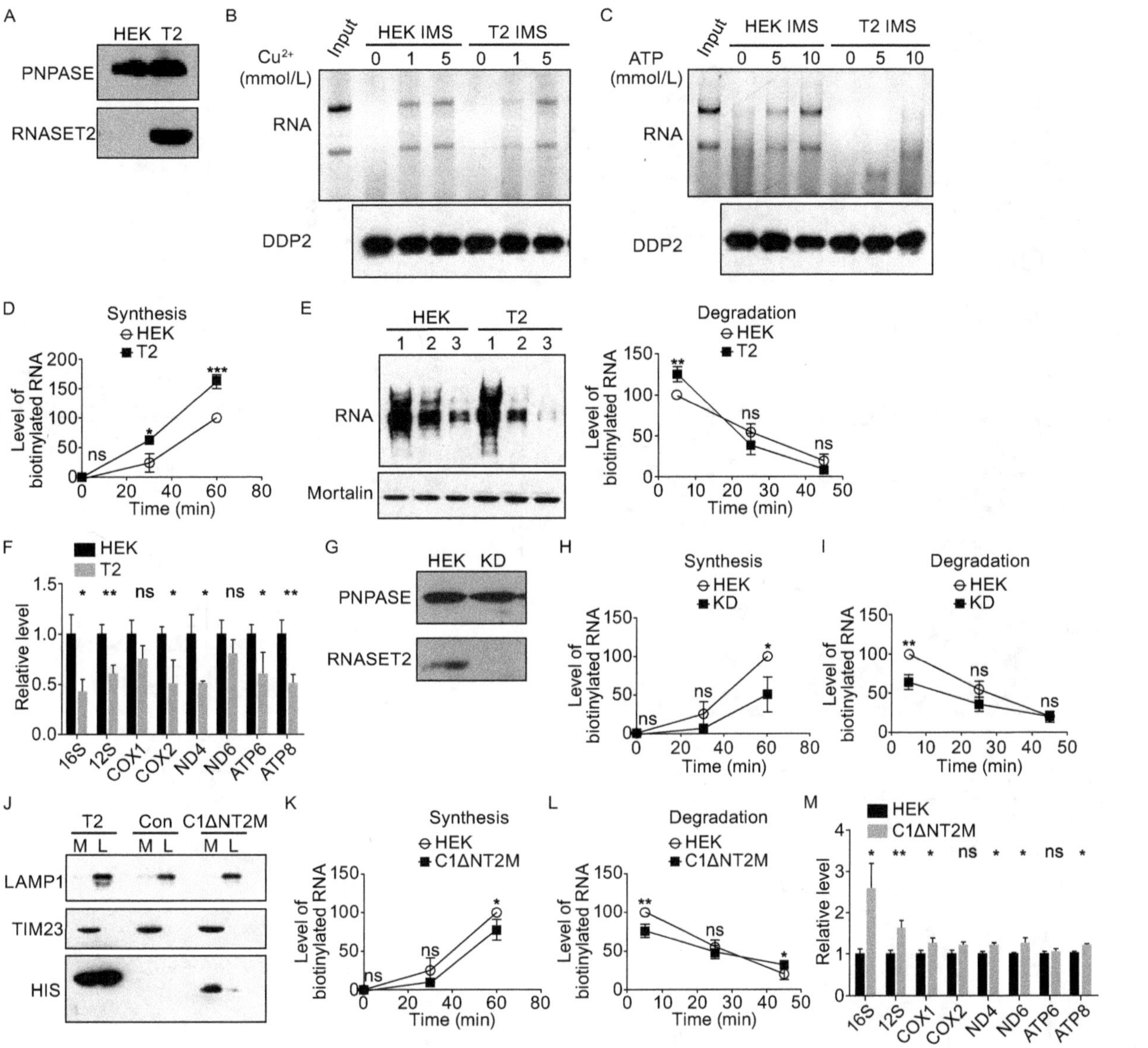

Figure 5. RNASET2 functions in mtRNA degradation. (A) Immunoblots of HEK and RNASET2-overexpressing (T2) cell lysates. PNPASE was used as a loading control. (B) IMS was isolated from HEK mitochondria or RNASET2-overexpressing (T2) mitochondria and the ribonuclease activity was tested for its sensitivity to different concentrations of Cu^{2+} (1 mmol/L and 5 mmol/L). Bottom panel is an immunoblot of IMS protein DDP2 showing the relative amount of IMS used for each *in vitro* degradation assay. (C) IMS was isolated from HEK mitochondria or RNASET2-overexpressing (T2) mitochondria and the ribonuclease activity was tested for its sensitivity to different concentrations of ATP (5 mmol/L and 10 mmol/L). (D) In organello mtRNA synthesis in control mitochondria (HEK) and mitochondria overexpressing RNASET2 (T2). (E) In organello mtRNA degradation in control mitochondria (HEK) and mitochondria overexpressing RNASET2 (T2). (F) qRT-PCR of mRNAs in HEK and RNASET2-overexpressing cells (T2) after H_2O_2 treatment for 2 h. (G) Immunoblots of HEK or RNASET2-knockdown (KD) cell lysates. PNPASE was used as a loading control. (H) In organello mtRNA synthesis in control mitochondria (HEK) and RNASET2-knockdown mitochondria (KD). (I) In organello mtRNA degradation in control mitochondria (HEK) and RNASET2-knockdown mitochondria (KD). (J) Immunoblots of mitochondria (M) and lysosomes (L) isolated from RNASET2-overexpressing cells (T2), control HEK cells (Con), and cells expressing C1-ΔN-RNASET2 (C65R, C118R) mutant (C1ΔNT2M). LAMP1 was used as a lysosomal marker, and TIM23 as a mitochondrial marker. (K) In organello mtRNA synthesis in control mitochondria (HEK) and mitochondria expressing C1-ΔN-RNASET2 (C65R, C118R) mutant (C1ΔNT2M). (L) In organello mtRNA degradation in control mitochondria (HEK) and mitochondria expressing C1-ΔN-RNASET2 (C65R, C118R) mutant (C1ΔNT2M). (M) qRT-PCR of mRNAs in HEK and C1-ΔN-RNASET2 (C65R, C118R) mutant expressing cells (C1ΔNT2M) after H_2O_2 treatment for 2 h. Statistical comparisons are performed using unpaired *t*-tests (n = 3 if not specified); $^*P < 0.05$, $^{**}P < 0.01$, $^{***}P < 0.001$, $^{****}P < 0.0001$. Data are presented as mean ± standard error of the mean (s.e.m.).

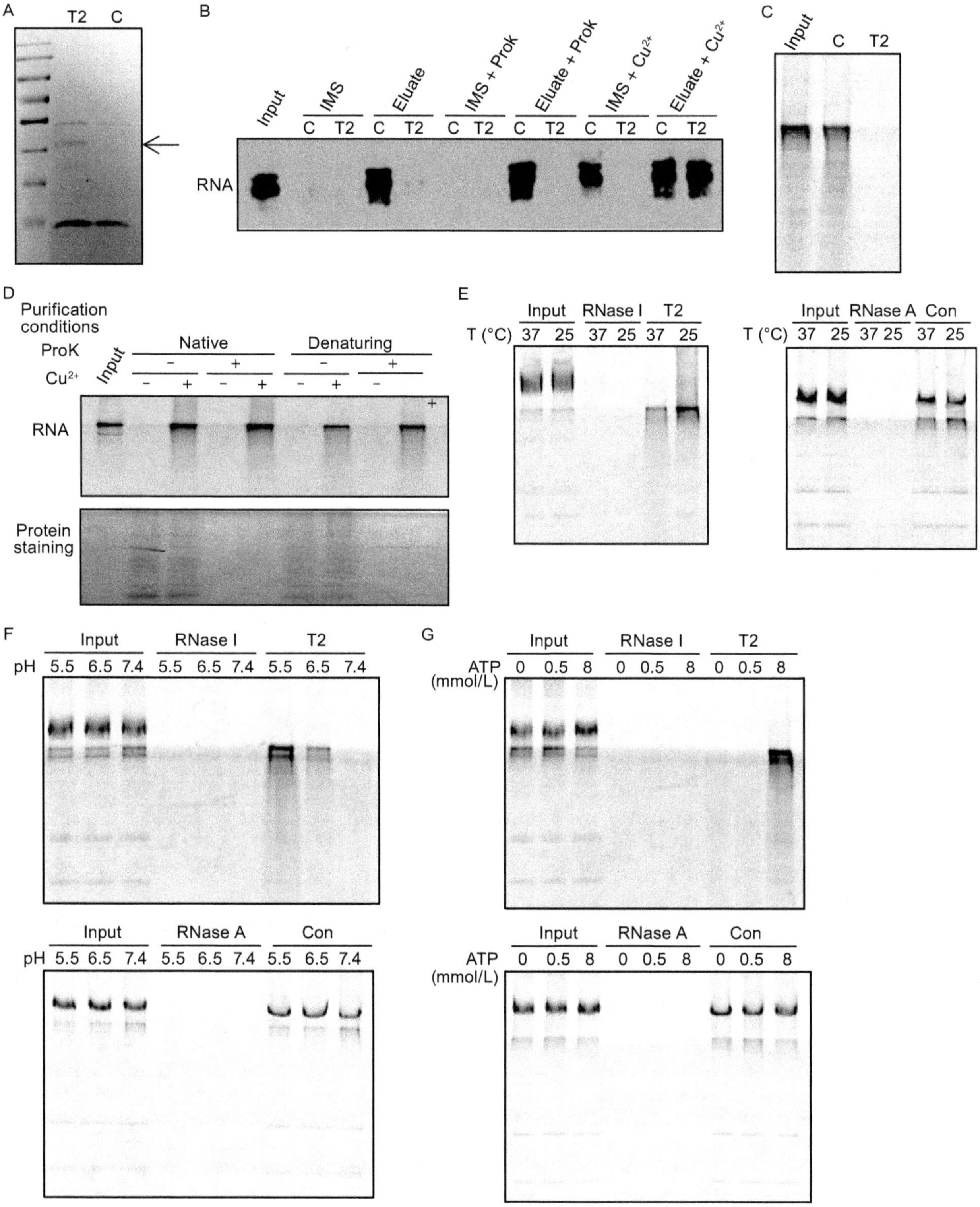
A
T2 C
B
Input
IMS
Eluate
IMS + Prok
Eluate + Prok
IMS + Cu2+
Eluate + Cu2+
C T2 C T2 C T2 C T2 C T2 C T2
RNA
C
Input
C T2
D
Purification conditions
ProK
Cu2+
Input
Native
Denaturing
RNA
Protein staining
E
Input RNase I T2
T (°C) 37 25 37 25 37 25
Input RNase A Con
T (°C) 37 25 37 25 37 25
F
Input RNase I T2
pH 5.5 6.5 7.4 5.5 6.5 7.4 5.5 6.5 7.4
Input RNase A Con
pH 5.5 6.5 7.4 5.5 6.5 7.4 5.5 6.5 7.4
G
Input RNase I T2
ATP (mmol/L) 0 0.5 8 0 0.5 8 0 0.5 8
Input RNase A Con
ATP (mmol/L) 0 0.5 8 0 0.5 8 0 0.5 8

◀ **Figure 6. Characterization of RNASET2 purified from HEK mitochondria.** (A) Dual-tag purification of RNASET2 (His and HA). Purification was performed using IMS from control HEK cells (C) or RNASET2-overexpressing cells (T2) under native condition. (B) Ribonuclease activity were examined in IMS samples and the purification samples (Eluate) from control HEK cells (C) or RNASET2-overexpressing cells (T2) using biotinylated *UCP2* mRNA as a substrate. The sensitivity of these activities to Cu^{2+} (0.5 mmol/L) and proteinase K (ProK) was also tested. (C) RNASET2 was purified from RNASET2-overexpressing mitochondria under denaturing condition and checked for ribonuclease activity using RNA purified from isolated mitochondria as substrates; C (control pulldown from HEK mitochondria) and T2 (RNASET2). (D) RNASET2s purified under native conditions and denaturing conditions had the same responses to proteinase K (ProK) and Cu^{2+} treatment. Lower panel is a coomassie staining gel of mitochondrial lysates as a positive control for proteinase K treatment. (E) Effect of temperature on RNASET2 purified from HEK mitochondria, RNaseI, and RNaseA (50 ng). Degradation was performed at 37°C (the temperature used for the other experiments if not specified) or 25°C. RNAs purified from isolated mitochondria were used as substrates. Con (control pulldown from HEK mitochondria). (F) Effect of pH on RNASET2 purified from HEK mitochondria, RNaseI, and RNaseA. Degradation was performed at pH 7.4 (the pH used for the other experiments if not specified), pH 6.5 or pH 5.5. (G) Effect of ATP on RNASET2 purified from HEK mitochondria, RNaseI and RNaseA. (H) Effect of Mg^{2+} and Cu^{2+} on RNASET2 purified from HEK mitochondria, RNaseI and RNaseA.

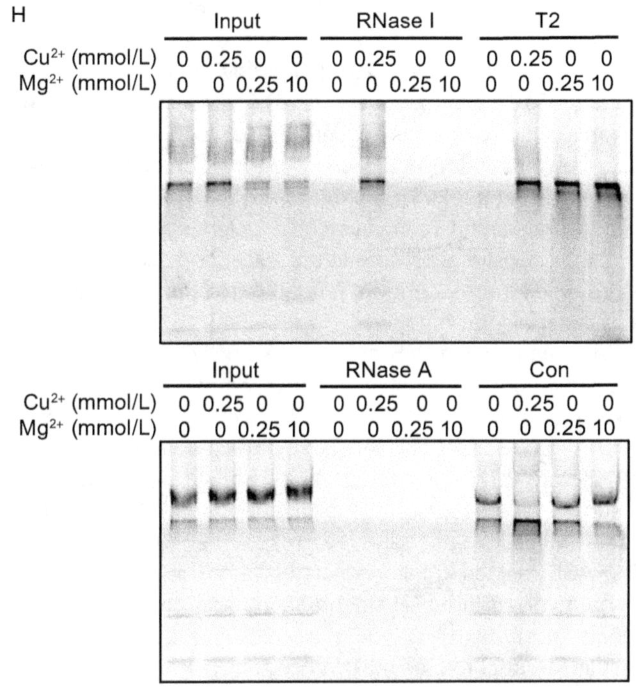

Figure 6. continued.

Another IMS localized nuclease is Endo G (Cote and Ruiz-Carrillo, 1993; Ohsato et al., 2002; Zhou et al., 2016). To examine whether it has similar characteristics as in organello mitochondrial RNA degradation and can potentially be also involved in the process, we purified human Endo G from HEK mitochondria and *E. coli* (Fig. S6A–E). Unlike RNASET2, Endo G was sensitive to proteinase K and the responses to the conditions tested were vastly different from those of RNASET2 or in organello mitochondrial RNA degradation. Thus, a direct involvement of Endo G on mitochondrial RNA degradation seems unlikely. That the human RNASET2 purified from mammalian cells and bacteria showed responses to so many conditions almost identical to those of the in organello mtRNA degradation together with its IMS localization further verified it as the enzyme responsible for mtRNA degradation.

DISCUSSION

For decades, the identity of the ribonuclease responsible for degradation of mammalian mitochondrial RNAs (mtRNAs) had been a great mystery. Our attempt at unraveling this mystery uncovered a mitochondrial IMS-localized ribonuclease activity that degrades mtRNAs and is sensitive to many cellular conditions such as pH and Ca^{2+}. We have done extensive studies to rule out the possibility that the action on the substrates was due to a cytosolic contamination or a mitochondrial leak. Localization of such a ribonuclease activity in the mitochondrial IMS guarantees a faster change of mitochondrial biogenesis and quick regulation of mitochondrial functions in response to signals outside of mitochondria.

At first glance, degradation of mtRNA in the IMS instead of the matrix where transcription and translation occur seems a little radical. However, this is not the first time a nuclease is reported in the IMS. REXO2, a 3' to 5' exonuclease specific for small oligomers also has an IMS localization (Bruni et al., 2013). It could act on oligonucleotides yielded by RNASET2 and recycle the NTPs for new mtRNA synthesis. Another example is Endo G, a nuclease reported to primarily localize in mitochondrial IMS and function on degradation of both nuclear and mitochondrial DNA under special circumstances (Cote and Ruiz-Carrillo, 1993; Ohsato et al., 2002; Zhou et al., 2016). Proteinase K insensitivity of RNASET2 is also a surprising finding. However, it is not a unique feature. Same characteristic is shared by ribonucleases in *Leishmania tarentolae* mitochondria (Alfonzo et al., 1998; Simpson et al., 1992).

Degradation of mtRNAs in the mitochondrial IMS means there are active transports of RNAs from the matrix across the mitochondrial inner membrane. Both mitochondrial RNA import and export have been previously reported. Mitochondria import a wide array of RNAs from the cytosol, including tRNAs, 5S rRNA and other non-coding RNAs, and the import pathways have been partially characterized (Chang and Clayton, 1989; Duchene et al., 2009; Mercer et al., 2011; Noh et al., 2016; Smirnov et al., 2011; Wang et al., 2010; Zhang et al., 2014). Mitochondrion-derived MitomiRs could also be exported to cytosol to function in post-transcriptional regulation of gene expression (Bienertova-Vasku et al., 2013; Duarte et al., 2015). These existing import and export pathways could potentially be the pathways for substrate delivery to the IMS ribonuclease activity. More studies are needed to understand substrate selection and the transport pathways.

The mammalian IMS ribonuclease activity shown here is quite different from yeast mtRNA degradosome and bacterium degradosome (Dziembowski et al., 1998; Miczak et al., 1996; Szczesny et al., 2012). Yeast degradosome has its ribonuclease component Dss1 in the matrix, and bacterium degradosome ribonuclease PNPASE is absent in yeast (Miczak et al., 1996; Wang et al., 2010). Mammalian PNPASE resides in mitochondrial IMS. It has been reported to be involved in mtRNA degradation and processing (Clemente et al., 2015; Daoud et al., 2012), but based on our results, a direct role as mtRNA ribonuclease seems unlikely. We have shown that mammalian mitochondrial membranes have no apparent ribonuclease activity (Fig. 2A and 2B), and PNPASE, a mitochondrial membrane bound protein, has no ribonuclease activity (Fig. 1A–C). A known role for mammalian PNPASE is importing nucleus-encoded RNAs into mitochondria (Sato et al., 2017; Vedrenne et al., 2012; von Ameln et al., 2012; Wang et al., 2010), so it could have trafficking role in substrate transport for the IMS ribonuclease activity. Evidence that mammalian PNPASE is not the ribonuclease for mtRNA degradation but could be involved in their transport also comes from RNAi silencing of PNPASE expression. In some cases, a decrease of mtRNA level was observed; while in others, there were no significant changes (Slomovic and Schuster, 2008).

An important question that arose from this study is how the RNA degradation activity is coordinated with the transcriptional activity in mitochondria. Mitochondria appear to be capable of quickly up-regulate transcriptional activity when the IMS RNA degradation activity is up, and vice versa, hence maintaining relatively stable RNA levels (Fig. 5). A signaling pathway or pathways are clearly needed for such regulation. The responses of both RNASET2 activity and in organello mitochondrial RNA degradation to conditions such as ATP and pH also suggest coupling and a feedback circuit between mitochondrial RNA degradation and ETC activity/ATP synthesis.

Interestingly, RNASET2 has been shown to be involved in cancer suppression and loss of function mutation causes cystic leukoencephalopathy (Acquati et al., 2011; Henneke et al., 2009). An enzymatic dead mutant without the N-terminus also has tumor suppression activity (Nesiel-Nuttman et al., 2015). However, we have shown that truncation of the N-terminus led to mislocalization of the protein, so possible involvement of its ribonuclease activity in tumor suppression has not been rule out yet. Since mammalian RNASET2 has little ribonuclease activity at low pH, the function of lysosomal pool remains to be elucidated. More work is also needed to understand its physiological roles.

MATERIALS AND METHODS

Cell lines

Cell lines used include HEK293, HeLa, TM6, and stable cell lines generated on these cell lines. See Supplemental Experimental Procedures for details.

Plasmids

Plasmids used include *PQCXIP-RNASET2-HAHisPC*, *PQCXIP-C1-ΔN-RNASET2* (*C65R*, *C118R*)*-HAHisPC*, *PQCXIP-RNASET2-GFP*, *PQCXIP-RNASET2* (*C184R*)*-GFP*, *PET28A-RNASET2-HisPC*, *PQCXIP-EndoG-HAHis*, *PET28A-EndoG-HAHis*, *PQCXIP-APEX-HisPC*, *PQCXIP-RNASET2-APEX-HisPC*, *PQCXIP-MIA40-APEX-HisPC*, *PQCXIP-PNPASE-HAHis*, *PQCXIP-MIA40-Avi-HisFlag*, *PQCXIP-RNASET2-Avi-HisFlag*, and *PQCXIP-TOM22-BirA-HAHis*. See Supplemental Experimental Procedures for construction details.

Isolation of crude mammalian mitochondria and cytosol

Cells were washed once with PBS buffer, resuspended in ice-cold mitoprep buffer (0.225 mol/L manifold, 0.075 mol/L sucrose, and 20 mmol/L HEPES pH 7.4), and broken in a glass-Teflon homogenizer on ice with 30 strokes. Nuclei and unbroken cells were pelleted at 700 ×*g* for 5 min, and the homogenization repeated once. Supernatants from both times were centrifuged again at 700 ×*g*. Crude mitochondria were pelleted from second-round supernatants at 11,000 ×*g* for 5 min, washed once with mitoprep buffer and resuspended in mitoprep buffer of desired volume for further use. Post-mitochondrial supernatant was spun at 21,000 ×*g* for 10 min and the supernatant was collected as cytosol.

Sub-mitochondrial fractionation

Hypotonic treatment was performed by incubating mitochondria for 20 min on ice by diluting mitoprep buffer with 10 volumes of 20 mmol/L HEPES pH 7.4 with one gentle vortexing at 10 min. Mitoplasts were separated from IMS by centrifugation at 15,000 ×*g* for 4 min. Salt wash was performed by adding 300 mmol/L of NaCl (from 5 mol/L stock) into mitoprep buffer (in cases of intact mitochondria) or mitoplasting mixture (in cases of mitoplasts) after 20 min on ice for another 5 min. Mitochondrial matrix was isolated by sonicating the salt-washed mitoplast in mitoprep buffer and separating the soluble fraction (matrix) from membrane by 10 min 21,000 ×*g* centrifugation. Total soluble was isolated by sonicating mitochondria in mitoprep buffer and separating the soluble fraction (total soluble) from membranes by 10 min 21,000 ×*g* centrifugation. The pellet was washed twice with mitoprep buffer and used as total membranes.

Mitoplasting after in organello mtRNA synthesis was carried out by proteinase K treatment. 500 µg mitochondria in 200 µL in organello synthesis mixture were pelleted by 12,000 ×*g* 4 min spin, washed with 1 mL mitoprep buffer, and resuspended in 300 µL mitoprep buffer with 2 μg proteinase K. The mixture was incubated on ice for 15 min with one vortexing at 8 min. 1 mL mitoprep buffer was added after the incubation and mitoplasts were pelleted at 15,000 ×*g* for 4 min, washed once with 1 mL mitoprep buffer, and proceeded to in organello degradation. Control mitochondria samples skipped the last two spins, as spinning and resuspending at this stage ruptures the outer membrane. This new method was used to ensure degradation of some inner membrane bound proteins that might be involved in mtRNA export from the matrix.

In vitro degradation assay

In vitro degradation assay for PNPASE was performed as previously described (Wang et al., 2010). All the other assays were performed in 20 mmol/L HEPES pH 7.4 at 37°C for 10 min if not otherwise specified. Mitochondrial lysates that contain membrane fractions had 0.5% Triton X. For experiments with membrane samples, equal amount of Triton X was added into all samples, normally the final concentration not exceeding 0.1% because of dilution. ~2 μg of IMS, ~2 μg of Matrix or ~10 μg of membrane was used for each 20 μL reaction. For purified protein samples, RNaseI (Thermo) and RNaseA (Thermo), 50 ng was used for each reaction. Reactions with bacterial samples were incubated for 20 min instead of 10 min. Substrates included 1 ng biotinylated *UCP2*, and 300 ng mtRNA. For effects of metal ions on the degradation, different concentrations of $MgCl_2$, $CuSO_4$, $MnCl_2$, or $ZnCl_2$ were used. Reaction was stopped by adding equal volume of SDS-Urea-EDTA buffer (2× SDS loading buffer with 8 mol/L urea and 15 mmol/L EDTA) and incubating at 90°C for 5 min. Samples were cooled to room temperature and 0.5 μg of proteinase K was added for a 5 min incubation at 37°C. Biotinylated samples were run on SDS-PAGE, transferred to a nylon membrane, and detected with nucleic acid detection kit (Thermo). mtRNA samples were run either on SDS-PAGE or agarose gels and stained with EtBr.

In organello RNA synthesis

In organello RNA synthesis was performed in 200 μL mitoprep buffer containing 4 mmol/L ATP pH 7.4, 20 mmol/L succinate, 1 mmol/L $CaCl_2$ and 1 μL Biotin RNA Labeling Mix (Roche) with 500 μg mitochondria at 37°C. For each time point (0 min, 30 min, and 60 min), 60 μL reaction mix was taken out and mitochondria were pelleted at 18,000 ×*g* for 2 min. Pellets were stored at −80°C for at least 15 min before next preparation step. For loading, samples were taken out of −80°C, quickly dissolved in 30 μL SDS-Urea-EDTA buffer (SDS loading buffer with 8 mol/L urea and 15 mmol/L EDTA) preheated to 90°C, and incubated at 90°C for 5 min. Samples were then cooled to room temperature and 0.5 μg of proteinase K was added for a 5 min incubation at 37°C. Biotinylated samples were run on SDS-PAGE, transferred to a nylon membrane (400 mA for 1.5 h), and detected with nucleic acid detection kit (Thermo). These synthesis conditions are less inhibitory to mtRNA degradation, and were used for the in organello degradation assays that examined their responses to mitoplasting, temperature, pH, ATP, and metal ions.

For comparison of in organello synthesis between different cell lines, more inhibitory synthesis conditions were used to ensure the yield of newly synthesized RNA more representing the real synthesis rate. Therefore, in organello mtRNA synthesis were carried out in buffer containing 4 mmol/L ATP pH 7.4, 20 mmol/L succinate, 0.5 mmol/L $CaCl_2$, 10 mmol/L $MgCl_2$, and 1 mg/mL HEK cytosol.

In organello mtRNA degradation

For in organello degradation assays that examined their responses to mitoplasting, temperature, pH, ATP, and metal ions, 500 μg mitochondria first underwent mtRNA synthesis in 200 μL mitoprep buffer containing 4 mmol/L ATP pH 7.4, 20 mmol/L succinate, 1 mmol/L $CaCl_2$, and 1 μL Biotin RNA Labeling Mix (Roche) at 37°C for 45 min. Mitochondria were pelleted at 12,000 ×*g* for 4 min at 4°C, washed with 1 mL ice cold mitoprep buffer, resuspended in 150 μL mitoprep buffer, and incubated on ice for 15 min with one vortexing at 8 min. 50 μL ice cold cocktail containing 4 mmol/L UTP, 40 mmol/L Ca^{2+}, 1 μL Ribolock RNase inhibitor (Thermo), and 1 μg RNaseI (Thermo) (original buffer exchanged to mitoprep buffer) in mitoprep buffer was added to the sample (RNase inhibitor was used to eliminate cytosolic interference and RNaseI was used to quickly degrade the RNA from broken mitochondria). The samples were shifted to 37°C. For each time point (0 min, 30 min, and 60 min), 60 μL reaction mix was taken out and mitochondria were pelleted at 18,000 ×*g* for 2 min. Pellets were stored at −80°C for at least 15 min before next preparation step. For loading, samples were taken out of −80°C, quickly dissolved in 30 μL SDS-Urea-EDTA buffer (SDS loading buffer with 8 M urea and 15 mmol/L EDTA) preheated to 90°C, and incubated at 90°C for 5 min. Samples were then cooled to room temperature and 0.5 μg of proteinase K was added for a 5 min incubation at 37°C. Biotinylated samples were run on SDS-PAGE, transferred to a nylon membrane (400 mA for 1.5 h), and detected with nucleic acid detection kit (Thermo).

For comparison of in organello mtRNA degradation between different cell lines, in organello mtRNA synthesis were carried out in buffer containing 4 mmol/L ATP pH 7.4, 20 mmol/L succinate, 0.5 mmol/L $CaCl_2$, and 1 mg/mL HEK cytosol for 45 min at 37°C before the following wash and degradation steps. Mitochondria should be isolated from stable cells lines grown to less than 5 doubling times from the initial frozen stocks.

Biotin-phenol labeling with APEX in live cells

500 μmol/L biotin-phenol (ApexBio) was added to cell medium for 30 min. Then 1 mmol/L of H_2O_2 was added and the plates were gently agitated for 1 min. The reaction was quickly quenched by replacing the medium with DPBS containing 5 mmol/L Trolox (abcam) and 10 mmol/L sodium ascorbate (Solarbio). Cells were then washed with the same solution three times. For Ni-NTA (Qigen) and HA bead (Thermo Scientific) enrichment of PNPASE, mitochondria were isolated from the biotin-labeled cells and PNPASE was first purified from the lysate under denaturing condition using Ni-NTA beads and then using HA beads under native conditions according to the manufactures' instructions.

In vivo biotinylation of Avi-tagged protein by spatially localized biotin ligase (BirA)

HEK293 cells co-transfected with the Avi-tagged protein expressing plasmid and BirA fusion protein expressing plasmid were gown in medium containing 25 μmol/L D-Biotin (Amresco) for 12 h. The cells were harvested, washed with PBS time times, lysed in 1× SDS protein loading buffer, and run on SDS-PAGE for biotin detection.

Additional procedures

Additional procedures include protein purification, Western blotting, mtRNA isolation, qPCR, Optiprep gradient centrifugation, identification of IMS ribonuclease, fluorescence microscopy and image acquisition, and *in vitro* transcription. See Supplemental Experimental Procedures for details.

ACKNOWLEDGEMENTS

We thank Haiteng Deng for the help on Mass Spec and Zhi Lu and Hongwei Wang for discussion. This research was supported by the Priority Research Program of the Ministry of Science and Technology 2017YFA0504600, the National Natural Science Foundation of China (Grant Nos. 31371439 and 91649103), and Ministry of Education 1000 youth program.

ABBREVIATIONS

APEX, ascorbate peroxidase; BirA, biotin ligase; IMS, intermembrane space; mtRNAs, mitochondrial RNAs

COMPLIANCE WITH ETHICS GUIDELINES

Peipei Liu, Jinliang Huang, Qian Zheng, Leiming Xie, Xinping Lu, Jie Jin, and Geng Wang declare that they have no conflict of interest. All institutional and national guidelines for the care and use of laboratory animals were followed.

REFERENCES

Acquati F, Bertilaccio S, Grimaldi A, Monti L, Cinquetti R, Bonetti P, Lualdi M, Vidalino L, Fabbri M, Sacco MG et al (2011) Microenvironmental control of malignancy exerted by RNASET2, a widely conserved extracellular RNase. Proc Natl Acad Sci USA 108:1104–1109

Alfonzo JD, Thiemann OH, Simpson L (1998) Purification and characterization of MAR1. A mitochondrial associated ribonuclease from Leishmania tarentolae. J Biol Chem 273:30003–30011

Anderson S, Bankier AT, Barrell BG, de Bruijn MH, Coulson AR, Drouin J, Eperon IC, Nierlich DP, Roe BA, Sanger F et al (1981) Sequence and organization of the human mitochondrial genome. Nature 290:457–465

Bienertova-Vasku J, Sana J, Slaby O (2013) The role of microRNAs in mitochondria in cancer. Cancer Lett 336:1–7

Bonawitz ND, Rodeheffer MS, Shadel GS (2006) Defective mitochondrial gene expression results in reactive oxygen species-mediated inhibition of respiration and reduction of yeast life span. Mol Cell Biol 26:4818–4829

Borowski LS, Dziembowski A, Hejnowicz MS, Stepien PP, Szczesny RJ (2013) Human mitochondrial RNA decay mediated by PNPase-hSuv3 complex takes place in distinct foci. Nucleic Acids Res 41:1223–1240

Bruni F, Gramegna P, Oliveira JM, Lightowlers RN, Chrzanowska-Lightowlers ZM (2013) REXO2 is an oligoribonuclease active in human mitochondria. PLoS ONE 8:e64670

Chang DD, Clayton DA (1989) Mouse RNAase MRP RNA is encoded by a nuclear gene and contains a decamer sequence complementary to a conserved region of mitochondrial RNA substrate. Cell 56:131–139

Chen HW, Rainey RN, Balatoni CE, Dawson DW, Troke JJ, Wasiak S, Hong JS, McBride HM, Koehler CM, Teitell MA et al (2006) Mammalian polynucleotide phosphorylase is an intermembrane space RNase that maintains mitochondrial homeostasis. Mole Cell Biol 26:8475–8487

Chujo T, Ohira T, Sakaguchi Y, Goshima N, Nomura N, Nagao A, Suzuki T (2012) LRPPRC/SLIRP suppresses PNPase-mediated mRNA decay and promotes polyadenylation in human mitochondria. Nucleic Acids Res 40:8033–8047

Clemente P, Pajak A, Laine I, Wibom R, Wedell A, Freyer C, Wredenberg A (2015) SUV3 helicase is required for correct processing of mitochondrial transcripts. Nucleic Acids Res 43:7398–7413

Cote J, Ruiz-Carrillo A (1993) Primers for mitochondrial DNA replication generated by endonuclease G. Science 261:765–769

Daoud R, Forget L, Lang BF (2012) Yeast mitochondrial RNase P, RNase Z and the RNA degradosome are part of a stable supercomplex. Nucleic Acids Res 40:1728–1736

Duarte FV, Palmeira CM, Rolo AP (2015) The emerging role of MitomiRs in the pathophysiology of human disease. Adv Exp Med Biol 888:123–154

Duchene AM, Pujol C, Marechal-Drouard L (2009) Import of tRNAs and aminoacyl-tRNA synthetases into mitochondria. Curr Genet 55:1–18

Dziembowski A, Malewicz M, Minczuk M, Golik P, Dmochowska A, Stepien PP (1998) The yeast nuclear gene DSS1, which codes for a putative RNase II, is necessary for the function of the mitochondrial degradosome in processing and turnover of RNA. Mol Gen Genet 260:108–114

Hallberg BM, Larsson NG (2014) Making proteins in the powerhouse. Cell Metab 20:226–240

Han S, Udeshi ND, Deerinck TJ, Svinkina T, Ellisman MH, Carr SA, Ting AY (2017) Proximity biotinylation as a method for mapping proteins associated with mtDNA in living cells. Cell Chem Biol 24:404–414

Henneke M, Diekmann S, Ohlenbusch A, Kaiser J, Engelbrecht V, Kohlschutter A, Kratzner R, Madruga-Garrido M, Mayer M, Opitz L et al (2009) RNASET2-deficient cystic leukoencephalopathy resembles congenital cytomegalovirus brain infection. Nat Genet 41:773–775

Irie M (1999) Structure-function relationships of acid ribonucleases: lysosomal, vacuolar, and periplasmic enzymes. Pharmacol Ther 81:77–89

Jan CH, Williams CC, Weissman JS (2014) Principles of ER cotranslational translocation revealed by proximity-specific ribosome profiling. Science 346:1257521

Khidr L, Wu G, Davila A, Procaccio V, Wallace D, Lee WH (2008) Role of SUV3 helicase in maintaining mitochondrial homeostasis in human cells. J Biol Chem 283:27064–27073

Kim DI, Roux KJ (2016) Filling the void: proximity-based labeling of proteins in living cells. Trends Cell Biol 26:804–817

Levy S, Allerston CK, Liveanu V, Habib MR, Gileadi O, Schuster G (2016) Identification of LACTB2, a metallo-beta-lactamase protein, as a human mitochondrial endoribonuclease. Nucleic Acids Res 44:1813–1832

Luhtala N, Parker R (2010) T2 Family ribonucleases: ancient enzymes with diverse roles. Trends Biochem Sci 35:253–259

Malecki M, Stepien PP, Golik P (2010) Assays of the helicase, ATPase, and exoribonuclease activities of the yeast mitochondrial degradosome. Methods Mol Biol 587:339–358

Margossian SP, Li H, Zassenhaus HP, Butow RA (1996) The DExH box protein Suv3p is a component of a yeast mitochondrial 3'-to-5' exoribonuclease that suppresses group I intron toxicity. Cell 84:199–209

Mercer TR, Neph S, Dinger ME, Crawford J, Smith MA, Shearwood AM, Haugen E, Bracken CP, Rackham O, Stamatoyannopoulos JA et al (2011) The human mitochondrial transcriptome. Cell 146:645–658

Miczak A, Kaberdin VR, Wei CL, Lin-Chao S (1996) Proteins associated with RNase E in a multicomponent ribonucleolytic complex. Proc Natl Acad Sci USA 93:3865–3869

Mishra P, Chan DC (2016) Metabolic regulation of mitochondrial dynamics. J Cell Biol 212:379–387

Nesiel-Nuttman L, Doron S, Schwartz B, Shoseyov O (2015) Human RNASET2 derivatives as potential anti-angiogenic agents: actin binding sequence identification and characterization. Oncoscience 2:31–43

Noh JH, Kim KM, Abdelmohsen K, Yoon JH, Panda AC, Munk R, Kim J, Curtis J, Moad CA, Wohler CM et al (2016) HuR and GRSF1 modulate the nuclear export and mitochondrial localization of the lncRNA RMRP. Genes Dev 30:1224–1239

Ohsato T, Ishihara N, Muta T, Umeda S, Ikeda S, Mihara K, Hamasaki N, Kang D (2002) Mammalian mitochondrial endonuclease G. Digestion of R-loops and localization in intermembrane space. Eur J Biochem FEBS 269:5765–5770

Portnoy V, Palnizky G, Yehudai-Resheff S, Glaser F, Schuster G (2008) Analysis of the human polynucleotide phosphorylase (PNPase) reveals differences in RNA binding and response to phosphate compared to its bacterial and chloroplast counterparts. RNA 14:297–309

Roux KJ, Kim DI, Raida M, Burke B (2012) A promiscuous biotin ligase fusion protein identifies proximal and interacting proteins in mammalian cells. J Cell Biol 196:801–810

Rubio MA, Rinehart JJ, Krett B, Duvezin-Caubet S, Reichert AS, Soll D, Alfonzo JD (2008) Mammalian mitochondria have the innate ability to import tRNAs by a mechanism distinct from protein import. Proc Natl Acad Sci USA 105:9186–9191

Sanchez MI, Mercer TR, Davies SM, Shearwood AM, Nygard KK, Richman TR, Mattick JS, Rackham O, Filipovska A (2011) RNA processing in human mitochondria. Cell Cycle 10:2904–2916

Sarkar D, Park ES, Emdad L, Randolph A, Valerie K, Fisher PB (2005) Defining the domains of human polynucleotide phosphorylase (hPNPaseOLD-35) mediating cellular senescence. Mol Cell Biol 25:7333–7343

Sato R, Arai-Ichinoi N, Kikuchi A, Matsuhashi T, Numata-Uematsu Y, Uematsu M, Fujii Y, Murayama K, Ohtake A, Abe T et al (2017) Novel biallelic mutations in the PNPT1 gene encoding a mitochondrial-RNA-import protein PNPase cause delayed myelination. Clin Genet. doi:10.1111/cge.13068

Schafer B, Hansen M, Lang BF (2005) Transcription and RNA-processing in fission yeast mitochondria. RNA 11:785–795

Simpson AM, Bakalara N, Simpson L (1992) A ribonuclease activity is activated by heparin or by digestion with proteinase K in mitochondrial extracts of Leishmania tarentolae. J Biol Chem 267:6782–6788

Slomovic S, Portnoy V, Yehudai-Resheff S, Bronshtein E, Schuster G (2008) Polynucleotide phosphorylase and the archaeal exosome as poly(A)-polymerases. Biochim et Biophys Acta 1779:247–255

Slomovic S, Schuster G (2008) Stable PNPase RNAi silencing: its effect on the processing and adenylation of human mitochondrial RNA. RNA 14:310–323

Smirnov A, Entelis N, Martin RP, Tarassov I (2011) Biological significance of 5S rRNA import into human mitochondria: role of ribosomal protein MRP-L18. Genes Dev 25:1289–1305

Szczesny RJ, Borowski LS, Malecki M, Wojcik MA, Stepien PP, Golik P (2012) RNA degradation in yeast and human mitochondria. Biochim et Biophys Acta 1819:1027–1034

Szczesny RJ, Wojcik MA, Borowski LS, Szewczyk MJ, Skrok MM, Golik P, Stepien PP (2013) Yeast and human mitochondrial helicases. Biochim et Biophys Acta 1829:842–853

Vedrenne V, Gowher A, De Lonlay P, Nitschke P, Serre V, Boddaert N, Altuzarra C, Mager-Heckel AM, Chretien F, Entelis N et al (2012) Mutation in PNPT1, which encodes a polyribonucleotide nucleotidyltransferase, impairs RNA import into mitochondria and causes respiratory-chain deficiency. Am J Hum Genet 91:912–918

von Ameln S, Wang G, Boulouiz R, Rutherford MA, Smith GM, Li Y, Pogoda HM, Nurnberg G, Stiller B, Volk AE et al (2012) A mutation in PNPT1, encoding mitochondrial-RNA-import protein PNPase, causes hereditary hearing loss. Am J Hum Genet 91:919–927

Wang G, Chen HW, Oktay Y, Zhang J, Allen EL, Smith GM, Fan KC, Hong JS, French SW, McCaffery JM et al (2010) PNPASE regulates RNA import into mitochondria. Cell 142:456–467

Williams CC, Jan CH, Weissman JS (2014) Targeting and plasticity of mitochondrial proteins revealed by proximity-specific ribosome profiling. Science 346:748–751

Zhang X, Zuo X, Yang B, Li Z, Xue Y, Zhou Y, Huang J, Zhao X, Zhou J, Yan Y et al (2014) MicroRNA directly enhances mitochondrial translation during muscle differentiation. Cell 158:607–619

Zhou Q, Li H, Li H, Nakagawa A, Lin JL, Lee ES, Harry BL, Skeen-Gaar RR, Suehiro Y, William D et al (2016) Mitochondrial endonuclease G mediates breakdown of paternal mitochondria upon fertilization. Science 353:394–399

6

Inhibition of retroviral Gag assembly by non-silencing miRNAs promotes autophagic viral degradation

Na Qu[1], Zhao Ma[1], Mengrao Zhang[1], Muaz N. Rushdi[1,2], Christopher J. Krueger[1,2], Antony K. Chen[1]✉

[1] Department of Biomedical Engineering, College of Engineering, Peking University, Beijing 100871, China
[2] Wallace H Coulter Department of Biomedical Engineering, Georgia Institute of Technology, Atlanta, GA 30332, USA
✉ Correspondence: chenak@pku.edu.cn (A. K. Chen)

ABSTRACT

We recently reported an unconventional mechanism by which miRNAs inhibit HIV-1 viral production. This occurs when miRNAs bind nonspecifically to the viral structural protein Gag, interfering with viral RNA-mediated Gag assembly at the plasma membrane. Consequently, misassembled viral complexes are redirected into the endocytic pathway where they are delivered to lysosomes for degradation. In this study, we demonstrate that autophagy is a critical mediator of the viral degradation pathway and that this pathway is not HIV-1 specific. Misassembled viral complexes were found to colocalize extensively with LC3 and p62 in late endosomes/lysosomes, demonstrating a convergence of autophagy with functional degradative compartments. Knocking down autophagosome formation machineries reduced this convergence, while treatment with autophagy-inducer rapamycin enhanced the convergence. Furthermore, similar autophagy-dependent nonspecific miRNA inhibition of murine leukemia virus (MLV) assembly was shown. Overall, these results reveal autophagy as a crucial regulator of the retroviral degradation pathway in host cells initiated by nonspecific miRNA-Gag interactions. These findings could have significant implications for understanding how cells may regulate retroviral complex assembly by miRNA expression and autophagy, and raise the possibility that similar regulations can occur in other biological contexts.

KEYWORDS microRNA, Gag protein, autophagy

INTRODUCTION

MicroRNAs (miRNAs) are small noncoding RNAs, 19–24 nucleotides in length, with gene-silencing functions critical to the regulation of numerous physiological and pathological processes (Filipowicz et al., 2008; Sharp, 2009). To achieve gene silencing, miRNAs associate with the RNA-induced silencing complex (RISC) to target specific mRNAs for degradation or translational repression. In addition to this role, emerging evidence has shown that miRNAs can also interact with other proteins, impacting cellular physiology via a variety of mechanisms independent of gene silencing (Chen et al., 2014; Eiring et al., 2010; Fabbri et al., 2012; Lehmann et al., 2012; Prud'homme et al., 2016; Ranganathan et al., 2017; Vickers et al., 2011; Yelamanchili et al., 2015). We previously discovered one such unconventional miRNA function, wherein miRNA acts as a potent inhibitor of the formation of HIV-1 virus particles (Chen et al., 2014).

Gag is the main retroviral structural protein that orchestrates the formation of HIV-1 virus particles. To drive viral assembly, thousands of Gag molecules must coalesce around the viral genome to form a highly-ordered Gag multimer at the plasma membrane (PM) (Briggs et al., 2004). To date, much evidence has shown that the process is strongly dependent on interaction between the nucleocapsid domain of Gag (NC) and the viral genome. One major interaction involves specific binding of NC to the psi-element in the viral genome, enabling its selective recruitment to the PM from the complex cytoplasmic environment (Aldovini and Young, 1990; Berkowitz et al., 1993; Gorelick et al., 1990). Another interaction involves nonspecific binding of NC to regions outside the psi-element, which is thought to facilitate Gag coalescence and multimerization at the PM (Kutluay and Bieniasz, 2010; Kutluay et al., 2014; Muriaux et al., 2001; Rulli et al., 2007). Recently, we showed that miRNAs can also bind to NC, forming miRNA-Gag complexes that block viral formation by disrupting viral RNA-mediated Gag assembly at the PM (Chen et al., 2014). This effect was more pronounced when the miRNAs did not participate in RNAi (nonsense miRNAs with no complementary target sites in endogenous mRNAs or the viral genome). The resulting misassembled viral complexes were then endocytosed and ultimately delivered to lysosomes for degradation. However, the question of why the misassembled viral complexes were routed to lysosomes, as opposed to other destinations, following endocytosis remains to be clarified.

Autophagy is a naturally-occurring catabolic process in which cellular homeostasis and survival are maintained through the removal and recycling of unwanted cellular materials in lysosomes (Klionsky, 2007; Mizushima, 2007). There are three main types of autophagy: macroautophagy, microautophagy, and chaperone-mediated autophagy. Of these, macroautophagy is the most primitive and well-studied form of autophagy (Feng et al., 2014). During this process, target substrates are enclosed within cytosolic double-membrane delivery vesicles termed autophagosomes. Subsequent fusion of autophagosomes with late endosomes and/or lysosomes enables the contents to be decomposed into macromolecular constituents that can be recycled. Macroautophagy has been observed in the clearance of surplus and damaged organelles (Liu and Czaja, 2013; Rambold and Lippincott-Schwartz, 2011), proteins and ribonucleoprotein aggregates (Frankel et al., 2017; Wong et al., 2012), and pathogens (Mizushima et al., 2008). These unique functions of macroautophagy thus prompted us to investigate its potential role in the miRNA-mediated viral degradation pathway. Here, we provide evidence for a pivotal role of macroautophagy in regulating the delivery of misassembled viral complexes to lysosomes. Additionally, we demonstrate a similar regulation by miRNAs and role of autophagy in disrupting the formation of murine leukemia virus (MLV), a retrovirus that belongs to a different genus from HIV-1. These findings linking miRNAs and autophagy could facilitate the understanding of how cellular virus resistance could be enhanced, benefiting the design of generalizable therapeutic approaches through miRNA expression or autophagy-induction.

RESULTS AND DISCUSSION

Non-silencing miRNAs mediate the redirection of HIV-1 Gag complexes from the plasma membrane to lysosomes

To demonstrate the ability of non-silencing miRNAs to alter intracellular redistribution of HIV-1 Gag complexes, we prepared HEK 293 cell lines stably expressing either the exogenous miRNA hsa-miR-146a (MiR^+) or an empty vector (Ctrl), and transfected both with pNL43ΔPΔE, an HIV-1 proviral clone construct (See MATERIALS AND METHODS). Since hsa-miR-146a has no known target sites in the viral genome or the endogenous mRNAs of HEK 293 cells, we hypothesized that this non-silencing miRNA could function to disrupt viral assembly through mechanisms other than gene silencing. Supporting this possibility, HIV-1 Gag transfected MiR^+ cells exhibited reduced capacity to form viral particles compared to Ctrl cells (Fig. 1A), assayed by calculating the amount of Gag in the supernatant relative to total Gag in cells and the supernatant. Additionally, total Gag expression was similar in MiR^+ and in Ctrl cells (Fig. 1B), suggesting that hsa-miR-146a could not cause silencing of viral genes in HEK 293 cells. Thus, we concluded that the observed reduction in virus release in MiR^+ cells was not caused by miRNA-mediated gene silencing.

The reduction in virus release, despite the similar total level of Gag expression in MiR^+ and Ctrl cells, raised the possibility that non-silencing miRNAs function to alter intracellular distribution of Gag complexes. Supporting this idea, electron micrographs and fluorescence microscopy both showed that HIV-1 Gag localized to large vacuoles within MiR^+ cells while appear to cluster at the PM and exhibit typical viral budding characteristics in Ctrl cells (Fig. 1C and 1D). Large Gag vacuoles, identified as Gag-containing vacuoles with diameters greater than 1 μm, were present in nearly 50% of MiR^+ cells but only in ~10% of Ctrl cells assessed at 48 h post-transfection of the viral constructs (Fig. 1E). The increased vacuole formation was driven in part by clathrin-mediated endocytosis, as inhibiting the endocytosis machinery using dominant-negative dynamin-K44A (Dyn-K44A) significantly reduced vacuole formation (Fig. 1F). Furthermore, immunofluorescence imaging with lysosome-associated membrane protein (LAMP)-specific antibodies identified the Gag-containing vacuoles as late endosomes and lysosomes (Fig. 1G). Similar results were obtained when analogous experiments were performed in a second cell line that stably expresses hsa-miR-888 (MiR^+888), a non-silencing miRNA that has no sequence homology with miR-146a (Figs. S1 and S2). Thus, under conditions where cells express non-silencing miRNAs, misassembled viral complexes are targeted from the PM to lysosomes via endocytosis.

Large vacuoles were not observed in MiR^+ cells when Gag was not expressed (Fig. S3). This suggests that vacuole formation results from interaction between the non-silencing miRNA and Gag. To test this possibility, we questioned if vacuoles could still form in cells expressing mutant Gag lacking NC (ΔNC-Gag). Results indicated that large Gag vacuoles could not form in MiR^+ or Ctrl cells (Fig. 2A and 2B), suggesting that miRNA-NC interactions are necessary for large Gag vacuole formation. Supporting this, fluorescence *in situ* hybridization (FISH) experiments showed that non-silencing miRNAs and HIV-1 viral RNAs could colocalize with Gag at the PM and in vacuoles in MiR^+

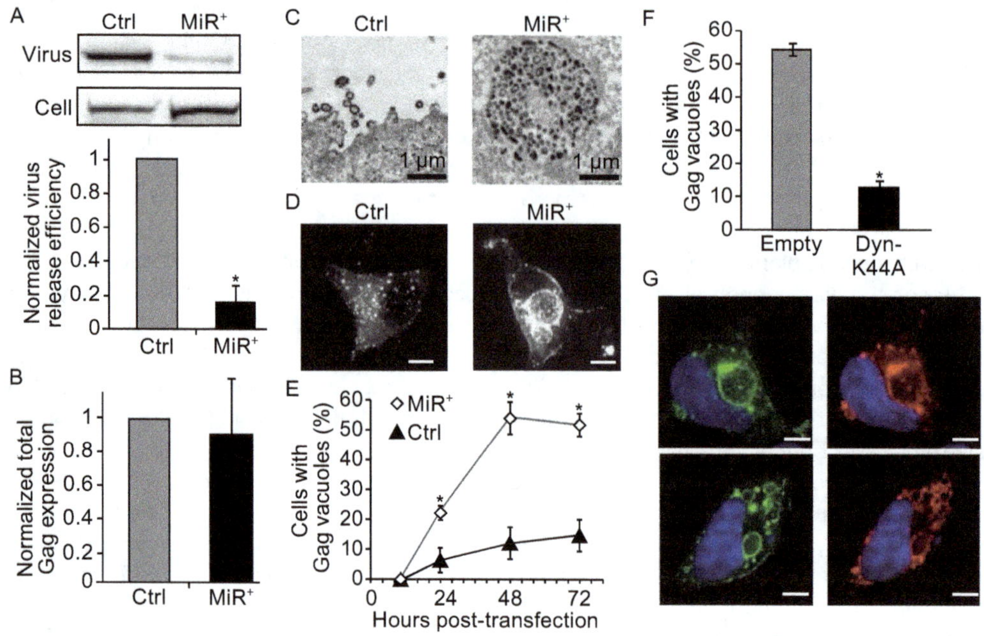

Figure 1. Overexpression of non-silencing miRNAs causes redistribution of HIV-1 Gag complexes from the PM to lysosomes through endocytosis. (A) The effect of overexpressing non-silencing miRNAs on virus particle release in MiR$^+$ and Ctrl cells at 48 h post-transfection of pNL43ΔPΔE. Western blot was performed with pooled Ig from HIV-1-infected patients (HIV-Ig) to detect Gag in viral particles and cells. Virus release efficiency was calculated as described in MATERIALS AND METHODS, normalized to virus release efficiency in Ctrl cells. (B) Total Gag expression levels (virus Gag plus cellular Gag) measured by Western blot in MiR$^+$ and Ctrl cells after transfection with pNL43ΔPΔE, normalized to total Gag expression in Ctrl cells. (C) Representative electron micrograph images of MiR$^+$ and Ctrl cells transfected with pNL43ΔPΔE. Characteristic Gag-enriched viral buds are observed at the PM of Ctrl cells, while MiR$^+$ cells tend to form large intracellular vacuoles. (D) Representative fluorescence images of Gag expression in MiR$^+$ and Ctrl cells through pNL43ΔPΔE-Gag-EGFP transfection. (E) Post-transfection time-course quantification of the percentage of cells with intracellular Gag-enriched vacuoles with diameter greater than 1 µm. (F) Co-transfection of pNL43ΔPΔE-Gag-EGFP with Dyn-K44A, a dominant-negative mutant that inhibits endocytosis, but not with empty vector, led to significant reduction in Gag vacuole formation. (G) Representative images of LAMP3 and LAMP1 in MiR$^+$ cells expressing pNL43ΔPΔE-Gag-EGFP. LAMP3 and LAMP1 colocalize extensively with Gag-EGFP in large vacuoles. Data represent mean ± SD of at least three replicates. Each replicate experiment was performed by examining at least 200 transfected cells. Unless otherwise noted, scale bar = 10 µm.

cells (Fig. 2C). Thus, together with the finding that Gag exhibited reduced capacity to form viral particles, these results confirm our previous finding that miRNAs and Gag can interact to disrupt viral assembly at the PM, with the misassembled viral complexes being sequestered in lysosomes following endocytosis.

Autophagy interacts with the endocytosis-mediated lysosomal degradation pathway to drive degradation of misassembled viral complexes

Following endocytosis, the observation that misassembled viral complexes are sequestered to lysosomes, as opposed to transiting to other destinations such as the trans-Golgi network or the extracellular milieu through exocytosis suggests that lysosomal delivery of misassembled viral complexes is a highly-regulated process. One possible regulation is through macroautophagy, which deliver unwanted materials to lysosomes for degradation. Increasing evidence suggests that macroautophagy could intersect with the endocytosis-mediated lysosomal degradation pathway (Sanchez-Wandelmer and Reggiori, 2013). Given that the misassembled Gag complexes appear to be highly sequestered in the late endosomes of MiR$^+$ cells, we sought to investigate whether this sequestration is autophagy-dependent.

To test this, we characterized autophagy activity in MiR$^+$ and Ctrl cells. Western blot analysis of LC3, an indicator of autophagy initiation and a marker of autophagosomes, showed that total LC3 levels and LC3 conversion capacity in MiR$^+$ and in Ctrl cells are similar (Fig. S4). This suggests overexpressing non-silencing miRNAs does not impact LC3 gene expression or conversion capacity. However, immunofluorescence with antibodies against LC3 showed different intracellular LC3 distribution patterns in Gag-expressing MiR$^+$ cells as compared with Ctrl cells (Figs. 3A and

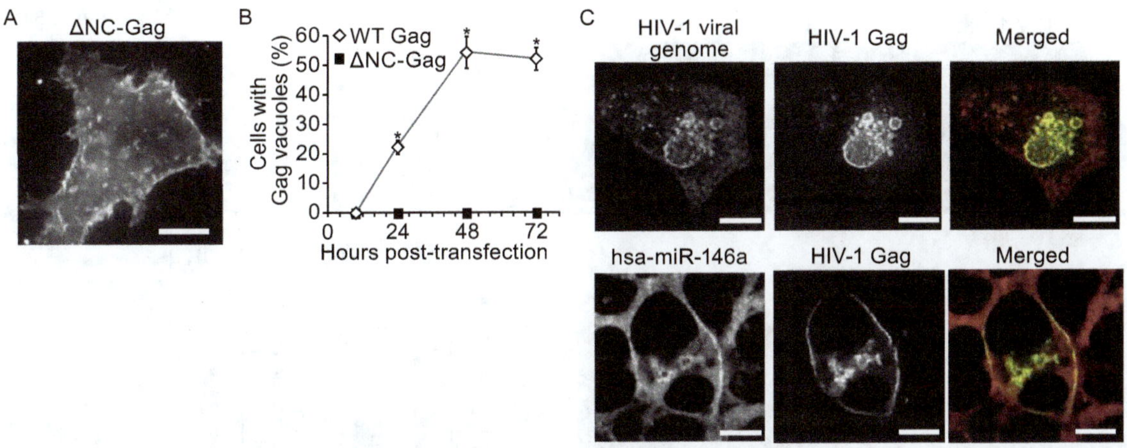

Figure 2. Interaction of miRNAs with HIV-1 Gag proteins drives intracellular Gag redistribution. (A) Representative fluorescence micrograph of ΔNC-Gag in MiR$^+$ cells. MiR$^+$ cells were transfected with pNL43ΔPΔEΔNC-Gag-EGFP and imaged at 48 h post-transfection. Note that ΔNC-Gag does not form intracellular vacuoles. (B) Time-course quantification of Gag vacuole formation by ΔNC-Gag as compared with wildtype Gag in MiR$^+$ cells. Data represent mean ± SD of at least three replicates. Each replicate experiment was performed by examining at least 200 transfected cells. (C) Representative images of hsa-miR-146a or HIV-1 viral genome (detected by FISH) and Gag (detected by Gag-EGFP) in MiR$^+$ cells transfected with pNL43ΔPΔE-Gag-EGFP. Scale bar = 10 μm.

S5). In Ctrl cells, LC3 and HIV-1 Gag proteins both exhibited a dispersedly-distributed and punctate staining pattern in the cytosol but did not colocalize with each other. In MiR$^+$ cells, LC3 localized extensively at the periphery and in the lumen of the Gag-enriched late endosomes/lysosomes. Thus, it appears that under conditions where misassembled viral complexes are internalized via endocytosis, autophagosomes are redistributed to the endocytosed complexes and participate in virus degradation.

To further investigate the dependence on autophagy in directing the endocytosed complexes to lysosomes for degradation, we tested if altering autophagic activity through pharmacological treatment or shRNA knockdown could impact the formation of intracellular vacuoles enriched in Gag, LC3, and LAMP proteins. Treatment with rapamycin, an autophagy enhancer, led to an increase in vacuole formation (Fig. 3B). Conversely, knockdown of ATG5, part of the machinery which drives autophagosome formation, caused a significant reduction in LC3 conversion, in the number of the Gag-enriched vacuoles, and an overall increase in Gag levels in MiR$^+$ cells (Fig. 3C and 3D), as expected if inhibiting autophagy activity reduces viral degradation. Furthermore, knockdown of ULK1, a key regulator of autophagy induction and progression, also led to a reduction in vacuole formation, confirming large Gag vacuoles formed through autophagosome-mediated pathways (Fig. S6). Additional evidence showing that macroautophagy was indeed functional under conditions where misassembled viral complexes are delivered to lysosomes was obtained from immunofluorescence labeling experiments that showed sequestration and degradation of mitochondria in LC3-containing Gag vacuoles. Specifically, in cells containing large Gag vacuoles, mitochondria, immunostained with antibodies against mitochondrial ATP synthase submit unit beta (ATPB), appeared highly fragmented, with the total signal intensity significantly reduced as compared with that of cells not expressing the viral constructs (Fig. 3E and 3F). Additionally, mitochondria were observed within the large Gag vacuoles, and treatment with leupeptin to inhibit protease activity led to ATPB accumulation and increased Gag enrichment inside the lumen, not just at the vacuole periphery (Fig. S7), suggesting that the large vacuoles originated from macroautophagy and mediate rapid lysosomal proteolytic degradation of both cellular and viral substrates.

Treatment with bafilomycin A1, an inhibitor of vacuolar H^+-ATPase that blocks autophagosomal/autolysosomal acidification in cells, reduced the size of Gag vacuoles significantly as compared with DMSO-treated cells (Fig. 3G). This supports the idea that large Gag vacuole formation is driven by fusion of autophagosomes with late endosomes/lysosomes. Interestingly, large Gag vacuoles were labeled by immunostaining with antibodies against p62 (Fig. 3H), which interacts with LC3 and mediates autophagic clearance of protein aggregates (aggrephagy). Thus, misassembled viral complexes could potentially resemble aggregates of malformed proteins, which trigger their degradation through an aggrephagy-mediated pathway. Altogether, these findings indicate that under conditions where retroviral assembly at the PM is disrupted by miRNAs resulting in endocytosis of viral complexes, macroautophagy plays a pivotal role in directing the misassembled complexes to lysosomes for degradation. The observed direction of misassembled viral

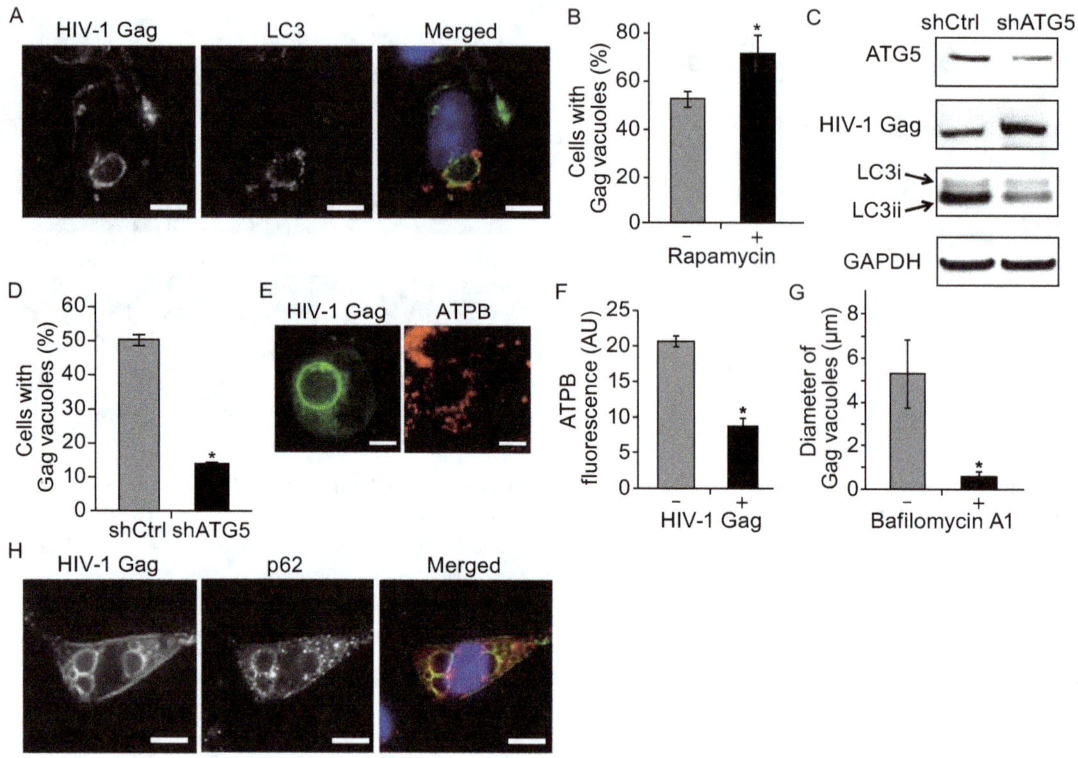

Figure 3. The role of autophagy in mediating lysosomal delivery of misassembled viral complexes. (A) Representative images of LC3 and HIV-1 Gag in MiR$^+$ cells. LC3 and HIV-1 Gag colocalize at the vacuoles. (B) Increase in Gag vacuole formation in response to rapamycin treatment. 24 h after MiR$^+$ cells were transfected with pNL43ΔPΔE-Gag-EGFP constructs, rapamycin (200 nmol/L) or DMSO control were added. Percentage of Gag-expressing MiR$^+$ cells containing at least one vacuole were assayed by fluorescence microscopy at 24 h following treatment. (C) The effect of ATG5 knockdown on HIV-1 Gag and LC3 conversion using shRNA against ATG5 (shATG5). At 48 h post-transfection, Western blot was performed to assess knockdown efficiency and its effect on Gag and LC3 conversion. shATG5 reduced the level of ATG5 by ~40%. This reduction was accompanied by ~50% reduction in LC3 conversion (calculated as the ratio of LC3ii to GAPDH) and a nearly two-fold increase in HIV-1 Gag accumulation in cells compared with cells transfected with the shCtrl plasmid. (D) Knockdown of ATG5 led to a reduction in Gag vacuole formation, assayed by fluorescence microscopy. (E) MiR$^+$ cells expressing pNL43ΔPΔE-Gag-EGFP were fixed and immunofluorescently-labeled with anti-ATPB antibodies to stain mitochondria. (F) Quantification of total mitochondrial ATPB signal per cell. (G) Decrease in Gag vacuole diameter in response to bafilomycin A1 treatment. 24 h after MiR$^+$ cells were transfected with pNL43ΔPΔE-Gag-EGFP constructs, 50 nmol/L bafilomycin A1 or DMSO control were added. Following 18 h incubation, cells were imaged for the presence of vacuoles. (H) Representative images of p62 and HIV-1 Gag in MiR$^+$ cells. p62 and HIV-1 Gag colocalize at the vacuoles. For (B) and (D), data represent mean ± SD of three replicates, with each replicative experiment performed by examining at least 200 transfected cells. For (F), data represent mean fluorescence intensity ± SD of 50 cells either untransfected or transfected with pNL43ΔPΔE-Gag-EGFP. For (G), data represent mean diameter ± SD of at least 200 individual vacuoles. Scale bar = 10 μm.

complexes to lysosomes through autophagy, as opposed to other possible routes, could potentially serve as an innate defense mechanism by which cells utilize non-silencing miRNAs to disrupt HIV-1 infection and dissemination.

Non-silencing miRNAs disrupt VLP formation and alter intracellular localization of MLV Gag and HIV-1 Gag through similar mechanisms

The ability of miRNAs to disrupt HIV-1 viral particle formation and cause viral degradation through autophagy led us to examine whether other retroviruses are regulated via similar mechanisms. Among retroviruses, MLV closely resembles HIV-1 in both assembly and budding, but belongs to a distinct viral genus (Jin et al., 2009; Jouvenet et al., 2006; Kutluay et al., 2014; Muriaux et al., 2004; Muriaux et al., 2001; Ono et al., 2005; Rein et al., 1994; Rulli et al., 2007). For example, Gag proteins of both viruses possess a nucleocapsid domain (NC) that is capable of interacting with RNAs nonspecifically (Kutluay et al., 2014; Muriaux et al., 2004; Rein et al., 1994). In addition, besides the viral genome, long-stranded cellular mRNAs can serve as scaffolds to facilitate Gag assembly, leading to formation of virus-like particles (VLPs) (Muriaux et al., 2001; Rulli et al., 2007).

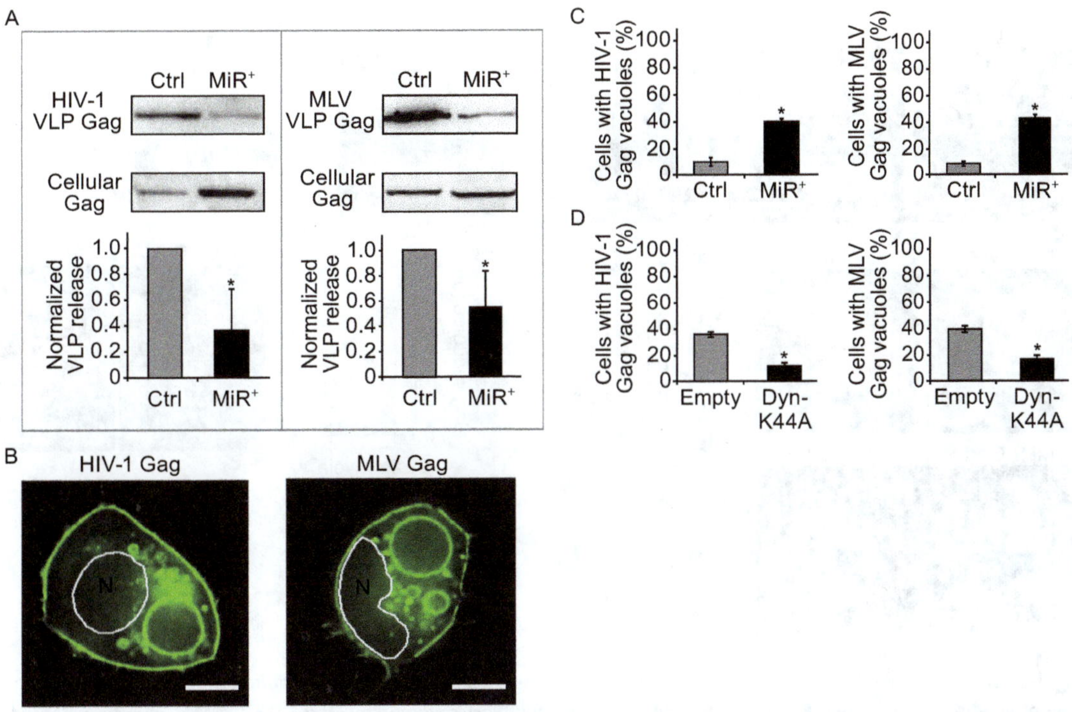

Figure 4. The effects of overexpressing non-silencing miRNAs on Gag assembly and intracellular localization of various retroviruses. (A) The effect of overexpressing non-silencing miRNAs on VLP production by MiR$^+$ and Ctrl cells transfected with CMV-driven plasmids encoding HIV-1 or MLV Gag proteins. Results were normalized to virus release efficiency in Ctrl cells. (B) Representative images of HIV-1 or MLV Gag (detected by Gag-EGFP) in MiR$^+$ cells. (C) Percentage of cells with large (diameter > 1 μm) intracellular vacuoles enriched in Gag after transfection of HIV-1 or MLV Gag. (D) Cells expressing HIV-1 or MLV Gag along with Dyn-K44A, a dominant-negative mutant that inhibits endocytosis, showed reduced Gag vacuole formation as compared with cells expressing an empty vector. Data represent mean ± SD of at least three replicates. For (C) and (D), each replicate experiment was performed by examining at least 200 transfected cells. Scale bar = 10 μm.

Furthermore, the PM is the major site of viral assembly in most cell types (Jin et al., 2009; Jouvenet et al., 2006; Ono et al., 2005). Therefore, since miRNAs can block formation of HIV-1 viral particles by interfering with Gag multimerization at the PM through interaction with NC and elicit autophagy-dependent viral degradation, we hypothesized that VLP formation and intracellular fate of MLV Gag complexes exhibit similar miRNA- and autophagy-mediated regulation. To test this possibility, MiR$^+$ and Ctrl cells were transfected with CMV-driven plasmid constructs encoding MLV Gag (See MATERIALS AND METHODS). For comparison, analogous experiments were performed with plasmids encoding CMV-driven HIV-1 Gag or simian foamy virus (SFV) Gag (See MATERIALS AND METHODS). The latter experiment serves as a negative control because SFV budding, unlike HIV-1 or MLV budding, requires envelope protein (Linial, 1999; Mullers, 2013; Shaw et al., 2003), and therefore should not occur in these experimental conditions.

VLP release efficiency, assayed at 48 h post-transfection of the CMV-driven constructs, showed that, similar to HIV-1 Gag, MLV Gag exhibited a reduced capacity to form VLPs in MiR$^+$ cells as compared with Ctrl cells (Fig. 4A). In contrast, no VLP release was detected in MiR$^+$ or Ctrl cells transfected with SFV Gag, as expected since envelope protein was not expressed (Fig. S8). Consistent with these findings, fluorescence microscopy imaging showed that both HIV-1 Gag and MLV Gag exhibited reduced puncta formation and an increased localization in large vacuoles in MiR$^+$ cells as compared with Ctrl cells (Fig. 4B and 4C). In contrast, no difference in the localization of SFV Gag was observed in MiR$^+$ or Ctrl cells. In both cells, SFV Gag formed intracellular aggregates, reflecting the inability to undergo viral budding at the PM in the absence of envelope protein (Fig. S9). Similar results were also obtained in MiR$^+$888 cells as compared with Ctrl cells (Figs. S10 and S11). Thus, it appears that over-expressing non-silencing miRNAs can disrupt MLV Gag assembly into VLPs at the PM and lead to internalization of the resulting misassembled Gag complexes. Supporting this possibility, co-expression of MLV Gag with Dyn-K44A significantly reduced large Gag vacuole formation (Figs. 4D and S12), which was also observed in cells expressing both HIV-1 Gag and Dyn-K44A. This suggests that the observed intracellular accumulation of MLV or

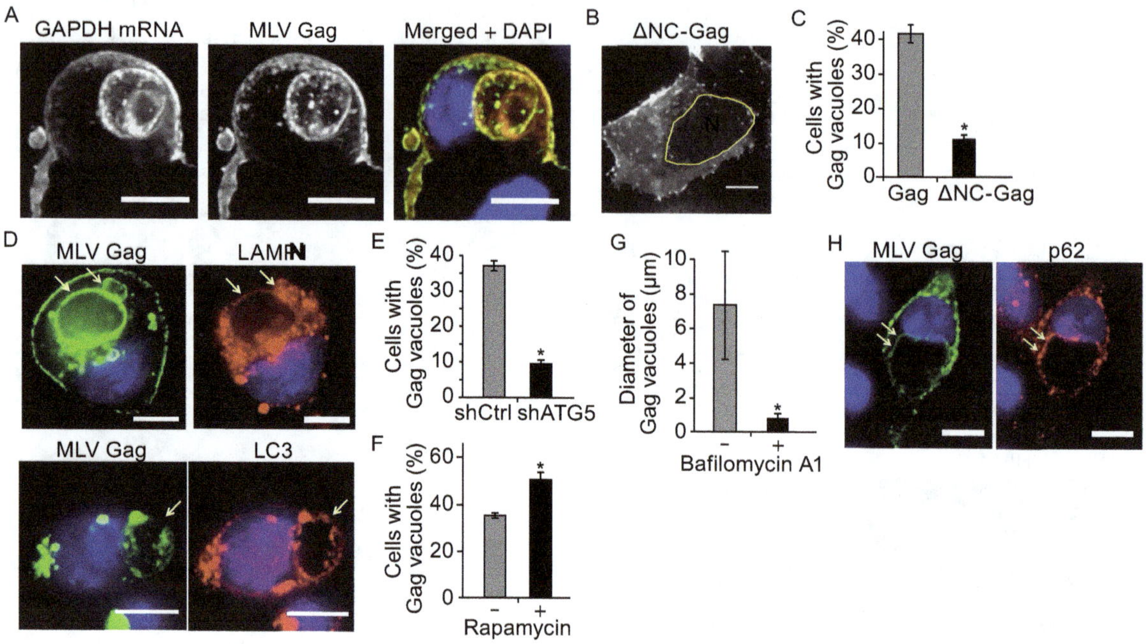

Figure 5. The effect of RNA-Gag interactions and autophagy on lysosomal delivery of misassembled MLV Gag complexes. (A) Representative images of GAPDH mRNA, detected by FISH, and Gag, detected by Gag-EGFP, in MiR$^+$ cells transfected with MLV Gag-EGFP. (B) Representative image of MLV-ΔNC-Gag, detected by MLV ΔNC-Gag-EGFP in MiR$^+$ cells. N indicates nucleus. (C) Percentage of MiR$^+$ cells with large (diameter > 1 μm) intracellular vacuoles at 48 h post-transfection of MLV Gag or MLV ΔNC-Gag. (D) Representative images of LC3 and LAMP1, detected by immunofluorescence, in MiR$^+$ cells expressing MLV Gag-EGFP. Arrows point to the vacuoles. (E) Knockdown of ATG5 led to a reduction in Gag vacuole formation. (F) Treatment with rapamycin (200 nmol/L) led to increased Gag vacuole formation. (G) Decrease in Gag vacuole diameter in response to bafilomycin A1 treatment. (H) Representative images of p62 and MLV Gag in MiR$^+$ cells. p62 and MLV Gag colocalize at the vacuoles. Arrows point to the vacuoles. For (C), (E), and (F), data represent mean ± SD of three replicates, with each replicative experiment performed by visually examining at least 200 transfected cells. For (G), data represent mean diameter ± SD of at least 200 individual vacuoles. Scale bar = 10 μm.

HIV-1 Gag was largely due to endocytosis of Gag complexes at the PM.

To further examine the similarity between the effects of non-silencing miRNAs on MLV and HIV-1 Gag, we performed experiments to characterize the nature of the MLV Gag vacuoles. FISH experiments showed GAPDH mRNA, an abundant endogenous mRNA, colocalized with Gag at the PM and at the periphery of the Gag-enriched vacuoles (Fig. 5A). These results support the idea that Gag can use cellular mRNAs as assembly scaffolds and suggest that miRNAs can disrupt this process to facilitate internalization of misassembled Gag complexes through endocytosis. Supporting this idea, expression of MLV Gag lacking NC did not lead to the formation of large vacuoles in MiR$^+$ cells (Fig. 5B and 5C), indicating RNA-Gag interaction is essential for the miRNA-mediated internalization of MLV Gag, as in HIV-1 Gag. Thus, these findings suggest that MLV Gag complexes, like those of HIV-1, undergo redistribution from the PM to cytoplasm when miRNAs compete with scaffolding RNAs for Gag binding. Furthermore, immunofluorescence experiments identified the vacuoles as autophagosomes and lysosomes (Fig. 5D), suggesting that macroautophagy converges with components of the endocytosis-mediated lysosomal degradation pathway. Lysosomal sequestration of MLV Gag was highly dependent on autophagy activity, as shRNA knockdown of ATG5 or ULK1 led to a significant reduction in large Gag vacuole formation, whereas treatment with rapamycin enhanced vacuole formation (Figs. 5E, 5F, and S13). Additionally, treatment with autophagy inhibitor bafilomycin A1 led to a reduction in the size of MLV Gag vacuoles (Fig. 5G), suggesting their formation requires maturation of autophagosomes and subsequent fusion with lysosomes. Finally, large Gag vacuoles are p62-positive, suggesting they result from aggrephagy-mediated degradation (Fig. 5H). Altogether, these findings suggest that non-silencing miRNAs can mediate the redistribution and clearance of retroviruses that rely on long-stranded RNAs as scaffolds for assembly and particle formation.

CONCLUSION

MicroRNAs are predominantly regarded as gene silencers involved in the regulation of various cellular processes, including embryogenesis, cellular differentiation, and pathogenesis. In these contexts, miRNAs, associated with proteins of the RISC, prevent the expression of specific genes

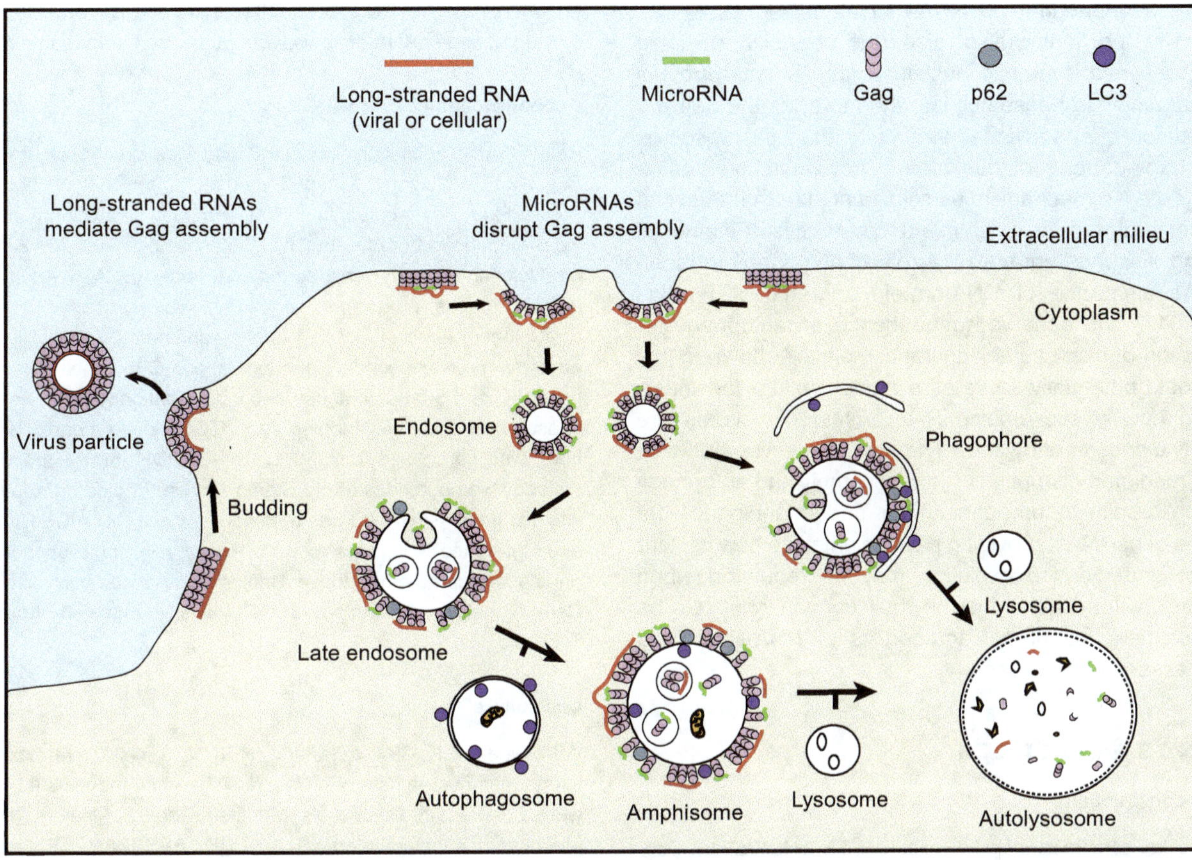

Figure 6. Schematic model of how microRNAs and macroautophagy function to block virus production. In the absence of Gag-miRNA complexes, Gag and scaffold RNAs (viral genome or long-stranded mRNAs) form stable complexes at the PM, resulting in viral budding. When Gag-miRNA complexes are present within the assembling complexes, particle formation is disrupted, resulting in internalization of misassembled complexes through endocytosis. The endocytosed Gag-miRNA complexes could be engulfed by phagophores, which mediates their autophagic degradation through a p62-dependent process. Additionally, fusion of autophagosomes to the late endosomes could form amphisomes, which can then mature into autolysosomes that degrade the enclosed misassembled viral complexes.

through hybridization with target mRNAs. Alternatively, increasing evidence has shown that miRNAs can also interact with proteins outside the RISC and that these interactions can influence cellular processes through mechanisms independent of gene silencing (Chen et al., 2014; Eiring et al., 2010; Fabbri et al., 2012; Lehmann et al., 2012; Prud'homme et al., 2016; Ranganathan et al., 2017; Vickers et al., 2011; Yelamanchili et al., 2015). In one such study, we showed that both exogenous and endogenous miRNAs, particularly those not involved in mediating gene silencing, can compete with viral RNA for HIV-1 Gag binding, forming miRNA-Gag complexes that prevent viral RNA-mediated Gag assembly into high-order multimers essential for viral budding. The misassembled viral complexes were ultimately redirected into the endocytic pathway where they were delivered to lysosomes for degradation (Chen et al., 2014).

In this study, we demonstrated that targeting of misassembled viral complexes from the PM to lysosomes for degradation is driven by macroautophagy (Fig. 6). The process involves p62, which potentially recognizes the endocytosed misassembled viral complexes as protein aggregates. Additionally, we showed that the assembly of MLV can also be regulated by miRNAs and autophagy in cells. The analogous effects of non-silencing miRNAs on the misassembly and degradation of HIV-1 and MLV Gag proteins suggest that the ability of Gag to interact nonspecifically with a diverse class of RNAs, as demonstrated previously (Campbell and Vogt, 1995; Chen et al., 2014; Jouvenet et al., 2009; Kutluay et al., 2014; Muriaux et al., 2001), is a crucial determinant for the observed miRNA-mediated viral blockage effect.

Our new findings linking miRNAs and autophagy could have significant implications for retroviral therapeutics and for understanding how cells regulate retroviral assembly and potentially other RNA-mediated protein assembly processes. For example, miRNAs, provided they are not otherwise preoccupied with gene silencing, might be utilized therapeutically as a general inhibitor to disrupt or, if present in sufficient excess, even completely block formation of various retroviruses dependent on RNA-mediated Gag assembly. Additionally, the finding also raises the possibility of using small, nonsense synthetic nucleic acids as anti-retroviral agents against viral assembly. Furthermore, as the delivery and retention of endocytosed viral complexes in lysosomes is autophagy-dependent, this dependency could serve as an internal defense mechanism by cells to ensure that the virus does not exit the cells to infect other cells. Finally, as emerging evidence unravels the role of autophagy in mediating the clearance of RNA-protein aggregates (Frankel et al., 2017), the observed dependence on autophagy for degradation of misassembled viral complexes shown in this study could potentially serve as a model system for understanding how misassembled RNA-protein complexes are generally disposed of by cells. The new connection between miRNA-mediated disruption of viral assembly and autophagy could be crucial in broadening our understanding of the functions of miRNAs, providing new insights into how cellular behavior and disease evolution may be regulated when miRNAs function, rather than in their conventional role as gene silencers, as potent antagonists of RNA-scaffolded protein assembly processes.

MATERIALS AND METHODS

Plasmid construction

HIV-1 proviral constructs including pNL43ΔPΔE, pNL43ΔPΔE-Gag-EGFP, pNL43ΔPΔEΔNC, and pNL43ΔPΔEΔNC-Gag-EGFP have been described previously (Chen et al., 2014). CMV-driven expression plasmids encoding HIV-1 Gag (pCR3.1-Gag) and HIV-1 Gag-mCherry (pCR3.1-Gag-mCherry) were kind gifts of Dr. Sanford Simons (Rockefeller University, New York, NY) and those that encode MLV Gag-EGFP (pcDNA3-MLV-Gag-EGFP) and SFV Gag-EGFP (pcDNA3-SFV-Gag-EGFP) were kind gifts of Dr. Stephan Gould (Johns Hopkins University, School of Medicine, Baltimore, MD). To generate pCR3.1-Gag-EGFP, the coding region of EGFP was first PCR amplified from pEGFP-N1 vector (Clontech) with forward primer 5′-attgcggccgcatggtgagcaagggcgagga-3′ and reverse primer 5′-atttctagattacttgtacagctcgtccatg-3′. The PCR product was then inserted into PCR3.1-Gag-mCherry backbone digested with *Not*I and *Xba*I to excise mCherry. To generate pcDNA3-MLV-Gag and pcDNA3-SFV-Gag, EGFP was first removed from pcDNA3-MLV-Gag-EGFP and pcDNA3-SFV-Gag-EGFP using *Bam*HI and *Xba*I, followed by end-filling the digested vector with Klenow and self-ligation of the resulting blunt-ended vectors. This operation leads to a stop codon in frame with the Gag sequences. To construct pcDNA3-MLV-ΔNC-Gag and pcDNA3-MLV-ΔNC-Gag-EGFP, plasmids that encode MLV Gag or MLV Gag-EGFP that harbor deletion of the nucleocapsid domain, respectively, the cDNA sequences corresponding to amino acids 16 to 23 in the NC domain of MLV Gag were deleted by PCR-mediated site-directed mutagenesis with forward primer 5′-GACCAGTGCGCCTACTGCAAGGAGAAG-3′ and reverse primer 5′-CTCGCCGCCCTGGCGGTCCTGCTTCTGGC-3′. shRNA constructs, including shRNA-ATG5 (TRC number: TRCN 0000150645; clone ID: NM_004849.1-1043s1c1), shRNA-ULK1 (TRC number: TRCN0000000837; clone ID: NM_003565.x-535s1c1), and shRNA-controls were purchased from Sigma Aldrich.

Antibodies and chemicals

Anti-LC3A/B rabbit polyclonal antibody was purchased from Cell Signaling Technology. Anti-p62 mouse monoclonal antibody and Alexa®Fluor 647-labeled Anti-LAMP1 mouse monoclonal antibody were purchased from Santa Cruz Biotechnology. Anti-ATPB mouse monoclonal antibody was purchased from ABCAM. Anti-LAMP3 (CD63) mouse monoclonal antibody was purchased from BD Pharmingen. Alexa®Fluor 594 donkey-anti-rabbit, Alexa®Fluor 594 goat-anti-mouse secondary antibodies, anti-GFP rabbit monoclonal antibody, and goat-anti-mouse secondary antibody HRP were purchased from Life Technologies. Anti-ATG5 mouse monoclonal antibody was purchased from MBL. Anti-GAPDH mouse monoclonal antibody was purchased from Sigma Aldrich. Pooled Ig from HIV-1-infected patients (HIV-Ig) was obtained from the NIH AIDS Research and Reference Reagent Program. Rapamycin and leupeptin were purchased from Sigma Aldrich. Bafilomycin A1 was purchased from Cell Signaling Technology. DAPI was purchased from Life Technologies.

Cell culture

Wild type HEK 293 cells and HEK 293 cells overexpressing exogenous human hsa-miR-146a (MiR$^+$) or hsa-miR-888 (MiR$^+$888) were cultured in Dulbecco's Modified Eagle's Medium (DMEM, Mediatech), supplemented with 10% (*v*/*v*) FBS (PAN^{TM} Biotech), 1× GlutaMAX^{TM} (Thermo Fisher) at 37°C, 5% (*v*/*v*) CO_2, and 90% relative humidity. Design of miRNA expression plasmids and generation of miRNA cell lines have been described previously (Chen et al., 2014). All experiments were performed with cells at passage numbers between 5 and 25.

Transfection

Transfection was performed with FuGENE® 6 (Promega) as per manufacturer's protocols when cells reached 50%–70% confluency. For fluorescence imaging studies where fluorescent proteins (FP) are used, cells were transfected with a mixture containing both FP-tagged and untagged viral constructs at a 1:3 ratio. For shRNA knockdown studies, shRNA constructs and viral constructs were transfected at 3:1 ratio.

Virus particle purification, Western blotting and analysis of viral release efficiency

To collect viral particles, culture supernatant harvested at 48 h post-transfection of viral constructs was centrifuged at 1,000 ×*g* for 10 min and then filtered through a 0.45 μm filter to remove cell debris and large aggregates. Subsequently, 20 μL of Dynabeads®280 streptavidin (Life Technologies), precleaned twice with 1× PBS, was added into every 10 mL of the supernatant in order to assist the visualization of pellet after ultracentrifugation (100,000 ×*g* for 1 h). The viral particles and the beads were resuspended in lysis buffer (0.5% Triton X-100, 50 mmol/L pH = 7.5 Tris-HCl, 300 mmol/L NaCl) containing protease inhibitor cocktail (1:100). The lysates and the beads were further separated by microcentrifugation. To collect cell lysates after removing culture supernatants, the cells were washed once in cold 1× PBS, trypsinized, pelleted, and finally lysed in lysis buffer containing protease inhibitor cocktail (1:100). Gag from both supernatant and the cell lysates were analyzed by SDS-PAGE on 10% acrylamide gels and transferred to Immobilon™-P membranes (Millipore). Immunoblotting was carried out with HIV-Ig or anti-GFP antibodies. Release efficiency was calculated (at 48 h post-transfection) as the ratio of supernatant Gag to total Gag, both determined by densitometry analysis of Western blot images using Fiji software.

Electron microscopy (EM)

Cells were washed twice with 0.1 mol/L sodium cacodylate buffer (pH = 7.4) followed by fixation with 2.5% glutaraldehyde in the same buffer for 2 h, and then post-fixed with 1% OsO_4 for 1 h at room temperature. After rinsing several times in cacodylate buffer and distilled water, the cells were incubated in 0.1% tannic acid (in cacodylate buffer) for 30 min, and stained in 1% uranyl acetate for 1 h. They were washed again in distilled water and dehydrated in a graded ethanol series and embedded in SPI-Pon 812 resin (SPI Supplies, PA, USA). Ultrathin (70 nm) sections were cut using an ultramicrotome (UC7, Leica Microsystem), and collected on copper grids with a single slot, stained with uranyl acetate and lead citrate. Then the sections were observed under a Tecnai G^2 20 TWIN electron microscope at 120 kV and photographed with an Eagle (4k×4k) digital camera (FEI, Oregon, USA).

Fluorescence microscopy

All fluorescent microscopy experiments were performed on an Olympus IX 83 motorized inverted fluorescence microscope equipped with a 20× UCPlanFL N 0.7NA or a 100× UPlanSApo 1.4NA objective lens, back-illuminated EMCCD camera (Andor), Sutter excitation and emission filter wheels and an MT-20E excitation source (Olympus). Images were acquired using the Olympus MT20 filter set for DAPI, EGFP and TAMRA and a filter set (ET620/60x, ET700/75m, T660lpxr, Chroma) for Cy5. All images were acquired using CellSens Dimension software. Three-dimensional image stacks were acquired with 0.25 μm increments in the z-direction. Results were analyzed with Fiji (Schindelin et al., 2012) or AutoQuant deconvolution software (MediaCybernetics).

Fluorescence *in situ* hybridization

Procedures for imaging fluorescence *in situ* hybridization (FISH) of hsa-miR-146a utilizing TSA Plus signal amplification (PerkinElmer) have been described before (Chen et al., 2014). FISH of unspliced HIV-1 viral RNA and GAPDH mRNA was performed as previously described, with modifications (Chen et al., 2014; Zhao et al., 2016). Specifically, cells were fixed in PBS solution containing 4% (*w*/*v*) paraformaldehyde for 20 min at room temperature, washed with 1× PBS, and permeabilized at 4°C in 70% (*v*/*v*) ethanol overnight. On the next day, the cells were washed thrice with wash buffer containing 2× SSC and 10% (*v*/*v*) formamide and then incubated in hybridization buffer (10% (*w*/*v*) dextran sulfate, 2× SSC, 10% (*v*/*v*) formamide) containing 250 nmol/L singly Quasar®570-labeled oligonucleotide probes against GAPDH mRNA (Cat# SMF-2026-1, Bioresearch) or 100 nmol/L singly TAMRA-labeled oligonucleotide probes against unspliced HIV-1 viral RNA (Chen et al., 2014) for 24 h at 37°C in a cell culture incubator. Prior to microscopy imaging, slides were washed thrice with wash buffer and then incubated in wash buffer for 30 min at 37°C, followed by two washes with 2× SSC and a final wash in 1× PBS to remove the unbound probe. Cells were incubated in 1× PBS for imaging.

Data analysis

All experiments were repeated at least three times unless otherwise stated. Statistics were performed using Student's *t*-test or one-way ANOVA with post-hoc testing of pairwise comparisons using Fischer's Protected Least Significant Difference. Significant difference (indicated by *) was taken at the $P < 0.05$ level.

ACKNOWLEDGEMENTS

This project was supported by grants from the National Key R&D Program of China (2016YFA0501603 and 2016YFA0100702), the National Natural Science Foundation of China (Grant No. 81371613), the Beijing Natural Science Foundation (7162114) and China's 1000 Young Talent Award program. We thank Dr. Ying-Chun Hu for her professional technical assistance in EM sample preparation and image analysis at the Core Facilities of College of Life Sciences, Peking University.

ABBREVIATIONS

LAMP, lysosome-associated membrane protein; miRNAs, microRNAs; MLV, murine leukemia virus; PM, plasma membrane; RISC, RNA-induced silencing complex; SFV, simian foamy virus; VLPs, virus-like particles

COMPLIANCE WITH ETHICS GUIDELINES

Na Qu, Zhao Ma, Mengrao Zhang, Muaz N. Rushdi, Christopher J. Krueger, and Antony K. Chen declare that they have no conflict of interest. This article does not contain any studies with human or animal subjects performed by the any of the authors.

REFERENCES

Aldovini A, Young RA (1990) Mutations of RNA and protein sequences involved in human immunodeficiency virus type 1 packaging result in production of noninfectious virus. J Virol 64:1920–1926

Berkowitz RD, Luban J, Goff SP (1993) Specific binding of human immunodeficiency virus type 1 gag polyprotein and nucleocapsid protein to viral RNAs detected by RNA mobility shift assays. J Virol 67:7190–7200

Briggs JA, Simon MN, Gross I, Krausslich HG, Fuller SD, Vogt VM, Johnson MC (2004) The stoichiometry of Gag protein in HIV-1. Nat Struct Mol Biol 11:672–675

Campbell S, Vogt VM (1995) Self-assembly in vitro of purified CA-NC proteins from Rous sarcoma virus and human immunodeficiency virus type 1. J Virol 69:6487–6497

Chen AK, Sengupta P, Waki K, Van Engelenburg SB, Ochiya T, Ablan SD, Freed EO, Lippincott-Schwartz J (2014) MicroRNA binding to the HIV-1 Gag protein inhibits Gag assembly and virus production. Proc Natl Acad Sci USA 111:E2676–E2683

Eiring AM, Harb JG, Neviani P, Garton C, Oaks JJ, Spizzo R, Liu S, Schwind S, Santhanam R, Hickey CJ et al (2010) miR-328 functions as an RNA decoy to modulate hnRNP E2 regulation of mRNA translation in leukemic blasts. Cell 140:652–665

Fabbri M, Paone A, Calore F, Galli R, Gaudio E, Santhanam R, Lovat F, Fadda P, Mao C, Nuovo GJ et al (2012) MicroRNAs bind to Toll-like receptors to induce prometastatic inflammatory response. Proc Natl Acad Sci USA 109:E2110–E2116

Feng Y, He D, Yao Z, Klionsky DJ (2014) The machinery of macroautophagy. Cell Res 24:24–41

Filipowicz W, Bhattacharyya SN, Sonenberg N (2008) Mechanisms of post-transcriptional regulation by microRNAs: are the answers in sight? Nat Rev Genet 9:102–114

Frankel LB, Lubas M, Lund AH (2017) Emerging connections between RNA and autophagy. Autophagy 13:3–23

Gorelick RJ, Nigida SM Jr, Bess JW Jr, Arthur LO, Henderson LE, Rein A (1990) Noninfectious human immunodeficiency virus type 1 mutants deficient in genomic RNA. J Virol 64:3207–3211

Jin J, Sherer NM, Heidecker G, Derse D, Mothes W (2009) Assembly of the murine leukemia virus is directed towards sites of cell-cell contact. Plos Biol 7:e1000163

Jouvenet N, Neil SJD, Bess C, Johnson MC, Virgen CA, Simon SM, Bieniasz PD (2006) Plasma membrane is the site of productive HIV-1 particle assembly. Plos Biol 4:2296–2310

Jouvenet N, Simon SM, Bieniasz PD (2009) Imaging the interaction of HIV-1 genomes and Gag during assembly of individual viral particles. Proc Natl Acad Sci USA 106:19114–19119

Klionsky DJ (2007) Autophagy: from phenomenology to molecular understanding in less than a decade. Nat Rev Mol Cell Biol 8:931–937

Kutluay SB, Bieniasz PD (2010) Analysis of the initiating events in HIV-1 particle assembly and genome packaging. PLoS Pathog 6: e1001200

Kutluay SB, Zang T, Blanco-Melo D, Powell C, Jannain D, Errando M, Bieniasz PD (2014) Global changes in the RNA binding specificity of HIV-1 gag regulate virion genesis. Cell 159:1096–1109

Lehmann SM, Kruger C, Park B, Derkow K, Rosenberger K, Baumgart J, Trimbuch T, Eom G, Hinz M, Kaul D et al (2012) An unconventional role for miRNA: let-7 activates Toll-like receptor 7 and causes neurodegeneration. Nat Neurosci 15:827–835

Linial ML (1999) Foamy viruses are unconventional retroviruses. J Virol 73:1747–1755

Liu K, Czaja MJ (2013) Regulation of lipid stores and metabolism by lipophagy. Cell Death Differ 20:3–11

Mizushima N (2007) Autophagy: process and function. Gene Dev 21:2861–2873

Mizushima N, Levine B, Cuervo AM, Klionsky DJ (2008) Autophagy fights disease through cellular self-digestion. Nature 451:1069–1075

Mullers E (2013) The foamy virus Gag proteins: what makes them different? Viruses-Basel 5:1023–1041

Muriaux D, Mirro J, Harvin D, Rein A (2001) RNA is a structural element in retrovirus particles. Proc Natl Acad Sci USA 98:5246–5251

Muriaux D, Costes S, Nagashima K, Mirro J, Cho E, Lockett S, Rein A (2004) Role of murine leukemia virus nucleocapsid protein in virus assembly. J Virol 78:12378–12385

Ono A, Waheed AA, Joshi A, Freed EO (2005) Association of human immunodeficiency virus type 1 gag with membrane does not require highly basic sequences in the nucleocapsid: use of a novel gag multimerization assay. J Virol 79:14131–14140

Prud'homme GJ, Glinka Y, Lichner Z, Yousef GM (2016) Neuropilin-1 is a receptor for extracellular miRNA and AGO2/miRNA complexes and mediates the internalization of miRNAs that modulate cell function. Oncotarget 7:68057–68071

Rambold AS, Lippincott-Schwartz J (2011) Mechanisms of mitochondria and autophagy crosstalk. Cell Cycle 10:4032–4038

Ranganathan P, Ngankeu A, Zitzer NC, Leoncini P, Yu XY, Casadei L, Challagundla K, Reichenbach DK, Garman S, Ruppert AS et al (2017) Serum miR-29a is upregulated in acute graft-versus-host disease and activates dendritic cells through TLR binding. J Immunol 198:2500–2512

Rein A, Harvin DP, Mirro J, Ernst SM, Gorelick RJ (1994) Evidence that a central domain of nucleocapsid protein is required for RNA packaging in murine leukemia-virus. J Virol 68:6124–6129

Rulli SJ Jr, Hibbert CS, Mirro J, Pederson T, Biswal S, Rein A (2007) Selective and nonselective packaging of cellular RNAs in retrovirus particles. J Virol 81:6623–6631

Sanchez-Wandelmer J, Reggiori F (2013) Amphisomes: out of the autophagosome shadow? EMBO J 32:3116–3118

Schindelin J, Arganda-Carreras I, Frise E, Kaynig V, Longair M, Pietzsch T, Preibisch S, Rueden C, Saalfeld S, Schmid B et al (2012) Fiji: an open-source platform for biological-image analysis. Nat Methods 9:676–682

Sharp PA (2009) The centrality of RNA. Cell 136:577–580

Shaw KL, Lindemann D, Mulligan MJ, Goepfert PA (2003) Foamy virus envelope glycoprotein is sufficient for particle budding and release. J Virol 77:2338–2348

Vickers KC, Palmisano BT, Shoucri BM, Shamburek RD, Remaley AT (2011) MicroRNAs are transported in plasma and delivered to recipient cells by high-density lipoproteins. Nat Cell Biol 13:423–433

Wong E, Bejarano E, Rakshit M, Lee K, Hanson HH, Zaarur N, Phillips GR, Sherman MY, Cuervo AM (2012) Molecular determinants of selective clearance of protein inclusions by autophagy. Nat Commun 3:1240

Yelamanchili SV, Lamberty BG, Rennard DA, Morsey BM, Hochfelder CG, Meays BM, Levy E, Fox HS (2015) MiR-21 in extracellular vesicles leads to neurotoxicity via TLR7 signaling in SIV neurological disease. PLoS Pathog 11:e1005032

Zhao D, Yang Y, Qu N, Chen M, Ma Z, Krueger CJ, Behlke MA, Chen AK (2016) Single-molecule detection and tracking of RNA transcripts in living cells using phosphorothioate-optimized 2'-*O*-methyl RNA molecular beacons. Biomaterials 100:172–183

7

Molecular dynamics and principal components of potassium binding with human telomeric intra-molecular G-quadruplex

Zhiguo Wang[1✉], **Ruping Chen**[1,2], **Ling Hou**[1], **Jianfeng Li**[1], **Jun-Ping Liu**[1,2,3,4✉]

[1] Institute of Aging Research, School of Medicine, Hangzhou Normal University, Hangzhou 311121, China
[2] School of Medicine, Shandong University, Jinan 250100, China
[3] Department of Immunology, Central Eastern Clinical School, Monash University, Melbourne, VIC 3004, Australia
[4] Center for Cancer Research, Monash Institute of Medical Research, Clayton, VIC 3168, Australia
✉ Correspondence: zhgwang@aliyun.com (Z. Wang), jun-ping.liu@monash.edu (J.-P. Liu)

ABSTRACT

Telomere assumes intra-molecular G-quadruplex that is a significant drug target for inhibiting telomerase maintenance of telomeres in cancer. Metal cations have been recognized as playing important roles in stabilizing G-quadruplex, but their binding processes to human telomeric G-quadruplex remain uncharacterized. To investigate the detailed binding procedures, molecular dynamics simulations were conducted on the hybrid [3 + 1] form-one human telomeric intra-molecular G-quadruplex. We show here that the binding of a potassium ion to a G-tetrad core is mediated by two alternative pathways. Principal component analysis illustrated the dominant concerted motions of G-quadruplex occurred at the loop domains. MM-PBSA calculations revealed that binding was energetically favorable and driven by the electrostatic interactions. The lower binding site was found more constructive favorable for binding. Our data provide useful information on a potassium-mediated stable structure of human telomeric intra-molecular G-quadruplex, implicating in ion disorder associated conformational changes and targeted drug design.

KEYWORDS G-quadruplex, potassium ion, molecular dynamics, principal component analysis, MM-PBSA

INTRODUCTION

Telomere G-quadruplex consists of guanine-rich nucleic acid sequences characteristic of a four-stranded structure at chromosomal ends (telomeres) of eukaryotic cells (Moyzis et al., 1988). G-quadruplex exists in several π-π stacked G-tetrad layers, with each having four coplanar guanines that are held together through eight Hoogsteen hydrogen bonds (Burge et al., 2006; Williamson, 1994). Human telomeric DNA is double-stranded for most of its length with a terminal single-stranded overhang of 100–200 bases at the extreme 3′-end that has the propensity to fold into intra-molecular G-quadruplex structures under physiological ionic conditions (Burge et al., 2006; Makarov et al., 1997; Neidle and Parkinson, 2003). The intra-molecular G-quadruplexes have been recognized as significant drug targets by inhibiting the activity of telomerase that is overexpressed and required in ~85% of cancers (Maizels, 2006; Neidle and Parkinson, 2003; Oganesian and Bryan, 2007; Paeschke et al., 2005). So, inhibition of telomerase by targeting the stabilization of intra-molecular telomeric G-quadruplex has become an attractive strategy for the development of novel anti-cancer drugs.

Information at the atomic level of the human telomeric intra-molecular G-quadruplex structures is essential for structure-based rational drug design. By applying NMR and X-ray crystallography techniques, human G-quadruplexes at telomeres, gene promoter regions of proto-oncogenes (c-myc, c-kit, bcl-2) and untranslated regions of mRNAs have been investigated (Bejugam et al., 2007; Dai et al., 2006; Dexheimer et al., 2006; Phat et al., 2007; Rankin et al., 2005; Siddiqui-Jain et al., 2002). For the human telomeric G-quadruplex, the single stranded G-rich overhang confers the intra-molecular G-quadruplex structures with different conformational topologies, stabilization of which is important in telomere-elicited signaling in aging and cancer.

Monovalent cations, mainly alkali metal ions and ammonium ions, are essential for G-quadruplex, because the ions are particularly sensitive to the structural folding of G-quadruplex involving in the formation of loop types and overall conformations (Gaynutdinov et al., 2008; Phan et al., 2006, 2007a, b; Pinnavaia et al., 1978; Rujan et al., 2005). KCl and NaCl associate with parallel and anti-parallel G-quadruplexes respectively (Gaynutdinov et al., 2008; Phan et al. 2007a, b). Folding of the hybrid [3 + 1] form-one and form-two of human telomeric G-quadruplexes is facilitated by K^+ (Bončina et al., 2012; Li et al., 2009), whereas unfolding is accompanied by potassium release from human telomeric DNA d[AG_3($TTAGGG)_3$] (Bončina et al., 2012). A hierarchy potencies of the monovalent cations in stabilizing G-quadruplex is $K^+ > NH_4^+ > Na^+$, suggesting that a weak stabilizer can be substituted by a strong stabilizer in solution to stabilize telomeres.

Besides NMR and X-ray crystallography, computer based molecular dynamics (MD) has proven to be a powerful tool to investigate the characteristics of telomeric G-quadruplex at the atomic level (Akhshi et al., 2012; Špačková et al., 2001; Zhu et al., 2013). Two MD studies on the binding process of d[G_4] and d[$G_4T_4G_4$] G-quadruplexes to cations have been reported using the AMBER 4.1 software and PARM94 all atom force field (Fadrná et al., 2004; Špačková et al., 1999). However, the force field appeared inaccurate apparently in predicting the nucleic loop conformations. In addition, multiple binding pathways of cations to a two-layered 15-mer thrombin-binding quadruplex aptamer have been shown with the use of a sum of microsecond molecular simulation (Reshetnikov et al., 2011). Although monovalent cations especially K^+ packed between G-tetrad layers have been accepted as a leading co-factor for the stability of G-quadruplex, little is known about the binding process of cations to any types of human telomeric intra-molecular G-quadruplex.

To study the cation binding process to human telomeric d[TAG_3($TTAGGG)_3$] that forms an intra-molecular antiparallel G-quadruplex, we carried out MD simulations on the interaction between potassium and G-quadruplex in solution. The G-quadruplex has three backbone strands in one direction and one strand in the opposite direction, and the hybrid [3 + 1] G-tetrads are in turn connected by the double-chain-reversal, edge-wise and edge-wise loops (Fig. 1). To discriminate from another hybrid G-quadruplex with altered loop types, the structure analyzed here was named as hybrid [3 + 1] form-one G-quadruplex. AMBER12 software and parmbsc0 combined FF99SB force field were applied, which has been demonstrated to provide closest G-quadruplex loops conformations and valued for DNA and G-quadruplex in many of nanoseconds state-of-the-art molecular dynamics simulations in aqueous solution (Fadrná et al., 2009; Pérez et al., 2007; Zhu et al., 2013). In the present work, using the popular molecular mechanic Poisson-Boltzmann surface area (MM-PBSA) and principal component analysis (PCA), we identified the preferred binding site and the most dominant motions during binding process, shading a fresh light on potential mechanisms of regulating the stabilities of telomeric intra-molecular G-quadruplex by physiological ions.

RESULTS

Binding process of the strong stabilizer K^+ with human telomeric hybrid [3 + 1] G-quadruplex was calculated in two independent 300 ns MD simulations (denoted as sim-K1 and sim-K2). The time scale is necessary for the monitoring of binding process and sampling conformations as previously suggested (Reshetnikov et al., 2011; Zhu et al., 2013).

Two pathways corresponding to the top and bottom G-tetrad gates in human hybrid [3 + 1] form-one G-quadruplex were recorded. The gates of central channel were constructed by electronegative O6 atoms of the top and bottom G-tetrad guanines (Fig. 1), and through these gated pathways cations migrated into the central channel of G-quadruplex. The top gate was capped by T1, A2 and the second edge-wise loop bases, and the bottom gate was half capped by bases of T12, T13 and A14. The loop caps prevented direct interactions between G-tetrads and K^+ ions from environment, suggesting that the loop bases blocking the pathways should realign to permit an access of cations to the central channel.

Potassium capture in sim-K1

Two K^+ ions were located in the channel of G-quadruplex. The first K^+ ion became initially bound at about 1 ns and reached quickly to the upper binding site between the top and central G-tetrads. While the second K^+ ion reached the lower binding site between the bottom and central G-tetrads through the bottom pathway, it bound at 159.3 ns and spent about 5.6 ns to accomplish the binding process. The key structures during the binding process were extracted and presented in Fig. 2. The first K^+ bound with the top cap nucleotides at 1.02 ns, and the base of T18 nucleotide in the second edge-wise loop tilted outward to accommodate the hydrated K^+ ion (Fig. 2A). K^+ became dehydrated and passed through the hindrance of cap residues to the space between the cap and top G-tetrad (Fig. 2B). In this process, there was a transient twist of the C3′-O3′ bond in the nucleotide T1 with the ε torsion angle changed from ~170° to 69°, and the bonds of the C1′-N1 between T1 and C1′-N9 in A2 were also twisted to create more space for the passage. The K^+ quickly became completely dehydrated and arrived at the upper binding site at 1.80 ns (Fig. 2C).

With one K^+ bound the G-quadruplex was stable, without apparent conformational changes detected till the second K^+ beginning binding to the bottom cap at 159.30 ns (Fig. 2D). The second K^+ approached to the binding site by the *re* face of the bottom cap nucleotides, where G5 was partly exposed to water bulk. In this process, two water molecules became dehydrated, the C3′-O3′ bond in T13 was twisted with the ε

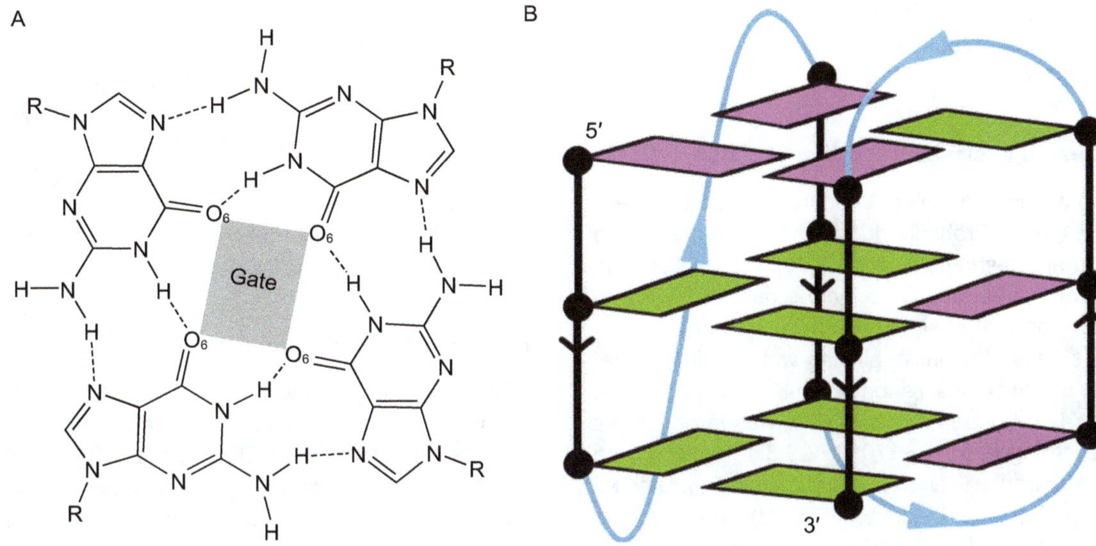

Figure 1. Structure of human telomeric hybrid [3 + 1] form-one G-quadruplex. (A) Top view of the top G-tetrad layer, Hoogsteen hydrogen bonds were denoted as dot lines; (B) side view of the whole schematic structure of G-quadruplex.

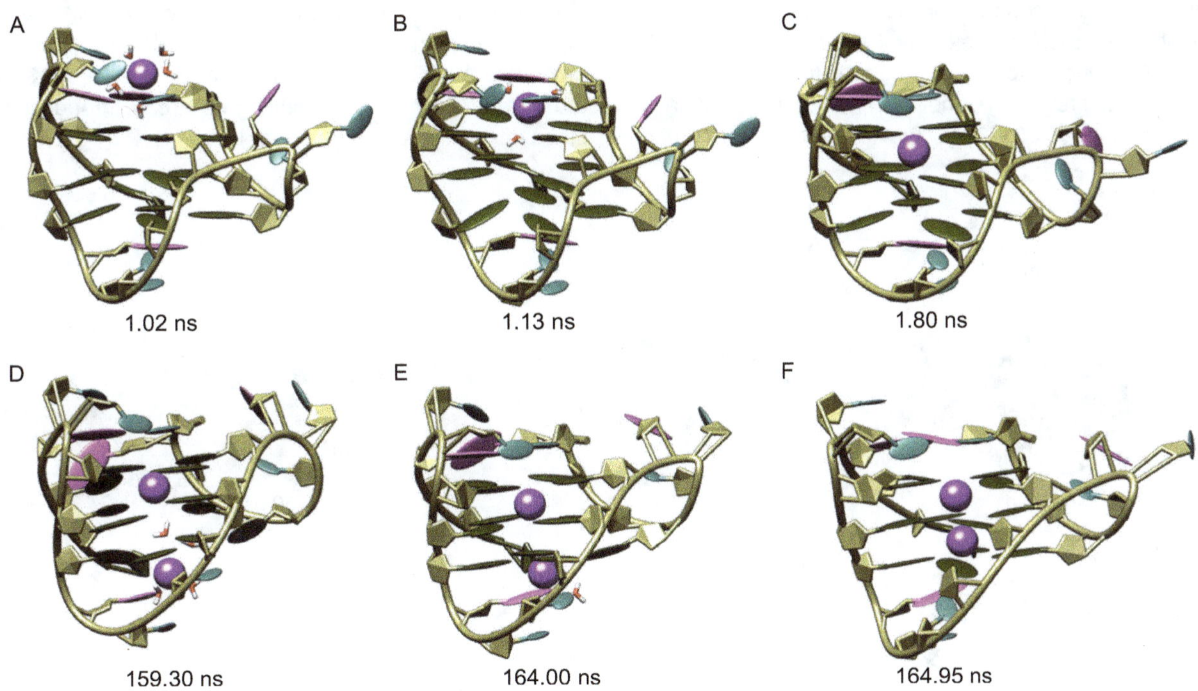

Figure 2. Binding process of potassium ions with human telomeric hybrid [3 + 1] form-one G-quadruplex in sim-K1. The bases of adenine, thymine and guanine nucleotides were colored orchid, cyan and olive green, respectively; potassium ions were represented as sphere and colored purple.

torsion angle changed from ~119° to 90.7°, and the base parts of G5 and T12 were temporarily rotated by the electrostatic interactions between K^+, O6 in G5 and O4 in T12. Bound K^+ became further dehydrated to get closer to the center of bottom G-tetrad gate (Fig. 2E), and finally arrived at the lower binding site at 164.95 ns by losing all of its coordinated waters. The resulting G-quadruplex with two K^+ ions bound seems to be quite stable benefited from internal electrostatic interactions.

Potassium capture in sim-K2

The binding process of K^+ with G-quadruplex in the sim-K2 was shown in Fig. 3. Probably due to the difference of the starting structures (Reshetnikov et al., 2011), the initial binding that occurred at 144.20 ns around the top pathway was significantly later than that in sim-K1 (Fig. 3A). The conformational changes caused by initial binding were similar to that caused by the binding of first K^+ ion in sim-K1, except that the twisting of C3′-O3′ bond in nucleotide T1 was larger (the ε torsion angle changed from ~170° to 20.3°) resulting in a bigger channel entrance to allow relatively less dehydrated K^+ ion to pass through the cap hindrance (Fig. 3B). This suggests the probable mechanism by which the bulky K^+ with five water molecules bound at the upper binding site at 145.10 ns. In addition, the cap entrance was kept open by the coordinated water molecules thereby generating a broad passage directly linking the top G-tetrad gate to environment.

The open state of the cap entrance lasted for about 4.8 ns, during which a second K^+ ion was found to be captured through the same pathway as the first one (Fig. 3C and 3D). However, it was unable to pass through the top gate that was held firmly by the first K^+ ion, and was finally compelled back to environment (Fig. 3E). Then the top pathway entrance was closed, the remaining K^+ with three coordinated waters oscillated between the upper binding site and the geometric center of central G-tetrad layer during the period from ~146 ns to ~187 ns. With time, two waters (the lower water and one of upper coordinated waters) escaped to the bulk from the bottom G-tetrad gate and the wide groove respectively. Finally, by passing through the central G-tetrad, the K^+ reached the lower binding site at 187.60 ns, a conformation achieved till the end of observation (Fig. 3F).

RMSD and G-tetrad gate area

Root mean square deviation (RMSD) of the G-quadruplex reveals the entire structural fluctuations of a nucleotide structure throughout the MD simulation. The RMSDs of two simulations were given in Fig. 4A and 4B. We observed no drastic fluctuations with long smooth flat RMSD curves over 125 ns, suggesting that the resulting G-quadruplexes are considerably stable with no significant changes. Fluctuations seem originated from the effects of water solvent and environmental K^+ ions and the bound K^+ ions that affect involving atoms during the binding process. To understand the directly involved residues in the binding process, the areas of top and bottom G-tetrad gates constructed by their respective guanine O6 atoms were monitored (Fig. 1), using the molsurf module in AMBER12. The presence or absence of K^+ inside the G-tetrad core affected the gate areas through electrostatic interactions (Fig. 4C and 4D).

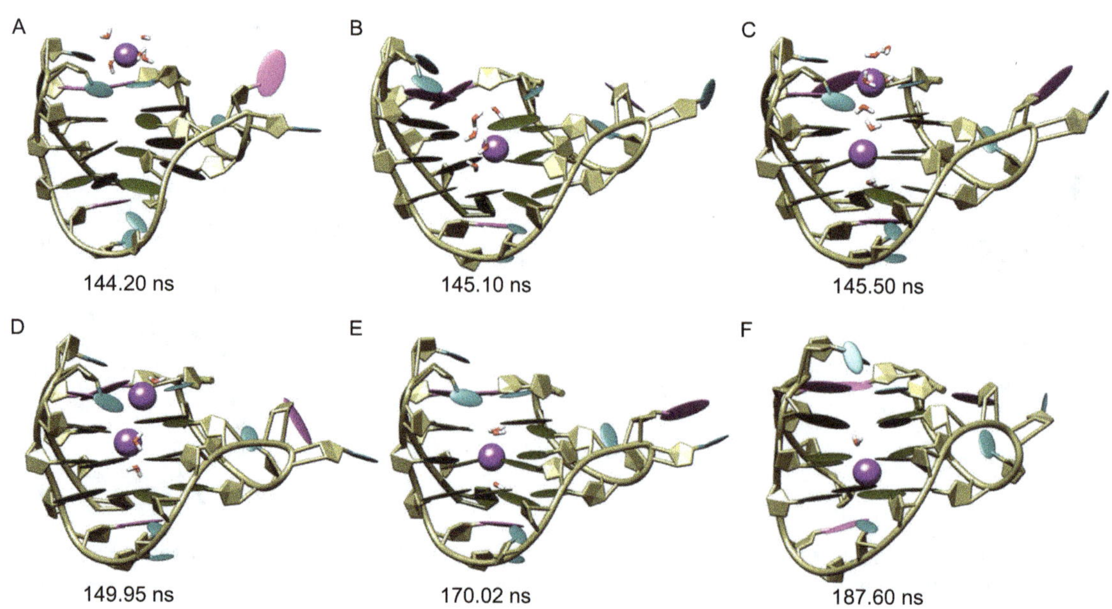

Figure 3. Binding process of potassium ions with human telomeric hybrid [3 + 1] form-one G-quadruplex in sim-K2. The bases of adenine, thymine and guanine nucleotides were colored orchid, cyan and olive green, respectively; potassium ions were represented as sphere and colored purple.

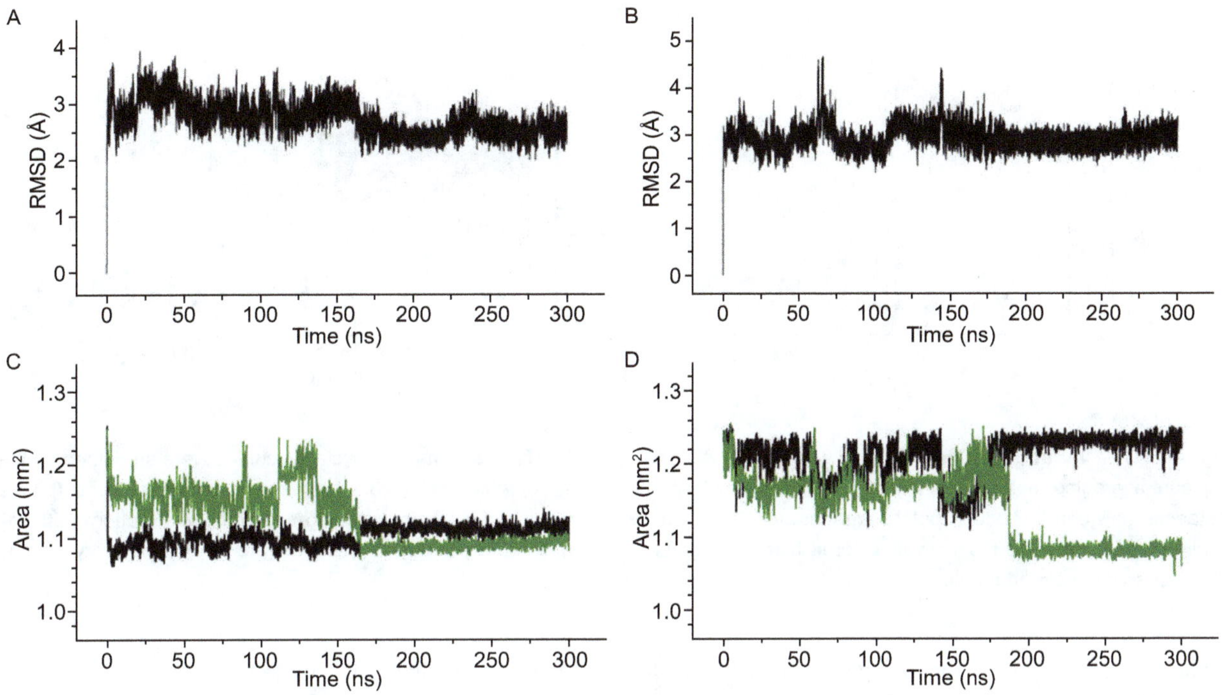

Figure 4. The RMSDs and the fluctuations of G-tetrad gate areas. (A and B) represent the RMSDs of sim-K1 and sim-K2, respectively; (C and D) represent the fluctuations of G-tetrad gate areas, in which the top and bottom gate area were shown as black and green lines, respectively.

In sim-K1, the area of top G-tetrad gate showed a sudden decrease at the starting time, since G-quadruplex get initial binding by the first K^+ ion at 1.02 ns. At the meantime, the bottom gate area also decreased slightly by the relatively weak electrostatic attraction (Fig. 4C). The area of top gate did not vary significantly till the second K^+ ion moved to the inside of bottom cap at 164 ns. While the top area became increased by about 0.04 nm^2 due to the upward shift of the first K^+ ion pushed by electrostatic repulsion from the second K^+ ion, the bottom G-tetrad gate area decreased significantly by the binding of the second K^+ ion through bottom pathway. The temporary increase in the bottom G-tetrad gate area between 111.64 ns and 136.59 ns corresponded to the inward tilt of G5 and G11 bases, the Hoogsteen hydrogen bonds between G5 and G11 bases destroyed in this period, so the structures were instable and reverted to the common form. No matter the G-quadruplex is in a free form or becomes completely bound by two K^+ ions, the top gate area appeared larger than the bottom one, suggesting that the top gate is a preferred K^+ binding pathway.

The large variations of gate areas before 144.20 ns in sim-K2 corresponded to transient binding of K^+ ions to the outside of G-quadruplex gates and subsequent quick dissociation. Accompanied with the effective binding of K^+ through the top pathway, the top gate area became drastically decreased at about 145 ns and achieved to a minimum at about 150 ns as the coordinated water dehydrated (Fig. 3D and 4D). Then the top gate area gradually increased and finally recovered close to the initial size in a synchronized manner with the location shift of bound K^+ ion from the upper binding site to the lower binding site (Fig. 3E and 3F). In contrast, the lower gate area increased by the tilt of G5 and G11 bases as the K^+ ion shifted, and a sudden decrease occurred when the K^+ achieved the lower binding site. The final area of the lower gate is 1.08 nm^2, near identical to that of sim-K1.

Role of coordinated water

Although water molecules in the cation binding process to nucleic acids play indispensable role, the roles of water molecules in K^+ binding to human telomeres have not been elucidated. We found that the coordinated water molecules of the bound K^+ ions participated in the whole binding processes in both MD simulations (Figs. 2 and 3). To unveil the effects of coordinated waters (within 3.5 Å of K^+ ion) (Reshetnikov et al., 2011), fluctuations of coordinated water numbers of bound K^+ were monitored as K^+ binding proceeded, and their impacts on conformation were analyzed (Fig. 5A). At the meantime, the geometric center of the four O6 atoms in the central G-tetrad (CCG) was set as a reference, and the distances between K^+ ions and CCG across

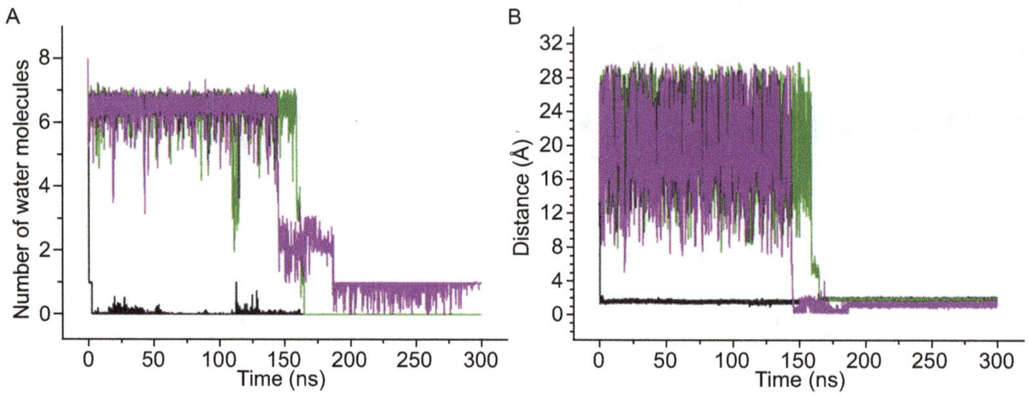

Figure 5. The variations of coordinated waters and distances in sim-K1 and sim-K2. For clarity, curves were smoothed by the adjacent-averaging method with a window of 10 points. (A) The number of waters within 3.5 Å of binding potassium ions; (B) the distances between the bound potassium ions and CCG. The variations in sim-K1 corresponding to the top and bottom pathway binding processes were respectively colored black and green, the variations in sim-K2 corresponding to the top pathway binding processes were colored purple.

the entire simulations (as a direct location indicator of binding K^+ ions) were presented in Fig. 5B.

In sim-K1, the coordinated waters of the first bound K^+ ion decreased significantly as the ion moved towards the CCG through the top pathway, with the number changed quickly from six to one at the beginning of MD simulation (Fig. 5A and 5B). This appeared in accordance with the binding process (Fig. 2), suggesting that cation binding and dehydration processes are synchronous. The last coordinated water molecule returned back to environment through the upside of the wide groove. As shown in Fig. 5A, the intermittent fluctuations of the curve (black) before 159 ns corresponded to one water molecule that occasionally accessed into the detection range. This water molecule formed weak hydrogen bonds with the top G-tetrad and was finally compelled back to environment leading to a more compact G-quadruplex structure. The binding of the second K^+ ion also started with water dehydration at about 159 ns (Fig. 2D). Within a period of about 6 ns, the number of coordinated water molecules quickly changed to four, then to one and zero. The starting time and duration of the dehydration process agreed with the binding process of K^+ ion. In sim-k2, the dehydration process was similar to that in simulation one, except for a much longer dehydration period (over 42 ns), and there was one coordinated water molecule retained in association with bound K^+ till the end of MD.

The variations in distance between bound K^+ ions and CCG were consistent with the binding and dehydration processes as shown in Fig. 5B. In sim-K1, the original distance between the upper binding site and CCG was ~1.58 Å, but when the second K^+ ion achieved the lower binding site, the distance became ~1.96 Å. The increased distance was near identical to the distance between the lower binding site and CCG, suggesting a symmetrical-like binding conformation along the central G-tetrad plane. In sim-K2, while there was only one bound K^+ ion, the traversing conformations corresponded to the zero values in distance (purple curve) (Fig. 3E). The subsequent distance between the binding site and CCG was ~1.39 Å.

Conformational alternations

Dominant concerted motions of G-quadruplex in both ensembles were investigated by PCA. The concerted motions among atoms are expressed by the eigenvectors of the covariance matrix, and the extent of concerted fluctuational motion is expressed by the corresponding eigenvalue. Previous studies have shown that overall fluctuations of macromolecule are accountable by a few low frequency eigenvectors with large eigenvalues (Zhu et al., 2013). The eigenvalues of the first 30 eigenvectors and their relative cumulative deviations of these two G-quadruplex ensembles were shown in Fig. 6A. The majority of conformational fluctuations (>75%) was due to the concerted motions specified by the first six eigenvectors. Contributions to the motions were specified by the first eigenvectors in sim-K1 and sim-K2 (31% and 32%, respectively) (Fig. 6A).

To further investigate the essential motions, we investigated two-dimensional projections defined by the first two eigenvectors of sim-K1 and sim-K2. Both projected trajectories were confined within certain regions (Fig. 6B), suggesting concerted motions. The main regions of the two projections superimposed well, indicative of G-quadruplex structures in similar dominant conformational alternations in the two independent MD ensembles (Fig. 6B). The directions and extents of conformational movements along the first eigenvectors of sim-K1 and sim-K2 were represented by arrows (Fig. 6C and 6D). The most flexible regions consisted of three connecting loops (colored purple), especially the

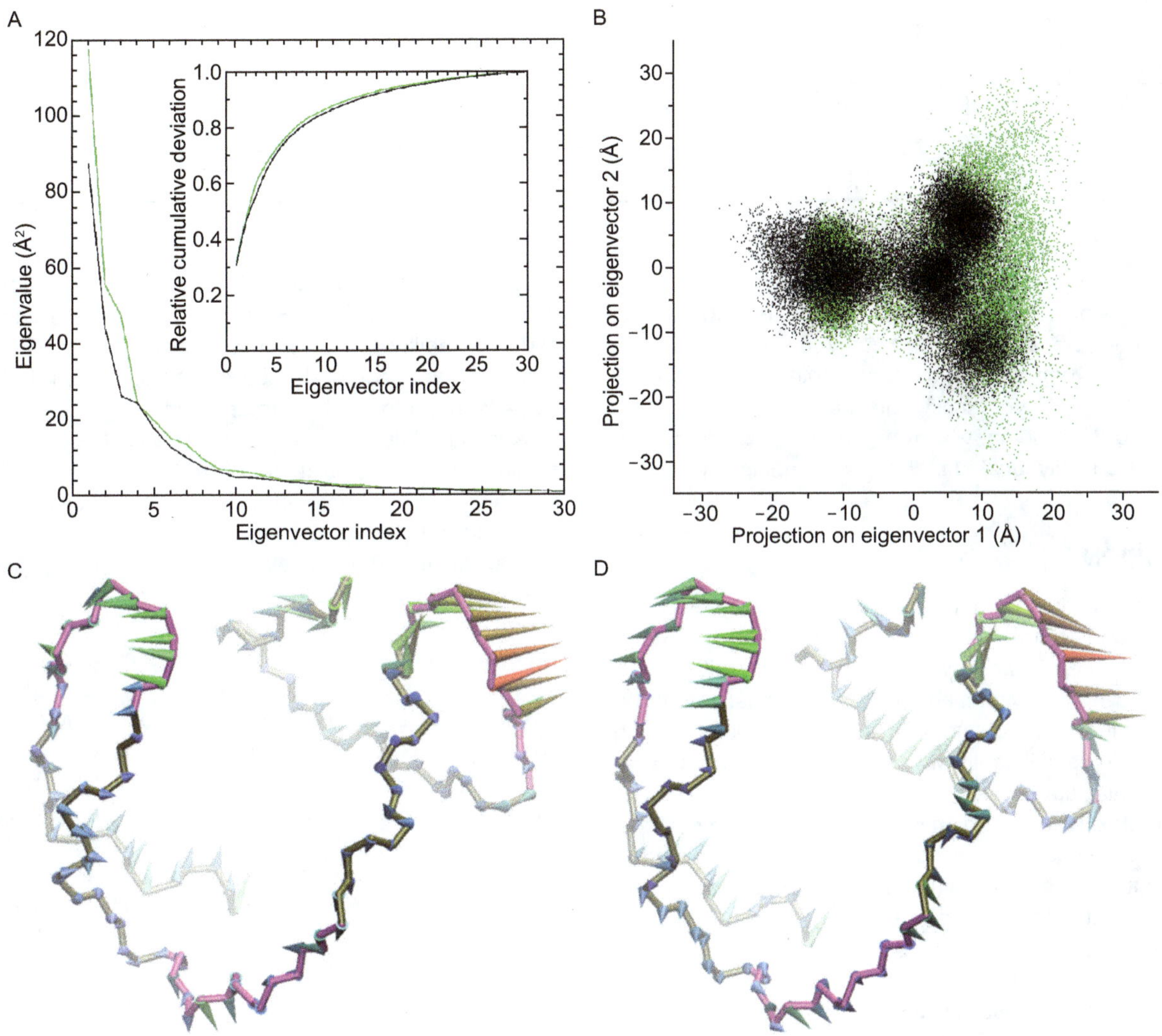

Figure 6. Principal component analysis. (A) Eigenvalue profiles constructed by the first 30 eigenvectors from PCA analysis of sim-K1 and sim-K2, which were represented by black and green lines, respectively. (B) two-dimensional projection of backbone atomic trajectories along the first two eigenvectors, projections in sim-K1 and sim-K2 were represented by black and green dots, respectively. (C and D) respectively show the dominant motions along the first eigenvector in sim-K1 and sim-K2 by porcupine plot.

double-chain-reversal and the second edge-wise loops (Fig. 6C and 6D). In both ensembles, the double-chain-reversal and the second edge-wise loops moved outwards, whereas the first edge-wise loop moved downwards (Fig. 6C and 6D). The similarity of the motion patterns suggests that G-quadruplexes in the two ensembles undergo comparable conformational changes, with the motions contributing to a compact structural conformation.

Binding free energies

The binding free energies of G-quadruplex and K^+ ions in sim-K1 and sim-K2 were summarized in Table 1. MM-PBSA method has been shown to be reliable in correlation with experimental results (Yang et al., 2011). Whereas the major driving force of binding energy is contributed by electrostatic interaction, polar solvation and van del Waals interactions build negative contributions. Consistently, our data showed that in sim-K1, the independent binding free energies from K^+ ions to the upper and lower binding sites were −73.79 and −76.58 kcal·mol^{-1}, respectively (Table 1). In sim-K2, the binding free energy was −35.01 kcal·mol^{-1}, where the negative values indicated that the binding processes are energetically favorable. The difference in large binding energy between sim-K1 and sim-K2 may be caused by their conformational distinctions as the G-quadruplex structure in sim-K1 was more compact than that in sim-K2. Another significant factor for the energy difference was that the

Table 1. Binding free energy calculations for the G-quadruplex-K^+ ion systems

Calculations	ΔE_{ele}	ΔE_{vdW}	$\Delta G_{PB,sol}$	$\Delta G_{np,sol}$	ΔE_{MM}	ΔG_{sol}	$T\Delta S$	ΔG_{bind}
sim1-K_{upper}	−743.60	14.98	643.81	−0.61	−728.62	643.20	−11.63	-73.79
sim1-K_{lower}	−748.20	14.15	647.31	−0.61	−734.05	646.70	−10.77	-76.58
sim2-K	−699.42	9.83	645.27	−0.61	−689.59	644.67	-9.92	-35.01

bound K^+ ion in sim-K2 was incompletely dehydrated, suggesting unfavorable electrostatic interactions between K^+ and G-tetrad O6 atoms. The energy difference of 2.79 kcal·mol^{-1} between lower and upper binding sites in sim-K1 indicated that the lower binding site was more energetically favorable for K^+ binding, which may be the reason that the K^+ in sim-K2 finally located at the lower binding site.

DISCUSSION

The present study shows the detailed dynamic binding processes of potassium ions and human telomeric G-quadruplex. Our results showed that G-quadruplex recruits one or two K^+ ions from environment and that, K^+ destines at the central channel of the G-tetrads. The starting time of binding process varied significantly, which could be attributed to the different initial structures. The initial structures consist of G-quadruplex, K^+ ions and environmental waters in both MD ensembles. Though the hybrid [3 + 1] form-one G-quadruplex structures share the same conformation, the potassium counterions added according to a rough Coulombic potential around G-quadruplex had different locations in sim-K1 and sim-K2, which leaded to the time difference of approaching to the binding sites. But once the K^+ reaches the space between the caps and G-tetrad gates, binding occurs and finishes quickly. So the rate-limiting step would be effective initial binding of potassium. This conclusion is in consistent with the recent findings of cation binding to 15-TBA DNA duplexes (Reshetnikov et al., 2011).

Coordinated water molecules have been found to play an important role in the whole process of K^+ binding to G-tetrads. The findings that partial dehydration occurred as the initial binding took place suggest that the spaces for binding between the caps and G-tetrads cannot accommodate fully hydrated K^+ ion and dehydration is a prerequisite for K^+ binding. In addition, we found that dehydration consistently started from the lateral coordinated waters of K^+ by checking the interval conformations, and the traversing process started with the bottom coordinated water of K^+ through the top pathway, or the top water through bottom pathway. So, the coordinated water molecules appear to play as precursors of K^+ binding in the interior traversing process.

In validating our results with two additional MD simulations using Na^+ as counterion named sim-Na1 and sim-Na2, the binding processes of G-quadruplex and Na^+ shown in Fig. S1 confirmed our findings in that the binding processes of Na^+ and K^+ are similar, and binding of Na^+ and K^+ produces comparable effects on G-quadruplex conformations (Figs. S2 and S3). However, we found that the higher hydration ability of Na^+ than that of K^+ makes Na^+ spend more time to accomplish the traverse process in both pathways from the top or bottom entry gates.

Recently Rajendran et al. succeeded in the detection and single-molecule visualization of the G-hairpin and G-triplex intermediates and have found that ionic concentrations are pivotal to the yield of G-quadruplex and intermediates (Rajendran et al., 2014). They showed that in the concurrent presence of 10 mmol/L Mg^{2+} and 100 mmol/L K^+, the maximum yields of G-triplex (57%) and G-quadruplex (76%) were obtained. In our simulations, the concentration of K^+ (~174 mmol/L), as calculated from the volume of octahedron box and the number of K^+ ions, should allow the full occupation of the G-tetrad core by two K^+ ions according to Rajendran's analysis. Though our MD simulations may cover most key conformational changes of G-quadruplex upon K^+ binding, the time scale limited by current computer hardware and affordable computing time cannot ensure that all simulations have reached their final state. So, our current MD simulations focused on the description of the binding process at atomic level. To recognize the overall functions of the K^+ and Mg^{2+} in the folding mechanism of G-quadruplex, a number of MD simulations form different starting conformations and further investigations on the interactions of metal cations and G-quadruplex intermediates are still needed.

Our MD simulations and MM-PBSA calculations suggest that the two K^+ binding sites are not equivalent and the lower binding site is energetically favorable. The major driving force in K^+ binding to G-quadruplex is electrostatic interactions of K^+ with G-tetrads. When designing G-quadruplex targeted drugs, molecules with respective specificities towards the top and bottom gates of K^+ entries in human telomeric G-quadruplex might be a desirable strategy for improved bioactivities. In addition, the findings in this work may provide some new insight into the structural characteristics of human telomeric intra-molecular G-quadruplex under conditions of ion disorder or electrolyte disturbances s of severe illness of metabolic disorders.

MATERIALS AND METHODS

Structure preparation

The initial structure of human telomeric hybrid [3 + 1] form-one G-quadruplex was retrieved from PDB data bank with the entry ID of

2JSM (Phan et al. 2007a, b). There are ten sets of coordinates for each atom of d[$TAG_3(TTAGGG)_3$], only the first set was retained since they were well superimposed with each other. No cations were existed in this NMR detected structure. In order to investigate the binding process of K^+ ions and keep system in electric neutrality, 22 potassium counterions were added around G-quadruplex by using the AMBER xleap module. The complex was further immersed into the center of a truncated octahedron box of TIP3P water molecules with a margin distance of 12.0 Å. According to this procedure, two independent models were created and subjected to molecular dynamics calculations. Moreover, models using Na^+ as counterion were also built in the same way.

Molecular dynamics simulations

In order to remove bad contacts between nucleotides and surrounding cations and water molecules, each model was energy minimized by steepest descent method for 2000 steps with AMBER sander module, during which K^+ ions and water molecules were set free while all nucleotides of G-quadruplex were restricted by a harmonic constraint of 100 $kcal \cdot mol^{-1} \cdot Å^{-2}$. The system energy was further minimized by conjugate gradient method for 4000 steps with no constraint. Then the system was gradually heated in the NVT ensemble from 0 K to 300 K over a period of 200 ps using Langevin thermostat with coupling coefficient of 1.0 ps and a weak constraint of 10 $kcal \cdot mol^{-1} \cdot Å^{-2}$ on nucleotides, and the system was subjected to an equilibrium simulation for 200 ps by removing all constraints. Finally, a 300 ns production simulation for each model was conducted under NPT ensemble, during which the temperature was kept 300 K by using Berendsen thermostat method with time constant of 1 ps, isotropic constant pressure boundary condition was applied by using Berendsen pressure coupling algorithm with a time constant of 1 ps. Periodic boundary conditions were applied to avoid surface effects. Covalent bonds containing hydrogen atoms were constrained using the SHAKE algorithm, and the long-range electrostatic interactions were treated by the Particle Mesh Ewald (PME) method. Both of the cutoff distances for long-range electrostatic and van der Waals interactions were set to 10.0 Å, time step was set to 2 fs in all of the simulations, coordinates were saved every 1 ps to record the molecular dynamics trajectory.

Principal component analysis

To identify essential degrees of freedom in molecular motions from molecular dynamics (MD) trajectories, principal component analysis was performed for production trajectory by removing the overall translational and rotational motions. Because PCA filters the main modes of collective motion from much more local fluctuations, which in many cases reflect the functions of proteins or nucleotides, PCA was also termed "essential dynamics". The principle of PCA has been described in detail elsewhere (Amadei et al., 1993; Zhu et al., 2013), and it has been widely used in the MD studies of protein and nucleotides.

PCA analysis was carried out by using the method of Interactive Essential Dynamics (IED) with the ptraj module in AMBER12 (Mongan, 2004). The G-quadruplex backbone atoms of the whole 300 ns trajectory were selected for calculation and analysis. Trajectory projections on certain eigenvectors were conducted to show the time-dependent motions that the atoms perform in the corresponding vibrational modes, they can also help determine coupled motions between different eigenvectors. Porcupine plot was generated by using VMD software to show a graphical summary of the motions along a particular eigenvector (Humphrey et al., 1996).

Binding free energy calculations

The binding free energies of potassium ions and G-quadruplex were obtained by using MM-PBSA method in AMBER12, snapshots of the last 100 ns with an interval of 200 ps were extracted for the calculation. The binding free energy (ΔG_{bind}) can be computed by the free energy difference between binding complex ($G_{complex}$) and the sum of G-quadruplex nucleotides (G_{quad}) and ions (G_{ion}) as follows:

$$\Delta G_{bind} = G_{complex} - (G_{quad} + G_{ion}) \quad (1)$$

Each free energy term can be calculated according to the equation:

$$G = E_{MM} + G_{solv} - TS \quad (2)$$

where E_{MM} is the molecular mechanics energy in gas phase, G_{solv} is the solvation free energy, and TS represents the entropy contribution. The E_{MM} term consists of internal strain energy (E_{int}), van der Waals energy (E_{vdW}) and electrostatic energy (E_{ele}):

$$E_{MM} = E_{int} + E_{vdW} + E_{ele} \quad (3)$$

The solvation free energy is divided into a polar part (G_{PB}) and a nonpolar part (G_{np}):

$$G_{solv} = G_{PB,sol} + G_{np,sol} \quad (4)$$

Therefore, according to above equations, ΔG_{bind} can be expressed as:

$$\Delta G_{bind} = \Delta E_{int} + \Delta E_{vdW} + \Delta E_{ele} + \Delta G_{PB,sol} + \Delta G_{np,sol} - T\Delta S \quad (5)$$

In equation 5, internal strain energy difference (ΔE_{int}) equals zero for our system since no contribution difference of bond, angle and torsion takes place upon ion bonding. The polar part of solvation $\Delta G_{PB,\ sol}$ is calculated by solving the Poisson-Boltzmann (PB) equations with the value of 1 and 80 for the interior and exterior dielectric constants respectively (Gilson et al., 1988). The nonpolar part of solvation $\Delta G_{np,\ sol}$ is determined by the solvent accessible surface area (SASA) using AMBER molsurf module according to the following equation

$$\Delta G_{np,sol} = \gamma \Delta SASA + \beta \quad (6)$$

where surface tension γ and offset β were set to the standard 0.00542 and 0.92 $kcal \cdot mol^{-1}$, respectively. The entropy term TΔS was calculated with normal mode analysis using the NMODE module in AMBER12 (Case et al., 2005).

ACKNOWLEDGEMENTS

This work was supported by grants from the the National Basic Research Program (973 Program) (Nos. 2012CB911200 and 2012CB910104) and National Natural Science Foundation of China (Grant No. 31100584).

ABBREVIATIONS

MD, molecular dynamics; MM-PBSA, molecular mechanic Poisson-Boltzmann surface area; PCA, principal component analysis; sim-K1, the first molecular dynamics simulation of G-quadruplex with potassium ions; sim-K2, the second molecular dynamics simulation of G-quadruplex with potassium ions; sim-Na1, the first molecular dynamics simulation of G-quadruplex with sodium ions; sim-Na2, the second molecular dynamics simulation of G-quadruplex with sodium ions.

COMPLIANCE WITH ETHICS GUIDELINES

Zhiguo Wang, Ruping Chen, Ling Hou, Jianfeng Li and Jun-Ping Liu declare that they have no conflict of interest. This article does not contain any studies with human or animal subjects performed by the any of the authors.

REFERENCES

Akhshi P, Acton G, Wu G (2012) Molecular dynamics simulations to provide new insights into the asymmetrical ammonium ion movement inside of the $[d(G3T4G4)]_2$ G-quadruplex DNA structure. J Phys Chem B 116:9363–9370

Amadei A, Linssen A, Berendsen HJ (1993) Essential dynamics of proteins. Proteins 17:412–425

Bejugam M, Sewitz S, Shirude PS, Rodriguez R, Shahid R, Balasubramanian S (2007) Trisubstituted isoalloxazines as a new class of G-quadruplex binding ligands: small molecule regulation of c-kit oncogene expression. J Am Chem Soc 129:12926–12927

Bončina M, Lah J, Prislan I, Vesnaver G (2012) Energetic basis of human telomeric DNA folding into G-quadruplex structures. J Am Chem Soc 134:9657–9663

Burge S, Parkinson GN, Hazel P, Todd AK, Neidle S (2006) Quadruplex DNA: sequence, topology and structure. Nucleic Acids Res 34:5402–5415

Case DA, Cheatham TE, Darden T, Gohlke H, Luo R, Merz KM, Onufriev A, Simmerling C, Wang B, Woods RJ (2005) The Amber biomolecular simulation programs. J Comput Chem 26:1668–1688

Dai J, Dexheimer TS, Chen D, Carver M, Ambrus A, Jones RA, Yang D (2006) An intramolecular G-quadruplex structure with mixed parallel/antiparallel G-strands formed in the human BCL-2 promoter region in solution. J Am Chem Soc 128:1096–1098

Dexheimer TS, Sun D, Hurley LH (2006) Deconvoluting the structural and drug-recognition complexity of the G-quadruplex-forming region upstream of the BCL-2 P1 promoter. J Am Chem Soc 128:5404–5415

Fadrná E, Špačková N, Štefl R, Koča J, Cheatham TE, Šponer J (2004) Molecular dynamics simulations of guanine quadruplex loops: advances and force field limitations. Biophys J 87:227–242

Fadrná E, Špačková N, Sarzynska J, Koca J, Orozco M, Cheatham TE III, Kulinski T, Sponer J (2009) Single stranded loops of quadruplex DNA as key benchmark for testing nucleic acids force fields. J Chem Theory Comput 5:2514–2530

Gaynutdinov TI, Neumann RD, Panyutin IG (2008) Structural polymorphism of intramolecular quadruplex of human telomeric DNA: effect of cations, quadruplex-binding drugs and flanking sequences. Nucleic Acids Res 36:4079–4087

Gilson MK, Sharp KA, Honig BH (1988) Calculating the electrostatic potential of molecules in solution: method and error assessment. J Comput Chem 9:327–335

Humphrey W, Dalke A, Schulten K (1996) VMD: visual molecular dynamics. J Mol Graph 14:33–38

Li H, Cao E, Gisler T (2009) Force-induced unfolding of human telomeric G-quadruplex: a steered molecular dynamics simulation study. Biochem. Biophys Res Commun 379:70–75

Maizels N (2006) Dynamic roles for G4 DNA in the biology of eukaryotic cells. Nat Struct Mol Biol 13:1055–1059

Makarov VL, Hirose Y, Langmore JP (1997) Long G tails at both ends of human chromosomes suggest a C strand degradation mechanism for telomere shortening. Cell 88:657–666

Mongan J (2004) Interactive essential dynamics. J Comput Aid Mol Des 18:433–436

Moyzis RK, Buckingham JM, Cram LS, Dani M, Deaven LL, Jones MD, Meyne J, Ratliff RL, Wu JR (1988) A highly conserved repetitive DNA sequence, (TTAGGG)n, present at the telomeres of human chromosomes. Proc Natl Acad Sci USA 85:6622–6626

Neidle S, Parkinson GN (2003) The structure of telomeric DNA. Curr Opin Struc Biol 13:275–283

Oganesian L, Bryan TM (2007) Physiological relevance of telomeric G-quadruplex formation: a potential drug target. Bioessays 29:155–165

Paeschke K, Simonsson T, Postberg J, Rhodes D, Lipps HJ (2005) Telomere end-binding proteins control the formation of G-quadruplex DNA structures in vivo. Nat Struct Mol Biol 12:847–854

Pérez A, Marchán I, Svozil D, Sponer J, Cheatham TE III, Laughton CA, Orozco M (2007) Refinement of the AMBER force field for nucleic acids: improving the description of alpha/gamma conformers. Biophys J 92:3817–3829

Phan AT, Luu KN, Patel DJ (2006) Different loop arrangements of intramolecular human telomeric (3+1) G-quadruplexes in K^+ solution. Nucleic Acids Res 34:5715–5719

Phan AT, Kuryavyi V, Burge S, Neidle S, Patel DJ (2007a) Structure of an unprecedented G-quadruplex scaffold in the human c-kit promoter. J Am Chem Soc 129:4386–4392

Phan AT, Kuryavyi V, Luu KN, Patel DJ (2007b) Structure of two intramolecular G-quadruplexes formed by natural human telomere sequences in K^+ solution. Nucleic Acids Res 35:6517–6525

Pinnavaia T, Marshall CL, Mettler CM, Fisk CL, Miles HT, Becker ED (1978) Alkali metal ion specificity in the solution ordering of a nucleotide, 5′-guanosine monophosphate. J Am Chem Soc 100:3625–3627

Rajendran A, Endo M, Hidaka K, Sugiyama H (2014) Direct and single-molecule visualization of the solution-state structures of G-hairpin and G-triplex intermediates. Angew Chem Int Ed 53:4107–4112

Rankin S, Reszka AP, Huppert J, Zloh M, Parkinson GN, Todd AK, Ladame S, Balasubramanian S, Neidle S (2005) Putative DNA quadruplex formation within the human c-kit oncogene. J Am Chem Soc 127:10584–10589

Reshetnikov RV, Sponer J, Rassokhina OI, Kopylov AM, Tsvetkov PO, Makarov AA, Golovin AV (2011) Cation binding to 15-TBA quadruplex DNA is a multiple-pathway cation-dependent process. Nucleic Acids Res 39:9789–9802

Rujan IN, Meleney JC, Bolton PH (2005) Vertebrate telomere repeat DNAs favor external loop propeller quadruplex structures in the presence of high concentrations of potassium. Nucleic Acids Res 33:2022–2031

Siddiqui-Jain A, Grand CL, Bearss DJ, Hurley LH (2002) Direct evidence for a G-quadruplex in a promoter region and its targeting with a small molecule to repress c-MYC transcription. Proc Natl Acad Sci USA 99:11593–11598

Špačková N, Berger I, Šponer J (1999) Nanosecond molecular dynamics simulations of parallel and antiparallel guanine quadruplex DNA molecules. J Am Chem Soc 121:5519–5534

Špačková N, Berger I, Šponer J (2001) Structural dynamics and cation interactions of DNA quadruplex molecules containing mixed guanine/cytosine quartets revealed by large-scale MD simulations. J Am Chem Soc 123:3295–3307

Williamson JR (1994) G-quartet structures in telomeric DNA. Annu Rev Biophys Biomol Struct 23:703–730

Yang Y, Shen Y, Liu H, Yao X (2011) Molecular dynamics simulation and free energy calculation studies of the binding mechanism of allosteric inhibitors with p38α MAP kinase. J Chem Inf Model 51:3235–3246

Zhu H, Xiao S, Liang H (2013) Structural dynamics of human telomeric G-quadruplex loops studied by molecular dynamics simulations. PLoS One 8:e71380

8

Structural insight into mechanisms for dynamic regulation of PKM2

Ping Wang[1], **Chang Sun**[1], **Tingting Zhu**[1], **Yanhui Xu**[1,2]✉

[1] Fudan University Shanghai Cancer Center and Institute of Biomedical Sciences, Shanghai Medical College of Fudan University, Shanghai 200032, China
[2] State Key Laboratory of Genetic Engineering, School of Life Sciences, Fudan University, Shanghai 200433, China
✉ Correspondence: xuyh@fudan.edu.cn (Y. Xu)

ABSTRACT

Pyruvate kinase isoform M2 (PKM2) converts phosphoenolpyruvate (PEP) to pyruvate and plays an important role in cancer metabolism. Here, we show that post-translational modifications and a patient-derived mutation regulate pyruvate kinase activity of PKM2 through modulating the conformation of the PKM2 tetramer. We determined crystal structures of human PKM2 mutants and proposed a "seesaw" model to illustrate conformational changes between an inactive T-state and an active R-state tetramers of PKM2. Biochemical and structural analyses demonstrate that PKM2^{Y105E} (phosphorylation mimic of Y105) decreases pyruvate kinase activity by inhibiting FBP (fructose 1,6-bisphosphate)-induced R-state formation, and PKM2^{K305Q} (acetylation mimic of K305) abolishes the activity by hindering tetramer formation. K422R, a patient-derived mutation of PKM2, favors a stable, inactive T-state tetramer because of strong intermolecular interactions. Our study reveals the mechanism for dynamic regulation of PKM2 by post-translational modifications and a patient-derived mutation and provides a structural basis for further investigation of other modifications and mutations of PKM2 yet to be discovered.

KEYWORDS pyruvate kinase M2, crystal structure, allosteric regulation, Warburg effect, post-translational modifications

Ping Wang and Chang Sun have contributed equally to this work.

INTRODUCTION

Most somatic cells produce energy predominantly through oxidative phosphorylation, whereas cancer cells produce energy mainly through the less efficient glycolytic pathway, followed by lactic acid production under aerobic conditions (Warburg, 1956). This aerobic glycolysis (also known as the Warburg effect) plays an important role in tumorigenesis (Vander Heiden et al., 2009; Koppenol et al., 2011; Chaneton and Gottlieb, 2012; Yang and Lu, 2013; Wong et al., 2015). Pyruvate kinase transfers one phosphate group from phosphoenolpyruvate (PEP) to adenosine di-phosphate (ADP), and thus generates one pyruvate molecule and one adenosine tri-phosphate (ATP) molecule. This step is one of the rate-limiting steps of glycolysis, and therefore pyruvate kinase is one of the primary mediators of glycolysis.

Four mammalian isoforms of pyruvate kinase-PKL, PKR, PKM1 and PKM2-are expressed in various types of cells and tissues. PKL is expressed in the liver, and PKR is expressed in the red blood cells. PKM1 and PKM2 are derived from alternative splicing of the *PKM* gene (Noguchi et al., 1986; Noguchi et al., 1987). PKM1 is a constitutively active isoform expressed in differentiated cells from various tissues. In contrast, PKM2 has low basal activity and is activated by an effector molecule fructose 1,6-bisphosphate (FBP). PKM2 is expressed in most proliferating cells (Christofk et al., 2008a). Furthermore, the regulation of PKM2 pyruvate kinase activity plays an essential role in cancer metabolism and is crucial for the growth and survival of cancer cells (Chaneton and Gottlieb, 2012; Yang and Lu, 2013; Wong et al., 2015). Recently, PKM2 has been reported to function as a nuclear protein kinase to regulate gene transcription and promote tumorigenesis (Gao et al., 2012; Yang et al., 2012a, b; Gao et al., 2013; Lv et al., 2013; Keller et al., 2014).

PKM2 exists in a dynamic population of monomer, dimer and tetramer and its pyruvate kinase activity relies on the formation of the tetramer. Upon formation of the tetramer, PKM2 can adopt the inactive T-state or active R-state conformation (Morgan et al., 2013). The pyruvate kinase activity of PKM2 is regulated by metabolic intermediates and post-translational modifications. For example, metabolic intermediates such as FBP and *N*-succinyl-5-aminoimidazole-4-carboxamide ribose-5′-phosphate (SAICAR) increase the pyruvate kinase activity of PKM2 (Dombrauckas et al., 2005; Chaneton et al., 2012; Keller et al., 2012). Synthetic compounds have been screened for their ability to activate PKM2 and, thus, suppress cancer growth (Boxer et al., 2010; Jiang et al., 2010; Anastasiou et al., 2012; Kung et al., 2012). Some of these compounds have been shown to promote tetramer formation of PKM2 and enhance its enzymatic activity (Anastasiou et al., 2012). In addition, serine and phenylalanine have been reported to bind to the allosteric site of PKM2 and activate or inhibit its activity, respectively (Chaneton et al., 2012; Morgan et al., 2013).

Post-translational modifications including phosphorylation, oxidation, hydroxylation and acetylation also regulate PKM2 activity to promote cancer proliferation. For example, acetylation of residue K305 inhibits pyruvate kinase activity of PKM2 *in vitro* (Lv et al., 2011) and acetylation of residue K433 affects FBP binding and prevents PKM2 activation (Lv et al., 2013). Phosphorylation of residue Y105 inhibits the tetramer formation and pyruvate kinase activity of PKM2 (Hitosugi et al., 2009). Furthermore, oxidation of residue C358 inhibits PKM2 activity and promotes the metabolic changes required for proliferation (Anastasiou et al., 2011). PKM2 hydroxylation of P403 and P408 promotes HIF-1 transactivation in cancer cells (Luo et al., 2011). The mutation R399E of PKM2 ($PKM2^{R399E}$) was shown to disrupt the tetramer formation on one of dimer interfaces, thereby producing dimers and decreasing its pyruvate kinase activity (Gao et al., 2012). In addition, mutations K422R and H391Y of PKM2 ($PKM2^{K422R}$ and $PKM2^{H391Y}$) were shown to decrease its pyruvate kinase activity in Bloom Syndrome (BS) patients, who are prone to cancer (Anitha et al., 2004; Akhtar et al., 2009; Gupta et al., 2010; Iqbal et al., 2014).

Although PKM2 has been studied for decades, how its activity is regulated remains poorly understood. In this study, we revealed how pyruvate kinase activity of PKM2 is regulated by post-translational modifications and a patient-derived mutation. On the basis of our observations, we propose a model for dynamic regulation of PKM2. Our study also provides a structural basis for further investigation of dynamic regulation of PKM2 by other post-translational modifications and mutations involved in cancer metabolism.

RESULTS

Effects of post-translational modifications and a patient-derived mutation on PKM2 activity

To investigate how the enzymatic activity of PKM2 is regulated, we purified wild-type PKM2 ($PKM2^{WT}$), $PKM2^{R399E}$ (a dimeric mutant) (Gao et al., 2012) and $PKM2^{K422R}$ (a patient-derived mutation). Given the difficulty to obtain phosphorylated or acetylated PKM2 proteins, we purified $PKM2^{Y105E}$ (a phosphorylation mimic of Y105) and $PKM2^{K305Q}$ (an acetylation mimic of K305) to mimic PKM2 containing the two modifications. The above four PKM2 mutants and $PKM2^{WT}$ were used for enzymatic activity assays and structural studies (Fig. S1A). We first measured the enzymatic activities for wild-type and mutants of PKM2 and calculated the Km values for PEP. The kinetic activities were calculated based on the PEP saturation curves in the absence or presence of FBP, an allosteric activator of PKM2 (Fig. 1A and 1B). We also calculated the normalized pyruvate kinase activity (kcat/Km, $(mol/L)^{-1}$ s^{-1}), as represented by the value of Vmax/Km of $PKM2^{WT}$, because an equal amount of protein was used for all activity assays (Fig. 1C). We could not detect protein kinase activity for human PKM2 under our experimental conditions using histone H3 as substrate (data not shown). Therefore, only pyruvate kinase activities (enzymatic activity hereafter if not specified) were measured for wild-type and mutants of PKM2. A high enzyme concentration in the reaction would greatly decrease the accuracy of calculated reaction rate. Because of the extremely fast reaction rate (less than 1.5 min), we measured the pyruvate kinase activity in nanomolar concentrations.

In the presence of FBP, $PKM2^{WT}$ showed a significant increase in enzymatic activity (compared to that in the absence of FBP) with a normalized kcat/Km of 5.6 and a decrease in Km to 0.2 mM (Fig. 1A–C and Table S1). $PKM2^{K305Q}$ showed barely detectable pyruvate kinase activity in the absence of FBP, and a normalized kcat/Km of 0.2 in the presence of FBP, indicating a significant decrease in activity. Compared to $PKM2^{WT}$, $PKM2^{R399E}$ showed a decrease in activity in the absence or presence of FBP, with a kcat/Km of 0.6 or 1.8, respectively. Interestingly, $PKM2^{Y105E}$ showed a slight decrease in activity in the presence, but not the absence, of FBP, with a kcat/Km of 1.2 in the absence of FBP and 4.3 in the presence of FBP. In contrast, $PKM2^{K422R}$ significantly decreased the enzymatic activity (kcat/Km of 0.2) and substrate binding affinity only in the absence of FBP. Taken together, these mutations showed different effects on pyruvate kinase activity of PKM2.

Enzymatic activities of PKM2 mutants are not positively correlated with PKM2 tetramer formation

Tetramer formation is essential for the pyruvate kinase activity of PKM2 (Dombrauckas et al., 2005). The gel-filtration analyses of wild-type and mutants of PKM2 show a mixed population of PKM2 in monomer, dimer and tetramer (Figs. 1D, 1E and S1B). $PKM2^{WT}$ appears to preferentially form a dimer in the absence of FBP and tends to form a tetramer in the presence of FBP. Compared to $PKM2^{WT}$, $PKM2^{Y105E}$ and $PKM2^{R399E}$ showed similar patterns of oligomerization in the absence of FBP. In contrast,

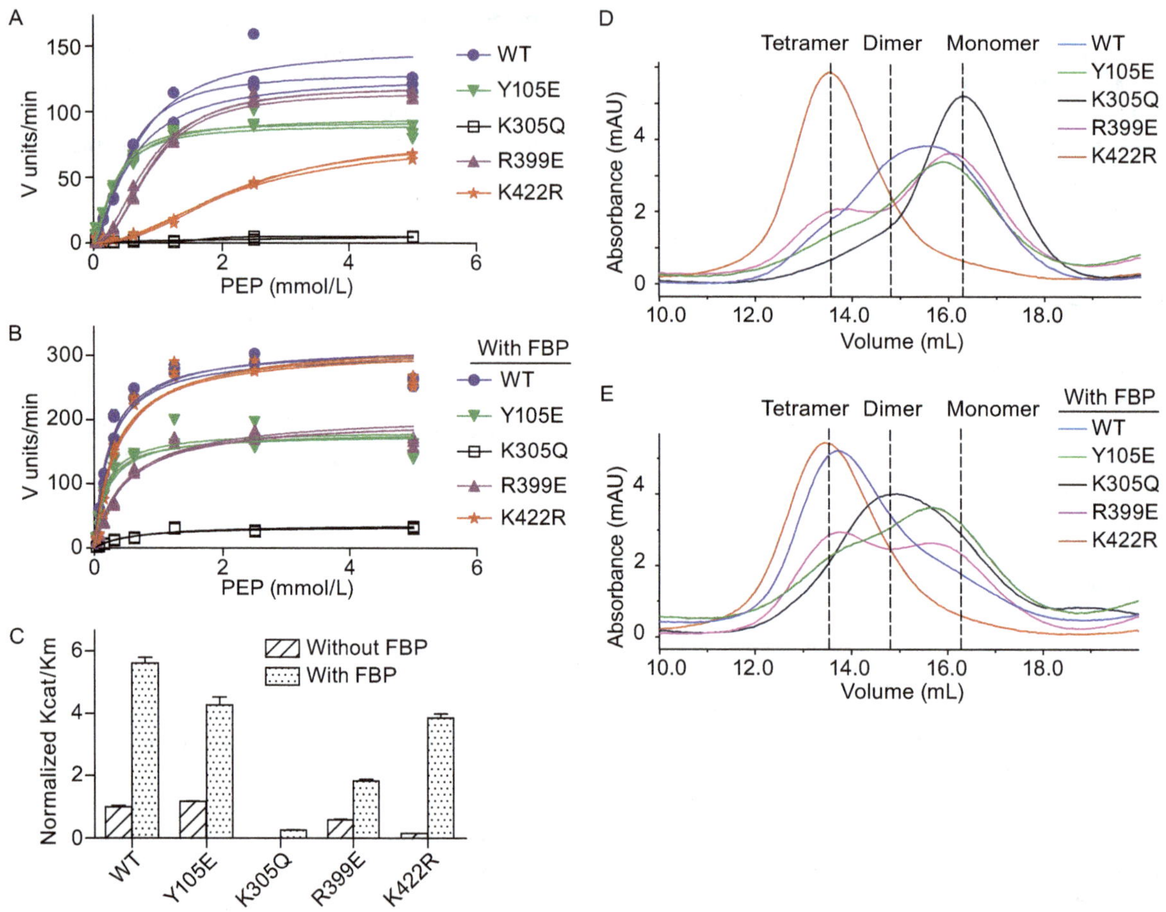

Figure 1. Activities and tetramer formation for wild-type PKM2 and mutants of PKM2. (A and B) Kinetic activities of PKM2 proteins in the absence (A) or presence (B) of FBP. The data were fit with the Allosteric Sigmoidal Equation (A) or the Michaelis-Menten Equation (B). (C) Normalized activities (kcat/Km) of wild-type PKM2 and mutants of PKM2 calculated according to the results from Fig. 1A and 1B with the value (kcat/Km) of PKM2^{WT} as a standard. The error bars represent mean ± SD for triplicate experiments. (D and E). Gel filtration of wild-type and mutants of PKM2 in the absence (D) or presence (E) of FBP. Peak positions of monomer, dimer and tetramer are indicated as dashed lines. The color scheme for PKM2 proteins is indicated. Superdex 200 (GE Healthcare, 10/300 GL) was used in gel-filtration analyses.

PKM2^{K305Q} tended to form a monomer whereas PKM2^{K422R} mainly formed a tetramer in the absence of FBP. In the presence of FBP, PKM2^{Y105E} and PKM2^{R399E} only slightly increased their tendency to form a tetramer, whereas PKM2^{K305Q} only formed a dimer; PKM2^{K422R} maintained its tendency to form the tetramer. The above observations are generally consistent with the prediction that the more tetramer that forms, the more active the protein is. However, exceptions exist. For example, in the absence of FBP, PKM2^{K422R} showed a significantly high level of tetramer formation, but was much less active than PKM2^{WT}. The results indicate that tetramer formation alone cannot explain how PKM2 activity is regulated by these mutations.

It was also noted that PKM2^{WT} formed a tetramer in a concentration- and time-dependent manner (Fig. S1C and S1D). As indicated by size exclusive chromatography, the higher concentration of protein used, the more tetramer formed (Fig. S1C). Interestingly, monomeric PKM2 eluted when either tetrameric or monomeric PKM2 was re-subjected to size exclusion chromatography under similar experimental conditions (Fig. S1D). To minimize concentration- and time-dependent confounding factors, we compared the oligomerization states of PKM2^{WT} and all the mutants under similar protein concentrations and experimental conditions.

Enzymatic activities of PKM2 mutants are not positively correlated with their thermal stability

The thermal stability of PKM2 was reported to reflect its enzymatic activity (Morgan et al., 2010). Thus, we next performed a thermal-shift assay to investigate the correlation between enzymatic activity and thermal stability of PKM2 proteins. $PKM2^{WT}$ and $PKM2^{Y105E}$ showed two peaks (Tm: ~44°C and ~60°C) in the absence of FBP and a single peak (Tm value of ~59°C) in the presence of FBP (Fig. S1E, Table S1). Compared to $PKM2^{WT}$, $PKM2^{K305Q}$ was less stable in the absence (Tm: ~44°C) and presence (Tm: ~54°C) of FBP, which is consistent with its lesser tendency toward tetramer formation and lower enzymatic activity. The Tm values of $PKM2^{R399E}$ (~ 56°C) and $PKM2^{K422R}$ (~ 61°C) were slightly enhanced by the addition of FBP, indicating that both proteins are relatively stable and FBP could not significantly enhance their thermal stabilities. These results indicate that thermal stability is not positively correlated with enzymatic activity or tetramer formation of the wild-type and mutants of PKM2 tested in this study. Taken together, the above results indicate that changes of the pyruvate kinase activity of PKM2 could not be predicted based on the state of oligomerization or thermal stability.

Crystal structures of human PKM2 mutants

To investigate the molecular mechanism for the regulation of PKM2 activity, we determined the crystal structures of $PKM2^{K305Q}$, $PKM2^{Y105E}$, $PKM2^{R399E}$ and $PKM2^{K422R}$ in the absence of FBP, in addition to $PKM2^{K422R}$ in the presence of FBP ($PKM2^{K422R\_FBP}$) (Fig. S2A and Table 1). Although $PKM2^{WT}$ and PKM2 mutants (except $PKM2^{K422R}$) were present in the form of a monomer in a certain percentage in solution with low protein concentration (less than 0.2 mg/mL) (Fig. 1D), all PKM2 proteins formed tetramer in the crystals. The results are consistent with the observation that higher protein concentration leads to more tetramer formation for PKM2 proteins in gel filtration (Fig. S1C). The crystal structures of wild-type PKM2 in the presence of oxalate ($PKM2^{Oxalate}$) (PDB: 3BJT) (Christofk et al., 2008b) and PKM2 in complex with phenylalanine ($PKM2^{Phe}$) (PDB: 4FXJ) (Morgan et al., 2013) were used for the following structural comparison. In addition, we drafted a model for four monomers in the tetramer structure to clarify the following description (Fig. S2A).

The first striking observation from the structural analyses is that $PKM2^{Oxalate}$, $PKM2^{K305Q}$ and $PKM2^{K422R\_FBP}$ adopted the R-state conformation, whereas all other PKM2 mutants adopted the T-state conformation (Fig. S2A). Notably, phenylalanine was previously reported to function as an inhibitor to lock PKM2 in an inactive T-state conformation (Morgan et al., 2013). $PKM2^{Oxalate}$ adopted an R-state conformation because the crystals were obtained in the presence of oxalate (an analog of PEP), which may induce an R-state conformation of PKM2 (Christofk et al., 2008b). Similarly, $PKM2^{K305Q}$ adopted an R-state conformation because the protein was crystallized in the presence of malonate (0.2 mol/L), which was observed in the active site of the $PKM2^{K305Q}$ structure (Fig. S2B). Both oxalate and malonate are analogs of PEP and may favor the formation of the R-state tetramer. These results suggest that wild-type PKM2 and mutants of the PKM2 tetramer tend to adopt the T-state conformation in the apo form but prefer an R-state conformation in the presence of either FBP or an analog of PEP.

A "seesaw" model in the R-/T-state transition of PKM2

Previous studies have demonstrated that a PKM2 tetramer in the R-state (active) is more active than that in the T-state (inactive). Structural comparison indicates that each individual monomer of PKM2 adopts similar fold with a root-mean-squared deviation (rmsd) of less than 0.6 Å for approximately 430 aligned Cα atoms (Fig. S2C). However, structure analysis indicates that the PKM2 tetramer undergoes significant changes between R- and T-state conformations with a rotation about the helix α9 of five degrees for each monomer (Fig. 2B and 2C). This observation is consistent with previous studies of the PKM protein in *Leishmania Mexicana* (Morgan et al., 2010). The PKM2 tetramer is formed through intermolecular interactions between four monomers on large (A-A′) and small (C-C′) interfaces (Fig. 2A). We propose a seesaw model for overall conformational changes during transitions between the R-state ($PKM2^{K305Q}$) and the T-state ($PKM2^{K422R}$) (Fig. 2B and 2C).

On the C-C′ interface, each monomeric PKM2 adopts a rigid conformation and rotates alone with strand β20 (Fig. 2B–D). When PKM2 adopts the T-state conformation, the FBP activation motif (FAM, residues G514–T522) and helix α18 on monomer A interact with FAM and α18 on monomer B on one side of two parallel seesaws. Helices α14/α15 on monomer A and α14/α15 on monomer B are apart from one another on the other side (Figs. 2D and S3). In a process of transitioning from the T-state to the R-state, FAM/α18 on monomer A and FAM/α18 on monomer B move away from one another, and helices α14/α15 on monomer A and α14/α15 on monomer B move closer to one another. Consistent with the above observations, the distance between residues M525 (β20) on monomer A and M525 (β20) on monomer B is approximately 5.0 Å in either the R- or T-state tetramer (Table S2). In contrast, the distance between residues L488 (in α18) on monomer A and L488 (in α18) on monomer B is ~25.9 Å in the T-state and ~27.4 Å in the R-state conformation, respectively; the distance between residues E397 (in α14) on monomer A and E397 (in α14) on monomer B is ~20.7 Å in T-state and ~14.0 Å in the R-state conformation.

On the A-A′ interface, each monomeric PKM2 rotates alone with helix α9 (Fig. 2E). When PKM2 adopts the R-state conformation, oxalate is recognized in the active site by residues R294, G295 and D296 of α8 on monomer A. The three residues are stabilized by residue R342 of α11 on monomer C through forming two intermolecular hydrogen bonds. In the process of an R- to T-state transition, helices α10/α11/α13 on monomer A move away from α10/α11/α13 on monomer C, whereas helices α14/α15/α16 on monomer A move closer to α14/α15/α16 on monomer C. The

Table 1. Crystallographic data and structure refinement statistics

Protein	$PKM2^{Y105E}$	$PKM2^{K305Q}$	$PKM2^{R399E}$	$PKM2^{K422R}$	$PKM2^{K422R\_FBP}$
Data collection					
Wavelength (Å)	0.97927	0.97927	0.97923	0.97923	0.97923
Resolution range (Å)	50–3.2 (3.31–3.20)[a]	50–2.3 (2.38–2.30)	50–2.4 (2.49– 2.40)	50–2.3 (2.38–2.30)	50–2.6 (2.69–2.60)
Space group	*P* 31	*P* 1 21 1	*P* 1 21 1	*P* 1 21 1	*P* 1 21 1
Unit cell (Å, °)	124.5, 124.5, 257.0, 90.0, 90.0, 120.0	94.8, 117.4, 110.3, 90.0, 113.2, 90.0	95.6, 71.2, 170.3, 90, 104.3, 90	97.9, 70.9, 169.5, 90.0, 100.2, 90.0	81.7, 152.6, 97.7, 90.0, 104.2, 90.0
Total reflections	610595	610809	513532	581223	411179
Unique reflections	72016	98003	87781	98424	70893
Multiplicity	8.5 (8.3)	6.2 (6.3)	5.8 (5.8)	5.9 (5.9)	5.6 (5.8)
Completeness (%)	98.72 (88.38)	99.52 (95.59)	98.30 (84.27)	96.72 (89.08)	98.9 (100)
Mean *I*/sigma (*I*)	5.56 (2.87)	11.35 (3.33)	7.34 (3.91)	7.47 (2.93)	35.0 (2.68)
R-sym	0.192 (0.768)	0.098 (0.552)	0.166 (0.83)	0.137 (0.676)	0.084 (0.854)
Structure refinement					
R-factor	0.2202	0.2131	0.2101	0.1974	0.2759
R-free[b]	0.2652	0.2601	0.2562	0.2397	0.3036
RMS (bonds)	0.010	0.002	0.002	0.002	0.005
RMS (angles)	1.39	0.63	0.57	0.60	0.87
Average B-factor	40.90	25.20	32.30	37.80	89.70
Ramachandran plot statistics					
Most favored regions (%)	84.2	91.8	92.2	90.7	93.5
Allowed regions (%)	14.1	7.9	7.2	8.6	5.8
Generously allowed regions (%)	1.4	0.3	0.3	0.4	0.4
Disallowed regions (%)	0.3	0	0.3	0.3	0.3

[a] The values for the data in the highest resolution shell are shown in parentheses.

[b] *R*free = ∑Test||*F*obs| − |*F*calc||/∑Test |*F*obs|, where "Test" is a test set of about 5% of the total reflections randomly chosen and set aside prior to refinement for the structure.

conformational change leads to a flip of the side chain on residue R342 on monomer C, which stabilizes residues R294/G295/D296 on monomer A in the R-state conformation. As a result, PKM2 forms an inactive T-state tetramer, in which oxalate is not well recognized in the active site (Fig. 2F). The mutation R342 W of PKM2 abolished its enzymatic activity, further supporting its significance for catalysis (Fig. S1G). Consistent with above observations, the distances between residues K311 (in α9) on monomer A and K311 (in α9) on monomer C are approximate 18.0 Å in the two states (Table S2). In contrast, the distance between residues E373 (in α13) on monomer A and E373 (in α13) on monomer C is ~37.8 Å in the T-state and ~31.1 Å in the R-state conformation; the distance between residues V414 (in α15) on monomer A and V414 (in α15) on monomer C is ~36.3 Å in the T-state and ~40.5 Å in the R-state conformation.

In support of the above conformational changes, the transition of PKM2 from the R-state to the T-state, the distance between β20 on monomer A and β20 on monomer C along with the C-C interface decreases from ~59.6 Å to ~53.6 Å, whereas the distance between α9 on monomer A and α9 on monomer B along with the A-A′ interface increases from ~43.4 Å to ~51.6 Å (Table S2). Regulation of

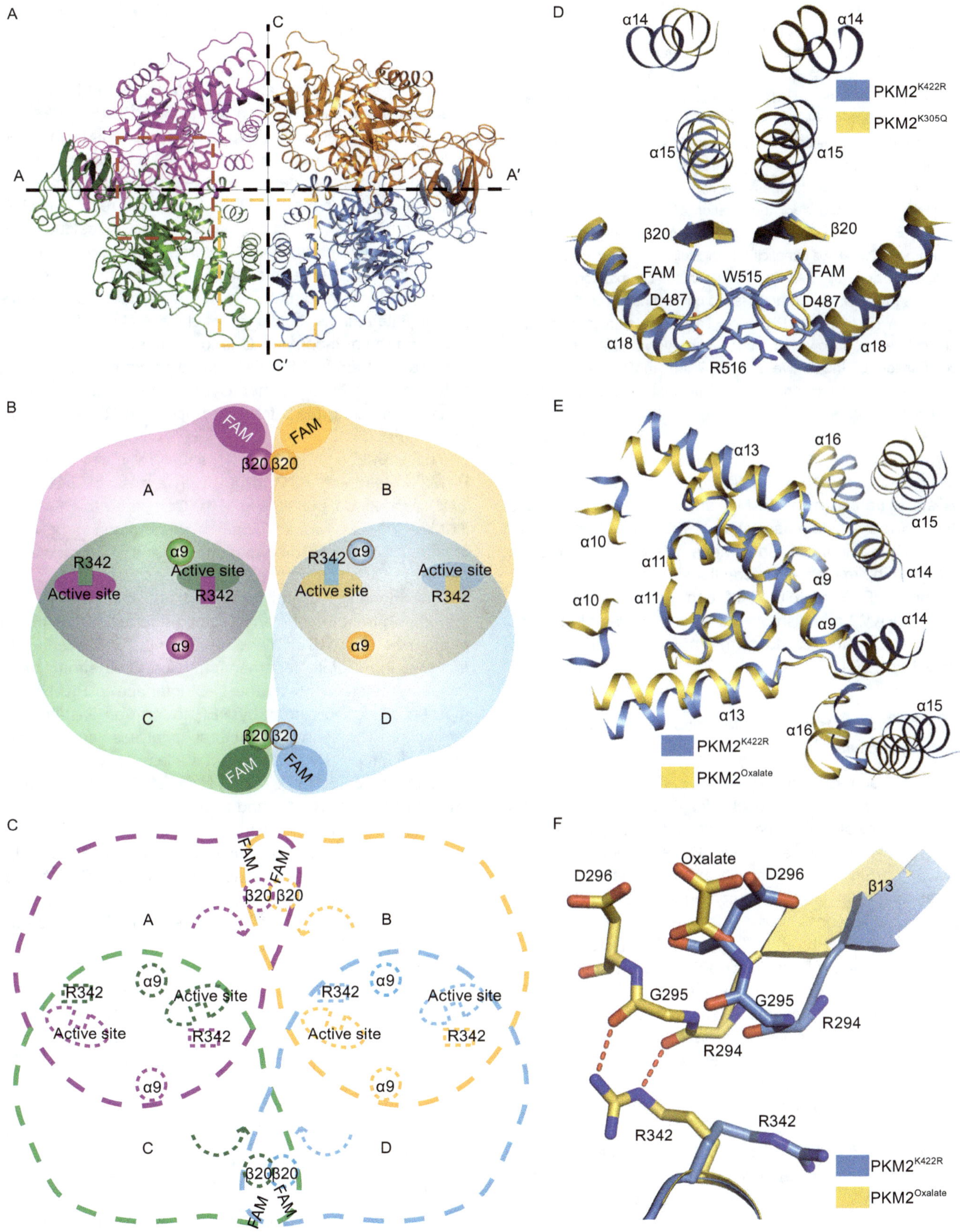
A
C
A
A′
C′
B
FAM
FAM
β20
β20
A
B
α9
R342
Active site
R342
Active site
R342
α9
Active site
Active site
R342
α9
α9
C
D
β20
β20
FAM
FAM
C
FAM
FAM
β20
β20
A
B
α9
R342
Active site
R342
α9
Active site
Active site
R342
Active site
R342
α9
α9
C
D
β20
β20
FAM
FAM
D
α14
α14
PKM2K422R
PKM2K305Q
α15
α15
β20
β20
FAM
W515
FAM
D487
D487
α18
α18
R516
E
α13
α16
α15
α10
α11
α9
α14
α10
α11
α9
α14
α13
α15
α16
PKM2K422R
PKM2Oxalate
F
Oxalate
D296
D296
β13
G295
G295
R294
R294
R342
R342
PKM2K422R
PKM2Oxalate

Figure 2. Crystal structure of human PKM2 and a "seesaw" model for conformational transitions of PKM2 tetramer. (A) Ribbon representation of the human PKM2 structure ($PKM2^{K422R}$) with A-A′ (red box) and C-C′ (yellow box) interfaces indicated as dashed lines. Four monomers are shown in different colors. (B and C) A "seesaw" model for the conformational transitions between the R-state (B) and T-state (C) conformations of the PKM2 tetramer. Critical elements for the conformational changes are indicated. Dashed arrows indicate the directions for the rotation of each monomer from the R- (B) to the T- (C) state. The monomers are indicated as monomer A to D for simplicity in the following description. (D and E) A close-up view for the structural comparison of the PKM2 structure in the R- and T-state conformations on the C-C′ (D) or A-A′ (E) interface. PKM2 in R- and T-state conformations are colored in yellow and blue, respectively. (F) A structural comparison of the active site of PKM2 in R- and T-state tetramers. PKM2 in R- and T-state conformations are colored in yellow and blue, respectively.

intermolecular interactions on either the A-A′ (PEP binding site) or C-C′ (α14/α15 or FAM/α18) interface will lead to an R-/T-state transition of PKM2 and affect its enzymatic activity. For example, association of PEP or its analog oxalate of PKM2 induces an R-state conformation through enhancing intermolecular interactions on the A-A′ interface. In addition, FBP recruits FAM back to the FBP binding pocket of PKM2 and disrupts the intermolecular interactions on the C-C′ interface to form an R-state conformation (Fig. 2B and 2C).

$PKM2^{K305Q}$ disrupts tetramer formation and impairs enzymatic activity

$PKM2^{K305Q}$ adopts a similar R-state conformation to that of $PKM2^{Oxalate}$ with a rmsd of 0.48 Å for 1750 aligned Cα atoms (Fig. 3). The existence of malonate in the condition of crystallization may lead to an active R-state conformation of $PKM2^{K305Q}$ (Fig. S2B). No significant conformational change was observed for the two compared structures (Fig. S2A). However, differences exist between $PKM2^{K305Q}$ and $PKM2^{Oxalate}$. In the $PKM2^{Oxalate}$ structure, two hydrogen bonds are formed between (1) residues K305 (on helix α9 of monomer C) and E384 (on helix α13) on monomer A and (2) K305 and I35 (close to helix α1) on monomer A (Fig. 3). The hydrogen bonds facilitate the intermolecular interactions on the A-A′ interface. No such contact was observed in $PKM2^{K305Q}$ structure.

The FBP binding pocket on the C-C′ interface of $PKM2^{K305Q}$ are structurally similar to that of $PKM2^{Oxalate}$. Consistently, the enzymatic activity of $PKM2^{K305Q}$ could be slightly elevated by FBP (Fig. 1C). Gel-filtration and thermal-shift assays demonstrated that FBP retains its ability to stabilize $PKM2^{K305Q}$ and promote its dimer formation (Figs. 1D, 1E and S1E). Altogether, mutation K305Q or acetylation of residue K305 inhibits PKM2 activity through hindering the tetramer formation on A-A′ interface.

$PKM2^{K422R}$ forms a tetramer in the T-state in the absence of FBP and the R-state in the presence of FBP

Different from $PKM2^{Oxalate}$ or $PKM2^{K305Q}$, $PKM2^{K422R}$ adopts a T-state conformation in the absence of FBP and an R-state conformation in the presence of FBP (Fig. S2A). In the $PKM2^{Oxalate}$ structure, two hydrogen bonds are formed between residue K422 (on monomer C) and residues P403 and Y444 (on monomer D), whereas FAM is flexible and invisible (Fig. 3). In the $PKM2^{K422R}$ structure, residue K422R on monomer C forms two hydrogen bonds with residue E418 on monomer D. Intermolecular interactions are mediated by hydrogen bonds between residue D487 on α18 of monomer C and W515 and R516 on FAM of monomer D, and hydrophobic interactions between residues W482, W515 and P517 on both monomer C and monomer D. The FBP association stabilizes the FAM and disrupts the interactions between FAM on monomer C and α18 on monomer D, therefore allowing the tetramer to adopt an R-state conformation (Fig. 3).

The $PKM2^{K422R}$ tetramer is structurally similar to the $PKM2^{Phe}$ tetramer (PDB: 4FXJ) with a rmsd of 0.41 Å for 1720 aligned Cα atoms (Fig. 4A). Compared to $PKM2^{Phe}$, $PKM2^{K422R}$ has two additional hydrogen bonds between residue K422R on α15 on monomer C and E418 on α15 on monomer D, which may lead to the dissociation of helices α14 on monomer C and helices α15 on monomer D, and therefore strengthen intermolecular interactions (α15 on monomer C and α15 on monomer D) on the C-C′ interface to facilitate the T-state tetramer formation. The strong intermolecular interactions on the C-C′ interface in $PKM2^{K422R}$ may be difficult to disrupt by PEP to form an R-state conformation. The above structural analyses support the observations that $PKM2^{K422R}$ forms a less active tetramer in solution (compared to $PKM2^{WT}$) but is stable in the thermal-shift assay (Figs. 1C, 1D and S1E).

Although $PKM2^{K422R}$ tends not to adopt a PEP-induced R-state conformation because of strong intermolecular interactions on the C-C′ interface (compared to $PKM2^{Oxalate}$), association of FBP did favor the R-state tetramer formation. Our structure of $PKM2^{K422R\_FBP}$ offers direct evidence for the R-state formation: the $PKM2^{K422R}$ tetramer adopts a similar fold to that of $PKM2^{WT}$ bound to FBP (PDB: 4B2D) (Fig. 4B). In support of the above observations, $PKM2^{K422R}$ showed pyruvate kinase activity comparable to that of $PKM2^{WT}$ in the presence of FBP, but significantly decreased activity and enhanced cooperativity in the absence of FBP (Fig. 1C and Table S1).

$PKM2^{R399E}$ tends to inhibit the formation of R-state tetramer

$PKM2^{R399E}$ also adopts the T-state conformation (Fig. S2A). No significant difference existed between the structures of $PKM2^{R399E}$ and $PKM2^{Phe}$ (Fig. 4A). Residue R399 of α14 on

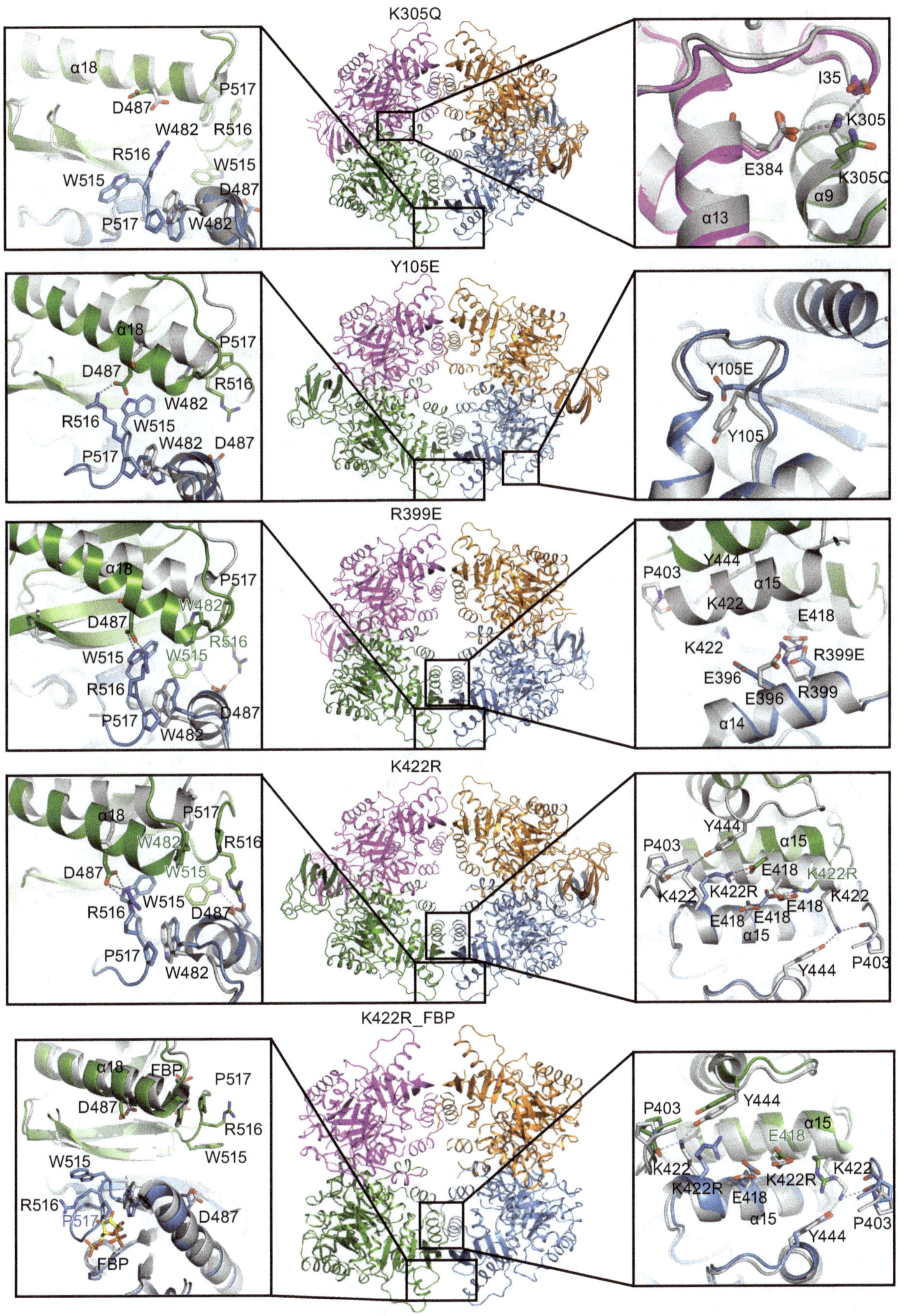
K305Q
α18
P517
D487
W482
R516
W515
I35
K305
E384
K305Q
α9
α13
Y105E
Y105
R399E
P403
Y444
α15
K422
E418
R399E
E396
R399
α14
K422R
K422R_FBP
FBP

Figure 3. Structural comparisons of PKM2 mutants and PKM2^{Oxalate} (in the R-state). Close-up views of PKM2 mutants with critical residues indicated as stick representations. PKM2^{Oxalate} (in the R-state) is colored in gray.

monomer C forms hydrogen bonds with residues E396 of α14 and E418 of α15 on monomer D on the C-C′ interface of the PKM2^{WT} structure. However, residue R399E on monomer C may not form the same hydrogen bonds in PKM2^{R399E} even in the R-state (in the presence of PEP or FBP) because the three negatively charged residues may repel one another (Fig. 3). The existence of R399E in PKM2 will inhibit the formation of the R-state conformation induced by FBP or PEP, and lead to a decrease of enzymatic activity (compared to PKM2^{WT}). Consistently, in the presence of FBP, PKM2^{R399E} is less active and has less of a tendency to form a tetramer compared to PKM2^{WT} (Fig. 1C and 1E). In contrast, no such significant difference in the extent of tetramer formation was observed in the absence of FBP.

PKM2^{Y105E} inhibits FBP association and formation of the FBP-induced R-state of PKM2

A previous study indicated that phosphorylation of residue Y105 inhibits FBP association and tetramer formation of PKM2, and thus decreases its enzymatic activity (Hitosugi et al., 2009). In the crystal structure, PKM2^{Y105E} binds to proline, which was derived from a crystallization condition and might have facilitated a T-state conformation (as did phenylalanine) (Fig. S2D). No significant conformational change around residue Y105E was observed between structures of PKM2^{Y105E} and PKM2^{Oxalate} (Fig. 3). The crystal structure could not reveal the mechanism for the inhibition of FBP association. However, it is possible that phosphorylation of Y105 functions during the dynamic transition, which could not be observed in crystal structure. In support of this conjecture, PKM2^{Y105E} has similar enzymatic activity and tetramer formation (compared to PKM2^{WT}) in the absence of FBP (Fig. 1C and 1D). The addition of oxalate leads to an increase in tetramer formation to a comparable extent for PKM2^{Y105E} and PKM2^{WT} (Fig. S1F). In contrast, the addition of FBP results in significantly less of an increase in the extent of tetramer formation for PKM2^{Y105E} (compared to PKM2^{WT}) (Fig. 1E). Taken together, the mutation Y105E, or phosphorylation of residue Y105, inhibits FBP association and prevents the FBP-induced R-state formation of PKM2.

DISCUSSION

A working model for the dynamic regulation of PKM2

Based on above analyses, here we propose a working model for the dynamic regulation of PKM2 (Fig. 5). PKM2 exists in a mixed population of monomer, dimer and tetramer, and prefers a T-state conformation if the tetramer forms in the absence of FBP. Upon FBP association, PKM2 undergoes a change from the T-state to the R-state conformation, which favors the recognition of PEP in the active site and enhances its pyruvate kinase activity. Regulation of the intermolecular interactions on the A-A′ or C-C′ interface plays an important role in conformational changes of PKM2. A "Rock and Lock" model has been proposed based on structural studies of pyruvate kinase from *L. Mexicana* (Morgan et al., 2010). However, the model was not sufficient to reveal the mechanism for dynamic regulation of human PKM2. In this work, a "seesaw" model was proposed based on the comparison of human PKM2 structures in T- and R-state tetramers (Fig. 2B and 2C). Our biochemical and structural studies revealed the mechanisms for dynamic regulation of the pyruvate kinase activity of PKM2 by inhibitor/activator, post-translational modifications and a patient-derived mutation (Fig. 5). The two layers of regulation provide precise control of pyruvate kinase activity of PKM2 in various biological and pathological conditions.

Post-translational modifications (such as acetylation and phosphorylation) and the observed patient-derived mutation regulate PKM2 pyruvate kinase activity through modulating tetramer formation or the transition between T- and R-state conformations (Fig. 5). PKM2 with an acetylated K305 mainly forms a monomer in solution and forms a dimer in the presence of FBP because the protein loses the intermolecular interactions on the A-A′ interface. The lack of tetramer formation leads to a significant decrease in the enzymatic activity. Phosphorylation of residue Y105 inhibits FBP association and prevents the FBP-induced R-state formation of PKM2, but experiences minor effects in the absence of FBP. Acetylation of residue K433 was reported to decrease PKM2 activity through inhibiting FBP association (Lv et al., 2013). PKM2^{R399E} tends to inhibit the formation of R-state tetramer (R399E on monomer A repels E396 on monomer A and E418 on monomer B on the C-C′ interface) and leads to a decrease of activity even in the presence of FBP. Interestingly, PKM2^{K422R} tends not to form the R-state conformation in the absence of FBP because the hydrogen bonds between residues K422R on monomer A and E418 on monomer B on the C-C′ interface favor a stable T-state tetramer formation. Moreover, FBP could induce the re-formation of the R-state tetramer of PKM2^{K422R}. Thus, PKM2^{K422R} shows low activity and high cooperativity in the absence of FBP but comparable activity to that of PKM2^{WT} in the presence of FBP. Notably, the cellular concentration of PEP ranges from 4 μmol/L to 140 μmol/L in human and rat tissues (Liebermeister, 2005). Under such conditions, PKM2^{K305Q} and PKM2^{K422R} show a significant decrease in pyruvate kinase activity (compared to PKM2^{WT}). Consistently, acetylation of K305 on PKM2 changes metabolic intermediates and promotes cell proliferation and tumor growth, and the K422R mutation of PKM2 enhances HeLa cell proliferation (Gupta et al., 2010; Lv et al., 2011).

The pyruvate kinase activity of PKM2 could be regulated by small molecules. FBP could recruit the FAM to the FBP binding pocket on the C-C′ interface and facilitate R-state

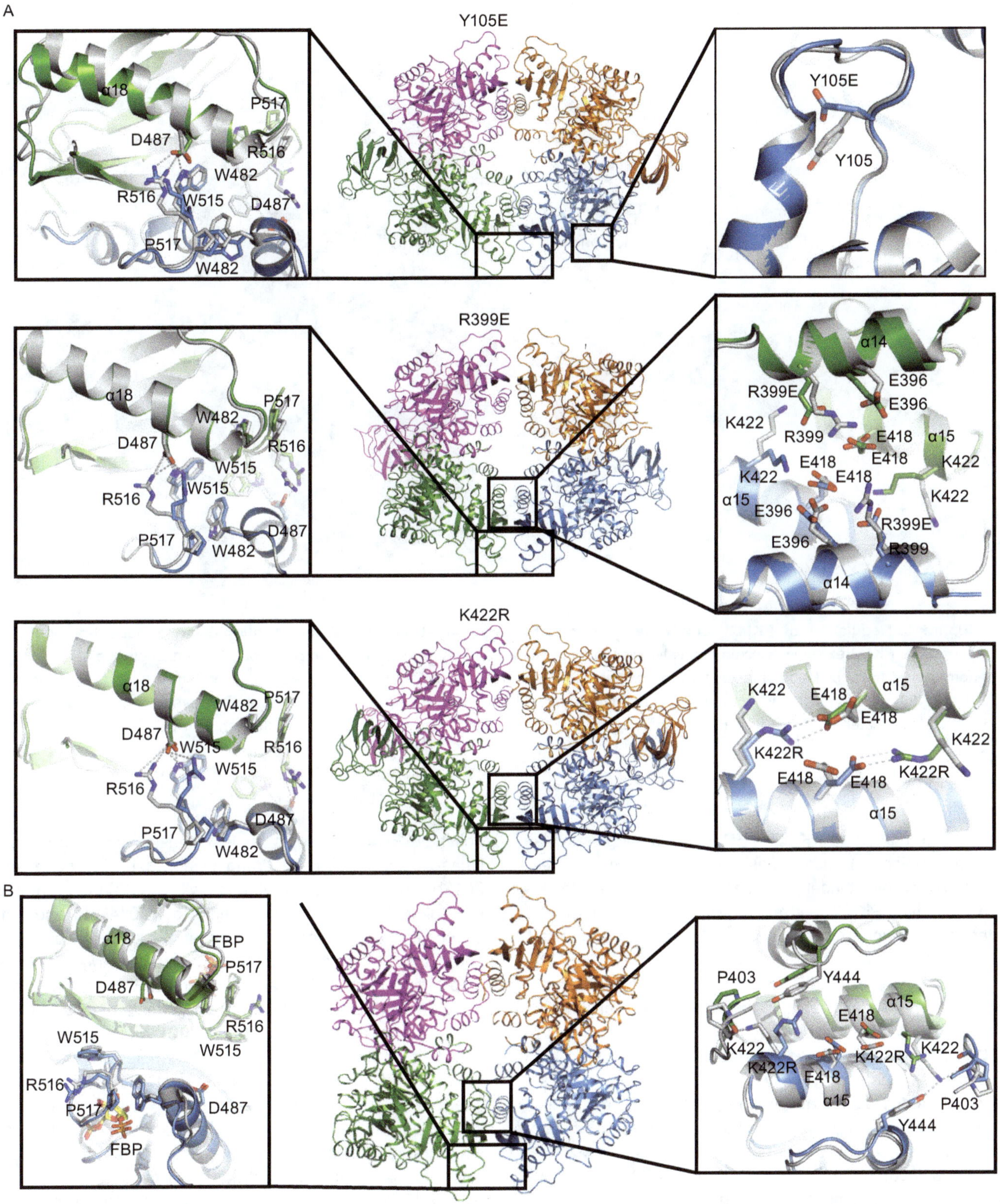

Figure 4. Structural comparisons of PKM2 mutants with PKM2^{Phe} (in T-state) and PKM2^{K422R_FBP} with PKM2 wild type bound with FBP. (A) Close-up views of PKM2 mutants with critical residues indicated as stick representation. PKM2^{Phe} (in T-state) is colored in gray. (B) Close-up views of PKM2^{K422R_FBP} with critical residues indicated as stick representation. PKM2^{FBP} (in R-state, PDB: 4B2D) is colored in gray.

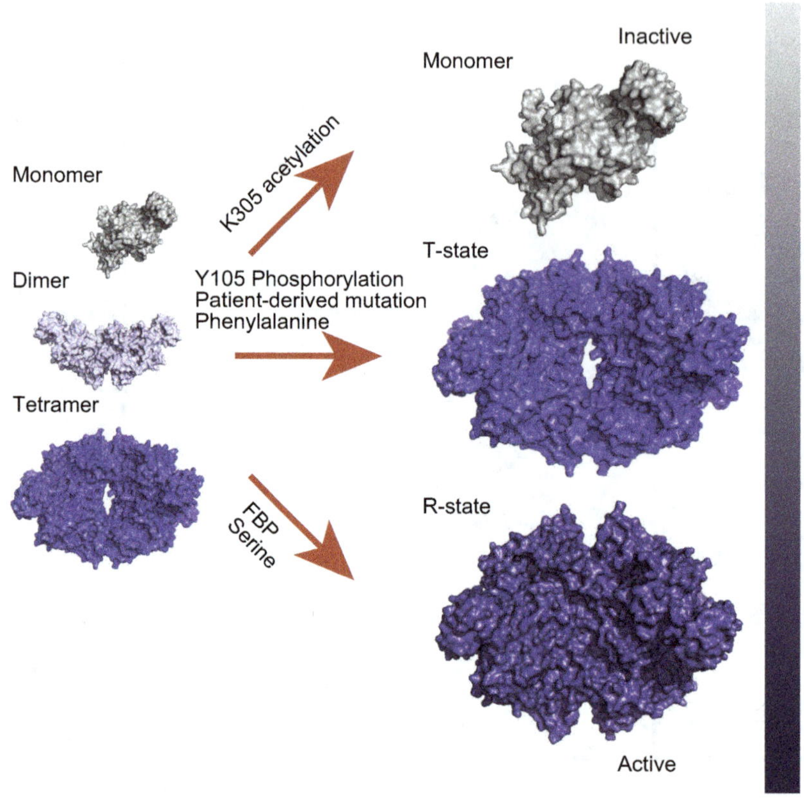

Figure 5. Regulation of PKM2 activity by inhibitor/activators, post-translational modifications and a patient-derived mutation. PKM2 exists in a mixed population of monomer, dimer and tetramer. Acetylation of residue K305 disrupts the tetramer formation. Phosphorylation of residue Y105, the patient derived mutation K422R and phenylalanine association induce an inactive T-state tetramer formation, whereas FBP and serine promote its active R-state formation. The level of pyruvate kinase activity of PKM2 is indicated by a colored bar, in which gray represents low activity and deep blue represents high activity.

tetramer formation of PKM2. PEP, or its analog (such as oxalate, malonate), associates with PKM2 on the A-A′ interface and allosterically induces R-state tetramer formation. Another PKM2 activator, TEPP-46, enhances the intermolecular interactions on the A-A′ interface and favors an R-state conformation (Anastasiou et al., 2012). The crystal structure of PKM2 bound to phenylalanine demonstrates that the amino acid could lock PKM2 in an inactive T-state (Morgan et al., 2013), while the crystal structure of PKM2 bound to serine demonstrates that the amino acid could also allosterically activate PKM2 (Chaneton et al., 2012). SAICAR is a metabolite abundant in proliferating cells and stimulates both pyruvate and protein kinase activities of PKM2 (Keller et al., 2012; Keller et al., 2014). $PKM2^{Q393K}$ (Q393 is located on helix α14) is a SAICAR-insensitive mutant, suggesting that the compound might regulate PKM2 through both A-A′ and C-C′ interfaces.

PKM2 has long been known to have pyruvate kinase activity. However, several studies recently indicated that PKM2 also possesses protein kinase activity, which plays an important role in tumorigenesis. For example, EGFR-activated ERK2 promotes nuclear translocation and protein kinase activity of PKM2 (Yang et al., 2012a; Yang et al., 2012b). SAICAR directly interacts with PKM2 and promotes its protein kinase activity (Keller et al., 2014). In addition, association of tyrosine phosphorylated peptide, acetylation of residue K433 or the mutation of R399E of PKM2 promotes its protein kinase activity and nuclear localization, which are correlated with predominant dimer formation (Gao et al., 2012; Gao et al., 2013; Lv et al., 2013). Interestingly, these dimers are formed with the loss of intermolecular interaction on the C-C′ interface, suggesting that regulation on this interface may be responsible for the switch from pyruvate kinase activity to protein kinase activity of PKM2. Further structural and biochemical studies may reveal the mechanisms for the switch of substrate specificity of PKM2.

In summary, our study provides structural insight into dynamic regulation of PKM2 by post-translational modifications and a patient-derived mutation. The "seesaw" model reveals the mechanism for the transition between T- and R-state conformations and provides a platform for further investigation of other modifications/mutations of PKM2.

ACCESSION NUMBERS

The atomic coordinates of the $PKM2^{Y105E}$, $PKM2^{K305Q}$, $PKM2^{R399E}$, $PKM2^{K422R}$, and $PKM2^{K422R\_FBP}$ have been deposited in the Protein Data Bank with the PDB codes 4QG6, 4QG8, 4QG9, 4QGC, and 4RPP, respectively.

MATERIALS AND METHODS

Protein crystallization

The ORF of human PKM2 was kindly gifted from Kun-liang Guan's lab at Fudan University. Procedures for mutagenesis and protein purification can be found in the Supplemental Materials. The crystals for PKM2 mutants were obtained at 18°C by the hanging-drop, vapor-diffusion method by mixing 1 μL protein solution (15–20 mg/mL) with 1 μL reservoir solution containing 0.2 mol/L malonate pH 5.0, 20% PEG3350 (for $PKM2^{K305Q}$); 0.2 mol/L L-proline, 0.1 mol/L HEPES pH 7.5, 24% PEG1500 (for $PKM2^{Y105E}$); 0.2 mol/L $CaAc_2$, 20% PEG3350 (for $PKM2^{R399E}$); 0.2 mol/L K_2SO_4, 20% PEG3350 (for $PKM2^{K422R}$); and 0.1 mol/L sodium chloride, 0.1 mol/L BIS-TRIS propane pH 9.0, 25% PEG 1500 for ($PKM2^{K422R\_FBP}$).

Data collection and structure determination

Crystals were mounted on nylon loops and flash-cooled in a cold nitrogen stream at 100 K. The data sets were collected at a single wavelength at the Shanghai Synchrotron Radiation Facility (SSRF) in China on beamline BL17U. Data were indexed, integrated and scaled using the HKL2000 program (Otwinowski and Minor 1997). The structures of PKM2 mutants were determined by molecular replacement using the monomer of PKM2 wild-type structure (3BJT.pdb) as a searching model (Christofk et al., 2008b). Details about structure determination and refinement can be found in Supplemental Materials.

Measurement of pyruvate kinase activity

A lactate dehydrogenase (LDH)-coupled pyruvate kinase activity assay was performed to measure the activity of wild-type PKM2 and mutants of PKM2 as previously described (Morgan et al., 2013). Because the reaction was sigmoidal, the reaction rates of PKM2 (in the absence of FBP) were calculated according to the Allosteric Sigmoidal Equation. For the reaction rates of PKM2 in the presence of FBP, we performed the calculation according to Michaelis-Menten Equation. Details for the pyruvate kinase assay can be found in Supplemental Materials.

Thermal shift assay and gel filtration assay

The thermal-shift assay and the gel-filtration (Superdex 200, GE healthcare, 10/300 GL) assay were performed to examine the thermal stability and tetramer formation of PKM2 proteins. Detailed descriptions can be found in Supplemental Materials.

ACKNOWLEDGMENTS

We thank staff members of beamline BL17U at SSRF (Shanghai Synchrotron Radiation Facility, China) for their assistance in data collection, and staff members of Biomedical Core Facility, Fudan University for their help on biochemical analyses. This work was supported by grants from the National Natural Science Foundation of China (Grant Nos. 31270779 and 31425008), the Basic Research Project of Shanghai Science and Technology Commission (12JC1402700), the Program of Shanghai Subject Chief Scientist (14XD1400500), and "ShuGuang" project (11SG06) supported by Shanghai Municipal Education Commission and Shanghai Education Development Foundation.

ABBREVIATIONS

ADP, adenosine di-phosphate; ATP, adenosine tri-phosphate; FBP, fructose 1,6-bisphosphate; PEP, phosphoenolpyruvate; PKM2, pyruvate kinase isoform M2.

COMPLIANCE WITH ETHICS GUIDELINES

Ping Wang, Chang Sun, Tingting Zhu and Yanhui Xu declare that they have no conflict of interest.

This article does not contain any studies with human or animal subjects performed by the any of the authors.

REFERENCES

Akhtar K, Gupta V, Koul A, Alam N, Bhat R, Bamezai RN (2009) Differential behavior of missense mutations in the intersubunit contact domain of the human pyruvate kinase M2 isozyme. J Biol Chem 284:11971–11981

Anastasiou D, Poulogiannis G, Asara JM, Boxer MB, Jiang JK, Shen M, Bellinger G, Sasaki AT, Locasale JW, Auld DS et al (2011) Inhibition of pyruvate kinase M2 by reactive oxygen species contributes to cellular antioxidant responses. Science 334:1278–1283

Anastasiou D, Yu Y, Israelsen WJ, Jiang JK, Boxer MB, Hong BS, Tempel W, Dimov S, Shen M, Jha A et al (2012) Pyruvate kinase M2 activators promote tetramer formation and suppress tumorigenesis. Nat Chem Biol 8:839–847

Anitha M, Kaur G, Baquer NZ, Bamezai R (2004) Dominant negative effect of novel mutations in pyruvate kinase-M2. DNA Cell Biol 23:442–449

Boxer MB, Jiang J, Vander Heiden MG, Shen M, Veith H, Cantley LC, Thomas CJ (2010) Identification of activators for the M2 isoform of human pyruvate kinase version 3. In: Probe reports from the NIH molecular libraries program. National Center for Biotechnology Information, Bethesda

Chaneton B, Gottlieb E (2012) Rocking cell metabolism: revised functions of the key glycolytic regulator PKM2 in cancer. Trends Biochem Sci 37:309–316

Chaneton B, Hillmann P, Zheng L, Martin AC, Maddocks OD, Chokkathukalam A, Coyle JE, Jankevics A, Holding FP, Vousden

KH et al (2012) Serine is a natural ligand and allosteric activator of pyruvate kinase M2. Nature 491:458–462

Christofk HR, Vander Heiden MG, Harris MH, Ramanathan A, Gerszten RE, Wei R, Fleming MD, Schreiber SL, Cantley LC (2008a) The M2 splice isoform of pyruvate kinase is important for cancer metabolism and tumour growth. Nature 452:230–233

Christofk HR, Vander Heiden MG, Wu N, Asara JM, Cantley LC (2008b) Pyruvate kinase M2 is a phosphotyrosine-binding protein. Nature 452:181–186

Dombrauckas JD, Santarsiero BD, Mesecar AD (2005) Structural basis for tumor pyruvate kinase M2 allosteric regulation and catalysis. Biochemistry 44:9417–9429

Gao X, Wang H, Yang JJ, Liu X, Liu ZR (2012) Pyruvate kinase M2 regulates gene transcription by acting as a protein kinase. Mol Cell 45:598–609

Gao X, Wang H, Yang JJ, Chen J, Jie J, Li L, Zhang Y, Liu ZR (2013) Reciprocal regulation of protein kinase and pyruvate kinase activities of pyruvate kinase M2 by growth signals. J Biol Chem 288:15971–15979

Gupta V, Kalaiarasan P, Faheem M, Singh N, Iqbal MA, Bamezai RN (2010) Dominant negative mutations affect oligomerization of human pyruvate kinase M2 isozyme and promote cellular growth and polyploidy. J Biol Chem 285:16864–16873

Hitosugi T, Kang S, Vander Heiden MG, Chung TW, Elf S, Lythgoe K, Dong S, Lonial S, Wang X, Chen GZ et al (2009) Tyrosine phosphorylation inhibits PKM2 to promote the Warburg effect and tumor growth. Sci Signal 2:ra73

Iqbal MA, Siddiqui FA, Chaman N, Gupta V, Kumar B, Gopinath P, Bamezai RN (2014) Missense mutations in pyruvate kinase M2 promote cancer metabolism, oxidative endurance, anchorage independence and tumor growth in a dominant negative manner. J Biol Chem 289:8098–8105

Jiang JK, Boxer MB, Vander Heiden MG, Shen M, Skoumbourdis AP, Southall N, Veith H, Leister W, Austin CP, Park HW et al (2010) Evaluation of thieno[3,2-b]pyrrole[3,2-d]pyridazinones as activators of the tumor cell specific M2 isoform of pyruvate kinase. Bioorg Med Chem Lett 20:3387–3393

Keller KE, Tan IS, Lee YS (2012) SAICAR stimulates pyruvate kinase isoform M2 and promotes cancer cell survival in glucose-limited conditions. Science 338:1069–1072

Keller KE, Doctor ZM, Dwyer ZW, Lee Y-S (2014) SAICAR induces protein kinase activity of PKM2 that is necessary for sustained proliferative signaling of cancer cells. Mol Cell 53:700–709

Koppenol WH, Bounds PL, Dang CV (2011) Otto Warburg's contributions to current concepts of cancer metabolism. Nat Rev Cancer 11:325–337

Kung C, Hixon J, Choe S, Marks K, Gross S, Murphy E, DeLaBarre B, Cianchetta G, Sethumadhavan S, Wang X et al (2012) Small molecule activation of PKM2 in cancer cells induces serine auxotrophy. Chem Biol 19:1187–1198

Liebermeister W (2005) Predicting physiological concentrations of metabolites from their molecular structure. J Comput Biol 12:1307–1315

Luo W, Hu H, Chang R, Zhong J, Knabel M, O'Meally R, Cole RN, Pandey A, Semenza GL (2011) Pyruvate kinase M2 is a PHD3-stimulated coactivator for hypoxia-inducible factor 1. Cell 145:732–744

Lv L, Li D, Zhao D, Lin R, Chu Y, Zhang H, Zha Z, Liu Y, Li Z, Xu Y et al (2011) Acetylation targets the M2 isoform of pyruvate kinase for degradation through chaperone-mediated autophagy and promotes tumor growth. Mol Cell 42:719–730

Lv L, Xu YP, Zhao D, Li FL, Wang W, Sasaki N, Jiang Y, Zhou X, Li TT, Guan KL et al (2013) Mitogenic and oncogenic stimulation of K433 acetylation promotes PKM2 protein kinase activity and nuclear localization. Mol Cell 52:340–352

Morgan HP, McNae IW, Nowicki MW, Hannaert V, Michels PA, Fothergill-Gilmore LA, Walkinshaw MD (2010) Allosteric mechanism of pyruvate kinase from *Leishmania mexicana* uses a rock and lock model. J Biol Chem 285:12892–12898

Morgan HP, O'Reilly FJ, Wear MA, O'Neill JR, Fothergill-Gilmore LA, Hupp T, Walkinshaw MD (2013) M2 pyruvate kinase provides a mechanism for nutrient sensing and regulation of cell proliferation. Proc Natl Acad Sci U S A 110:5881–5886

Noguchi T, Inoue H, Tanaka T (1986) The M1- and M2-type isozymes of rat pyruvate kinase are produced from the same gene by alternative RNA splicing. J Biol Chem 261:13807–13812

Noguchi T, Yamada K, Inoue H, Matsuda T, Tanaka T (1987) The L- and R-type isozymes of rat pyruvate kinase are produced from a single gene by use of different promoters. J Biol Chem 262:14366–14371

Otwinowski Z, Minor W (1997) Processing of X-ray diffraction data collected in oscillation mode. Methods Enzymol 276:307–326

Vander Heiden MG, Cantley LC, Thompson CB (2009) Understanding the Warburg effect: the metabolic requirements of cell proliferation. Science 324:1029–1033

Warburg O (1956) On the origin of cancer cells. Science 123:309–323

Wong N, Ojo D, Yan J, Tang D (2015) PKM2 contributes to cancer metabolism. Cancer Lett 356:184–191

Yang W, Lu Z (2013) Regulation and function of pyruvate kinase M2 in cancer. Cancer Lett 339:153–158

Yang W, Xia Y, Hawke D, Li X, Liang J, Xing D, Aldape K, Hunter T, Alfred Yung WK, Lu Z (2012a) PKM2 phosphorylates histone H3 and promotes gene transcription and tumorigenesis. Cell 150:685–696

Yang W, Zheng Y, Xia Y, Ji H, Chen X, Guo F, Lyssiotis CA, Aldape K, Cantley LC, Lu Z (2012b) ERK1/2-dependent phosphorylation and nuclear translocation of PKM2 promotes the Warburg effect. Nat Cell Biol 14:1295–1304

9

Evolution from genetics to phenotype: reinterpretation of NSCLC plasticity, heterogeneity, and drug resistance

Yingjiao Xue[1,2,3,5], Shenda Hou[1,2,3,5], Hongbin Ji[1,2,3,4,5], Xiangkun Han[1,2,3,5]✉

[1] Key Laboratory of Systems Biology, Shanghai Institutes for Biological Sciences, Chinese Academy of Sciences, Shanghai 200031, China
[2] Innovation Center for Cell Signaling Network, Shanghai Institutes for Biological Sciences, Chinese Academy of Sciences, Shanghai 200031, China
[3] CAS Center for Excellence in Molecular Cell Science, Institute of Biochemistry and Cell Biology, Shanghai Institutes for Biological Sciences, Chinese Academy of Sciences, Shanghai 200031, China
[4] School of Life Science and Technology, Shanghai Tech University, Shanghai 200120, China
[5] Institute of Biochemistry and Cell Biology, Shanghai Institutes for Biological Sciences, Chinese Academy of Sciences, Shanghai 200031, China
✉ Correspondence: xkhan@sibs.ac.cn (X. Han)

ABSTRACT

Lung cancer is the leading cause of cancer-related deaths worldwide. Targeted therapy is beneficial in most cases, but the development of drug resistance stands as an obstacle to good prognosis. Multiple mechanisms were explored such as genetic alterations, activation of bypass signaling, and phenotypic transition. These intrinsic and/or extrinsic dynamic regulations facilitate tumor cell survival in meeting the demands of signaling under different stimulus. This review introduces lung cancer plasticity and heterogeneity and their correlation with drug resistance. While cancer plasticity and heterogeneity play an essential role in the development of drug resistance, the manipulation of them may bring some inspirations to cancer prognosis and treatment. That is to say, lung cancer plasticity and heterogeneity present us with not only challenges but also opportunities.

KEYWORDS lung cancer, plasticity, heterogeneity, drug resistance, phenotypic transition

Yingjiao Xue and Shenda Hou have contributed equally to this work.

INTRODUCTION

Lung cancer is the most common cause of cancer-related deaths in both men and women with a stable worldwide 5-year survival rate of approximately 15% despite advanced medical efforts (Key et al., 2014). Primary therapeutic strategies include surgery, chemotherapy, radiotherapy, and targeted therapy. Although these treatments can initially reduce tumor burden, relapse occurs in most cases.

Drug resistance and metastasis are the two major causes of mortality, both of which are attributed to genetic heterogeneity, genetic evolution, and tumor plasticity. Together, these three aspects produce tumor heterogeneity and allow cancer cells to survive pressures induced by dynamic changes in the surrounding microenvironment and immune system. Plasticity refers to several processes including differentiation, dedifferentiation and transdifferentiation, portraying the dramatic capacity of cells to change their fates (Shoshani and Zipori, 2015). As for cancer cells, epithelial-mesenchymal transition (EMT), and subtype transdifferentiation best embody the property of tumor plasticity. As for the arising of secondary mutations and gene amplifications against tyrosine kinase inhibitor (TKI) treatment, genetic evolution and heterogeneity theories could provide reasonable explanation. Evidences from multiple works have highlighted characteristics of cancer cell plasticity and genetic heterogeneity as a major cause of drug resistance during pre-clinical and clinical studies (Sequist et al., 2011;

Shien et al., 2013). This review primarily focuses on lung cancer plasticity and heterogeneity. These facilitate cancer cell survival and bring considerable challenges to clinical treatment, yet they may also provide inspirations on cancer prognosis and potential therapies.

PLASTICITY OF LUNG CANCER IN EPITHELIAL TO MESENCHYMAL TRANSITION

An example of a long-known plasticity of epithelia is EMT. During cancer progression, plasticity is exemplified by morphological and phenotypical conversions such as the expression of mesenchymal markers and loss of epithelial markers (Park et al., 2015). EMT provides cancer cells with a more invasive and motile phenotype thus enhancing their invasive and metastatic potential (Thiery et al., 2009). Recent work has also demonstrated that EMT has a role in multidrug resistance (Xin et al., 2013; Sui et al., 2014).

The role of EMT in modulating cancer dissemination has been intensively addressed. Transcriptional repressors of the E-cadherin gene are activated during EMT, leading to the loss of the epithelial phenotype. The repressors can either directly repress the activity of the E-cadherin promoter or indirectly repress E-cadherin transcription (Nurwidya et al., 2012). Downstream molecules, such as Snail and ZEB, can also induce the expression of metalloproteases, which degrade the basement membrane thereby favoring invasion (Thiery et al., 2009).

EMT is also highly associated with EGFR-TKI induced drug resistance. Non-small cell lung cancer (NSCLC) cell lines sensitive to gefitinib express EMT markers such as E-cadherin, claudin-4 and claudin-7, but resistant cell lines have lost expression of these genes, while the mesenchymal marker vimentin was highly expressed in the resistant cell lines (Frederick et al., 2007). When the EMT-inducing ligand TGF-β was added to HCC4006 lung cancer cells *in vitro,* morphological changes and resistance to erlotinib were observed in two weeks. On the other hand, removal of TGF-β helped cells reverse their morphology as well as sensibility to erlotinib (Suda et al., 2011).

More specific mechanisms of EMT are still being uncovered. It has been reported that Forkhead box protein M1 (FOXM1), pyruvate dehydrogenase kinase 4 (PDK4) and Cx26 are important regulators of EMT and are associated with drug resistance (Kong et al., 2014; Sun et al., 2014; Yang et al., 2015). In a clinical study, 4 out of 9 patients with *EGFR*-mutant lung adenocarcinoma exhibited changed EMT status pre-and post-gefitinib treatments (Uramoto et al., 2010). EMT is a primary cause of metastasis, which is responsible for approximately 95% of cancer mortality. Therefore, more emphasis should be placed on understanding EMT, especially as associated with drug resistance.

PLASTICITY IN HISTOLOGICAL TRANSITION AS A NEW MECHANISM OF DRUG RESISTANCE

NSCLC, which accounts for 75% of all lung cancers, and small cell lung cancer (SCLC) are two subtypes of lung cancer that share distinct histological and genetic signatures. The histological heterogeneity of lung cancer makes treatment complicated. The two major subtypes of non-small cell lung cancer, lung adenocarcinoma (ADC) and squamous cell carcinoma (SCC), account for almost 80% of all subtypes of NSCLC. These subtypes can highly benefit from multiple lines of therapeutics. It is generally accepted that there are different cells-of-origin for the different types of lung cancer. For example, lung ADC is believed to be derived from type-II pneumocytes, while lung SCC descend from basal cells. High expression of neuroendocrine markers, such as calcitonin-gene related peptide (CGRP) and adhesion molecule (Ncam1) in SCLC, indicates a neuroendocrine (NE)-associated pulmonary cell-of-origin (Sutherland and Berns, 2010). To date, however, no direct evidence has identified the true cells-of-origin for all subtypes of lung cancer. Interestingly, recent clinical studies revealed that drug treatments could induce a histological transition from ADC to SCLC or SCC. These observations challenge our current knowledge on the cell-of-origin of lung cancer and further highlight the importance of cancer plasticity in lung cancer progression.

Transformation from NSCLC to SCLC

Clinical observations in which primary lung ADC developed into SCLC after prolonged drug treatment (more than 2 years) have been previously reported (Morinaga et al., 2007). Engelman et al had a similar discovery. Specifically, it was reported that five patients (14%) with lung ADC were found to have a diagnosis of SCLC in EGFR TKI-resistant tumor biopsies despite maintaining the original mutation and became sensitive to classical SCLC treatment (Sequist et al., 2011). A long remission following TKI treatment and conservation of the original activating *EGFR* mutation suggests a transformation rather than a coexistence of SCLC and NSCLC. Histological transformation further supports plasticity and is one of main causes of TKI-resistance (Shien et al., 2014). In three out of the five resistant patients, genetic mechanisms of resistance were lost in the absence of the continued selective pressure of EGFR inhibitor treatment, and these cancers were sensitive to a second round of treatment with EGFR inhibitors (Sequist et al., 2011). Transformed SCLCs have features of classical SCLC including universal alterations of the RB tumor suppressor, reduced EGFR expression, and a heightened sensitivity to BCL-2 family inhibition (Niederst et al., 2015). Transformed SCLCs have *EGFR* mutations, but do not express EGFR protein, which could potentially explain why they are no longer sensitive to EGFR inhibitors. However, additional molecular mechanisms of transformation remain to be further investigated (Fig. 1).

Emma Norkowski et al found that SCLC developing with ADC, either synchronously or asynchronously, appears to be associated with *EGFR* mutations independent of TKI treatment. Two cases of NSCLC to SCLC transformation were observed after drug treatment, and two cases of phenotypic

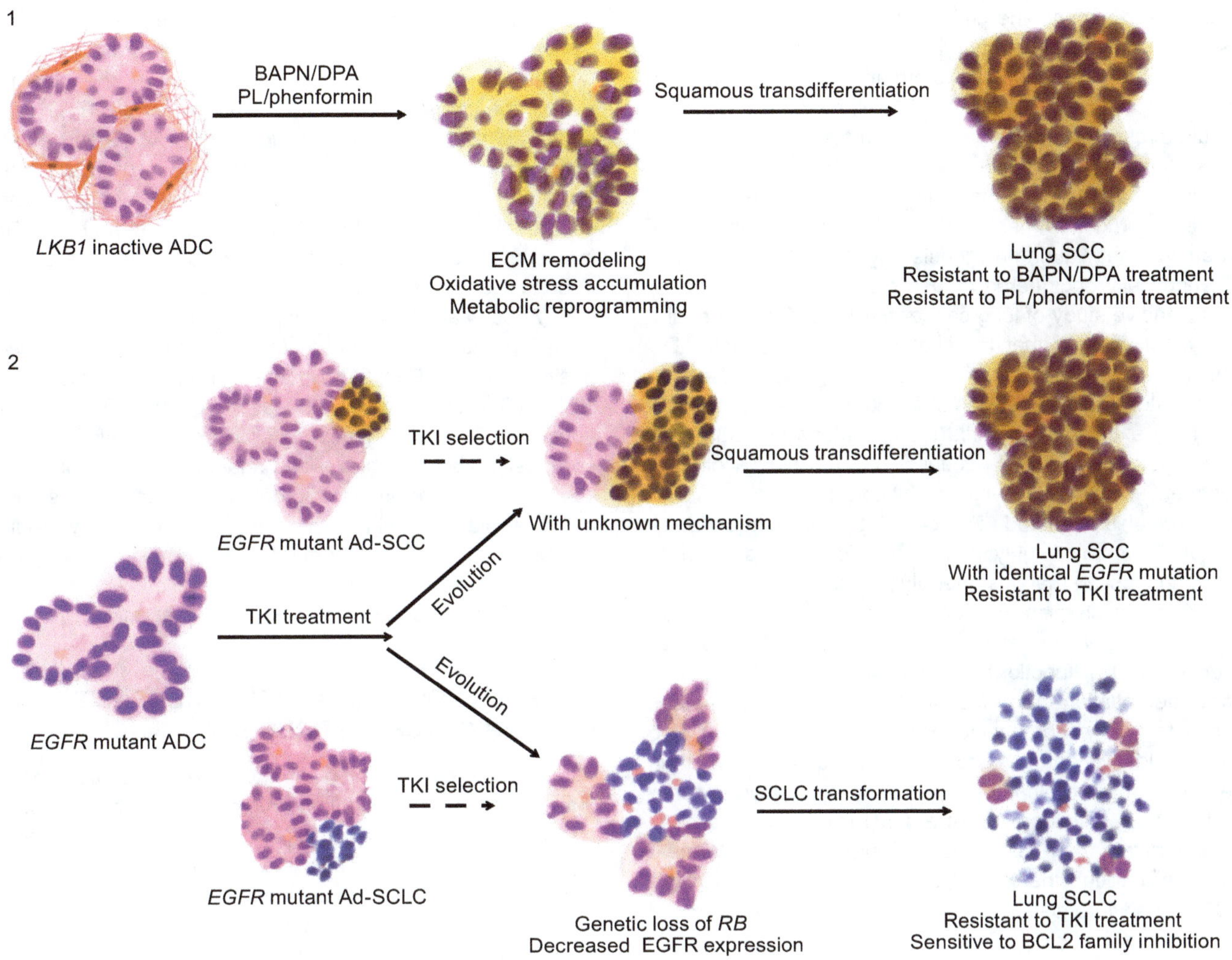

Figure 1. Histological transition and drug resistance. Lung ADC with *LKB1* deficiency treated with β-aminoproprionitrile (BAPN) or phenformin undergoes ECM remodeling, fibroblast losing, metabolic reprograming and oxidative stress accumulation. Along with these changes the histology gradually transdifferentiates from ADC to SCC. As a consequence, the tumor eventually becomes resistant to BAPN/DPA or piperlongumine (PL)/phenformin (1). In addition, histological transition in lung *EGFR* mutant tumor also contributes to drug resistance: on one hand, ADC can transdifferentiate into SCC which is resistant to EGFR inhibitor; on the other hand, ADC can also transform into SCLC with decreased EGFR expression and resistance to EGFR inhibitor. Although the tumor still harbors identical *EGFR* mutation after histological transition, the possibility of preexistent SCC or SCLC can not be excluded (2).

transition without TKI treatment were also observed. However, the possible coexistence of two cell types of origin could not be excluded (Norkowski et al., 2013). Although the spontaneous transformation of *EGFR* mutant ADC to SCLC may be possible, current evidence to support this are weak. Therefore, additional cases need to be analyzed to support this theory (Fig. 1).

Transdifferentiation from lung ADC to SCC

NSCLC can be further pathologically divided into three major subtypes: ADC, SCC and large cell carcinoma (Tuveson and Jacks, 1999; Jackson et al., 2001). There is also a mixed lung adenosquamous cell carcinoma subtype (Ad-SCC) that accounts for 4–10% of NSCLC subtypes. Identical genetic mutations between the adenomatous and squamous parts of a single Ad-SCC lesion suggests that the phenotypic transition between ADC and SCC occurs according to the cancer monoclonal theory (Hofmann et al., 1994; Toyooka et al., 2006; Kang et al., 2007; Ichinokawa et al., 2011).

Stress triggered by TKI treatment has been reported as a driver of phenotype transition in clinical studies. For example, ADCs with *EGFR* mutations may transform into SCCs following TKI treatment and eventually become resistant to TKI. In another case, a mutation in *PIK3CA* has been observed in a patient with an *EGFR* mutation following erlotinib and second-line chemotherapy. Histological analysis indicated a transformation into the SCC subtype (Kuiper et al., 2015). Furthermore, transformation into SCC has been documented in two cases of *EGFR* mutated ADC with acquired resistance to gefitinib treatment (Hsieh et al., 2015). These findings emphasize the need to understand the histological changes of human lung cancer subtypes before and after drug treatment and highlight the importance of repeated biopsy during diagnosis (Fig. 1).

Genetically engineered mouse models have allowed for the extensive study of lung cancer plasticity (Sugano et al.). Han et al showed that inactivation of *Lkb1* in lung ADC conferred plasticity that promoted a progressive transformation into SCC in mice (Han et al., 2014). This process was driven by extracellular matrix (ECM) remodeling caused by downregulation of lysyl oxidase (*Lox*). Pharmacological treatment using LOX enzymatic inhibitors in $Kras^{G12D}/Lkb1^{L/L}$ (*KL*) mice also promoted ADC-to-SCC transition. Importantly, the SCCs remain highly proliferative following drug treatment, suggesting that this phenotypic transition could be an underlying mechanism to drug resistance in pre-clinical trials. Additionally, the authors found that reactive oxygen species (ROS) functionally modulated the ADC-to-SCC transdifferentiation in the *KL* mouse model. The squamous transition was accompanied by metabolic reprogramming upon oxidative stress accumulation during phenformin treatment in pre-clinical trials, representing a novel mechanism of drug resistance (Li et al., 2015b). It has also been demonstrated that Yes-associated protein (YAP) overexpression, through the regulation of *P63I,* inhibited ADC-to-SCC transdifferentiation, while YAP knockdown promoted this process in *Lkb1*-deficient mouse models (Gao et al., 2014). *Lkb1* deficiency causes a robust accumulation of cellular oxidative stress, which confers lung cancer cells with higher plasticity and facilitates an adaptive metabolic reprogramming of cancer cell through phenotypic transition. These works provide insight into lung cancer plasticity and cell-fate determination during tumor progression. In conclusion, histological evolution from ADC to the less stress-sensitive SCC is a strategy for tumor cells to survive and confers drug resistance (Fig. 1).

PLASTICITY IN CHEMOTHERAPY RESISTANCE

Approximately 60% of NSCLC tumors do not harbor targetable driver mutations (Shea et al., 2016), and conventional cytotoxic chemotherapy remains the standard-of-care treatment for NSCLC patients. The most common cytotoxic agents used in treating NSCLC patients are alkylating agents, including cisplatin and carboplatin, which damage DNA by forming platinum-DNA adducts and thus disrupt DNA replication and transcription (Olaussen and Postel-Vinay, 2016). However, tumor cells adapt mechanisms to protect against cytotoxic agents, which remains an obstacle to successful treatment of NSCLC. A major mechanism of chemotherapy resistance is insufficient amount agent to reach the target DNA. For example, high expression of the detoxification protein Gluthatione S-transfer protein GSTP1 (Nakanishi et al., 1999), high expression of the plasma-membrane transporter Copper transporter 1 (CTR1) (Kim et al., 2014), and low expression of BCL-2 (Jeong et al., 2010) have been reported as mediators of resistance. Another major mechanism of resistance is failure to achieve cell death after platinum-DNA adduct formation. For instance, increased expression of ERCC1, which is a pivotal endonuclease in the nucleotide excision repair (NER) pathway (Zhou et al., 2004), and increased expression of survival signaling pathways, such as activation of the phosphatidylinositol 3 kinase (PI3K)/AKT pathway, and overexpression of HER2 (Calikusu et al., 2009) allow tumor cells to evade apoptosis. These epigenetic alterations also reflect cancer cell plasticity and enable cells to adapt to new environments caused by drug treatment. In order to enhance response and avoid resistance to chemotherapy, efforts to identify predictive biomarkers are highly needed. Moreover, selecting the appropriate chemotherapies in combination with targeted therapies or immunotherapies will be synergistic.

GENETIC EVOLUTION AND HETEROGENEITY OF LUNG CANCER IN DRUG RESISTANCE

In addition to cancer cell plasticity, genetic alterations are a major mechanism of acquired drug resistance that has been confirmed in patients. The target gene itself can be altered by mutation or amplification, thus limiting the ability of the drug to inhibit activity of the kinase that remains active and drives aberrant signaling. Another resistance mechanism is bypass signaling, which can be activated under pressure to circumvent the inhibited kinase. Regardless of mechanism, genetic alterations lead to tumor heterogeneity and provide advantages against drug treatment.

Drug resistance in *EGFR* mutant lung cancer

EGFR protein overexpression is frequently observed in NSCLC patients and is correlated with a poor prognosis (Hirsch et al., 2003). Hyperactivation of EGFR is closely associated with the development and progression of lung cancer. Mutation of the EGFR kinase domain accounts for the majority of EGFR activation. *EGFR* mutations occurred in 10% to 20% of NSCLC patients (Riely and Yu, 2015), so further exploration of EGFR as therapeutic target is highly valuable. Two predominant classes of EGFR inhibitors have been developed including monoclonal antibodies that target the extracellular domain of EGFR, such as cetuximab, and small molecule TKIs that target the catalytic domain of EGFR, such as gefitinib and erlotinib. TKIs are synthetic, mainly quinazoline-derived, low-weight molecules that interact with the intracellular tyrosine kinase domain of several receptors, including EGFR, and inhibit ligand-induced receptor phosphorylation by competing for the intracellular ATP-binding site (Ciardiello, 2000). TKIs show initial response rates of over 75% in mutant *EGFR*-NSCLC patients (Perez-Ramirez et al., 2015). In the Iressa Pan-Asia Study (IPASS), *EGFR* mutations were reported to be the strongest predictive biomarker for progression-free survival and objective response rate to first-line gefitinib versus carboplatin/paclitaxel (Fukuoka et al., 2011). However, after an initial marked response to EGFR TKIs, almost all patients inevitably relapse via development of acquired resistance within a median of 9–12 months (Riely and Yu, 2015; Santarpia et al., 2015). Because it was observed that TKIs induced drug resistance in *EGFR* mutant lung cancer, this indicates that genetic variation and tumor heterogeneity can act as intrinsic drivers for the acquisition of additional mutations upon drug treatment.

One of the greatest therapeutic challenges is gatekeeper mutation that change the accessibility of a hydrophobic pocket critical for the binding of small molecule TKIs (Zhang et al., 2009). For example, the secondary mutation T790M increases the affinity of EGFR for ATP and renders the EGFR kinase inhibitors gefitinib and erlotinib less effective in displacing this ATP (Yun et al., 2008). The T790M mutation is responsible for half of EGFR TKI resistance (Wang et al., 2015). Strategies were developed to overcome this resistance such as the generation of the quinazoline irreversible EGFR TKI afatinib (Pao et al., 2005); however, high concentration of this inhibitor is required to inhibit the T790M mutation and acquired resistance to afatinib was also observed in clinical studies (Hashida et al., 2015). Although intensive preclinical studies show impressive results for third-generation mutant-specific EGFR inhibitors such as AZD9291, rociletinib, and WZ4002, drug resistance has still been observed. New tertiary mutations were identified such as C797S in AZD9291-resistant T790M-positive tumors and L718Q, L844V, and C797S in WZ4002-resistant tumors (Thress et al., 2015; Yu et al., 2015). Sequentially detected mutations in NSCLC samples depict the genetic alterations in response to the modified inhibitors. A recent study revealed that the acquired resistance caused by T790M could occur either by selection of pre-existing T790M-positive clones or via genetic evolution from initially T790M-negative drug-tolerant cells (Hata et al., 2016). Although heterogeneity-based tumor cell selection stands as a new mechanism of drug resistance, it determines cell identity through genetic alterations in early stage (Fig. 2).

Moreover, alternative signaling pathways, or so-called bypass tracks, can be activated in resistant cells bypassing the need for signaling from the target. An increasing recognized drug resistance mechanism occurs through a bypass tracks module. For example, 15% to 20% of EGFR TKI–resistant cases are mediated by amplification of the *c-MET* receptor tyrosine kinase. Activation of signaling downstream of *c-MET* is independent of EGFR and allows resistant cells to grow despite EGFR inhibition (Bean et al., 2007; Engelman et al., 2007). To prevent *MET* amplification, combination therapies of EGFR TKI and MET inhibitors have been applied in clinical trials. However, cancer relapse inevitably occurred. It was discovered that ATP-binding cassette subfamily B member 1 (ABCB1) overexpression, which was associated with cancer stem cell (CSC) properties and EMT, was involved in the acquired resistance to MET inhibitors. ABCB1 inhibitors could be a potential drug to fight against this resistance (Sugano et al., 2015) although this remains to be investigated (Fig. 2).

Furthermore, systematic genetic and histological analyses have been performed on tumor biopsies from 37 patients with drug-resistant lung adenocarcinoma carrying *EGFR* mutations (Sequist et al., 2011). Other than *MET* amplification, some resistant cancers showed unexpected genetic alterations including mutations in *PIK3CA* and amplification of *HER2*. Aberrant activation of PI3K/AKT signaling in the presence of EGFR kinase inhibitor can result in drug resistance (Engelman and Janne, 2008). Takezawa et al found that *HER2* was amplified in 12% of tumors with acquired resistance and 1% of untreated lung adenocarcinomas (Takezawa et al., 2012) (Fig. 2).

Drug resistance of lung cancer with EML4-ALK fusion

Another example of drug resistance in lung cancer is anaplastic lymphoma kinase (ALK) target treatment. Echinoderm microtubule-associated protein-like 4 (EML4) and ALK (Choi et al. 2010) fusion protein is an oncogenic driver occurring in approximately 5% of NSCLC and leading to the expression of constitutively active ALK tyrosine kinase (Soda et al., 2007). Aberrant ALK activity activates mitogen-activated protein kinase kinase/extracellular signal-regulated protein kinase (MEK/ERK) and PI3K signaling pathways and promotes cell survival and proliferation (Shaw et al., 2013). Crizotinib is an ALK ATP-competitive inhibitor used as first-line therapy in the treatment of advanced NSCLC harboring *ALK* rearrangement. Unfortunately, responses to this agent are not long-lasting. Genetic alterations and activation of bypass tracks are two main mechanisms of acquiring resistance. Secondary mutations in *ALK* TK domain as well as *ALK* fusion gene amplification have been identified (Doebele et al., 2012; Katayama et al., 2012). L1196M, a gatekeeper mutation, has been detected, as well as L1152R, C1156Y, I1171T/N/S, F1174L/C/V, V1180L, G1202R, S1206Y, and G1269A mutations (Iams and Lovly, 2015). In crizotinib-resistant tumors, several distinct bypass tracks mediating resistance have been reported. Among them, activation of EGFR has been identified in several independent studies. As revealed by studies from the Massachusetts General Hospital, 5 out of 9 resistant patient samples have EGFR activation compared to sensitive samples. This finding has been further confirmed by *in vitro* experiments (Sasaki et al., 2011; Katayama et al., 2012; Yamaguchi et al., 2014). Katayama et al also found high-level of *c-KIT* gene amplification in 2 out of 18 resistant samples. Further studies showed that c-KIT overexpression required stem-cell factor to promote resistance and that resistance could be overcome by combined imatinib and crizotinib treatment

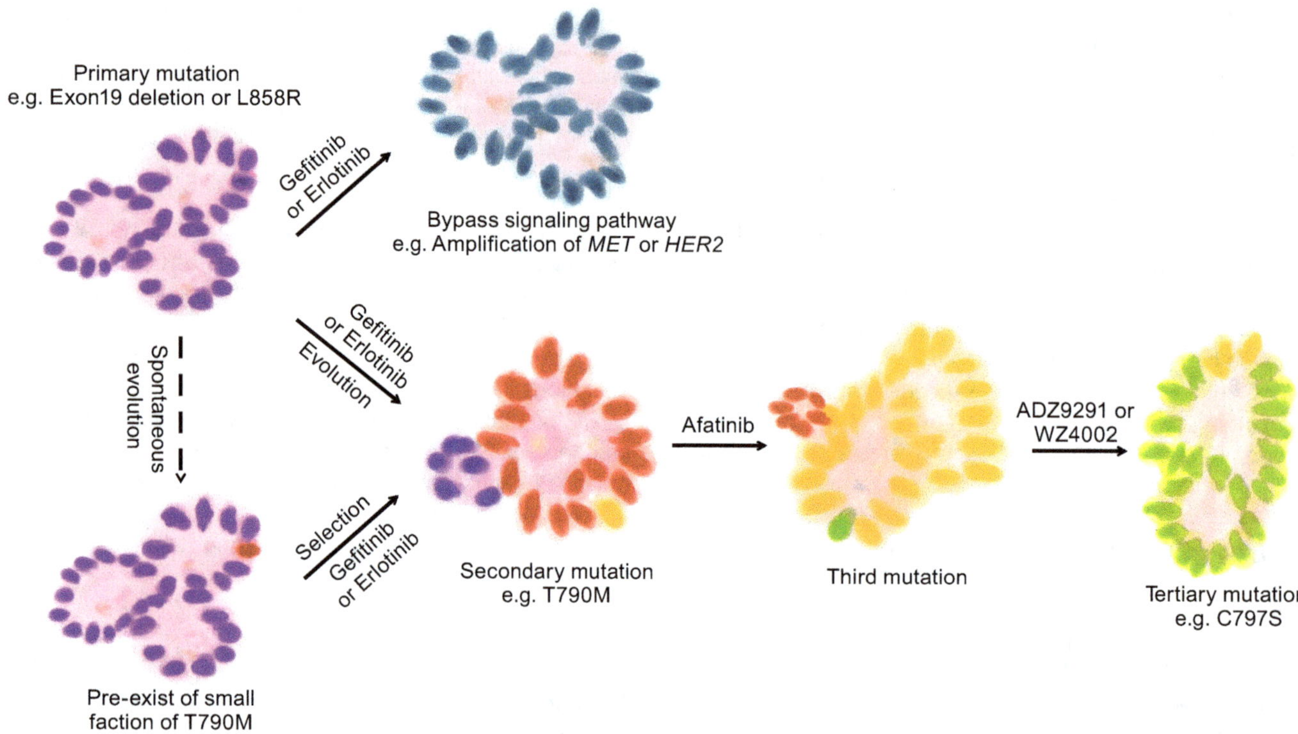

Figure 2. Drug resistance in *EGFR* mutant lung cancer. Lung ADC with *EGFR* mutation, such as exon 19 deletion or L858R substitution, can develop drug resistance after or even before drug treatment. In the first case, after being exposed to EGFR TKI, such as gefitinib or erlotinib, the tumor becomes resistant through bypass signaling pathway, such as amplification of *MET* or *HER2*. Also, under the pressure of inhibitor, the tumor can develop secondary mutation because of its genetic instability. As for the second case, the tumor has the capacity of spontaneous evolution before drug treatment. As a result, the tumor also develops secondary mutation. After drug treatment, the secondary mutation is subjected to selection and the cells with secondary mutation become a majority of the tumor. Admittedly, it is hard to diagnose in each tumor whether the secondary mutation has already existed before drug treatment or not. However in both cases, the tumor subsequently treated with the drug, which is targeted to the secondary or third mutation, finally develops drug resistance, manifesting as holding the third or tertiary mutation.

(Katayama et al., 2012). Upregulation of insulin like growth factor receptor 1R (IGF-1R) and SRC activity have also been detected in ALK TKI resistance (Crystal et al., 2014; Lovly et al., 2014). Furthermore, P2Y receptors were reported to mediate ALK inhibitor resistance and protein kinase C played an important role in this process (Wilson et al., 2015) (Fig. 3).

Based on the exploration of the resistance mechanisms, multiple second-generation ALK inhibitors functioning as first-line therapies and crizotinib resistance therapies are currently applied in clinic, such as ceritinib, or are in clinical trials, such as alectinib, brigatinib, X-396, PF-06463922, and TSR-011 (Iams and Lovly, 2015; Wilson et al., 2015). However, resistance to second-generation ALK inhibitors has also been observed. Katayama et al found a novel V1180L gatekeeper mutation from a cell line model and a novel I1171T secondary mutation from a patient who developed resistance to alectinib (Katayama et al., 2014). Another study suggested that inhibitors of heat shock protein 90 (Hsp90), such as 17-DMAG, ganetespib, and IPI-504, may overcome ligand-triggered resistance to second-generation ALK inhibitors (Tanimoto et al., 2014). However, resistance to Hsp90 inhibition has also been discovered. Expression of P-glycoprotein has been reported to be associated with resistance to 17-DMAG (Kim et al., 2015) (Fig. 3).

Genetic evolution and heterogeneity

It is generally accepted that tumors are monoclonal in origin. That is, only a single cell can cross over the border from normalcy to malignancy to become the ancestor of the cells in a tumor mass. As a result, all cells in a tumor are expected to share common genetic or biochemical markers (Weinberg, 2013). However, it is also known that tumors are composed of several subpopulations with genotypic and phenotypic heterogeneity. At initiation, the tumor is a relatively homogeneous collection of cells that soon become heterogeneous due to the continuous acquisition of new mutations. This acquisition is mostly due to the development of genetic instability. The clonal evolution model suggests

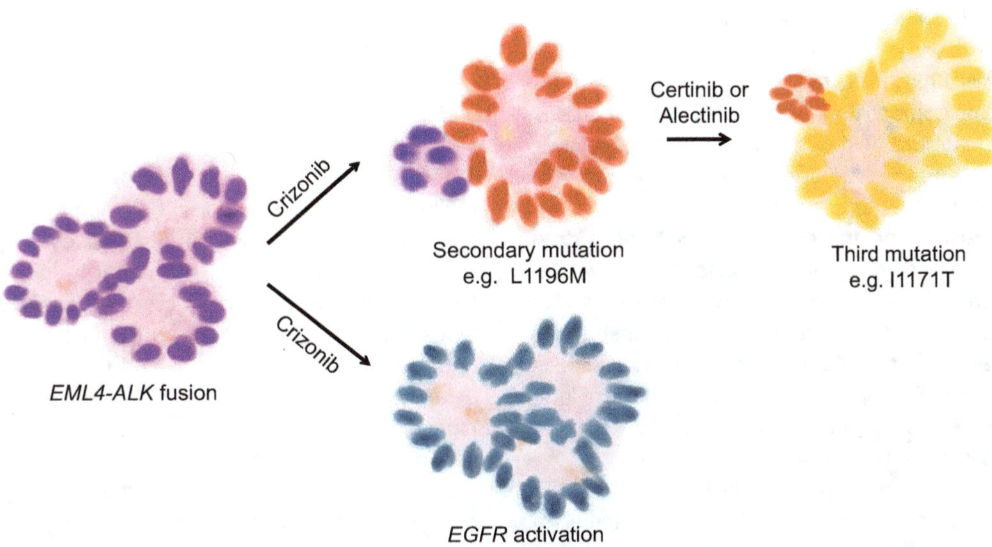

Figure 3. EML4-ALK fusion and TKI resistance. Similar to EGFR TKI resistance, lung ADC with *EML4-ALK* fusion gene also becomes resistant to ALK TKI in two major ways. One way is due to secondary mutation after crizonib treatment; and cells with third mutation dominate the tumor after being treated with second-generation ALK inhibitor. Another way is through EGFR activation in order to compensate the inhibitory effect of ALK inhibitor.

that the genetic evolution among cancer cells is responsible for cancer progression and heterogeneity in phenotype, function, and response to therapy (Hockel and Vaupel, 2001).

Tumor initiation and progression are dependent on rare, stochastic mutations that create genetic variability in a cell population. Most of the spontaneous mutations are neutral or deleterious, while disadvantage mutations will be discarded, and neutral mutations can still retain. Mutations that confer competitive advantages, such as increased proliferation and survival in the restrictive microenvironment, can be selected for by Darwinian forces. As a consequence, these cells have an advantage in outgrowth and eventual dominance in the local tissue environment. Normally, a majority of the tumor cells undergo several successive mutations and Darwinian selection after the initiating mutation. This process is called clonal succession, on which the clonal evolution model is based (Weinberg, 2013). However, a minority of tumor cells may harbor only one or two advantageous or neutral mutations. This provides genetic heterogeneity within a tumor. Additionally, cells within different parts of the tumor could experience different selective pressures before drug treatment, such as differential interactions with the extracellular matrix, gradients of oxygen, nutrition, growth factors, and blood supply. This differential exposure could be another cause of tumor heterogeneity (Marusyk and Polyak, 2010).

Following drug treatment, the primary stress afflicting tumor cells is caused by the drug and genetic heterogeneity might be a reasonable explanation of drug resistance. Although it is accepted that resistant properties occur after pharmacological treatment, researchers have found that some resistance-related mutations exist prior to treatment (Denis et al., 2015; Gurden et al., 2015). There is also the possibility that previously advantageous mutations may no longer benefit the tumor proliferation or survival, but that previously neutral mutations harbored by a very small fraction of the tumor cells become beneficial under drug exposure. Drug treatment becomes the catalyst for the accumulation of certain clones that then become detectable. The preexistence of EGFR TKI resistant cells have been validated in cell lines as well as in primary human lung tumors through high-throughput FISH analyses and peptide nucleic acid-clamping PCR or ultra-sensitive droplet digital PCR (Turke et al., 2010; Oh et al., 2011; Watanabe et al., 2015). Darwinian selection will allow tumor cell survival in patients with pre-existing resistant cells, even if it is a minimal subgroup, when drugs are administered (Gerlinger et al., 2012; Turtoi et al., 2015). On the other hand, if there are no pre-existing advantageous mutations against the drug treatment, the tumor will undergo regression. However, during this period of drug treatment, residual cells may still mutate. Tumor cells that adapt to drug treatment by acquiring additional mutations will dominate the tumor cell population rapidly. Both pre-existing and later evolving mutations could lead to detectable genetic alterations. Changes in the cell population embody the characteristic of cancer plasticity, which could explain how genetic alterations contribute to EGFR or ALK resistance.

MODULATING PLASTICITY AND HETEROGENEITY: AN OPPORTUNITY FOR CANCER TREATMENT

Cell intrinsic and cell extrinsic features contribute to tumor heterogeneity, which aids cancer cells in acquiring functional capabilities that allow the malignant cells to survive, proliferate, and disseminate through genetic alterations, epigenetic modifications or subtype transdifferentiation. Cell extrinsic properties arise from factors in the microenvironment including the composition of the extracellular matrix and factors sequestered to its constituents, the tumor's ability to recruit adequate blood supply, and the recruitment of stromal cells that aid in tumor growth (Marjanovic et al., 2013). These diverse aspects help cancer cells fight drug treatment and eventually develop resistance. However, studying tumor plasticity and heterogeneity provides us with more opportunities to fully understand the processes and molecular mechanisms that underlie drug resistance. It is therefore reasonable to be optimistic about the future of lung cancer treatment benefiting from an increased knowledge of plasticity and heterogeneity (Fig. 4).

Opportunities brought by cancer plasticity can be manifested in treatment. Loss of stem properties in tumor initiating cells (TICs) (Lovly et al.) can be beneficial for drug treatment with fewer concerns regarding clinical relapse. Cancer stem cell differentiation is a new hot spot for such studies. All-trans retinoic acid (Miller et al. 2004) is an agent that can induce differentiation and may be effective against residual tumor cells that are resistant to drug treatment. Moro et al found that several cycles of all-trans retinoic acid (ATRA) and cisplatin treatment are able to stably reduce lung TIC and dissemination (Moro et al., 2015). The histone deacetylase inhibitors senistein and C-phycocyanin have been reported to strengthen the anti-cancer affects of ATRA and reduce the metastatic potential of lung cancer (Zhou et al., 2012; Cheng et al., 2015; Greve et al., 2015; Li et al., 2015a). ATRA also works in the development of drug-resistant neuroblastoma both *in vitro* and *in vivo* (Matthay et al., 1999; Merrill et al., 2004). In advanced hepatocellular carcinoma, ATRA also helps against resistance to sorafenib, which is considered as the first-line treatment of the malignant disease (Guan et al., 2015). These studies suggest that depriving stem cell function might be universal and effective in cancer treatments, including in drug resistance in lung cancer (Fig. 4).

Moreover, the inhibition or interference of the transdifferentiation process may strengthen the effect of TKI treatment *EGFR* mutant lung ADC. In a *Lkb1*-deficient lung cancer mouse model, the quenching of ROS has been validated to functionally inhibit the ADC-to-SCC transdifferentiation (Li et al., 2015b). Although validation of phenotypic transition in human NSCLC is challenging, the potential for the combination treatment of ROS scavengers with transition inhibitors to overcome drug resistance is high. Additionally, impairing the ability of tumor cells to eliminate accumulated oxidative stress during metabolically controlled phenotypic transitions may sensitize tumors to ROS inducing reagent. Awareness of lung cancer plasticity may affect treatment regiments to maximize pathological response. If transdifferentiation can be detected early, corresponding changes to therapy may benefit patients. On the other hand, a forced transdifferentiation to a complete transformation may make tumors more sensitive to chemotherapy and radiation than the original tumor. In this manner, patients may achieve longer period of disease control and survival via effective chemotherapy and radiotherapy (Fig. 4).

Plasticity may allow cancer cells to revert back to normal cells. The discovery of the reversal phenomenon dates back to the 1970s. It was observed that the injection of blastocysts into malignant mouse embryonal carcinoma cells reverted them back into normalcy (Mintz and Illmensee, 1975). This shows the impact of the microenvironment on cancer progression. Cancer cell reversal to non-malignant cells has been observed in various types of cancers through transplantation into a normal adult or embryonic tissue (Lee et al., 2005; Krause et al., 2010; Bischof et al., 2013; Brock et al., 2015). This highlights a potential new approach to cancer treatment helping cancer cells transit back into the normal state. Recently, the development of systems biology has allowed for a more efficient exploration of phenotype switching and identification of essential molecular targets (Brock et al., 2015). While there are no current applications of cancer-normalcy-transition treatment, new mechanisms are continuously being uncovered. The study of normalcy reversal could lead to the development of novel cancer therapies one day.

Increasing evidence has shown that inhibition of EMT may be a possible mechanism of drug resistance. Suppression of the master EMT-inducer zinc finger E-box-binding homeobox 1 (ZEB1) significantly enhanced the chemosensitivity of docetaxel-resistant ADC *in vitro* and *in vivo,* and ectopic expression of ZEB1 increased chemoresistance (Ren et al., 2013). Furthermore, through the screening of established small anti-cancer molecules in the *EGFR* mutant NSCLC HCC827 cell line and a corresponding mesenchymal derivative cell line, Wilson et al. found that dasatinib, an Abl/Src inhibitor, was a potential candidate for treatment of erlotinib resistance associated with EMT (Wilson et al., 2014). A forced EMT experiment also helps finding of agents with epithelial CSC-specific toxicity. The trial is based on the study of epithelial CSCs which are enriched in EMT process. The candidate agent Salinomycin can lead to the loss of expression of breast cancer stem cell genes in patients (Gupta et al., 2009). However, recent research shows that non-cancer stem cells have the potential to dedifferentiate into cancer stem cells (Chaffer et al., 2011). A major limitation to the use of agents with CSC-specific toxicity is the possibility for the non-stem cells to convert into cancer stem cells. However, further research may identify new strategies to overcome these conflicts, such as finding new agents targeting both cancer stem cells and non-stem cells, or drugs specifically point to non-cancer stem cells and cancer stem cells respectively (Gupta et al., 2009).

Heterogeneity, particularly intratumoral heterogeneity, increases the challenges of cancer subtype classification. It may lead to the biased analysis of single tumor-biopsy samples since the sample may represent only a part of the tumor genomic landscape (Gerlinger et al., 2012). This represents a potential obstacle to successful biomarker discovery, validation, and application (Dibben et al., 2012;

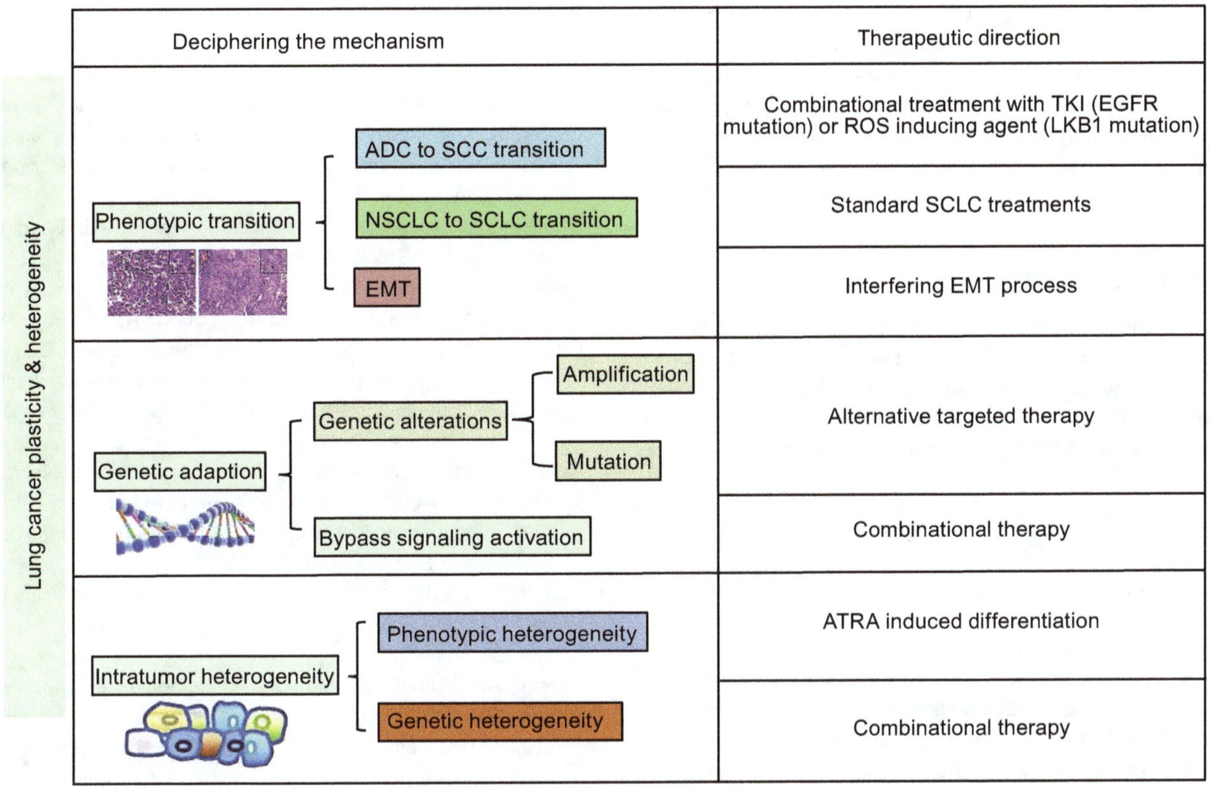

Figure 4. Summary of mechanisms of lung cancer plasticity, heterogeneity and the correlated therapeutics. This model illustrates the guidance of lung cancer plasticity and heterogeneity in clinical use, especially in drug resistance upon target therapies. Different therapeutic strategies depend on the unique mechanisms. Histologically, transition from ADC to SCC and switch from NSCLC to SCLC have been found broadly. Complete transition leads to the using of transited tumor therapy while combinational treatment is applied to partial transformation. EMT is a special aspect of phenotype alteration. It shows morphological plasticity of lung cancer with the switch of biological markers. Interfering EMT process can be a bright way to inhibit certain kind of drug resistance. In addition to histological transition, genetic adaption provides an evident proof for lung cancer heterogeneity. As the intensive study of cell and molecular mechanism, gene alteration and abnormal activation have been revealed in the drug-surviving cancer cells. More target-pointed drugs are needed, substitutionally or combinationally. Intratumor heterogeneity reinforces the composition of lung cancer that changes for adapting to the environment. A typical illustration is the acquirement of stem cell-like property, against which ATRA is an effective therapy. In terms of targeting genetic heterogeneity, identification of advantageous clones and application of combinational therapy could provide a better outcome for lung cancer patients.

Gerlinger et al., 2012). Heterogeneity may also partly contribute to the failure of personalized medicine. This highlights the need for biomarker studies, macro-dissections, and increased number of biopsies per patient (Dibben et al., 2012). Although no consensus has been reached, optimized biopsies may lead to better-targeted therapies or combined treatments (Fig. 4).

CONCLUSION

In the clinical treatment of lung cancer, drug resistance remains a frustrating issue. In this review, we attribute this problem to lung caner plasticity and genetic heterogeneity at both the histological and genetic levels. More specifically, these mechanisms include secondary mutations, bypass signaling pathway, histological transformation from NSCLC to SCLC, transdifferentiation from ADC to SCC, cancer stem cell differentiation, and the EMT program. Cancer cell plasticity and genetic alterations further contribute to the tumor heterogeneity of lung cancer, serving as selective material under restrictive conditions, such as drug exposure, tumor microenvironmental and metabolic stress. In these situations, only more adaptive cells can survive and accumulate in a tumor. As a result, drug resistance eventually develops. However, we can hope to be inspired by understanding and exploiting the mechanisms of plasticity. For instance, we can utilize oxidative stress scavengers to inhibit transdifferentiation, apply oxidative stress-inducing agents to ROS sensitive tumors, and use certain agents to limit the process of EMT and cancer stem cell differentiation. Moreover, plasticity-induced heterogeneity emphasizes the importance of the collection of more representative biopsies and the discovery of more optimized biomarkers for pathological diagnosis. Lung cancer plasticity and heterogeneity remains a significant challenge, but there is a real potential to exploit it in the fight against drug resistance.

ACKNOWLEDGEMENTS

This work was sponsored by grants from the National Natural Science Foundation of China (Grant No. 81402276), Science and Technology Commission of Shanghai Municipality (Grant No. 15XD1504000), and the 2014 Young Scientist fellowship from International Association for the Study of Lung Cancer (IASLC).

ABBREVIATIONS

17-DMAG, 17-(Dimethylaminoethylamino)-17-demethoxygeldanamycin; ADC, adenocarcinoma; Ad-SCC, adenosquamous cell carcinoma; ATRA, all-trans retinoic acid; CSC, cancer stem cell; ECM, extracellular matrix; EGFR, epidermal growth factor receptor; EML4-ALK, echinoderm microtubule-associated protein-like 4 and anaplastic lymphoma kinase; EMT, epithelial to mesenchymal transition; ERK, extracellular signal-regulated protein kinase; FOXM1, Forkhead box protein M1; Hsp90, the heat shock protein 90; IGF-1R, insulin like growth factor receptor 1R; Lox, lysyl oxidase; NSCLC, non-small cell lung cancer; PDK4, pyruvate dehydrogenase kinase 4; PI3K, phosphatidylinositol 3 kinase; ROS, reactive oxygen species; SCC, squamous cell carcinoma; SCLC, small cell lung cancer; TKI, tyrosine kinase inhibitor; TIC, tumor initiating cell.

COMPLIANCE WITH ETHICS GUIDELINES

Yingjiao Xue, Shenda Hou, Hongbin Ji and Xiangkun Han declare that they have no conflict of interest.

This article does not contain any studies with human or animal subjects performed by any of the authors.

REFERENCES

Bean J, Brennan C, Shih JY, Riely G, Viale A, Wang L, Chitale D, Motoi N, Szoke J, Broderick S et al (2007) MET amplification occurs with or without T790M mutations in EGFR mutant lung tumors with acquired resistance to gefitinib or erlotinib. Proc Natl Acad Sci USA 104:20932–20937

Bischof AG, Yuksel D, Mammoto T, Mammoto A, Krause S, Ingber DE (2013) Breast cancer normalization induced by embryonic mesenchyme is mediated by extracellular matrix biglycan. Integr Biol (Camb) 5:1045–1056

Brock A, Krause S, Ingber DE (2015) Control of cancer formation by intrinsic genetic noise and microenvironmental cues. Nat Rev Cancer 15:499–509

Calikusu Z, Yildirim Y, Akcali Z, Sakalli H, Bal N, Unal I, Ozyilkan O (2009) The effect of HER2 expression on cisplatin-based chemotherapy in advanced non-small cell lung cancer patients. J Exp Clin Cancer Res 28:97

Chaffer CL, Brueckmann I, Scheel C, Kaestli AJ, Wiggins PA, Rodrigues LO, Brooks M, Reinhardt F, Su Y, Polyak K et al (2011) Normal and neoplastic nonstem cells can spontaneously convert to a stem-like state. Proc Natl Acad Sci USA 108:7950–7955

Cheng J, Qi J, Li XT, Zhou K, Xu JH, Zhou Y, Zhang GQ, Xu JP, Zhou RJ (2015) ATRA and Genistein synergistically inhibit the metastatic potential of human lung adenocarcinoma cells. Int J Clin Exp Med 8:4220–4227

Choi YL, Soda M, Yamashita Y, Ueno T, Takashima J, Nakajima T, Yatabe Y, Takeuchi K, Hamada T, Haruta H et al (2010) EML4-ALK mutations in lung cancer that confer resistance to ALK inhibitors. N Engl J Med 363:1734–1739

Ciardiello F (2000) Epidermal growth factor receptor tyrosine kinase inhibitors as anticancer agents. Drugs 60(Suppl 1) 25–32; discussion 41–22

Crystal AS, Shaw AT, Sequist LV, Friboulet L, Niederst MJ, Lockerman EL, Frias RL, Gainor JF, Amzallag A, Greninger P et al (2014) Patient-derived models of acquired resistance can identify effective drug combinations for cancer. Science 346:1480–1486

Denis MG, Vallee A, Theoleyre S (2015) EGFR T790M resistance mutation in non small-cell lung carcinoma. Clin Chim Acta 444:81–85

Dibben SM, Holt RJ, Davison TS, Wilson CL, Taylor J, Paul I, McManus K, Kelly PJ, Proutski V, Harkin DP et al (2012) Implications for powering biomarker discovery studies. J Mol Diagn 14:130–139

Doebele RC, Pilling AB, Aisner DL, Kutateladze TG, Le AT, Weickhardt AJ, Kondo KL, Linderman DJ, Heasley LE, Franklin WA et al (2012) Mechanisms of resistance to crizotinib in patients with ALK gene rearranged non-small cell lung cancer. Clin Cancer Res 18:1472–1482

Engelman JA, Janne PA (2008) Mechanisms of acquired resistance to epidermal growth factor receptor tyrosine kinase inhibitors in non-small cell lung cancer. Clin Cancer Res 14:2895–2899

Engelman JA, Zejnullahu K, Mitsudomi T, Song Y, Hyland C, Park JO, Lindeman N, Gale CM, Zhao X, Christensen J et al (2007) MET amplification leads to gefitinib resistance in lung cancer by activating ERBB3 signaling. Science 316:1039–1043

Frederick BA, Helfrich BA, Coldren CD, Zheng D, Chan D, Bunn PA Jr, Raben D (2007) Epithelial to mesenchymal transition predicts gefitinib resistance in cell lines of head and neck squamous cell carcinoma and non-small cell lung carcinoma. Mol Cancer Ther 6:1683–1691

Fukuoka M, Wu YL, Thongprasert S, Sunpaweravong P, Leong SS, Sriuranpong V, Chao TY, Nakagawa K, Chu DT, Saijo N et al (2011) Biomarker analyses and final overall survival results from a phase III, randomized, open-label, first-line study of gefitinib versus carboplatin/paclitaxel in clinically selected patients with advanced non-small-cell lung cancer in Asia (IPASS). J Clin Oncol 29:2866–2874

Gao Y, Zhang W, Han X, Li F, Wang X, Wang R, Fang Z, Tong X, Yao S, Li F et al (2014) YAP inhibits squamous transdifferentiation of Lkb1-deficient lung adenocarcinoma through ZEB2-dependent DNp63 repression. Nat Commun 5:4629

Gerlinger M, Rowan AJ, Horswell S, Larkin J, Endesfelder D, Gronroos E, Martinez P, Matthews N, Stewart A, Tarpey P et al (2012) Intratumor heterogeneity and branched evolution revealed by multiregion sequencing. N Engl J Med 366:883–892

Greve G, Schiffmann I, Lubbert M (2015) Epigenetic priming of non-small cell lung cancer cell lines to the antiproliferative and differentiating effects of all-trans retinoic acid. J Cancer Res Clin Oncol 141:2171–2180

Guan DX, Shi J, Zhang Y, Zhao JS, Long LY, Chen TW, Zhang EB, Feng YY, Bao WD, Deng YZ et al (2015) Sorafenib enriches epithelial cell adhesion molecule-positive tumor initiating cells and exacerbates a subtype of hepatocellular carcinoma through TSC2-AKT cascade. Hepatology 62:1791–1803

Gupta PB, Onder TT, Jiang G, Tao K, Kuperwasser C, Weinberg RA, Lander ES (2009) Identification of selective inhibitors of cancer stem cells by high-throughput screening. Cell 138:645–659

Gurden MD, Westwood IM, Faisal A, Naud S, Cheung KM, McAndrew C, Wood A, Schmitt J, Boxall K, Mak G et al (2015) Naturally occurring mutations in the MPS1 gene predispose cells to kinase inhibitor drug resistance. Cancer Res 75:3340–3354

Han X, Li F, Fang Z, Gao Y, Li F, Fang R, Yao S, Sun Y, Li L, Zhang W et al (2014) Transdifferentiation of lung adenocarcinoma in mice with Lkb1 deficiency to squamous cell carcinoma. Nat Commun 5:3261

Hashida S, Yamamoto H, Shien K, Miyoshi Y, Ohtsuka T, Suzawa K, Watanabe M, Maki Y, Soh J, Asano H et al (2015) Acquisition of cancer stem cell-like properties in non-small cell lung cancer with acquired resistance to afatinib. Cancer Sci 106:1377–1384

Hata AN, Niederst MJ, Archibald HL, Gomez-Caraballo M, Siddiqui FM, Mulvey HE, Maruvka YE, Ji F, Bhang HC, Krishnamurthy Radhakrishna V et al (2016) Tumor cells can follow distinct evolutionary paths to become resistant to epidermal growth factor receptor inhibition. Nat Med 5:24

Hirsch FR, Varella-Garcia M, Bunn PA Jr, Di Maria MV, Veve R, Bremmes RM, Baron AE, Zeng C, Franklin WA (2003) Epidermal growth factor receptor in non-small-cell lung carcinomas: correlation between gene copy number and protein expression and impact on prognosis. J Clin Oncol 21:3798–3807

Hockel M, Vaupel P (2001) Tumor hypoxia: definitions and current clinical, biologic, and molecular aspects. J Natl Cancer Inst 93:266–276

Hofmann HS, Knolle J, Neef H (1994) The adenosquamous lung carcinoma: clinical and pathological characteristics. J Cardiovasc Surg (Torino) 35:543–547

Hsieh MS, Jhuang JY, Hua SF, Chou YH (2015) Histologic evolution from adenocarcinoma to squamous cell carcinoma after gefitinib treatment. Ann Thorac Surg 99:316–319

Iams WT, Lovly CM (2015) Anaplastic lymphoma kinase as a therapeutic target in non-small cell lung cancer. Cancer J 21:378–382

Ichinokawa H, Ishii G, Nagai K, Yoshida J, Nishimura M, Hishida T, Suzuki K, Ochiai A (2011) Clinicopathological characteristics of primary lung adenocarcinoma predominantly composed of goblet cells in surgically resected cases. Pathol Int 61:423–429

Jackson EL, Willis N, Mercer K, Bronson RT, Crowley D, Montoya R, Jacks T, Tuveson DA (2001) Analysis of lung tumor initiation and progression using conditional expression of oncogenic K-ras. Genes Dev 15:3243–3248

Jeong SH, Jung JH, Han JH, Kim JH, Choi YW, Lee HW, Kang SY, Hwang YH, Ahn MS, Choi JH et al (2010) Expression of Bcl-2 predicts outcome in locally advanced non-small cell lung cancer patients treated with cisplatin-based concurrent chemoradiotherapy. Lung Cancer 68:288–294

Kang SM, Kang HJ, Shin JH, Kim H, Shin DH, Kim SK, Kim JH, Chung KY, Kim SK, Chang J (2007) Identical epidermal growth factor receptor mutations in adenocarcinomatous and squamous cell carcinomatous components of adenosquamous carcinoma of the lung. Cancer 109:581–587

Katayama R, Shaw AT, Khan TM, Mino-Kenudson M, Solomon BJ, Halmos B, Jessop NA, Wain JC, Yeo AT, Benes C et al (2012) Mechanisms of acquired crizotinib resistance in ALK-rearranged lung Cancers. Sci Transl Med 4:120ra17

Katayama R, Friboulet L, Koike S, Lockerman EL, Khan TM, Gainor JF, Iafrate AJ, Takeuchi K, Taiji M, Okuno Y et al (2014) Two novel ALK mutations mediate acquired resistance to the next-generation ALK inhibitor alectinib. Clin Cancer Res 20:5686–5696

Key J, Kim YS, Tatulli F, Palange AL, O'Neill B, Aryal S, Ramirez M, Liu X, Ferrari M, Munden R et al (2014) Opportunities for nanotheranosis in lung cancer and pulmonary metastasis. Clin Transl Imaging 2:427–437

Kim ES, Tang X, Peterson DR, Kilari D, Chow CW, Fujimoto J, Kalhor N, Swisher SG, Stewart DJ, Wistuba II et al (2014) Copper transporter CTR1 expression and tissue platinum concentration in non-small cell lung cancer. Lung Cancer 85:88–93

Kim HJ, Lee KY, Kim YW, Choi YJ, Lee JE, Choi CM, Baek IJ, Rho JK, Lee JC (2015) P-glycoprotein confers acquired resistance to 17-DMAG in lung cancers with an ALK rearrangement. BMC Cancer 15:553

Kong FF, Zhu YL, Yuan HH, Wang JY, Zhao M, Gong XD, Liu F, Zhang WY, Wang CR, Jiang B (2014) FOXM1 regulated by ERK pathway mediates TGF-beta1-induced EMT in NSCLC. Oncol Res 22:29–37

Krause S, Maffini MV, Soto AM, Sonnenschein C (2010) The microenvironment determines the breast cancer cells' phenotype: organization of MCF7 cells in 3D cultures. BMC Cancer 10:263

Kuiper JL, Ronden MI, Becker A, Heideman DA, van Hengel P, Ylstra B, Thunnissen E, Smit EF (2015) Transformation to a squamous cell carcinoma phenotype of an EGFR-mutated NSCLC patient after treatment with an EGFR-tyrosine kinase inhibitor. J Clin Pathol 68:320–321

Lee LM, Seftor EA, Bonde G, Cornell RA, Hendrix MJ (2005) The fate of human malignant melanoma cells transplanted into zebrafish embryos: assessment of migration and cell division in the absence of tumor formation. Dev Dyn 233:1560–1570

Li B, Gao MH, Lv CY, Yang P, Yin QF (2015a) Study of the synergistic effects of all-transretinoic acid and C-phycocyanin on the growth and apoptosis of A549 cells. Eur J Cancer Prev 25:97–101

Li F, Han X, Li F, Wang R, Wang H, Gao Y, Wang X, Fang Z, Zhang W, Yao S et al (2015b) LKB1 inactivation elicits a redox imbalance to modulate non-small cell lung cancer plasticity and therapeutic response. Cancer Cell 27:698–711

Lovly CM, McDonald NT, Chen H, Ortiz-Cuaran S, Heukamp LC, Yan Y, Florin A, Ozretic L, Lim D, Wang L et al (2014) Rationale for co-targeting IGF-1R and ALK in ALK fusion-positive lung cancer. Nat Med 20:1027–1034

Marjanovic ND, Weinberg RA, Chaffer CL (2013) Cell plasticity and heterogeneity in cancer. Clin Chem 59:168–179

Marusyk A, Polyak K (2010) Tumor heterogeneity: causes and consequences. Biochim Biophys Acta 1805:105–117

Matthay KK, Villablanca JG, Seeger RC, Stram DO, Harris RE, Ramsay NK, Swift P, Shimada H, Black CT, Brodeur GM et al (1999) Treatment of high-risk neuroblastoma with intensive chemotherapy, radiotherapy, autologous bone marrow transplantation, and 13-cis-retinoic acid. Children's Cancer Group. N Engl J Med 341:1165–1173

Merrill RA, Ahrens JM, Kaiser ME, Federhart KS, Poon VY, Clagett-Dame M (2004) All-trans retinoic acid-responsive genes identified in the human SH-SY5Y neuroblastoma cell line and their regulated expression in the nervous system of early embryos. Biol Chem 385:605–614

Miller VA, Kris MG, Shah N, Patel J, Azzoli C, Gomez J, Krug LM, Pao W, Rizvi N, Pizzo B et al (2004) Bronchioloalveolar pathologic subtype and smoking history predict sensitivity to gefitinib in advanced non-small-cell lung cancer. J Clin Oncol 22:1103–1109

Mintz B, Illmensee K (1975) Normal genetically mosaic mice produced from malignant teratocarcinoma cells. Proc Natl Acad Sci USA 72:3585–3589

Morinaga R, Okamoto I, Furuta K, Kawano Y, Sekijima M, Dote K, Satou T, Nishio K, Fukuoka M, Nakagawa K (2007) Sequential occurrence of non-small cell and small cell lung cancer with the same EGFR mutation. Lung Cancer 58:411–413

Moro M, Bertolini G, Pastorino U, Roz L, Sozzi G (2015) Combination treatment with all-trans retinoic acid prevents cisplatin-induced enrichment of CD133+ tumor-initiating cells and reveals heterogeneity of cancer stem cell compartment in lung cancer. J Thorac Oncol 10:1027–1036

Nakanishi Y, Kawasaki M, Bai F, Takayama K, Pei XH, Takano K, Inoue K, Osaki S, Hara N, Kiyohara C (1999) Expression of p53 and glutathione S-transferase-pi relates to clinical drug resistance in non-small cell lung cancer. Oncology 57:318–323

Niederst MJ, Sequist LV, Poirier JT, Mermel CH, Lockerman EL, Garcia AR, Katayama R, Costa C, Ross KN, Moran T et al (2015) RB loss in resistant EGFR mutant lung adenocarcinomas that transform to small-cell lung cancer. Nat Commun 6:6377

Norkowski E, Ghigna MR, Lacroix L, Le Chevalier T, Fadel E, Dartevelle P, Dorfmuller P, Thomas de Montpreville V (2013) Small-cell carcinoma in the setting of pulmonary adenocarcinoma: new insights in the era of molecular pathology. J Thorac Oncol 8:1265–1271

Nurwidya F, Takahashi F, Murakami A, Takahashi K (2012) Epithelial mesenchymal transition in drug resistance and metastasis of lung cancer. Cancer Res Treat 44:151–156

Oh JE, An CH, Yoo NJ, Lee SH (2011) Detection of low-level EGFR T790M mutation in lung cancer tissues. APMIS 119:403–411

Olaussen KA, Postel-Vinay S (2016) Predictors of chemotherapy efficacy in Non-Small Cell Lung Cancer: a challenging landscape. Ann Oncol mdw321

Pao W, Miller VA, Politi KA, Riely GJ, Somwar R, Zakowski MF, Kris MG, Varmus H (2005) Acquired resistance of lung adenocarcinomas to gefitinib or erlotinib is associated with a second mutation in the EGFR kinase domain. PLoS Med 2:e73

Park DH, Jeon HS, Lee SY, Choi YY, Lee HW, Yoon S, Lee JC, Yoon YS, Kim DS, Na MJ et al (2015) MicroRNA-146a inhibits epithelial mesenchymal transition in non-small cell lung cancer by targeting insulin receptor substrate 2. Int J Oncol 75:186

Perez-Ramirez C, Canadas-Garre M, Jimenez-Varo E, Faus-Dader MJ, Calleja-Hernandez MA (2015) MET: a new promising biomarker in non-small-cell lung carcinoma. Pharmacogenomics 16:631–647

Ren J, Chen Y, Song H, Chen L, Wang R (2013) Inhibition of ZEB1 reverses EMT and chemoresistance in docetaxel-resistant human lung adenocarcinoma cell line. J Cell Biochem 114:1395–1403

Riely GJ, Yu HA (2015) EGFR: the paradigm of an oncogene-driven lung cancer. Clin Cancer Res 21:2221–2226

Santarpia M, Gil N, Rosell R (2015) Strategies to overcome resistance to tyrosine kinase inhibitors in non-small-cell lung cancer. Expert Rev Clin Pharmacol 8:461–477

Sasaki T, Koivunen J, Ogino A, Yanagita M, Nikiforow S, Zheng W, Lathan C, Marcoux JP, Du J, Okuda K et al (2011) A novel ALK secondary mutation and EGFR signaling cause resistance to ALK kinase inhibitors. Cancer Res 71:6051–6060

Sequist LV, Waltman BA, Dias-Santagata D, Digumarthy S, Turke AB, Fidias P, Bergethon K, Shaw AT, Gettinger S, Cosper AK et al (2011) Genotypic and histological evolution of lung cancers acquiring resistance to EGFR inhibitors. Sci Transl Med 3:75

Shaw AT, Hsu PP, Awad MM, Engelman JA (2013) Tyrosine kinase gene rearrangements in epithelial malignancies. Nat Rev Cancer 13:772–787

Shea M, Costa DB, Rangachari D (2016) Management of advanced non-small cell lung cancers with known mutations or rearrangements: latest evidence and treatment approaches. Ther Adv Respir Dis 10:113–129

Shien K, Toyooka S, Yamamoto H, Soh J, Jida M, Thu KL, Hashida S, Maki Y, Ichihara E, Asano H et al (2013) Acquired resistance to EGFR inhibitors is associated with a manifestation of stem cell-like properties in cancer cells. Cancer Res 73:3051–3061

Shien K, Yamamoto H, Soh J, Miyoshi S, Toyooka S (2014) Drug resistance to EGFR tyrosine kinase inhibitors for non-small cell lung cancer. Acta Med Okayama 68:191–200

Shoshani O, Zipori D (2015) Stress as a fundamental theme in cell plasticity. Biochim Et Biophys Acta-Gene Regul Mech 1849:371–377

Soda M, Choi YL, Enomoto M, Takada S, Yamashita Y, Ishikawa S, Fujiwara S, Watanabe H, Kurashina K, Hatanaka H et al (2007) Identification of the transforming EML4-ALK fusion gene in non-small-cell lung cancer. Nature 448:561–566

Suda K, Tomizawa K, Fujii M, Murakami H, Osada H, Maehara Y, Yatabe Y, Sekido Y, Mitsudomi T (2011) Epithelial to mesenchymal transition in an epidermal growth factor receptor-mutant lung cancer cell line with acquired resistance to erlotinib. J Thorac Oncol 6:1152–1161

Sugano T, Seike M, Noro R, Soeno C, Chiba M, Zou F, Nakamichi S, Nishijima N, Matsumoto M, Miyanaga A et al (2015) Inhibition of ABCB1 overcomes cancer stem cell-like properties and acquired resistance to MET inhibitors in non-small cell lung cancer. Mol Cancer Ther 14:2433–2440

Sui H, Zhu L, Deng W, Li Q (2014) Epithelial-mesenchymal transition and drug resistance: role, molecular mechanisms, and therapeutic strategies. Oncol Res Treat 37:584–589

Sun Y, Daemen A, Hatzivassiliou G, Arnott D, Wilson C, Zhuang G, Gao M, Liu P, Boudreau A, Johnson L et al (2014) Metabolic and transcriptional profiling reveals pyruvate dehydrogenase kinase 4 as a mediator of epithelial-mesenchymal transition and drug resistance in tumor cells. Cancer Metab 2:20

Sutherland KD, Berns A (2010) Cell of origin of lung cancer. Mol Oncol 4:397–403

Takezawa K, Pirazzoli V, Arcila ME, Nebhan CA, Song X, de Stanchina E, Ohashi K, Janjigian YY, Spitzler PJ, Melnick MA et al (2012) HER2 amplification: a potential mechanism of acquired resistance to EGFR inhibition in EGFR-mutant lung cancers that lack the second-site EGFRT790M mutation. Cancer Discov 2:922–933

Tanimoto A, Yamada T, Nanjo S, Takeuchi S, Ebi H, Kita K, Matsumoto K, Yano S (2014) Receptor ligand-triggered resistance to alectinib and its circumvention by Hsp90 inhibition in EML4-ALK lung cancer cells. Oncotarget 5:4920–4928

Thiery JP, Acloque H, Huang RY, Nieto MA (2009) Epithelial-mesenchymal transitions in development and disease. Cell 139:871–890

Thress KS, Paweletz CP, Felip E, Cho BC, Stetson D, Dougherty B, Lai Z, Markovets A, Vivancos A, Kuang Y et al (2015) Acquired EGFR C797S mutation mediates resistance to AZD9291 in non-small cell lung cancer harboring EGFR T790M. Nat Med 21:560–562

Toyooka S, Yatabe Y, Tokumo M, Ichimura K, Asano H, Tomii K, Aoe M, Yanai H, Date H, Mitsudomi T et al (2006) Mutations of epidermal growth factor receptor and K-ras genes in adenosquamous carcinoma of the lung. Int J Cancer 118:1588–1590

Turke AB, Zejnullahu K, Wu YL, Song Y, Dias-Santagata D, Lifshits E, Toschi L, Rogers A, Mok T, Sequist L et al (2010) Preexistence and clonal selection of MET amplification in EGFR mutant NSCLC. Cancer Cell 17:77–88

Turtoi A, Blomme A, Castronovo V (2015) Intratumoral heterogeneity and consequences for targeted therapies. Bull Cancer 102:17–23

Tuveson DA, Jacks T (1999) Modeling human lung cancer in mice: similarities and shortcomings. Oncogene 18:5318–5324

Uramoto H, Iwata T, Onitsuka T, Shimokawa H, Hanagiri T, Oyama T (2010) Epithelial-mesenchymal transition in EGFR-TKI acquired resistant lung adenocarcinoma. Anticancer Res 30:2513–2517

Wang H, Guo R, Zhang L (2015) TKI Resistance for T790M Mutation. Zhongguo Fei Ai Za Zhi 18:245–250

Watanabe M, Kawaguchi T, Isa S, Ando M, Tamiya A, Kubo A, Saka H, Takeo S, Adachi H, Tagawa T et al (2015) Ultra-sensitive detection of the pretreatment EGFR T790M mutation in non-small cell lung cancer patients with an EGFR-activating mutation using droplet digital PCR. Clin Cancer Res 21:3552–3560

Weinberg R (2013) The biology of cancer. Garland science

Wilson C, Nicholes K, Bustos D, Lin E, Song Q, Stephan JP, Kirkpatrick DS, Settleman J (2014) Overcoming EMT-associated resistance to anti-cancer drugs via Src/FAK pathway inhibition. Oncotarget 5:7328–7341

Wilson FH, Johannessen CM, Piccioni F, Tamayo P, Kim JW, Van Allen EM, Corsello SM, Capelletti M, Calles A, Butaney M et al (2015) A functional landscape of resistance to ALK inhibition in lung cancer. Cancer Cell 27:397–408

Xin H, Kong Y, Jiang X, Wang K, Qin X, Miao ZH, Zhu Y, Tan W (2013) Multi-drug-resistant cells enriched from chronic myeloid leukemia cells by Doxorubicin possess tumor-initiating-cell properties. J Pharmacol Sci 122:299–304

Yamaguchi N, Lucena-Araujo AR, Nakayama S, de Figueiredo-Pontes LL, Gonzalez DA, Yasuda H, Kobayashi S, Costa DB (2014) Dual ALK and EGFR inhibition targets a mechanism of acquired resistance to the tyrosine kinase inhibitor crizotinib in ALK rearranged lung cancer. Lung Cancer 83:37–43

Yang J, Qin G, Luo M, Chen J, Zhang Q, Li L, Pan L, Qin S (2015) Reciprocal positive regulation between Cx26 and PI3K/Akt pathway confers acquired gefitinib resistance in NSCLC cells via GJIC-independent induction of EMT. Cell Death Dis 6:e1829

Yu HA, Tian SK, Drilon AE, Borsu L, Riely GJ, Arcila ME, Ladanyi M (2015) Acquired resistance of EGFR-mutant lung cancer to a T790M-specific EGFR inhibitor: emergence of a third mutation (C797S) in the EGFR tyrosine kinase domain. JAMA Oncol 1:982–984

Yun CH, Mengwasser KE, Toms AV, Woo MS, Greulich H, Wong KK, Meyerson M, Eck MJ (2008) The T790M mutation in EGFR kinase causes drug resistance by increasing the affinity for ATP. Proc Natl Acad Sci USA 105:2070–2075

Zhang J, Yang PL, Gray NS (2009) Targeting cancer with small molecule kinase inhibitors. Nat Rev Cancer 9:28–39

Zhou W, Gurubhagavatula S, Liu G, Park S, Neuberg DS, Wain JC, Lynch TJ, Su L, Christiani DC (2004) Excision repair cross-complementation group 1 polymorphism predicts overall survival in advanced non-small cell lung cancer patients treated with platinum-based chemotherapy. Clin Cancer Res 10:4939–4943

Zhou RJ, Yang XQ, Wang D, Zhou Q, Xia L, Li MX, Zeng LL, Wang G, Yang ZZ (2012) Anti-tumor effects of all-trans retinoic acid are enhanced by genistein. Cell Biochem Biophys 62:177–184

10

MiR-130a regulates neurite outgrowth and dendritic spine density by targeting MeCP2

Yunjia Zhang[1,2], Mengmeng Chen[1,2], Zilong Qiu[3], Keping Hu[4], Warren McGee[5], Xiaoping Chen[5], Jianghong Liu[1], Li Zhu[1]✉, Jane Y. Wu[1,5]✉

[1] State Key Laboratory for Brain & Cognitive Science, Institute of Biophysics, Chinese Academy of Sciences, Beijing 100101, China
[2] University of Chinese Academy of Sciences, Beijing 100049, China
[3] Laboratory of Molecular Basis of Neural Plasticity, Institute of Neuroscience, Chinese Academy of Sciences, Shanghai 200031, China
[4] Institute of Medicinal Plant Development (IMPLAD), Chinese Academy of Medical Sciences, Beijing 100193, China
[5] Department of Neurology, Center for Genetic Medicine, Lurie Cancer Center, Northwestern University Feinberg School of Medicine, Chicago, IL 60611, USA
✉ Correspondence: zhuli@moon.ibp.ac.cn (L. Zhu), jane-wu@northwestern.edu (J. Y. Wu)

ABSTRACT

MicroRNAs (miRNAs) are critical for both development and function of the central nervous system. Significant evidence suggests that abnormal expression of miRNAs is associated with neurodevelopmental disorders. MeCP2 protein is an epigenetic regulator repressing or activating gene transcription by binding to methylated DNA. Both loss-of-function and gain-of-function mutations in the *MECP2* gene lead to neurodevelopmental disorders such as Rett syndrome, autism and *MECP2* duplication syndrome. In this study, we demonstrate that miR-130a inhibits neurite outgrowth and reduces dendritic spine density as well as dendritic complexity. Bioinformatics analyses, cell cultures and biochemical experiments indicate that miR-130a targets *MECP2* and down-regulates MeCP2 protein expression. Furthermore, expression of the wild-type MeCP2, but not a loss-of-function mutant, rescues the miR-130a-induced phenotype. Our study uncovers the *MECP2* gene as a previous unknown target for miR-130a, supporting that miR-130a may play a role in neurodevelopment by regulating MeCP2. Together with data from other groups, our work suggests that a feedback regulatory mechanism involving both miR-130a and MeCP2 may serve to ensure their appropriate expression and function in neural development.

KEYWORDS miR-130a, MECP2, neurite outgrowth, dendritic spines, dendrite morphology

INTRODUCTION

MicroRNAs (miRNAs) are small non-coding RNAs of 16–28 nucleotides (nt) in lengths. They are transcribed into pri-miRNA, which are then cleaved by Drosha to form pre-miRNA in the nucleus. After being transported to the cytoplasm and then cleaved by Dicer, mature miRNAs incorporate with AGO proteins to form RNA-induced silencing complexes (RISCs) to silence target mRNAs (Finnegan and Pasquinelli, 2013). MicroRNAs play important roles in the formation and the function of the nervous system (McNeill and Van Vactor, 2012). Aberrant expression of miRNAs may lead to neurodevelopmental and neurodegenerative diseases (Im and Kenny, 2012). We identified microRNA-130a (miR-130a) in a search for microRNAs regulated by the Tar-DNA binding protein-43 (TDP-43) (The detail will be described in a separate paper). It has been reported that miR-130a regulates neurogenesis by inhibiting the synthesis of substance P (Greco and Rameshwar, 2007), a neuropeptide that can increase neurite outgrowth of cultured chick dorsal root ganglia (Shigehiko and Takeshi, 1978). In *Xenopus laevis*, miR-130a is expressed in anterior neural tissues, eyes and branchial arches (Walker and Harland, 2008), and down-regulation of miR-130a led to smaller eyes (Gessert et al., 2010). In mice, miR-130a is expressed in developing somites, neural tube and restricted regions of the brain after embryonic day 9.5 (Hoesel et al., 2010). Cerebral expression of miR-130a then decreases during the postnatal period throughout the rest of development and is detected only at a low level in the adult mouse cortex (Eda et al., 2011; Søe et al., 2011). Several studies suggest that dysregulation of miR-130a may be associated with neurodevelopmental disorders. For example, the expression of miR-130a is up-regulated in the serum of children with autism (Vasu et al., 2014). Analyses of the copy number variation (CNV) of miR-130a have led to the identification of 2 patients diagnosed with autism spectrum disorder carrying duplications of the locus (11q12.1) containing miR-130a (Rosenfeld et al., 2010). These reports suggest a potential role for miR-130a in neural development. However, the target genes and mechanisms by which miR-130a functions in neurons remain to be investigated.

MECP2 (Methyl-CpG-binding Protein 2) is an X-linked gene, encoding a methylated DNA-binding protein and has been identified as a causative gene in Rett syndrome (Amir et al., 1999; Lewis et al., 1992; reviewed in Lombardi et al., 2015). The MeCP2 protein has multiple biological functions, including regulating RNA transcription, involving in RNA splicing and miRNA biogenesis (Chahrour et al., 2008; Cheng et al., 2014; Maunakea et al., 2013; Nan et al., 1998; Young et al., 2005; Lombardi et al., 2015). Loss and gain of function mutations of MeCP2 lead to Rett syndrome and *MECP2* duplication syndrome, both of which are progressive neurological disorders characterized by intellectual disability, autism and developmental regression (Lombardi et al., 2015). Postmortem brain pathology analyses, MeCP2 mouse models and *in vitro* experiments suggest that MeCP2 regulates axon outgrowth, dendritic spine formation and dendritic arbor complexity (Chao et al., 2007; Chapleau et al., 2009; Cheng et al., 2014; Jiang et al., 2013; Jugloff et al., 2005; Lombardi et al., 2015). Neuronal expression of MeCP2 at an appropriate level is crucial for neural development. However, the mechanisms by which how MeCP2 is regulated remain to be elucidated.

The mammalian *MECP2* gene has a highly conserved long 3′ untranslated region (3′UTR) which contains multiple binding sites for microRNAs (Coy et al., 1999; McGowan and Pang, 2015). It was proposed that microRNAs binding to the 3′UTR of *MECP2* and regulating its expression might contribute to neuronal maturation and neurodevelopmental diseases (Hansen et al., 2010). In the present study, we examined the interaction between miR-130a and MeCP2. Our data demonstrate that miR-130a inhibits neurite outgrowth and dendritic spine formation by regulating MeCP2.

RESULTS

MiR-130a is expressed in developing cerebral cortex and is predicted to regulate neuronal function

The mature sequence of miR-130a is conserved across various species including mammals, birds, amphibians and fish (Fig. S1). To study the function of miR-130a in neurodevelopment, we first examined the expression of miR-130a in developing rat brains. Primary and mature rno-miR-130a were detected in the rat cerebral cortices at both embryonic (E18) and postnatal (P1) stages (Fig. 1), which is consistent with previous results in mice (Hoesel et al., 2010). There were no dramatic changes of the level of miR-130a between the late embryonic and early postnatal stages.

To begin to explore biological function of miR-130a, we employed a bioinformatics approach using three miRNA target prediction tools—TargetScan, miRanda and PicTar—to predict candidate target genes for miR-130a (Agarwal et al., 2015; Krek et al., 2005; Betel et al., 2008). The pathway enrichment analysis (KOBAS 2.0; Xie et al., 2011) revealed potential target genes involved in multiple pathways important for neuronal function, such as axon guidance, synaptic vesicle formation and axonal transport. A number of the predicted miR-130a target genes are also associated with neurodevelopmental disorders, such as autism, schizophrenia and hereditary spastic paraplegia (Table 1). These results, together with its expression in brain tissues, suggest that miR-130a may play an important role in regulating expression of genes crucial for the function of the nervous system.

MiR-130a inhibits neurite outgrowth

Neuronal differentiation is an early and fundamental event in neurodevelopment. The sprouting and outgrowth of neurites

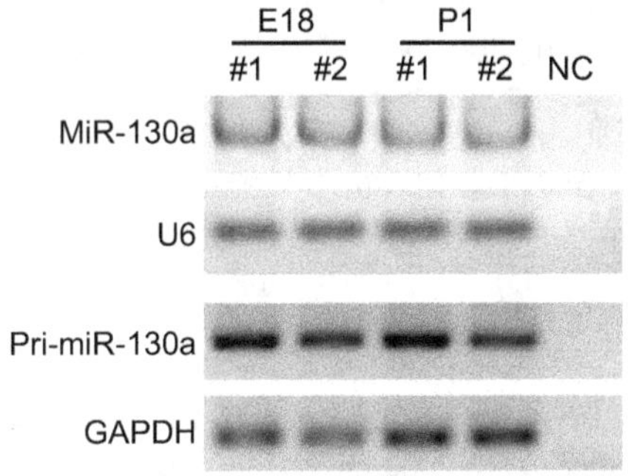

Figure 1. Expression of miR-130a in embryonic and postnatal rat brains. Rat cortical neurons were harvested from brains at the stages of E18 and P1. The levels of mature miR-130a (with U6 as an internal control) (upper panel) and primary miR-130a (with GAPDH as an internal control) (lower panel) were detected by RT-PCR. The last lane in each panel contained the negative control (NC) in which reverse transcriptase was omitted from the cDNA synthesis reactions.

Table 1. Candidate target protein-coding genes for miR-130a

Diseases and pathways		Database	MiR-130a target genes	Gene set size	*P*-value
Neuronal function	Axon guidance	KEGG PATHWAY	9	129	0.0102891
	Synaptic vesicle formation	KEGG PATHWAY	5	63	0.0321619
	Lissencephaly gene (LIS1) in neuronal migration and development	PID	3	19	0.0403738
	SNARE interactions in vesicular transport	KEGG PATHWAY	3	36	0.0807329
Nervous system disorders	Hereditary spastic paraplegia (SPG)	KEGG DISEASE	4	21	0.0058905
	Autistic spectrum disorder	FunDO	6	68	0.013854
	Schizophrenia	FunDO	10	172	0.0239614
	Lissencephaly	KEGG DISEASE	2	6	0.0218963
	Non-syndromic X-linked mental retardation	KEGG DISEASE	3	22	0.0368521
	Congenital disorders of development	KEGG DISEASE	13	244	0.0426269
	Isolated orofacial clefts	KEGG DISEASE	2	14	0.0803631
	Spinocerebellar ataxias	FunDO	2	12	0.0520478

Target genes for miR-130a were predicted using TargetScan, miRanda and PicTar. The intersection of the results from all three tools was accepted as the potential targets. A total of 475 potential target genes were identified by the bioinformatics analyses (see MATERIALS AND METHODS). Shown in the table are the pathways most enriched with protein-coding genes predicted to be targets of miR-130a ("MiR-130a target genes" column) and corresponding raw *P*-values. The "Gene set size" is determined based on the protein coding genes in the genome that have been annotated with the specific pathways by the respective database..

followed by the formation of axons and dendrites is an initial critical process in the early stage of development of the nervous system (Lefebvre, Sanes and Kay, 2015; Takano et al., 2015 and references within). To investigate the role of miR-130a in neuronal development, we examined the effect of miR-130a on neurite outgrowth using primary rat neuronal cultures. Cortical neurons were transfected with a control vector or a plasmid expressing pri-miR-130a that also expressed GFP to mark transfected neurons. Seventy-two hours following transfection, the neurite length was measured by tracing the longest neurite of randomly-selected transfected neurons and quantified using Image J, as published previously (Gao et al., 2010 and reference within). Neurons transfected with miR-130a showed shorter neurite length, approximately 65% of that in the control neurons (Fig. 2A and 2C). To confirm the effect of miR-130a on neurite outgrowth, a specific RNA inhibitor for miR-130a was used. Rat primary cortical neurons were transfected with a scrambled RNA control or the miR-130a inhibitor together with a plasmid expressing GFP. Neurons transfected with the miR-130a inhibitor showed greater neurite length, approximately 150% of that in neurons transfected with the control scrambled RNA (Fig. 2B and 2C). This effect of miR-130a on the neurite length was not due to its effect on neuronal survival, because miR-130a did not affect neuronal death as shown by a Terminal deoxyribonucleotidyl transferase-mediated dUTP-digoxigenin nick end labeling (TUNEL) assay and nuclear staining (Fig. S2). We also examined whether miR-130 affected the number of neurite branches per neuron (Fig. S3A). Quantitative analyses of distribution of neurite branch numbers did not show any significant difference between the control and miR-130a groups, indicating that neurite numbers were not affected by miR-130a (Fig. S3B).

MiR-130a expression decreases dendritic complexity

Dendritic formation and maturation is critical for the formation and maintenance of the nervous system. To study the effect of miR-130a on dendrite formation, rat E18 primary cortical neurons were cultured for 7 days and then transfected with the control or miR-130a, and then fixed at DIV18 for examining dendritic morphology. The dendrites were traced and analyzed using NeuronJ (ImageJ) and NeuronStudio with neuronal morphology quantified using Sholl analysis as described previously (Kutzing et al., 2010; Langhammer et al., 2010). Dendritic complexity was decreased in miR-130a transfected neurons as compared with the control neurons, including the number of branch points and terminal points per neuron, as well as the number of intersections (Fig. 3A–C). Further analyses revealed that miR-130a decreased the number of secondary and tertiary branches, but did not affect primary branches (Fig. 3D and 3E). The lack of effect on primary branches by miR-130a is consistent with what was observed with the number of neurites (Fig. S3). In addition, the total length of dendrites was decreased in miR-130a expressing neurons as compared with the control group, although the average length of segments did not change (Fig. 3F and 3G). These data together

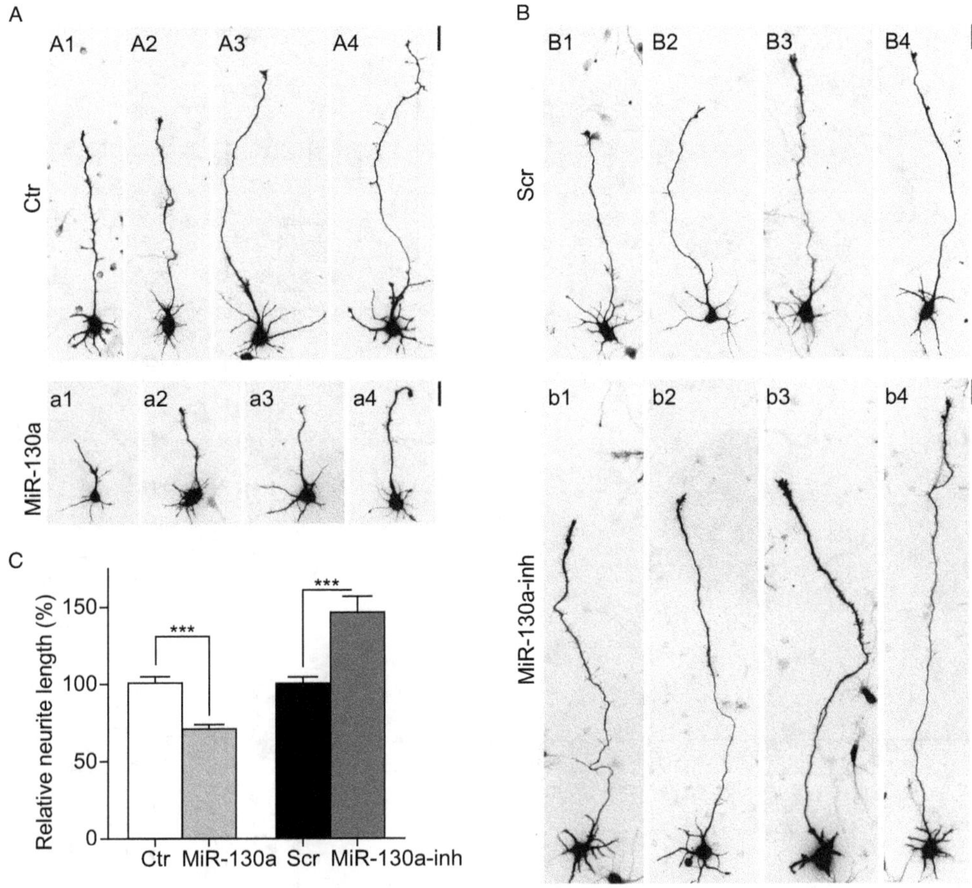

Figure 2. MiR-130a decreases the length of neurites in cultured primary cortical neurons. (A) Cortical neurons isolated from E18 rat embryos were electroporated with the vector control (Ctr; upper panels A1–A4) or the miR-130a expression plasmid (lower panels a1–a4) and cultured for 3 days. Representative images of neurons are shown from either control or miR-130a groups to illustrate the range of neurite outgrowth. Scale bar: 25 μm. (B) Cortical neurons from E18 embryos were electroporated with the scrambled RNA control (Scr; upper panels B1–B4) or the miR-130a inhibitor (MiR-130a-inh; lower panels b1–b4) and cultured for 3 days. Representative images of neurons are shown for the corresponding groups to illustrate the range of neurite outgrowth. Scale bar: 25 μm. (C) Quantification of the average lengths of the longest neurites (mean values ± SEM) in experiments shown in panels (A) and (B). At least 100 neurons in each group were randomly selected and scored in each experiment. ***, $P < 0.001$, Mann-Whitney test. Data represent three independent experiments.

show that miR-130a expression reduces dendritic complexity, in particular, inhibits the formation of the secondary and tertiary branches.

MECP2 is a target of miR-130a

To investigate mechanism of miR-130a function, we sought to identify candidate target genes for miR-130a. Among the potential target gene pathways predicted by our bioinformatics analyses (Table 1), we decided to first examine candidate genes involved in neurodevelopment and associated with neurodevelopmental diseases. *MECP2* is one of these predicted target genes for miR-130a. The human *MECP2* gene has a long conserved 3′UTR (~8 kb) with containing 4 predicted miR-130a binding sites. The most upstream miR-130a binding site located ~240 nucleotide downstream of the stop codon is highly conserved, from human to rodents (Fig. 4A). To validate that *MECP2* is a target gene of miR-130a, a dual luciferase reporter assay in HEK293 cells was performed. A luciferase reporter plasmid was constructed containing a 415-bp fragment of the 3′UTR of the human *MECP2* gene including the most highly conserved miR-130a binding site (the most upstream binding site in the 3′UTR; Fig. 4A). When co-transfected with miR-130a, luciferase activity of the *MECP2* 3′UTR reporter was significantly reduced, whereas mutating the predicted miR-130a binding site in the *MECP2* reporter (3′UTRmt) reversed the miR-130a inhibition, indicating that the 3′UTR of the human *MECP2* gene is responsive to miR-130a expression (Fig. 4B). The mRNA and protein levels of MeCP2 were

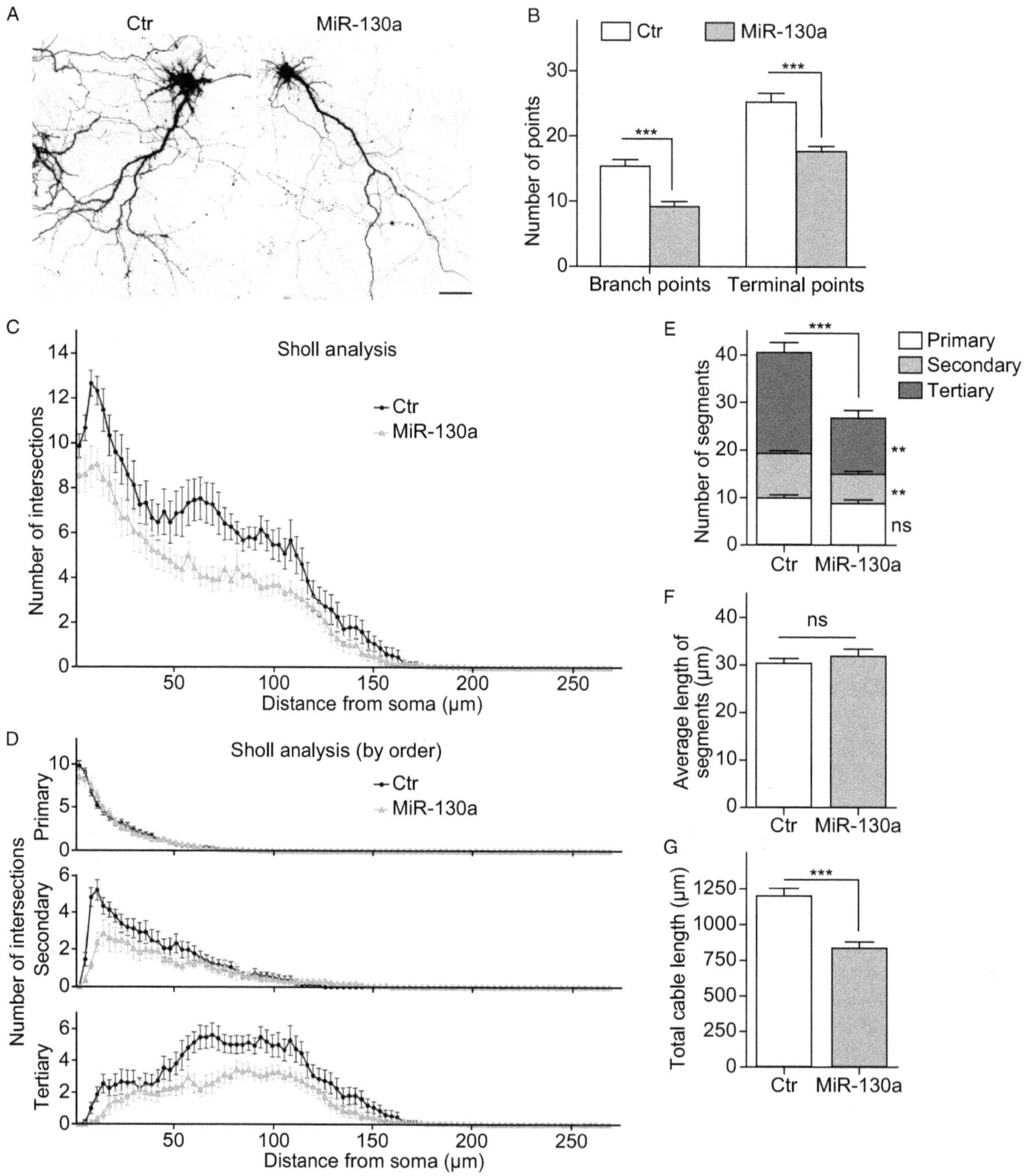

Figure 3. MiR-130a decreases the complexity of dendrites. (A) Cortical neurons were cultured for 7 days and transfected with control or miR-130a. Cells were fixed at 18 DIV. Images were taken using Leica confocal SP8. Image of 40× microscopy were demonstrated. Scale bar: 25 μm. (B) The number of branch points and terminal points per cell. (C) The Sholl analysis of dendritic complexity as measured by number of intersections. (D) The Sholl analysis of dendritic complexity by branch order-primary, secondary and tertiary branches. (E) The number of segments in total and by branch order. (F) Average length of segments per cell. (G) The total length of branches per cell. **, $P < 0.01$; ***, $P < 0.001$, Mann-Whitney test. Fifteen neurons were scored in each group in each experiment. The bar diagram represents mean values ± SEM. Data represent 3 independent experiments.

examined following transfection of pri-miR-130a or miR-130a inhibitor. Western blotting of transfected rat cortical neurons showed expression of miR-130a decreased MeCP2 protein expression to ~40% of that in the control group, whereas the expression of the miR-130a inhibitor increased the MeCP2 protein level to ~3 fold of that in the scramble control group (Fig. 4C and 4E). RT-PCR assays showed that expression of neither miR-130a nor miR-130a inhibitor altered the mRNA

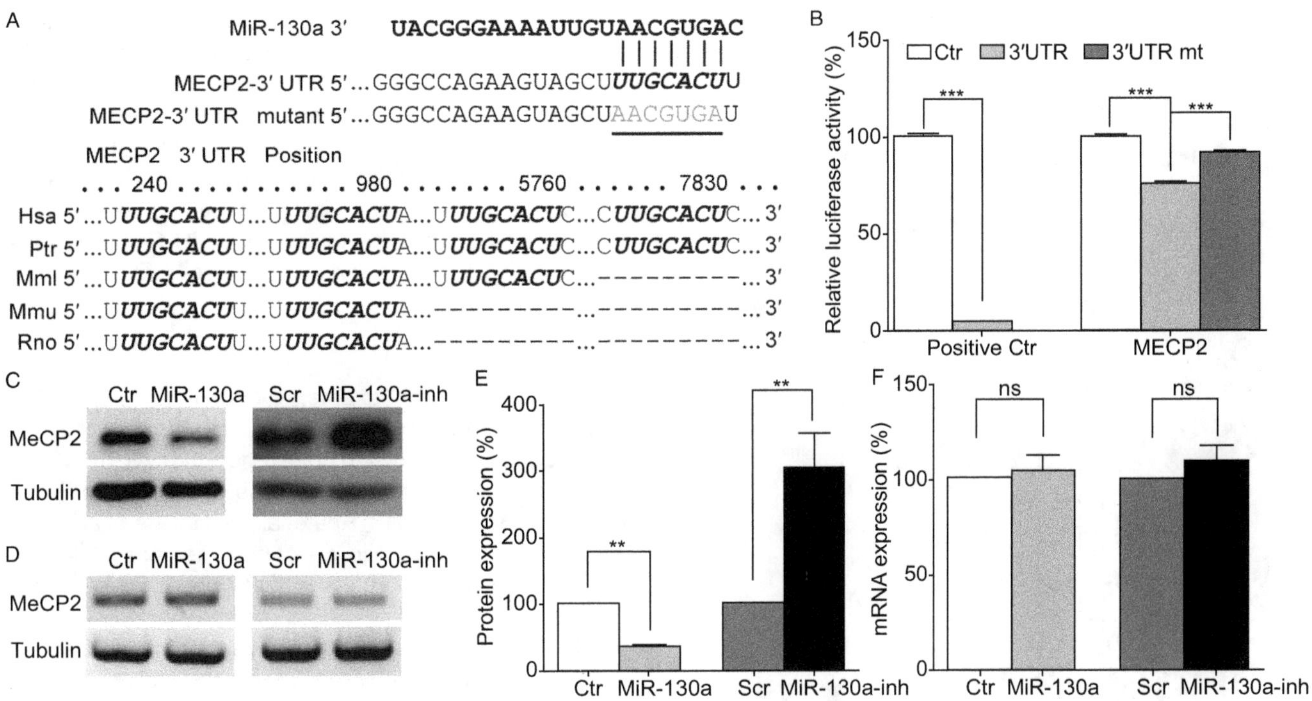

Figure 4. The *MECP2* gene is a target of miR-130a. (A) Alignment of the binding sites of miR-130a on MECP2 mRNA 3′UTR. The seed regions of miR-130a binding sites are in italics. Luciferase reporter plasmids containing the human MECP2-3′UTR regions, either the wild type UTR (MECP2-3′UTR) or the mutant UTR with the most highly conserved binding site for miR-130a mutated as shown (MECP2-3′UTR mutant). (B) A dual luciferase reporter assay. Relative luciferase activity was measured in HEK293 cells at 48 h post-transfection. The wild-type MECP2-3′UTR is referred to as "3′UTR", whereas the mutant described in panel (A) is indicated as "3′UTRmt". (C) The MeCP2 protein levels as detected by Western blotting. Rat E18 cortical neurons were electroporated with the control vector, miR-130a, scramble RNA (Scr) or miR-130a inhibitor (miR-130a-inh). β-Tubulin was used as an internal control. (D) The MeCP2 mRNA levels as assayed by RT-PCR with β-tubulin as an internal control. (E and F) Quantification and statistical analysis of data in (C) and (D). The bar diagram represents mean values ± SEM. Data represent three independent experiments. *, $P < 0.05$; **, $P < 0.01$; ***, $P < 0.001$. Mann-Whitney test.

levels of MeCP2 (Fig. 4D and 4F). Because miRNAs may regulate protein expression by translational inhibition or mRNA degradation or both mechanisms (Hausser and Zavolan, 2014; Vidigal and Ventura, 2015; Wilczynska and Bushell, 2015), these results suggest that miR-130a may decrease the MeCP2 protein level by affecting translation efficiency rather than degrading MeCP2 mRNA. Future studies are necessary to elucidate the precise mechanisms by which miR-130a regulates MeCP2 protein production.

Restoring MeCP2 expression rescues miR-130a-mediated inhibition of neurite outgrowth

It was reported that MeCP2 promotes axon outgrowth in mouse primary cortical neurons (Jugloff et al., 2005). The phosphorylation of MeCP2 at Ser80 is necessary for MeCP2 chromatin binding ability and neurological function (Tao et al., 2009). When the Ser80 was mutated to alanine ($MeCP2^{S80A}$), the S80A-mutant MeCP2 protein cannot be phosphorylated, acting as a loss-of function mutant. We co-transfected the wild-type MeCP2 or $MeCP2^{S80A}$ together with miR-130a into cortical neurons and examined if MeCP2 could rescue the effect of miR-130a on the neurite outgrowth. The average neurite length was measured as described above. Neurons transfected with MeCP2 alone showed longer neurite length, approximately 2 fold of that of control neurons, whereas neurons transfected with $MeCP2^{S80A}$ mutant did not show any changes in the average neurite length (Fig. 5A and 5B). Expression of MeCP2 in miR-130a transfected neurons significantly increased neurite length, ~30% longer than that of neurons transfected with miR-130a alone. However, expression of the mutant form of MeCP2, $MeCP2^{S80A}$, ablated this rescue effect. Our data show that miR-130a negatively regulates neurite length in part by down-regulating MeCP2.

MiR-130a decreases dendritic spine density and MeCP2 rescues the miR-130a-induced phenotype

Dendritic spines are membrane protrusions on dendrites and play critical roles in the formation of synapses and neural circuits (Ethell and Pasquale, 2005). We examined the effect of miR-130a on dendritic spine formation. In this set of

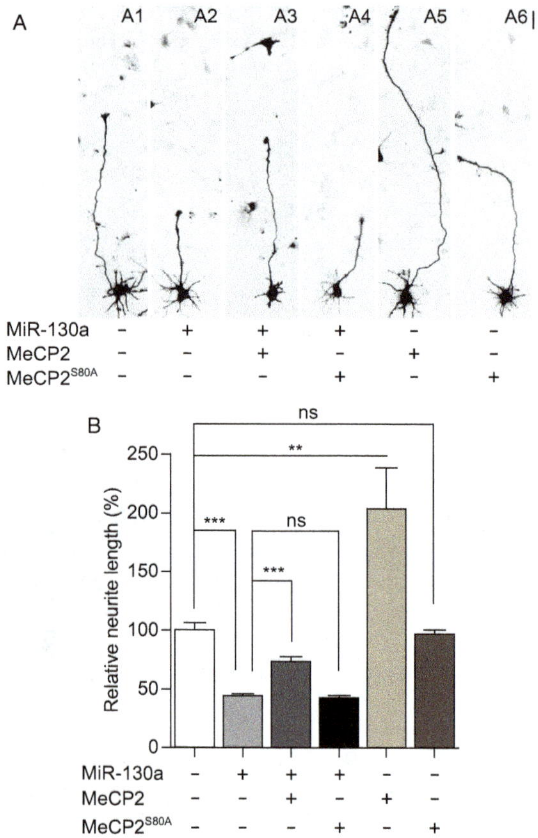

Figure 5. Expression of miR-130a decreases the length of neurites, a phenotype partially rescued by wild type MeCP2 but not by a loss-of-function MeCP2 mutant. (A) Rat E18 cortical neurons were electroporated with the vector control, miR-130a, MeCP2, MeCP2^{S80A} or miR-130a together with MeCP2 or MeCP2^{S80A} and cultured for 3 days before imaging. Scale bar: 25 µm. (B) The average length of the longest neurites in (A) were measured and statistically analyzed from three independent experiments (mean values ± SEM; at least 100 neurons in each group were scored in each experiment). **, $P < 0.01$; ***, $P < 0.001$. Mann-Whitney test.

experiments, rat E18 primary cortical neurons were cultured for 21 days *in vitro* and then transfected using Lipofectamine with the control or with miR-130a. Cells were then fixed 3 days post-transfection. The density of dendritic spines on the secondary and tertiary dendritic branches was determined. Neurons transfected with miR-130a showed fewer dendritic spines as compared with the control neurons (Fig. 6A, panel a1 and a2). Both gain-of- and loss-of-function mutations of the *MECP2* gene lead to neurodevelopment disorders with dendritic spine morphological abnormalities, suggesting the proper level of functional MeCP2 is critical for neural development (Jiang et al., 2013; Xu et al., 2014). Indeed, overexpression of MeCP2 alone reduced dendritic spine density (Fig. 6A, panel a5), consistent with previous data from hippocampal slice cultures (Xu et al., 2014).

We then asked whether the phenotype of decreased spine density by miR-130a was mediated by regulating MeCP2. Neurons co-expressing miR-130a and MeCP2 showed a partial rescue effect, with a higher density than neurons transfected with either miR-130a or MeCP2 alone (Fig. 6A, panel a3). This rescue effect was not observed when S80A mutant MeCP2 was co-transfected with miR-130a (Fig. 6A, panel a4). Taken together, our data suggest that miR-130a decreases dendritic spine density in part by regulating MeCP2.

DISCUSSION

In this study we sought to determine the role of miR-130a in neuronal development, and examined whether effects of miR-130a were mediated by regulating *MECP2* gene expression. Consistent with previous studies carried out in mice (Hoesel et al., 2010; Eda et al., 2011; Søe et al., 2011), miR-130a is expressed in rat cortical neurons at the late embryonic and early postnatal stages (Fig. 1). Bioinformatics analyses suggest that miR-130a may be involved in neurogenesis and associated with neurodevelopmental disorders such as schizophrenia, lissencephaly and autism (Table 1) (Poluch and Juliano, 2015; Selemon and Zecevic, 2015; Wang and Baraban, 2007; Wegiel et al., 2010). Neurogenesis is not only essential for the formation of the nervous system during development, but also critical for the maintenance and neural repair in the adult brain (Feliciano et al., 2015; Zhao et al., 2008). Defects in neurogenesis are associated with a large number of neurological disorders, including neurodevelopmental diseases and neurodegenerative disorders such as Alzheimer's Disease and Parkinson's Disease (Merson and Bourne, 2014; Schoenfeld and Cameron, 2015; Winner and Winkler, 2015).

Our bioinformatics analyses suggest that miR-130a may be involved in many pathways critical for neurodevelopment and associated with pathogenesis of neurodevelopmental diseases (Table 1). Using primary neuronal cultures, we provided experimental evidence that miR-130a plays a role in regulating both neurite outgrowth and dendritic formation. The outgrowth of neurites, which will differentiate into axons and dendrites, is an important event in neural development (Takano et al., 2015; Lefebvre, Sanes, Kay 2015). MiR-130a expression reduces neurite outgrowth, whereas miR-130a inhibitor increases neurite outgrowth (Fig. 2). Expression of miR-130a also reduces dendritic complexity without affecting the number of primary dendritic branches (Fig. 3). Finally, miR-130a reduces the density of dendritic spines (Fig. 6). Overall, these results suggest that miR-130a may play a complex role in regulating neurogenesis and neuronal differentiation. The expression of miR-130a was reported to be decreased in the cerebral cortex of mouse brain but increased in hippocampus from birth to adulthood (Eda et al., 2011). This dynamic expression pattern of miR-130a suggests that the function of miR-130a in the nervous system may be different depending on the specific regions and developmental stages. It should be noted that it has been

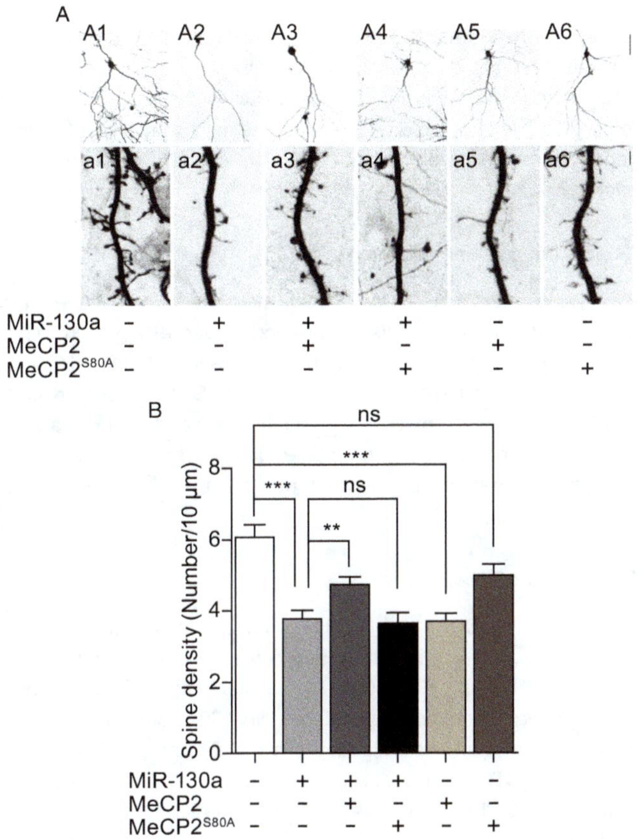

Figure 6. Expression of miR-130a decreased the density of dendritic spines, a phenotype that can be rescued by functional MeCP2 but not by MeCP2 mutant. (A) Rat E18 cortical neurons were cultured for 21 days and then transfected with vector control, miR-130a, MeCP2, MeCP2^{S80A} or miR-130a mixed with MeCP2 or MeCP2^{S80A}. Upper panels (A1–A6): image of 20× microscopy demonstrated the healthy morphology of neurons. The boxed areas by dotted lines in the top panels are enlarged and shown in lower panels and used for quantification of the secondary and tertiary dendrites. Scale bar: 75 μm. Lower panels (a1–a6): high magnifications of image (100× lens; zoom 4.0×) demonstrated the morphology of dendritic spines. Scale bar: 2.5 μm. (B) The average density of dendritic spines. Spines on the secondary and tertiary dendrites were measured. The bar diagram represents mean values ± SEM. Data represent three independent experiments (Ten neurons were scored in each experiment). **, $P < 0.01$; ***, $P < 0.001$, Mann-Whitney test.

observed that the expression of a number of miRNAs was affected in MeCP2-deficient mice, including the down-regulation of miR-130a (Urdinguio et al., 2010). Together with our data, these results suggest that miR-130a together with MeCP2 may participate in a feedback regulatory loop that maintain the appropriate levels of both miR-130a and MeCP2 to ensure their proper function of neural development in a spatially and temporally regulated manner. It is obvious that further studies are necessary for elucidating the role of miR-130a in neurodevelopment.

Our data support that miR-130a inhibits neurite outgrowth, at least in part, by regulating *MECP2*, a predicted target gene by three bioinformatics tools. MiR-130a expression decreased the protein level of MeCP2 but did not affect its mRNA level (Fig. 4). This can be explained by the possible activity of miR-130a in regulating translation efficiency rather than mRNA stability, a mechanism that has been described for a subset of miRNAs (Baek et al., 2008). The human *MECP2* gene has at least 4 transcripts of different lengths, among which the 10 kb mRNA contains the longest 3′UTR (~8 kb). This transcript is predominantly expressed in the brain, whereas the short transcripts are mainly expressed in the lung and liver (Pelka et al., 2005). Only the longest 3′UTR contains multiple miR-130a binding sites in different species. In addition, the 3′UTR of *MECP2* is highly conserved in both sequence and RNA secondary structure (Coy et al., 1999). These all suggest that miR-130a may regulate MeCP2 in a conserved manner and the regulation may be relatively specific to the nervous system. Variants in the *MECP2* 3′UTR were found among patients affected by autism and Rett syndrome, although none of the variants is predicted to be pathogenic to Rett syndrome (Santos et al., 2008; Shibayama et al., 2004). However, MeCP2 mRNA levels in four autism patients carrying conserved 3′UTR alterations were lower (Coutinho et al., 2007). These findings suggest that mutations in the 3′UTR might impact the expression of MeCP2 (McGowan and Pang, 2015).

It has been reported previously that DNA methylation and histone modification are important to neurite outgrowth in PC12 cells (Futamura et al., 1995; Persengiev and Kilpatrick, 1996). MeCP2 is a methylated DNA binding protein regulating gene transcription, playing important roles in histone deacetylation and chromatin remodeling by recruiting histone deacetylase (Chahrour et al., 2008; Jones et al., 1998; Nan et al., 1998). MeCP2 was reported to promote axon outgrowth in mouse cortical neurons, and the suppression of MeCP2 in PC12 cells inhibited the neurite extension (Cusack et al., 2004; Jugloff et al., 2005). MeCP2^{S80A} could not be phosphorylated and lacked the chromatin binding ability (Tao et al., 2009). In addition, it was found that S80-phosphorylated MeCP2 could suppress nuclear miRNA processing by binding Drosha, the initial processing complex for generating pre-miRNAs, and regulate Drosha function in an activity-dependent manner (Cheng et al., 2014). In our study, the wild type MeCP2 promoted neurite outgrowth and effectively rescued the phenotype induced by miR-130a, whereas MeCP2^{S80A} lost this activity (Fig. 5). Because MeCP2 also regulates microRNA biogenesis, including that of miR-130a (Urdinguio et al., 2010), it might be difficult to interpret complex results if the rescue experiments were performed in MeCP2-null background. These results suggest that miR-130a inhibits neurite outgrowth by targeting MeCP2, and that MeCP2 may affect neurite outgrowth via its activity in regulating chromatin structure and/or modulating miRNA processing. Much more work is necessary to further understand the underlying mechanisms, especially the reciprocal relationship between MeCP2 and miR-130a.

Reduced dendritic spine density is a common feature in individuals with Rett syndrome (Phillips and Pozzo-Miller, 2015; Xu et al., 2014), knockdown of the endogenous MeCP2 also reduced dendritic spine density (Chapleau et al., 2009). Consistent with this, expression of MeCP2 rescued the decreased spine density caused by miR-130a. It has been reported that dendritic spine density in terminal dendritic branches of young transgenic mice expressing MeCP2 (a *MECP2* duplication syndrome mouse model) was initially higher than the control group and then decreased after postnatal week 12 (Jiang et al., 2013). The dendrite outgrowth phenotype in the MeCP2 overexpressing mice during the early stage (See Fig. 1 of Jiang et al., 2013) is consistent with our observation. On the other hand, it has also been reported that expression of MeCP2 inhibited dendritic spine development in rat hippocampal slice cultures (Cheng et al., 2014). Data from different studies suggest that both MeCP2 overexpression and MeCP2 loss of function result in complex phenotypes in dendritic development (Na et al., 2013; Zhou et al., 2006; Jiang et al., 2013; Cheng et al., 2014). Therefore, the observation in our neuronal culture system that MeCP2 increased dendritic spine density in miR-130a expressing neurons may reflect a delicate balance and dynamic changes in the combinatorial effects of MeCP2 and microRNAs during dendritic development.

Neurons require proper homeostasis of MeCP2 to function normally (Lombardi et al., 2015). Loss of function of MeCP2 leads to severe neurodevelopmental diseases such as Rett syndrome, autism and schizophrenia; whereas the gain of function of MeCP2 may cause *MECP2* duplication syndrome (Lombardi et al., 2015; Van Esch, 2011). Phenotypes of Rett syndrome and *MECP2* duplication syndrome in mouse models can be reversed if MeCP2 level is properly restored (Robinson et al., 2012; Sztainberg et al., 2015). Taken together, our data support that miR-130a exhibits an inhibitory role on neurite outgrowth, dendritic branching and dendritic spine maturation, partially by targeting MeCP2. Seminal reviews on miRNAs have discussed about miRNAs as multi-facet regulators acting beyond simple repression of gene expression: buffering noise in the expression of their targets or setting a threshold-linear response (Bartel, 2009; Hausser and Zavolan, 2014). In this way, miR-130a could serve to buffer changes in MeCP2 protein expression and may help maintain MeCP2 at an appropriate level, which is essential for its function. Future studies are necessary to elucidate the complex mechanisms controlling all these events.

MATERIALS AND METHODS

Prediction of target genes for miR-130a by bioinformatics analyses

Target genes for miR-130a were predicted using TargetScan, miRanda and PicTar. The intersection of the results from all three tools was accepted as the potential targets. The Entrez Gene IDs of potential targets were downloaded from Ensembl and entered into KOBAS2.0 (with default setting) for annotation of gene pathways, disease associated genes and GeneOntology (GO) terms. The result of annotation was then used to identify statistically significantly enriched pathways, diseases and GO terms using protein-coding genes from the whole human genome set (20,192 genes) as background.

Cell culture and transfection

Embryonic day 18 (E18) rat cortical neurons were cultured and transfected following published protocols (Gao et al., 2010; Guo et al., 2011). Briefly, following electroporation (Amaxa Nucleofector II), neurons were seeded onto poly-D-lysine coated coverslips in 12-well plates. In the experiments for analyses of dendrite morphology and spine density, Lipofectamine2000 was used for transfecting neurons at 7 days and 21 days *in vitro* (DIV).

Plasmids and RNA inhibitors

MeCP2 and MeCP2^{S80A} plasmids were as described previously (Cheng et al., 2014). MiR-130a plasmid was generated using pCDH-CMV-MCS-EF1-copGFP (CD511B-1, System Biosciences) according to the manufacturer's instructions. MiR-130a genomic sequence was amplified with the following primers: forward, 5′-TGCTCTAGAGGTCATCTGAGAGTGTTGCCT-3′; reverse, 5′-CCGGGATCCTGACCCTCAGTTTTTCATCCA-3′. The miR-130a inhibitor was a methylation-modified RNA oligonucleotide complementary to miR-130a (GenePharma).

Luciferase reporter assay

Plasmids expressing wild-type or mutant human *MECP2* 3′UTR reporter genes containing a 415-bp fragment including the most conserved miR-130a binding site (shown as the "240 site" in Fig. 4A) were constructed using psiCHECK2 vector (Promega). The PCR amplification primers were as follows, wild-type MECP2-3′UTR primers: forward primer, 5′-CCGCTCGAGCGGAGCGGATTGCAAAGC-3′; and reverse primer, 5′-ATTTGCGGCCGCTGTAGACGGGGCACTGATGG-3′. The mutant forward primer is designed across the first miR-130a binding site, introducing a mutation that scrambles the entire seed region (underlined portion, see Fig. 4A): MECP2-3′ UTR mutant primers: forward, 5′-GGCCAGAAGTAGCT<u>AACGTGAT</u> TTCTA AACTAGGCTC-3′; and reverse, 5′-GCCTAGTTTAGA AATCACGTTAGCTACTTCTGGCCC-3′. The positive control was the same vector inserted with the complementary sequence of miR-130a. Luciferase assay was performed using Dual-Luciferase® Reporter Assay System (Promega). MiR-130a and MECP2 3′UTR were co-transfected to HEK293 cells at the ratio of 36.5:1. 48 h following transfection, cells were lysed with the activity of Renilla luciferase measured according to the manufacturer's instructions.

RNA extraction and RT-PCR

Total RNA was extracted using TriZol reagent (Invitrogen). Genomic DNA was removed with RNase-free DNase (NEB). Total cDNA was reverse-transcribed using SuperScript™ III Reverse Transcriptase

(Invitrogen). PCR primers of MeCP2 are: forward, 5′-AGAGGAA GTCTGGTCGCTCT-3′, reverse, 5′-CCAGGCTTTCTACCCCGTTT-3′. Primary miR-130a: forward, 5′-AGGATGAGAGGAAGGCTGTG-3′, reverse, 5′-AGAAAACAGTGACGCTGAGG-3′; GAPDH: forward, 5′-CCCCCAATGTATCCGTTGTG-3′, reverse, 5′-TAGCCCAGGATG CCCTTTAGT-3′; U6: forward, 5′-CTCGCTTCGGCAGCACA-3′, reverse, 5′-AACGCTTCACGAATTTGCGT-3′.

RT-PCR detection of mature miR-130a was performed using poly (T) adaptor-PCR as described previously (Shi et al., 2012). Poly (T) adaptor: 5′-AAGCAGTGGTATCAACGCAGAGTGC$(T)_{30}$VN-3′; poly(T) adaptor reverse primer: 5′-CTCACACGACTCACGACAGG GCAAGCAGTGGTATCAACGCAGAGTG-3′; miR-130a specific forward primer: 5′-TGCGGCAGTGCAATGTTAAAAGGGCAT-3′.

TUNEL assay

E18 rat cortical neurons were dissociated and electroporated with control vector or miR-130a plasmid. Cells were cultured for 3 days. Terminal deoxyribonucleotidyl transferase (TDT)-mediated dUTP-digoxigenin nick end labeling (TUNEL) was performed using In Situ Cell Death Detection Kit (Roche) as described in (Zhu et al., 2014).

Western blotting analysis and Immunostaining

MeCP2 antibody was prepared as previously described (Hu et al., 2006). Tubulin antibody and secondary antibodies were from ProteinTech Group. Protein lysates were prepared from rat brain tissues or cultured cells for Western blotting analyses using corresponding antibodies following our previously published protocols with a chemiluminescence HRP detection kit (Millipore) (Guo et al., 2011; Deng et al., 2015).

For immunostaining, cells were washed in pre-warmed phosphate-buffered saline (PBS) and fixed in 4% paraformaldehyde (PFA). The fixed cells were permeabilized and blocked with 10% normal goat serum (NGS) plus 0.3% Triton X-100 in PBS, incubated with primary antibody: rabbit anti-beta tubulin antibody (1:70) (PTG). After washing 5 times, the cells were incubated with secondary antibody conjugated to Alexa Fluor® 594 (1:500) (Invitrogen) and stained with 4',6'-diamidino-2-phenylindole (DAPI; 1 μg/mL). The coverslips were mounted with PermaFluor aqueous mounting medium (Thermo).

Morphological analyses of axons and dendrites

The images for neurite outgrowth were taken using an inverted fluorescence microscope (Nikon) following a previously published protocols (Gao et al., 2010). The images for dendrite morphology and dendritic spines were taken using a confocal microscope (Leica SP8); image analyses were carried out following the published protocols (Langhammer et al., 2010; Srivastava et al., 2011).

Statistical analyses

The data was normalized to the control group. Statistical analyses were performed using Mann-Whitney test, between individual groups and two-way ANOVA for comparing distributions. *, $P < 0.05$; **, $P < 0.01$; ***, $P < 0.001$. The bar diagram represents mean values ± SEM.

ACKNOWLEDGMENTS

We thank members of the Wu lab for helpful suggestions. We thank Dr. Yan Zhu and members of the Wu lab for discussions, suggestions and critical reading of the manuscript. MC, LZ, JL are supported by grants from the National Basic Research Program (973 Program) (No. 2013CB917803) and the National Natural Science Foundation of China (Grant No. 91132710). ZQ is supported by the National Natural Science Foundation of China (Grant Nos. 91432111 and 81527901). KH is supported by the National Natural Science Foundation of China (Grant Nos. 81070907 and 81271255). WM is supported by NIH (F30 NS090893). JYW is supported by NIH (RO1AG033004).

ABBREVIATIONS

MiR-130a, microRNA-130a; miR-130a-inh, microRNA-130a-inhibitor; MECP2, methyl-CpG-binding protein 2; TUNEL, TdT-mediated dUTP nick end labeling; Tuj1, neuronal class III β-tubulin.

COMPLIANCE WITH ETHICS GUIDELINES

All authors declare that there is no conflict of interest. All national and institutional guidelines for the care and use of laboratory animals were followed. This work did not involve any studies with human subjects performed by the any of the authors.

REFERENCES

Agarwal V, Bell GW, Nam J-W, Bartel DP, Izaurralde E (2015) Predicting effective microRNA target sites in mammalian mRNAs. eLife 4:e05005

Amir RE, Van den Veyver IB, Wan M, Tran CQ, Francke U, Zoghbi HY (1999) Rett syndrome is caused by mutations in X-linked MECP2, encoding methyl-CpG-binding protein 2. Nat Genet 23:185–188

Baek D, Villén J, Shin C, Camargo FD, Gygi SP, Bartel DP (2008) The impact of microRNAs on protein output. Nature 455:64–71

Bartel DP (2009) MicroRNAs: target recognition and regulatory functions. Cell 136:215–233

Betel D, Wilson M, Gabow A, Marks DS, Sander C (2008) The microRNA.org resource: targets and expression. Nucleic Acids Res 36:D149–D153

Chahrour M, Jung SY, Shaw C, Zhou X, Wong ST, Qin J, Zoghbi HY (2008) MeCP2, a key contributor to neurological disease, activates and represses transcription. Science 320:1224–1229

Chao H-T, Zoghbi HY, Rosenmund C (2007) MeCP2 controls excitatory synaptic strength by regulating glutamatergic synapse number. Neuron 56:58–65

Chapleau CA, Calfa GD, Lane MC, Albertson AJ, Larimore JL, Kudo S, Armstrong DL, Percy AK, Pozzo-Miller L (2009) Dendritic spine pathologies in hippocampal pyramidal neurons from Rett syndrome brain and after expression of Rett-associated MECP2 mutations. Neurobiol Dis 35:219–233

Cheng T-L, Wang Z, Liao Q, Zhu Y, Zhou W-H, Xu W, Qiu Z (2014) MeCP2 suppresses nuclear microRNA processing and dendritic growth by regulating the DGCR8/Drosha complex. Dev Cell 28:547–560

Coutinho AM, Oliveira G, Katz C, Feng J, Yan J, Yang C, Marques C, Ataíde A, Miguel TS, Borges L (2007) MECP2 coding sequence and 3′-UTR variation in 172 unrelated autistic patients. Am J Med Genet B 144:475–483

Coy JF, Sedlacek Z, Bächner D, Delius H, Poustka A (1999) A complex pattern of evolutionary conservation and alternative polyadenylation within the long 3′-untranslated region of the methyl-CpG-binding protein 2 gene (MeCP2) suggests a regulatory role in gene expression. Hum Mol Genet 8:1253–1262

Cusack SM, Rohn TT, Medeck RJ, Irwin KM, Brown RJ, Mercer LM, Oxford JT (2004) Suppression of MeCP2β expression inhibits neurite extension in PC12 cells. Exp Cell Res 299:442–453

Deng J, Yang M, Chen Y, Chen X, Liu J, Sun S, Cheng H, Li Y, Bigio EH, Mesulam M, Xu Q, Du S, Fushimi K, Zhu L, Wu JY (2015) FUS interacts with HSP60 to promote mitochondrial damage. PloSGenetics 11(9):e1005357

Eda A, Takahashi M, Fukushima T, Hohjoh H (2011) Alteration of microRNA expression in the process of mouse brain growth. Gene 485:46–52

Ethell IM, Pasquale EB (2005) Molecular mechanisms of dendritic spine development and remodeling. Prog Neurobiol 75:161–205

Feliciano DM, Bordey A, Bonfanti L (2015) Noncanonical sites of adult neurogenesis in the mammalian brain. Cold Spring Harbor Perspect Biol 7:a018846

Finnegan EF, Pasquinelli AE (2013) MicroRNA biogenesis: regulating the regulators. Crit Rev Biochem Mol Biol 48:51–68

Futamura M, Monden Y, Okabe T, Fujita-Yoshigaki J, Yokoyama S, Nishimura S (1995) Trichostatin A inhibits both ras-induced neurite outgrowth of PC12 cells and morphological transformation of NIH3T3 cells. Oncogene 10:1119–1123

Gao X, Joselin AP, Wang L, Kar A, Ray P, Bateman A, Goate AM, Wu JY (2010) Progranulin promotes neurite outgrowth and neuronal differentiation by regulating GSK-3β. Protein Cell 1:552–562

Gessert S, Bugner V, Tecza A, Pinker M, Kühl M (2010) FMR1/FXR1 and the miRNA pathway are required for eye and neural crest development. Dev Biol 341:222–235

Greco SJ, Rameshwar P (2007) MicroRNAs regulate synthesis of the neurotransmitter substance P in human mesenchymal stem cell-derived neuronal cells. Proc Natl Acad Sci 104:15484–15489

Guo W, Chen Y, Zhou X, Kar A, Ray P, Chen X, Rao EJ, Yang M, Ye H, Zhu L, Liu J, Xu M, Yang Y, Wang C, Zhang D, Bigio EH, Mesulam M, Shen Y, Xu Q, Fushimi K, Wu JY (2011) An ALS-associated mutation in the TDP-43 gene enhances protein aggregation, fibril formation and neurotoxicity. Nat Struct Mol Biol 18:822–830

Hansen KF, Sakamoto K, Wayman GA, Impey S, Obrietan K (2010) Transgenic miR132 alters neuronal spine density and impairs novel object recognition memory. PLoS ONE 5:e15497

Hausser J, Zavolan M (2014) Identification and consequences of miRNA–target interactions—beyond repression of gene expression. Nat Rev Genet 15:599–612

Hoesel B, Bhujabal Z, Przemeck GK, Kurz-Drexler A, Weisenhorn DMV, Angelis MHD, Beckers J (2010) Combination of in silico and in situ hybridisation approaches to identify potential Dll1 associated miRNAs during mouse embryogenesis. Gene Expr Patterns 10:265–273

Hu K, Nan X, Bird A, Wang W (2006) Testing for association between MeCP2 and the brahma-associated SWI/SNF chromatin-remodeling complex. Nat Genet 38(9):962–964

Im H-I, Kenny PJ (2012) MicroRNAs in neuronal function and dysfunction. Trends Neurosci 35:325–334

Jiang M, Ash RT, Baker SA, Suter B, Ferguson A, Park J, Rudy J, Torsky SP, Chao H-T, Zoghbi HY (2013) Dendritic arborization and spine dynamics are abnormal in the mouse model of MECP2 duplication syndrome. J Neurosci 33:19518–19533

Jones PL, Veenstra GCJ, Wade PA, Vermaak D, Kass SU, Landsberger N, Strouboulis J, Wolffe AP (1998) Methylated DNA and MeCP2 recruit histone deacetylase to repress transcription. Nat Genet 19:187–191

Jugloff DG, Jung BP, Purushotham D, Logan R, Eubanks JH (2005) Increased dendritic complexity and axonal length in cultured mouse cortical neurons overexpressing methyl-CpG-binding protein MeCP2. Neurobiol Dis 19:18–27

Krek A, Grün D, Poy MN, Wolf R, Rosenberg L, Epstein EJ, MacMenamin P, da Piedade I, Gunsalus KC, Stoffel M et al (2005) Combinatorial microRNA target predictions. Nat Genet 37:495–500

Kutzing MK, Langhammer CG, Luo V, Lakdawala H, Firestein BL (2010) Automated Sholl analysis of digitized neuronal morphology at multiple scales. J Vis Exp. 45:2354

Langhammer CG, Previtera ML, Sweet ES, Sran SS, Chen M, Firestein BL (2010) Automated Sholl analysis of digitized neuronal morphology at multiple scales: whole cell Sholl analysis versus Sholl analysis of arbor subregions. Cytom. A 77:1160–1168

Lefebvre JL, Sanes JR, Kay JN (2015) Development of dendritic form and function. Annu Rev Cell Dev Biol 31:741–777

Lewis JD, Meehan RR, Henzel WJ, Maurer-Fogy I, Jeppesen P, Klein F, Bird A (1992) Purification, sequence, and cellular localization of a novel chromosomal protein that binds to methylated DNA. Cell 69:905–914

Lombardi LM, Baker SA, Zoghbi HY (2015) MECP2 disorders: from the clinic to mice and back. J Clin Investig 125:2914

Maunakea AK, Chepelev I, Cui K, Zhao K (2013) Intragenic DNA methylation modulates alternative splicing by recruiting MeCP2 to promote exon recognition. Cell Res 23:1256–1269

McGowan H, Pang ZP (2015) Regulatory functions and pathological relevance of the MECP2 3′-UTR in the central nervous system. Cell Regen 4:1

McNeill E, Van Vactor D (2012) MicroRNAs shape the neuronal landscape. Neuron 75:363–379

Merson TD, Bourne JA (2014) Endogenous neurogenesis following ischaemic brain injury: insights for therapeutic strategies. Int J Biochem Cell Biol 56:4–19

Na ES, Nelson ED, Kavalali ET, Monteggia LM (2013) The impact of MeCP2 loss-or gain-of-function on synaptic plasticity. Neuropsychopharmacology 38:212–219

Nan X, Ng H-H, Johnson CA, Laherty CD, Turner BM, Eisenman RN, Bird A (1998) Transcriptional repression by the methyl-CpG-binding protein MeCP2 involves a histone deacetylase complex. Nature 393:386–389

Pelka GJ, Watson CM, Christodoulou J, Tam PP (2005) Distinct expression profiles of Mecp2 transcripts with different lengths of 3′-UTR in the brain and visceral organs during mouse development. Genomics 85:441–452

Persengiev SP, Kilpatrick DL (1996) Nerve growth factor induced differentiation of neuronal cells requires gene methylation. NeuroReport 8:227–231

Phillips M, Pozzo-Miller L (2015) Dendritic spine dysgenesis in autism related disorders. Neurosci Lett 601:30–40

Poluch S, Juliano SL (2015) Fine-tuning of neurogenesis is essential for the evolutionary expansion of the cerebral cortex. Cereb Cortex 25:346–364

Robinson L, Guy J, McKay L, Brockett E, Spike RC, Selfridge J, De Sousa D, Merusi C, Riedel G, Bird A (2012) Morphological and functional reversal of phenotypes in a mouse model of Rett syndrome. Brain 135:2699–2710

Rosenfeld JA, Ballif BC, Torchia BS, Sahoo T, Ravnan JB, Schultz R, Lamb A, Bejjani BA, Shaffer LG (2010) Copy number variations associated with autism spectrum disorders contribute to a spectrum of neurodevelopmental disorders. Genet Med 12:694–702

Santos M, Yan J, Temudo T, Oliveira G, Vieira JP, Fen J, Sommer S, Maciel P (2008) Analysis of highly conserved regions of the 3′-UTR of MECP2 gene in patients with clinical diagnosis of Rett syndrome and other disorders associated with mental retardation. Dis Mark 24:319–324

Schoenfeld TJ, Cameron HA (2015) Adult neurogenesis and mental illness. Neuropsychopharmacology 40:113–128

Selemon L, Zecevic N (2015) Schizophrenia: a tale of two critical periods for prefrontal cortical development. Transl Psychiatry 5: e623

Shi R, Sun Y-H, Zhang X-H, Chiang VL (2012) Poly (T) adaptor RT-PCR. In next-generation MicroRNA expression profiling technology. Methods Mol Biol 822:53–66

Shibayama A, Cook EH, Feng J, Glanzmann C, Yan J, Craddock N, Jones IR, Goldman D, Heston LL, Sommer SS (2004) MECP2 structural and 3′-UTR variants in schizophrenia, autism and other psychiatric diseases: A possible association with autism. Am J Med Genet B 128:50–53

Shigehiko N, Takeshi F (1978) Stimulatory effects of substance P and nerve growth factor (NGF) on neurite outgrowth in embryonic chick dorsal root ganglia. Neuropharmacology 17:73–76

Søe MJ, Møller T, Dufva M, Holmstrøm K (2011) A sensitive alternative for microRNA in situ hybridizations using probes of 2-O-methyl RNA+ LNA. J Histochem Cytochem 59:661–672

Srivastava DP, Woolfrey KM, Penzes P (2011) Analysis of dendritic spine morphology in cultured CNS neurons. J Vis Exp. 53:e2794

Sztainberg Y, Chen H-M, Swann JW, Hao S, Tang B, Wu Z, Tang J, Wan Y-W, Liu Z, Rigo F, Zoghbi HY (2015) Reversal of phenotypes in MECP2 duplication mice using genetic rescue or antisense oligonucleotides. Nature 528(7580):123–126

Takano T, Xu C, Funahashi Y, Namba T, Kaibuchi K (2015) Neuronal polarization. Development. 142(12):2088–2093

Tao J, Hu K, Chang Q, Wu H, Sherman NE, Martinowich K, Klose RJ, Schanen C, Jaenisch R, Wang W, Sun YE (2009) Phosphorylation of MeCP2 at Serine 80 regulates its chromatin association and neurological function. Proc Natl Acad Sci 106:4882–4887

Urdinguio RGI, Fernandez AF, Lopez-Nieva P, Rossi S, Huertas D, Kulis M, Liu CG, Croce CM, Calin GA, Esteller M (2010) Disrupted microRNA expression caused by Mecp2 loss in a mouse model of Rett syndrome. Epigenetics. 5(7):656–663

Van Esch H (2011) MECP2 duplication syndrome. Mol Syndromol 2:128–136

Vasu MM, Anitha A, Thanseem I, Suzuki K, Yamada K, Takahashi T, Wakuda T, Iwata K, Tsujii M, Sugiyama T (2014) Serum microRNA profiles in children with autism. Molecular autism 5:40

Vidigal JA, Ventura A (2015) The biological functions of miRNAs: lessons from in vivo studies. Trends Cell Biol 25:137–147

Walker JC, Harland RM (2008) Expression of microRNAs during embryonic development of *Xenopus tropicalis*. Gene Expression Patterns 8:452–456

Wang Y, Baraban SC (2007) Granule cell dispersion and aberrant neurogenesis in the adult hippocampus of an LIS1 mutant mouse. Dev Neurosci 29:91–98

Wegiel J, Kuchna I, Nowicki K, Imaki H, Wegiel J, Marchi E, Ma SY, Chauhan A, Chauhan V, Bobrowicz TW (2010) The neuropathology of autism: defects of neurogenesis and neuronal migration, and dysplastic changes. Acta Neuropathol 119:755–770

Wilczynska A, Bushell M (2015) The complexity of miRNA-mediated repression. Cell Death Differ 22:22–33

Winner B, Winkler J (2015) Adult Neurogenesis in Neurodegenerative Diseases. Cold Spring Harbor Perspect Biol 7:a021287

Xie C, Mao X, Huang J, Ding Y, Wu J, Dong S, Kong L, Gao G, Li C-Y, Wei L (2011) KOBAS 2.0: a web server for annotation and identification of enriched pathways and diseases. Nucleic Acids Res 39:W316–W322

Xu X, Miller EC, Pozzo-Miller L (2014) Dendritic spine dysgenesis in Rett syndrome. Frontiers in neuroanatomy 8

Young JI, Hong EP, Castle JC, Crespo-Barreto J, Bowman AB, Rose MF, Kang D, Richman R, Johnson JM, Berget S (2005) Regulation of RNA splicing by the methylation-dependent transcriptional repressor methyl-CpG binding protein 2. Proc Natl Acad Sci USA 102:17551–17558

Zhao C, Deng W, Gage FH (2008) Mechanisms and functional implications of adult neurogenesis. Cell 132:645–660

Zhou Z, Hong EJ, Cohen S, Zhao W-N, Ho HYH, Schmidt L, Chen WG, Lin Y, Savner E, Griffith EC (2006) Brain-specific phosphorylation of MeCP2 regulates activity-dependent Bdnf transcription, dendritic growth, and spine maturation. Neuron 52:255–269

Zhu L, Xu M, Yang M, Yang Y, Li Y, Deng J, Ruan L, Liu J, Du S, Liu X, Feng W, Fushimi K, Bigio EH, Mesulam M, Wang C, Wu JY (2014) An ALS-mutant TDP-43 neurotoxic peptide adopts an anti-parallel β-structure and induces TDP-43 redistribution. Hum Mol Genet. 23(25):6863–6877

11

Genetic approach to track neural cell fate decisions using human embryonic stem cells

Xuemei Fu[1,2]✉, Zhili Rong[2], Shengyun Zhu[1,2], Xiaocheng Wang[2], Yang Xu[2], Blue B. Lake[2]✉

[1] Shenzhen Children's Hospital, Shenzhen 518026, China
[2] Section of Molecular Biology, Division of Biological Sciences, University of California, San Diego, 9500 Gilman Drive, La Jolla, CA 92093-0322, USA
✉ Correspondence: fxmzj2004@163.com (X. Fu), b1lake@ucsd.edu (B. B. Lake)

ABSTRACT

With their capability to undergo unlimited self-renewal and to differentiate into all cell types in the body, human embryonic stem cells (hESCs) hold great promise in human cell therapy. However, there are limited tools for easily identifying and isolating live hESC-derived cells. To track hESC-derived neural progenitor cells (NPCs), we applied homologous recombination to knock-in the *mCherry* gene into the *Nestin* locus of hESCs. This facilitated the genetic labeling of Nestin positive neural progenitor cells with mCherry. Our reporter system enables the visualization of neural induction from hESCs both *in vitro* (embryoid bodies) and *in vivo* (teratomas). This system also permits the identification of different neural subpopulations based on the intensity of our fluorescent reporter. In this context, a high level of mCherry expression showed enrichment for neural progenitors, while lower mCherry corresponded with more committed neural states. Combination of mCherry high expression with cell surface antigen staining enabled further enrichment of hESC-derived NPCs. These mCherry$^+$ NPCs could be expanded in culture and their differentiation resulted in a down-regulation of mCherry consistent with the loss of Nestin expression. Therefore, we have developed a fluorescent reporter system that can be used to trace neural differentiation events of hESCs.

Xuemei Fu and Zhili Rong contribute equally to this work.

Author Contribution Conceived and designed the experiments: XF ZR YX. Performed the experiments: XF ZR SZ XCW BBL Analyzed the data: XF ZR YX BBL. Wrote the paper: XF YX BBL.

KEYWORDS Nestin, knock-in, human embryonic stem cells, neural progenitor cells

INTRODUCTION

Human embryonic stem cells (hESCs) undergo unlimited self-renewal and retain pluripotency to differentiate into all cell types of the body, making them invaluable tools for development of cellular therapeutics (Fu and Xu, 2011). Modeling human disease has further become possible with advances in genetic modification in hESCs either through homologous recombination (Song et al., 2010; Hockemeyer et al., 2011) or gene knockdown (Ordonez et al., 2012), as well as reprogramming of cells having disease-causing mutations into an ESC-like state (human induced pluripotent stem cells or hiPSCs) (Unternaehrer and Daley, 2011). Given the limitation of using mouse systems to develop human therapeutics, the use of these approaches can provide more physiologically relevant insight into disease pathology and help to identify more efficacious drug targets. These approaches additionally permit disease modeling of tissues that are previously unattainable or that show pathologies too advanced in donor tissues. This includes studies on the mechanisms underlying neuronal degeneration associated with central nervous system (CNS) diseases (e.g. Parkinson's, Alzheimer's or Amyotrophic lateral sclerosis).

Studies using human pluripotent stem cell derived neurons to model these diseases or generate neural tissue replacements rely upon the effectiveness of protocols that recapitulate normal embryonic development to direct the differentiation of hESCs into the neural cell types afflicted by the disease (Aas et al., 1996). A key step in this process is to differentiate hESCs into neural stem cells (NSCs) that can be further differentiated into neurons, astrocytes and glia. While significant progress has been made in improving the differentiation of hESCs into NSCs (Reubinoff et al., 2001; Itsykson et al., 2005; Wu et al., 2007; Chambers et al., 2009), this step remains an imperfect process with resulting cultures comprising a mixture of undifferentiated ESCs, NSCs, neurons, glia and non-CNS cell types including neural crest. Subsequent differentiation of these mixed cultures would fail to generate single pure lineages needed for transplantation, modeling or population based gene expression studies (e.g. RNA-seq).

To circumvent these issues, several groups have had success in using combinatorial cell surface marker staining to enrich for NSCs (Yuan et al., 2011; Israel et al., 2012). However this method requires further sorting strategies for continued monitoring of resultant cell differentiation states. Promoter-based viral reporter systems for the neural progenitor marker *Nestin* has also been used to specifically label NSCs within the human fetal brain and could be used to purify these cells for expansion or for further differentiation both in culture and following transplantation (Keyoung et al., 2001). Several groups have used *Nestin* reporter systems to identify and track neural progenitors both *in vivo* and *in vitro* (Sawamoto et al., 2001; Lenka et al., 2002; Mignone et al., 2004; Noisa et al., 2010). These transgenic or viral based strategies, however, involve integration of the exogenous DNA into random genomic loci that can modulate the promoter activity, depending on the epigenetic state around the integration site. To address this issue, we decided to generate knock-in cell lines using a *Nestin*-based fluorescent reporter system that would permit not only a simpler sorting strategy for isolating NPCs/NSCs but also would enable their live tracking during expansion or differentiation. Unlike transgenic approaches (Placantonakis et al., 2009), this system would ensure reporter expression specifically driven by the endogenous *Nestin* promoter and/or enhancer, and would not introduce any other genetic alterations that arise from random integration strategies. This would permit specific mCherry expression during neural differentiation of hESCs and the ability to generate enriched NSC cultures where multi-potentiality could be easily monitored. These cells will provide a unique opportunity for studies requiring purer neural cultures and live cell tracking, including transplantation, disease modeling or whole transcriptome analyses (Peljto and Wichterle, 2011).

RESULTS

Generation of a knock-in Nestin reporter cell line

We have recently developed a bacterial artificial chromosome (BAC) based targeting strategy that enables a high efficiency of homologous recombination in hESCs (Song et al., 2010). To knock-in the *mCherry* coding region into the *Nestin* locus, we constructed a BAC targeting vector in which the ATG of one endogenous *Nestin* allele was replaced with that of *mCherry* (Fig. 1A and 1B). In order to screen for homologous recombinants, one homologous arm of the BAC vector was shortened to 6.5 kb. Recombinants were then screened by Southern blotting to identify both the 8.8 kb germline band and the 11.2 kb mutant band (Fig. 1B). The LoxP-flanked selection marker cassette (*CAG-Neo*) was subsequently excised from the genome (Fig. 1C) through transient expression of the Cre enzyme (Song et al., 2007). Knock-in hESCs showed normal karyotypes (Fig. 2A) and retained the potential to differentiate into all three germ layers including neuroectoderm (Fig. 2B and 2C). Furthermore, mCherry-positive neural-like cells were found to migrate from teratoma tissues when they were cultured in neural progenitor medium (Fig. 2F–H). Therefore, our knock-in cell line may provide a means for monitoring the formation of neural lineages during early hESC differentiation.

mCherry fluorescence follows Nestin expression dynamics

To further validate whether our genetic tagging of the *Nestin* locus could allow tracing of neural differentiation from hESCs, *Nestin:mCherry* hESCs were differentiated with an established embryoid body (EB) neural induction protocol (Reubinoff et al., 2001; Zhang et al., 2001). mCherry positive cells could be observed soon after EB treatment with neural induction medium coincident with the expected emergence of neural progenitors (Fig. 3A). Further, mCherry positive cells were found to migrate out from attached EBs (Fig. 3B and 3C) and rapidly expanded to confluence similarly to that observed for teratoma tissues (Fig. 2F–H). This confirms that our *Nestin:mCherry* reporter system can monitor neural specification from both *in vitro* and *in vivo* differentiated hESCs.

Following out-migration from EB or teratoma tissues, mCherry positive cells displayed properties consistent with neural progenitor cells. They could be further expanded in culture for several passages (data not shown); showed significant proliferation as indicated by Ki-67 staining (Fig. 3D); and co-expressed endogenous Nestin (Fig. 3E). Furthermore, high mCherry expression was localized to neural rosette-like structures, as was previously reported for a transgenic *Nestin* reporter (Noisa et al., 2010), and co-localized with the neural stem cell marker Sox2 (Fig. 3F). Therefore, mCherry expression can faithfully label Nestin positive cycling progenitors during differentiation of hESCs into the neural lineage both *in vitro* and *in vivo*.

mCherry levels dropped during NPC differentiation

While our results indicate that the knock-in of *mCherry* into the *Nestin* locus recapitulates its endogenous expression, we next sought to define whether mCherry specifically labels neural progenitors. To do address this, mCherry positive cells were characterized using flow cytometry (Figs. 4A and 5). Interestingly, two different fractions showing low and high mCherry expression were revealed with global passaging of neural cultures coincident with an emerging population of

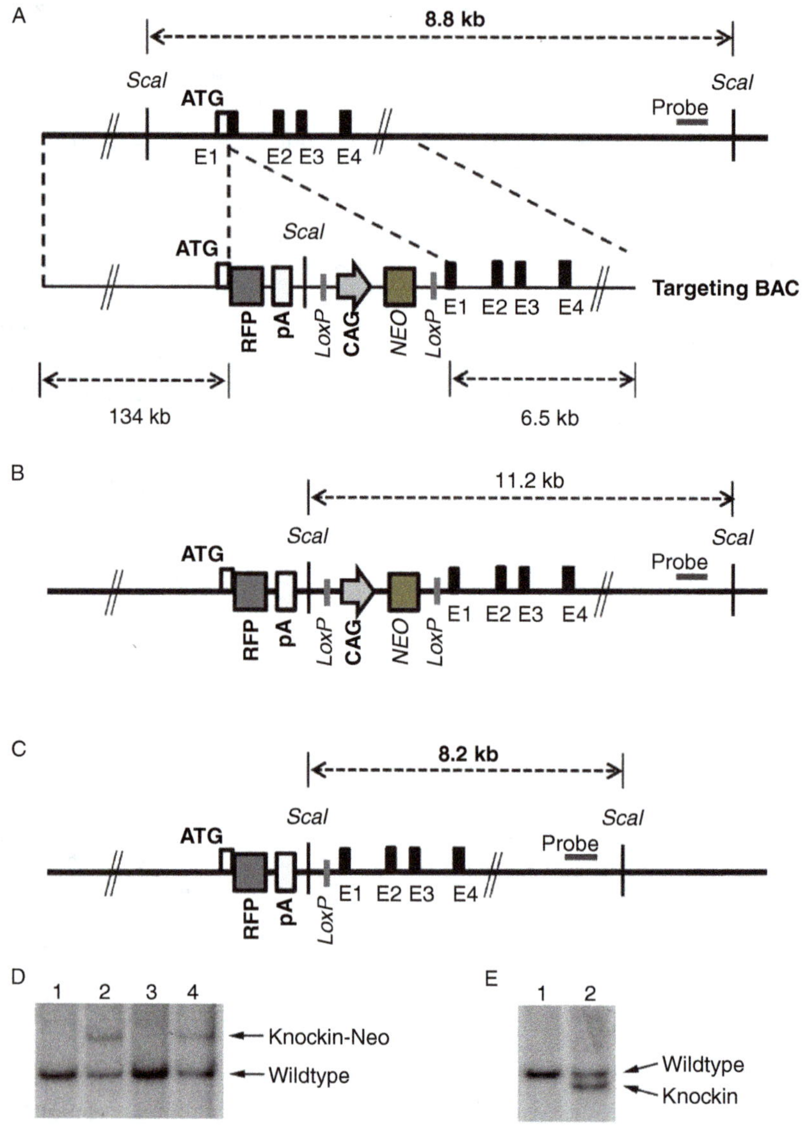

Figure 1. *Nestin:mCherry* knock-in strategy. (A) The germline configuration of the *Nestin* locus in hESCs showing: the exon containing the initiation ATG; the germline Scal fragment; the probe used for Southern blot analysis; and the BAC targeting vector containing the *mCherry* (RFP) expression cassette together with the LoxP-flanked selection marker. (B) Insertion of this cassette into the *Nestin* exon replaced the initiation ATG of *mCherry* with that of the *Nestin* gene and generated an altered Scal fragment size corresponding to the targeted allele. (C) The transient expression of Cre led to the excision of the selection marker allowing normal expression of the *mCherry* according to the endogenous *Nestin* gene. (D) Southern blot analysis of the targeted hESCs. Genomic DNA was digested with Scal and hybridized to the probe indicated in (A and B). The location of the targeted and wild-type fragments are indicated. Lanes 1 and 3: hESCs with random integration of the targeting construct. Lanes 2 and 4: homologous recombinants. (E) Southern blot analysis of the knock-in hESCs showing that the selection marker was excised from the targeted allele by LoxP/Cre-mediated deletion. Genomic DNA was digested with Scal and probed with the same targeting probe shown in (A, C).

CD44 positive cells (Fig. 4A) representing more differentiated neural cell types (Liu et al., 2004). Continued culture in this manner resulted in the progressive loss of mCherry expression over several passages (data not shown). Our results are consistent with $mCherry^{Lo}$ cells potentially representing more committed neural cell types as has been previously reported for a transgenic hESC system (Noisa et al., 2010). In fact, mouse Nestin reporters exhibited gradient levels of reporter activity during differentiation, with high levels associated with neural stem cells, lower levels observed in more differentiated yet dividing neuronal progenitors and finally little to no activity in fully differentiated neurons (Sawamoto et al., 2001; Lenka et al., 2002; Mignone et al., 2004). Consistently, mCherry expression was lost in terminally differentiated neurons that were labeled with a synapsin-GFP lentiviral reporter (Fig. 4B and 4C). In

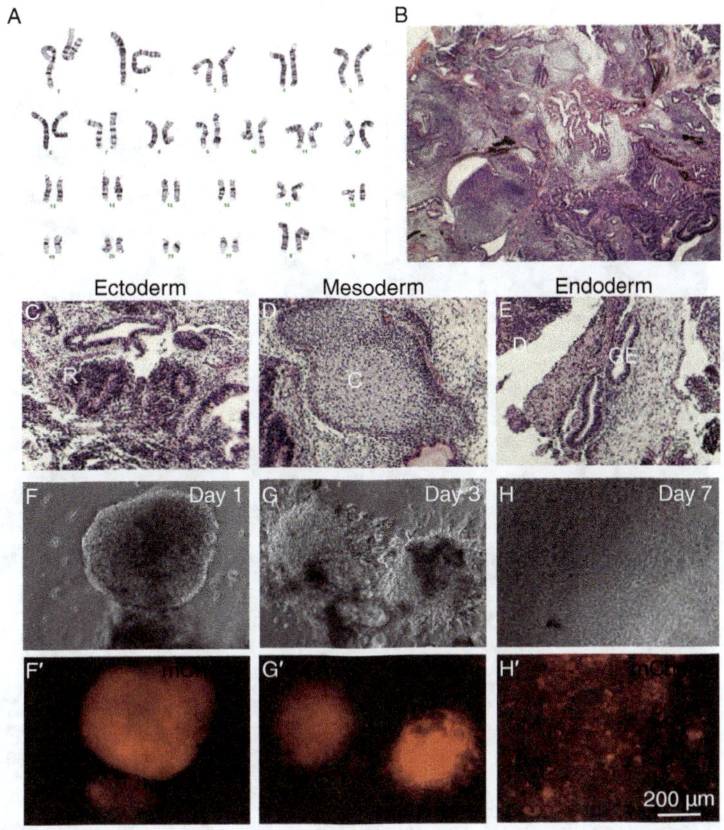

Figure 2. *Nestin-mCherry* knock-in hESCs. (A) Twenty metaphase cells of the knock-in hESCs were examined and exhibited normal karyotypes. (B) Knock-in hESCs formed well-differentiated teratomas in SCID mice. Cells of each of the three germ layers were identified in the teratomas including: (C) ectoderm (R, neural rosette); (D) mesoderm (C, cartilage); (E) endoderm (GE, gut-like epithelium). Out migration of mCherry positive neural cells were observed from teratomas formed by knock-in hESCs on day 1 (F–F′), day 3 (G–G′) and day 7 (H–H′) following attachment in NPC medium.

addition, $mCherry^{Hi}$ NPCs could be differentiated into both $GFAP^{+}$ astrocytes and $Tuj1^{+}$ neurons (Fig. 4D and 4E), confirming their multipotency. Therefore, our results demonstrate that mCherry levels correspond with endogenous *Nestin* expression, specifically labeling neural progenitors and lost upon their further differentiation.

One significant advantage to the labeling of NPCs with mCherry would be the potential for sorting and enriching this population. Furthermore, given the potential heterogeneity of neural progenitors in terms of stage of commitment and potentiality, it would be advantageous to specifically isolate or enrich bone fide neural stem cells. To determine the specificity of mCherry to NSCs or NPCs compared to more committed cell lineages, we used cell surface markers that highlight these different neural subpopulations. Previous reports have shown the use of multiple surface markers to identify NSCs based on a $CD184^{+}CD271^{-}CD44^{-}CD24^{+}$ staining signature (Yuan et al., 2011). Both CD271 and CD44 identify undifferentiated hESCs, as well as neural contaminants including neural crest stem cells, peripheral neural cells, more differentiated neuronal cell types, astrocyte progenitors and glia (Morrison et al., 1999; Luo et al., 2002; Liu et al., 2004; Lee et al., 2007). Alternatively, both CD184 and CD24 have been shown to be expressed on the surface of NSCs. Consistent with mCherry high expressers representing a more neural stem cell state, the $CD184^{+}CD44^{-}CD24^{+}$ population localized more within the $mCherry^{Hi}$ expressing cells in contrast to the $CD184^{-}CD24^{-}$ population (Fig. 5). Furthermore, the $CD271^{-}CD44^{-}CD24^{+}$ cells also mapped more to the $mCherry^{Hi}$ expressers unlike a similar $CD24^{-}$ population (Fig. 5). These results indicate that mCherry high expression is in part predicative of a stem cell state, but is not specific to this population (Fig. 5). This is likely the result of continued high Nestin expression in more committed neural progenitors as well as a potential lag in the mCherry expression compared to endogenous Nestin that is a result of their differing protein stabilities.

mCherry sorting strategy for NSC enrichment

While high mCherry expression occurs in neural stem cells (Fig. 3F), it in itself is not sufficient to fully purify this fraction from the global population. In order to achieve this, $mCherry^{Hi}$ expression would need to be combined with cell

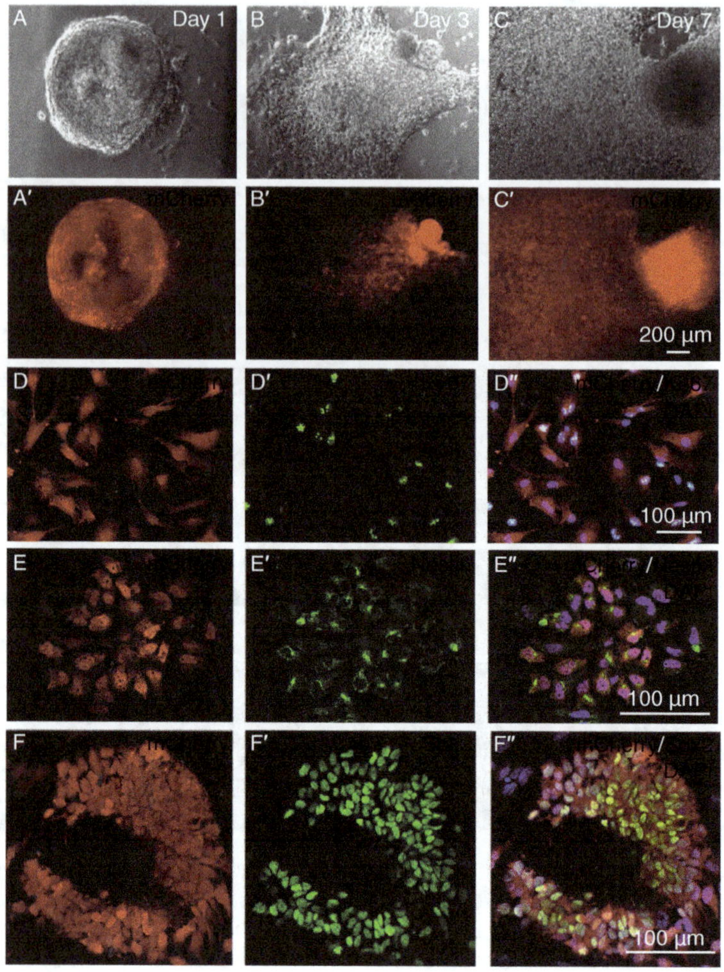

Figure 3. mCherry labeled Nestin positive neural progenitors derived from knock-in hESCs. Neural induction of embryoid bodies (EB) and corresponding mCherry expression on day 1 (A–A′), day 3 (B–B′) and day 7 (C–C′) following EB attachment. mCherry positive neural progenitor cells cultured in NPC medium showed active proliferation as indicated by Ki-67 staining (D–D″), co-expressed Nestin (E–E″) and co-localized with Sox2 in rosette-like structures (F–F″).

surface marker staining. Interestingly, neural cultures that were gated on the mCherry^{Hi} expression and that were additionally either CD271^{-}CD44^{-} or CD44^{-}CD184^{+} (Fig. 6A) were almost entirely CD24^{+} when neural cells were collected directly from induced EBs. This indicates a redundancy for this marker at this stage in neural differentiation. However, this trend was not necessarily observed after passaging of the total population, likely due to the spontaneous differentiation of NPCs (data not shown). Therefore, these results indicate that mCherry^{Hi} can substitute for the CD24^{+} staining when isolating NSCs directly from induced EBs, enabling a simpler sorting strategy for these cells. Consistently, the sorted CD184^{+}CD44^{-}CD271^{-} population, which was found to map to a mCherry^{Hi} population (Fig. 6B), showed significant further enrichment of the NSC markers *Sox1*, *Sox2* and *Pax6* over the mCherry^{Hi} population alone (Fig. 6C). Use of dual SMAD inhibitors Noggin and SB431542 during neural induction (Chambers et al., 2009) was found to significantly increase the number of NSCs sorted and showed a further increase in *Sox1* and *Pax6* expression levels in the enriched mCherry^{Hi} population (data not shown). Therefore, these results indicate that the combination of mCherry^{Hi} with a more limited surface marker sorting strategy that excludes CD24 can be sufficient to significantly enrich for NSCs. Furthermore, substituting CD24 for mCherry expression provides the distinct advantage of being able to further monitor the differentiation status of the NPC culture and permits tracking of terminal differentiation events.

DISCUSSION

Our studies demonstrate a knock-in approach that can be used together with existing strategies of purifying NPCs/NSCs derived from hESCs while additionally providing a means of tracking these cells live, both *in vitro* and *in vivo*.

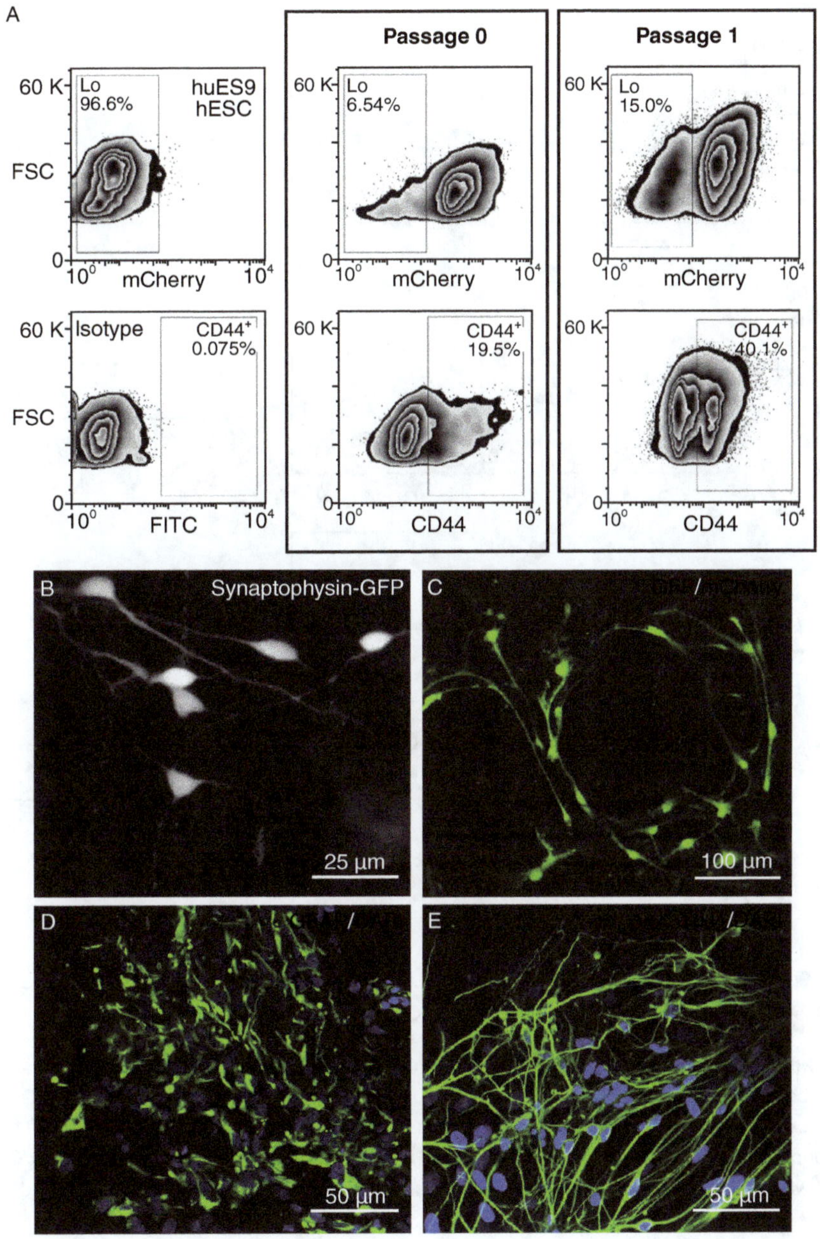

Figure 4. mCherry intensity is consistent with *Nestin* expression dynamics. (A) FACS analysis for mCherry expression levels in Hues9 hESCs or neural cultures derived from *Nestin:mCherry* knock-in Hues9 hESCs. Neural cultures that were out-migrating from EBs (Passage 0) or were passaged one time (Passage 1) were either stained with FITC isotype control or FITC labeled anti-CD44 antibody. Both mCherry low expressing fractions and CD44 positive fractions are indicated. (B) $mCherry^{Hi}$ population differentiates into synapsin positive neurons as shown by *Syn:GFP* reporter lentivirus. (C) mCherry expression is lost in terminally differentiated neurons identified by the *Syn:GFP* reporter. $mCherry^{Hi}$ NPCs can be differentiated into $GFAP^{+}$ astrocytes (D) and $Tuj1^{+}$ neurons (E).

Transgenic or viral *Nestin* reporters have been widely used to label neural progenitors (Keyoung et al., 2001; Sawamoto et al., 2001; Lenka et al., 2002; Mignone et al., 2004; Noisa et al., 2010) and Nestin positivity remains an important marker identifying hESC-derived NSCs (Reubinoff et al., 2001; Zhang et al., 2001; Itsykson et al., 2005; Joannides et al., 2007; Chambers et al., 2009). Our transgene-free *Nestin:mCherry* reporter hESC line provides the unique ability to monitor the differentiation status of our cultures with the sensitivity to distinguish high and low subpopulations of Nestin expression and without requiring Nestin immunostaining. While $mCherry^{Lo}$ cells likely correspond to more committed neural lineages or non-neural contaminants, the $mCherry^{Hi}$ population appears to enrich for neural progenitors with multi-lineage capabilities. Furthermore, the combination of $mCherry^{Hi}$ with CD184, CD44 and CD271 surface marker staining permitted significant enrichment of NSC-specific markers (*Sox1*, *Sox2* and

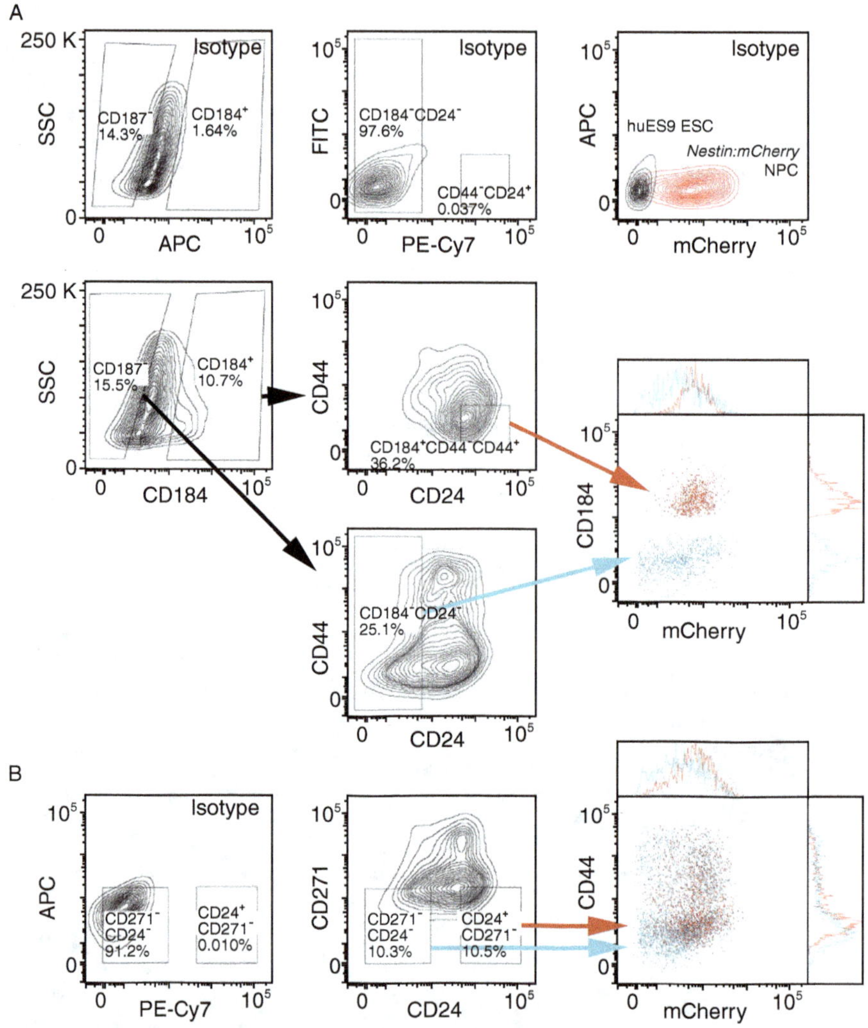

Figure 5. NSC markers map to the mCherry^{Hi} Population. *Nestin:mCherry* NPCs at P0 were stained with isotype control antibodies or FITC labeled anti-CD44, PE-Cy7 labeled anti-CD24, Alexa-647 anti-CD271 or APC labeled anti-CD184 antibodies. huES9 ES cells were used as an mCherry negative control. (A) CD184$^+$ cells were gated on CD44$^-$CD24$^+$, while CD184$^-$ were gated on CD24$^-$ and both populations were shown relative to the CD184 and mCherry distributions. (B) CD271$^-$CD24$^-$ and CD271$^-$CD24$^+$ populations were shown with respect to CD44 and mCherry distributions.

Pax6). These results indicate that our reporter system can be combined with additional strategies for specifically isolating or enriching NSCs including and not limited to: surface marker staining; additional knock-in strategies conferring NSC or lineage-specific drug resistance; or knock-in fluorescent reporters of additional neural lineage markers. Furthermore, our knock-in line was able to achieve significant enrichment of NSCs with a smaller subset of surface markers (CD184$^+$CD44$^-$CD271$^-$) as has been previously used (Yuan et al., 2011), providing a clear advantage over wild-type hESCs. The potential to replace the CD24 surface marker with intrinsic fluorescent reporter activity further provides a means to monitor the NSC cultures and track terminal differentiation. Our lineage-specific knock-in approach will enable further studies tracing the commitment of neural progenitors derived from hESCs; to perform population based studies; and to improve current protocols for directed differentiation of cell types needed for cell replacement therapy.

MATERIALS AND METHODS

Construction of BAC-based targeting vector

The *Nestin* BAC clone was purchased from Life Technologies Inc. and the targeting vector was constructed by recombineering as we previously described (Song et al., 2010).

Cell culture of hESCs

The HUES9 hESCs were cultured as described (Cowan et al., 2004). Briefly, hESCs were cultured on mouse embryonic fibroblast

(MEFs) feeder layer in knockout medium supplemented with 10% knockout serum replacement, 10% plasmanate, 1 mmol/L glutamine, 0.1 mmol/L nonessential amino acids, 10 ng/mL bFGF, and 0.1 μmol/L β-mercaptoethanol. To passage hESCs, confluent culture was washed with phosphate buffered saline (PBS), dissociated for 5 min with TrypLE and resuspended into single cells with culture medium. All tissue culture reagents were purchased from Life Technologies Inc. except where indicated otherwise.

Differentiation of hESCs into neural progenitors

The neural differentiation of hESCs was performed with established protocols (Reubinoff et al., 2001; Zhang et al., 2001). Briefly, hESC colonies were gently scraped from the plate and cultured in suspension as embryoid bodies (EB) on ultra-low cluster plates in EB formation medium (DMEM/F12-Glutamax, 10% characterized FBS, 1% Pen-Strep, 100 μmol/L β-mercaptoethanol). The media was changed to NPC induction medium (NPC medium: DMEM/F12-Glutamax, 0.5% N2, 1% B27, 1% Penicillin-Streptomycin, 10 ng/mL bFGF). After two days, the EBs were replated onto poly-L-ornithine/laminin coated plates. After neural cells migrated away from the EBs and proliferated to confluence, the culture was treated with 0.05% trypsin for 1–2 min at room temperature, resuspended into single cells, and either sorted or re-plated onto poly-L-ornithine/laminin coated dishes in NPC medium. In one sort experiment, BMP inhibitor Noggin (500 ng/mL, R&D Systems) and TGF-beta inhibitor SB431542 (10 μmol/L, Sigma-Aldrich) were included during EB differentiation stages prior to their attachment onto poly-L-ornithine/laminin coated dishes.

Teratoma formation of hESCs in SCID mice

Two million hESCs were suspended in the knockout medium supplemented with 30% matrigel (BD Biosciences) and injected subcutaneously into severe combined immunodeficiency (SCID) mice. After six to eight weeks, teratomas were recovered. For histological assessment, teratomas were fixed in 10% buffered formalin, embedded in paraffin, sectioned and stained with hematoxylin and eosin. To derive NPCs from teratomas, teratomas were chopped into small pieces and cultured in NPC medium. After several passages, the adherent culture became homogenous with mCherry positive cells.

Differentiation of hESC-derived NPCs into mature neurons

The hESC-derived neural progenitors were transduced with lentivirus harboring synapsin promoter-driven EGFP as described (Kim et al., 2011; Lake et al., 2012). The NPCs were cultured in the neuronal differentiation medium containing DMEM/F12 supplemented with N2/B27, BDNF (10 ng/mL, Peprotech), GDNF (10 ng/mL, Peprotech), Penicillin-Streptomycin (Invitrogen), glutamax (Invitrogen), cyclic AMP (1 μmol/L, Sigma Aldrich). The cells were cultured in differentiation media for two to three weeks with 50% of the medium replaced every other day.

Quantitative real-time PCR analysis

Total RNA from hESCs and hESC-derived neural cultures were isolated using RNAeasy Mini Kit (QIAGEN) or ZR RNA MicroPrep Kit (Zymo Research). Total RNA was reverse transcribed using the High Capacity cDNA Reverse Transcription Kit (Applied Biosystems) and analyzed on a StepOnePlus™ quantitative real-time PCR system (Applied Biosystems) as previously described (Song et al., 2007). The primers used are as following: *Sox1* (F-GAGTGGAAGGT CATGTCCGAGG; R-CCTTCTTGAGCAGCGTCTTGGT), *Sox2* (F-ATGCACCGCTACGACGTGA; R-CTTTTGCACCCCTCCCATTT), *Pax6* (F-TGTCCAACGGATGTGTGAGT; R-TTTCCCAAGCAAAGATGGAC), *GAPDH* (F-GATGACATCAAGAAGGTGGTGA; R-GTC TACATGGCAACTGTGAGGA). Data analysis was performed using Microsoft Excel.

Immunofluorescence staining

Neural progenitor cells grown on poly-L-ornithine/laminin coated 4-well chamber plates were fixed in 4% paraformaldehyde for 15 min at room temperature, permeabilized with 0.3% Triton-X-100 in Triton Buffer Solution (TBS) for 10 min, blocked with 2% FBS in TBS-0.05% Tween-20 for 20 min and stained with primary antibodies in blocking buffer at 4°C overnight. Primary antibodies included: anti-Nestin (611658; BD Pharmingen); anti-SOX2 (561469; BD Pharmingen); anti-RFP (600-401-379; Rockland); anti-ki67-Alexa fluor-488 (51-9007231; BD Stemflow Human neural Lineage analysis Kit); TuJ1 (MMS-435P, Covance); GFAP (ab7779, Abcam). The next day, the cells were stained with secondary antibodies (Alexa fluor 568 donkey anti-rabbit; Alexa fluor 488 donkey anti-mouse; Life Technologies, Inc.) in blocking buffer for 45 min and nuclei were counterstained with DAPI. Images were acquired using an Olympus FV1000 confocal microscope. Live culture images were also acquired using an Inverted Axioscope and AxioCam MRm (Carl Zeiss, Inc.). Image assembly was performed using Adobe Photoshop CS5 (Adobe Systems, Inc.).

Flow cytometry

hESC-derived neural cultures were washed with PBS, trypsinized with 0.05% trypsin for 1–2 min at room temperature and resuspended into single cells using DMEM/F12 with Glutamax containing 10% FBS. For cell surface staining, cells were incubated with CD184-APC (560936; BD Pharmingen), CD24-PE-Cy7 (561646; BD Pharmingen), CD44-PE (51-9007231; BD Pharmingen) and/or CD271-Alexa fluor 647 or BV421 (560877 or 562562; BD Pharmingen) antibodies in PBS containing 1% bovine serum albumen (BSA) according to manufacturer's protocol. Stained and unstained cells were analyzed using either a BD LSR-II or LSRFortessa flow cytometry machine (BD Biosciences). Control hES9 hESCs or hESC-derived NPCs were used for mCherry control and IgG-PE-Cy7, IgM-FITC, IgG-BV421 and IgG-APC (BD Pharmingen) were used for isotype controls. Cell sorting was performed using either BD FACSAria II or BD Influx cell sorter. Gating for all sorts was defined by isotype control staining. Flow cytometry data analysis was performed using FlowJo software (Tree Star, Inc.).

ACKNOWLEDGEMENTS

We thank Drs. Anirvan Ghosh, Ji-Eun Kim, Evan Snyder and Jean-Pyo Lee for help with neural differentiation of human ES cells. This

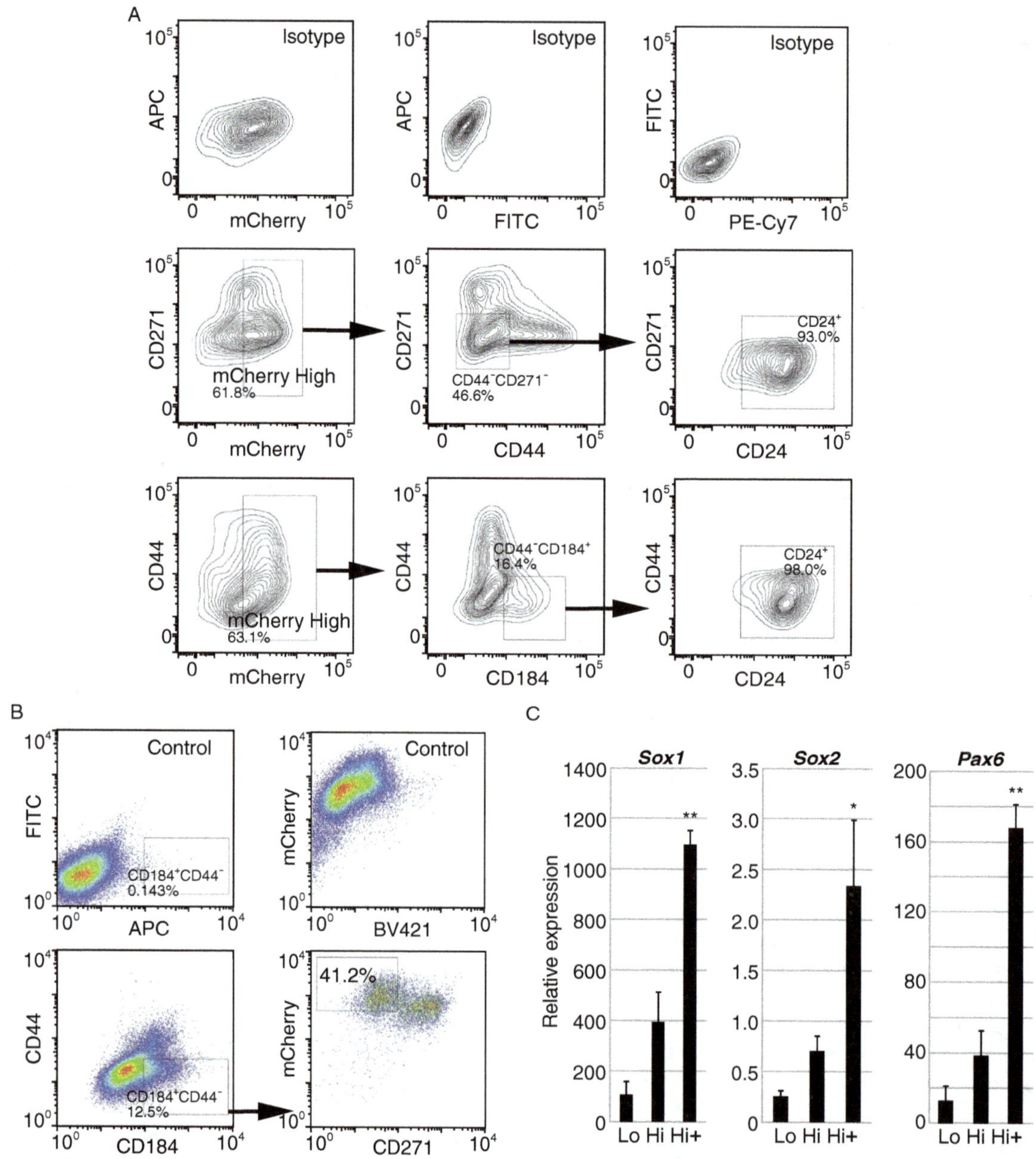

Figure 6. mCherry^{Hi}-based sorting strategy significantly enriches NSCs. (A) *Nestin:mCherry* NPCs at P0 were stained with isotype control antibodies or FITC labeled anti-CD44, PE-Cy7 labeled anti-CD24, Alexa-647 anti-CD271 or APC labeled anti-CD184 antibodies. mCherry^{Hi} cells were gated on CD271$^-$CD44$^-$ or CD44$^-$CD184$^+$ and shown with respect to CD44 and CD24 distributions. (B) *Nestin:mCherry* NPCs at P0 were either unstained (Control) or stained with FITC labeled anti-CD44, BV421 labeled anti-CD271 and APC labeled anti-CD184 antibodies. The CD184$^+$CD44$^-$ cells (all gating based on isotype control staining) were shown with respect to mCherry and CD271 distributions. The mCherry^{Hi}CD271$^-$ population was sorted for gene expression analyses. (C) mCherry^{Lo} (Lo), mCherry^{Hi} (Hi), mCherry^{Hi} CD184$^+$CD271$^-$CD44$^-$ (Hi +) populations were sorted from P0 *Nestin:mCherry* NPCs and real time PCR analysis was performed for *Sox1, Sox2* and *Pax6* expression relative to undifferentiated *Nestin:mCherry* huES9 hESCs. Results shown are from two independent experiments. $^*P < 0.05$ and $^{**}P < 0.01$ (two-tailed *T*-test relative to the mCherry^{Lo} condition).

work was supported by a grant from California Institute for Regenerative Medicine (RC1-00148) to YX and a grant from the National Natural Science Foundation of China (Grant No. 81172828) to XF. Histology was carried out by the UCSD Histology and Immunohistochemistry Shared Resource; confocal microscopy was performed within the UCSD Neuroscience Microscopy Shared Facility (Grant: P30 NS047101). FACS sorting was performed within UCSD Human Embryonic Stem Cell Core Facility.

COMPLIANCE WITH ETHICS GUIDELINES

All hESC work in this study has been approved by the INSTITUTIONAL EMBRYONIC STEM CELL RESEARCH OVERSIGHT COMMITTEE (ESCRO) of University of California, San Diego.

All animal work in this study has been approved by Institutional Animal Care and Use Committee (IACUC) of University of California, San Diego. Xuemei Fu, Zhili Rong, Shengyun Zhu, Xiaocheng Wang, Yang Xu, and Blue B. Lake declare that they have no conflict of interest.

REFERENCES

Aas T, Borresen AL, Geisler S, Smith-Sorensen B, Johnsen H, Varhaug JE, Akslen LA, Lonning PE (1996) Specific P53 mutations are associated with de novo resistance to doxorubicin in breast cancer patients. Nat Med 2:811–814

Chambers SM, Fasano CA, Papapetrou EP, Tomishima M, Sadelain M, Studer L (2009) Highly efficient neural conversion of human ES and iPS cells by dual inhibition of SMAD signaling. Nat Biotech 27:275–280

Cowan CA, Klimanskaya I, McMahon J, Atienza J, Witmyer J, Zucker JP, Wang S, Morton CC, McMahon AP, Powers D et al (2004) Derivation of embryonic stem-cell lines from human blastocysts. N Engl J Med 350:1353–1356

Fu X, Xu Y (2011) Self-renewal and scalability of human embryonic stem cells for human therapy. Regen Med 6:327–334

Hockemeyer D, Wang H, Kiani S, Lai CS, Gao Q, Cassady JP, Cost GJ, Zhang L, Santiago Y, Miller JC et al (2011) Genetic engineering of human pluripotent cells using TALE nucleases. Nat Biotechnol 29:731–734

Israel MA, Yuan SH, Bardy C, Reyna SM, Mu Y, Herrera C, Hefferan MP, Van Gorp S, Nazor KL, Boscolo FS et al (2012) Probing sporadic and familial Alzheimer's disease using induced pluripotent stem cells. Nature 482:216–220

Itsykson P, Ilouz N, Turetsky T, Goldstein RS, Pera MF, Fishbein I, Segal M, Reubinoff BE (2005) Derivation of neural precursors from human embryonic stem cells in the presence of noggin. Mol Cell Neurosci 30:24–36

Joannides AJ, Fiore-Heriche C, Battersby AA, Athauda-Arachchi P, Bouhon IA, Williams L, Westmore K, Kemp PJ, Compston A, Allen ND et al (2007) A scaleable and defined system for generating neural stem cells from human embryonic stem cells. Stem Cells 25:731–737

Keyoung HM, Roy NS, Benraiss A, Louissaint A Jr, Suzuki A, Hashimoto M, Rashbaum WK, Okano H, Goldman SA (2001) High-yield selection and extraction of two promoter-defined phenotypes of neural stem cells from the fetal human brain. Nat Biotechnol 19:843–850

Kim J-E, O'Sullivan ML, Sanchez CA, Hwang M, Israel MA, Brennand K, Deerinck TJ, Goldstein LSB, Gage FH, Ellisman MH et al (2011) Investigating synapse formation and function using human pluripotent stem cell-derived neurons. Proc Natl Acad Sci USA 108:3005–3010

Lake BB, Fink J, Klemetsaune L, Fu X, Jeffers JR, Zambetti GP, Xu Y (2012) Context-dependent enhancement of induced pluripotent stem cell reprogramming by silencing puma. Stem Cells 30:888–897

Lee G, Kim H, Elkabetz Y, Al Shamy G, Panagiotakos G, Barberi T, Tabar V, Studer L (2007) Isolation and directed differentiation of neural crest stem cells derived from human embryonic stem cells. Nat Biotechnol 25:1468–1475

Lenka N, Lu ZJ, Sasse P, Hescheler J, Fleischmann BK (2002) Quantitation and functional characterization of neural cells derived from ES cells using nestin enhancer-mediated targeting in vitro. J Cell Sci 115:1471–1485

Liu Y, Han SS, Wu Y, Tuohy TM, Xue H, Cai J, Back SA, Sherman LS, Fischer I, Rao MS (2004) CD44 expression identifies astrocyte-restricted precursor cells. Dev Biol 276:31–46

Luo Y, Cai J, Liu Y, Xue H, Chrest FJ, Wersto RP, Rao M (2002) Microarray analysis of selected genes in neural stem and progenitor cells. J Neurochem 83:1481–1497

Mignone JL, Kukekov V, Chiang AS, Steindler D, Enikolopov G (2004) Neural stem and progenitor cells in nestin-GFP transgenic mice. J Comp Neurol 469:311–324

Morrison SJ, White PM, Zock C, Anderson DJ (1999) Prospective identification, isolation by flow cytometry, and in vivo self-renewal of multipotent mammalian neural crest stem cells. Cell 96:737–749

Noisa P, Urrutikoetxea-Uriguen A, Li M, Cui W (2010) Generation of human embryonic stem cell reporter lines expressing GFP specifically in neural progenitors. Stem Cell Rev 6:438–449

Ordonez MP, Roberts EA, Kidwell CU, Yuan SH, Plaisted WC, Goldstein LS (2012) Disruption and therapeutic rescue of autophagy in a human neuronal model of Niemann Pick type C1. Hum Mol Genet 21:2651–2662

Peljto M, Wichterle H (2011) Programming embryonic stem cells to neuronal subtypes. Curr Opin Neurobiol 21:43–51

Placantonakis DG, Tomishima MJ, Lafaille F, Desbordes SC, Jia F, Socci ND, Viale A, Lee H, Harrison N, Tabar V et al (2009) BAC transgenesis in human embryonic stem cells as a novel tool to define the human neural lineage. Stem Cells 27:521–532

Reubinoff BE, Itsykson P, Turetsky T, Pera MF, Reinhartz E, Itzik A, Ben-Hur T (2001) Neural progenitors from human embryonic stem cells. Nat Biotech 19:1134–1140

Sawamoto K, Yamamoto A, Kawaguchi A, Yamaguchi M, Mori K, Goldman SA, Okano H (2001) Direct isolation of committed neuronal progenitor cells from transgenic mice coexpressing spectrally distinct fluorescent proteins regulated by stage-specific neural promoters. J Neurosci Res 65:220–227

Song H, Hollstein M, Xu Y (2007) p53 gain-of-function cancer mutants induce genetic instability by inactivating ATM. Nat Cell Biol 9:573–580

Song H, Chung S-K, Xu Y (2010) Modeling disease in human ESCs using an efficient BAC-based homologous recombination system. Cell Stem Cell 6:80–89

Unternaehrer JJ, Daley GQ (2011) Induced pluripotent stem cells for modelling human diseases. Philos Trans R Soc Lond B Biol Sci 366:2274–2285

Wu H, Xu J, Pang ZP, Ge W, Kim KJ, Blanchi B, Chen C, Sudhof TC, Sun YE (2007) Integrative genomic and functional analyses reveal neuronal subtype differentiation bias in human embryonic stem cell lines. Proc Natl Acad Sci USA 104:13821–13826

Yuan SH, Martin J, Elia J, Flippin J, Paramban RI, Hefferan MP, Vidal JG, Mu Y, Killian RL, Israel MA et al (2011) Cell-surface marker signatures for the isolation of neural stem cells, glia and neurons derived from human pluripotent stem cells. PLoS One 6: e17540

Zhang SC, Wernig M, Duncan ID, Brustle O, Thomson JA (2001) In vitro differentiation of transplantable neural precursors from human embryonic stem cells. Nat Biotechnol 19:1129–1133

12

miR-27b inhibits gastric cancer metastasis by targeting NR2F2

Qingzhao Feng[1✉], Xionglin Wu[5], Fuchao Li[2], Beibei Ning[4], Xiaofeng Lu[1], Yin Zhang[1], Ying Pan[3], Wenxian Guan[1✉]

[1] Department of General Surgery, The Affiliated Drum Tower Hospital of Medical School of Nanjing University, Nanjing 210008, China
[2] Department of Gerontology, The Affiliated Drum Tower Hospital of Medical School of Nanjing University, Nanjing 210008, China
[3] Department of Orthopaedics, The Affiliated Drum Tower Hospital of Medical School of Nanjing University, Nanjing 210008, China
[4] The Children's Hospital of Zhejiang University School of Medicine, Hangzhou 310003, China
[5] Department of Emergency, The Yiwu Affiliated Hospital of Wenzhou Medical University, Yiwu 322000, China
✉ Correspondence: izaobao@vip.qq.com (Q. Feng), guan-wx@163.com (W. Guan)

ABSTRACT

Increasing attention is focused on the down-regulation of miRNAs in cancer process. Nuclear receptor subfamily 2 (NR2F2, also known as COUP-TFII) is involved in the development of many types of cancers, but its role in gastric cancer remains elusive. In this experiment, oncomine and Kaplan-meier database revealed that NR2F2 was up-regulated in gastric cancer and that the high NR2F2 expression contributed to poor survival. MicroRNA-27b was targeted and down-regulated by NR2F2 in human gastric cancer tissues and cells. The ectopic expression of miR-27b inhibited gastric cancer cell proliferation and tumor growth *in vitro* and *in vivo*. Assays suggested that the overexpression of miR-27b could promote MGC-803 cells' migration and invasion and retard their metastasis to the liver. In addition, down-regulation of miR-27b enhanced GES-1 cells' proliferation and metastasis *in vitro*. These findings reveal that miR-27b is a tumor suppressor in gastric cancer and a biomarker for improving patients' survival.

KEYWORDS miR-27b, NR2F2, gastric cancer, tumor metastasis

INTRODUCTION

Gastric cancer (GC), the second leading cause of cancer-induced death, has a high incidence in countries like China or Japan (Kang et al., 2014; Yan et al., 2015). Various innate and acquired factors, including *H. pylori* infection, genetic, epigenetic and molecular alterations, are involved in gastric tumorigenesis via their impact on genetic expression and signaling pathways (Milne et al., 2009). However, the mechanism of GC oncogenesis remains unknown. Nuclear receptor subfamily 2 (NR2F2, group F, member 2, also known as COUP-TFII or ARP1) is a key molecule in mesenchymal-epithelial interaction during organogenesis (Tsai and Tsai, 1997). Meanwhile, Kieback et al. find that NR2F2 is expressed in tumor cell lines of human endometrial (Kieback et al., 1996), pancreatic (Qin et al., 2010), colorectal (Shin et al., 2009), and breast cancers (Prahalad et al., 2010). The expression of NR2F2 in terminally differentiated epithelial cells functions in mesenchymal-endothelial interactions, angiogenesis, tumor growth and metastasis by inhibiting TGF-β-induced growth (Qin et al., 2013).

MicroRNAs (miRs), as small single-stranded noncoding RNAs, are key post-transcriptional negative regulators that completely or partially bind to complementary sites in the 3′-untranslated-region (3′UTR) of target mRNAs. Recent studies show that miRNAs can regulate tumor growth,

Qingzhao Feng, Xionglin Wu, Fuchao Li contributed equally to this work.

metastasis and progression (Ma et al., 2007; Ruan et al., 2009; Aleckovic and Kang, 2015). A single miRNA can downregulate the expression of multiple target genes and thereby inhibit tumor metastasis. Therefore, miRNAs can be targeted to modulate the invasion-metastasis cascade (Lin et al., 2015). MiR-27b is a strain of intronic miRNA that regulates chondrosarcoma (Huang et al., 2016), cervical carcinogenesis (Yao et al., 2016) and neuroblastom (Lee et al., 2012).

Interestingly, a recent report demonstrates that the lower level of miR-27b expression correlates with gastric cancer proliferation (Tao et al., 2015). However, little is known about the function of miR-27b in gastric cancer metastasis. In our study, miR-27b was downregulated in gastric cancer tissues, with an inverse correlation with lymph node metastasis. Meanwhile, the miR-27b overexpression inhibited the proliferation and invasion of gastric cancer cells *in vitro* and suppresses tumor growth and liver metastasis of gastric cancer cells *in vivo*. Conversely, miR-27b inhibitors significantly enhanced the proliferation and invasion of gastric cancer cell *in vitro*. Our study concludes that miR-27b plays a suppressive role in gastric cancer metastasis.

RESULTS

Survival time shortened by high NR2F2 expression in gastric cancer

The expression of NR2F2 was measured in gastric cancer patients by oncomine database. NR2F2 was significantly upregulated in gastric cancer tissues compared with normal tissues (Fig. 1A). Q-RT found that NR2F2 expression was significantly higher in gastric cancer tissues than in normal tissues (Fig. 1B). Analysis of immunohistochemical staining and Western blot also support this result (Fig. S1). We subsequently used oncomine database to determine the correlation between the NR2F2 level and the survival time. Although the survival rate had no significant difference (Fig. 1C), the high level of NR2F2 brought a shorter survival time than the low level of NR2F2 did. Then we used Kaplan-meier piotter database to determine the influence of NR2F2 on the survival of gastric cancer patients. The results told that the high level of NR2F2 caused poor clinical survival of patients. (Fig. 1D). All results conclude that the up-regulated NR2F2 level in gastric tissues is negatively correlated with patients' survival.

Suppressive role of miR-27b in gastric cancer

Enough evidence demonstrates that miR-27b, like miRNA, can suppress the proliferation of gastric cancer by targeting ROR1 (Tao et al., 2015). Currently, the function of miR-27b in gastric cancer metastasis remains unclear. To determine its exact function, Q-RT was performed to detect the expression level of miR-27b in gastric cancer tissues and cell lines. As is shown in Fig. 2A, miR-27b expression was significantly decreased in gastric cancer samples compared to matched normal tissues. Of all the patients, 89.47% (17/19) had a lower expression of miR-27b in tumor tissues than that in the adjacent mucosa (Fig. 2B). We further tested the correlation between the level of miR-27b expression and the metastasis of gastric cancer, finding that both were negatively associated (Fig. 2C). We also observed that miR-27b expression was lower in gastric cancer cell lines MGC-803 than in GES-1 cells (human immortalized gastric epithelial cell line) (Fig. 2D). These results prove that the miR-27b level is down-regulated in gastric cancer tissues and gastric cancer cell lines, and is negatively correlated with the metastasis of gastric cancer.

Cell proliferation, migration and invasion suppressed by miR-27b in gastric cancer

To evaluate the biological function of miR-27b in gastric cancer, the cell line MGC-803 with low expression of endogenous miR-27b was stably transfected with miR-27b by lentiviral infection (Fig. S2A). MTT assays showed that the ectopic overexpression of miR-27b suppressed the proliferation of MGC-803 cells (Fig. 3A). Moreover, soft agar formation assays revealed that overexpression miR-27b could inhibit the expanding of MGC-803 cellular colonies (Fig. 3B). The effects of miR-27b on the migration and invasion of gastric cancer cells were further examined by Transwell assays. The ectopic overexpression of miR-27b in MGC-803 cells dramatically inhibited cell migration (Fig. 3C). Also, we used a Matrigel transmembrane invasion assay to study the invasive properties of MGC-803 cells. Overexpression of miR-27b reduced the invasiveness of gastric cancer cells. Consistent with these phenotypes, then we detected that mRNA levels of MMP2, MMP9, Cyclin D1, c-Myc and the protein levels of c-Myc and Cyclin D1 were downregulated with overexpressed miR-27b (Fig. 3D and 3E). These results suggest that miR-27b can suppress the proliferation, migration and invasion of gastric cancer cells and inhibit the growth and metastasis of gastric cancer.

Proliferation, invasion and metastasis of GES-1 cells promoted by the knockdown of miR-27b Given that the overexpression of miR-27b inhibited the proliferation, migration and invasion of gastric cancer cells, we performed loss-of-function experiments to verify the function of miR-27b in GES-1 cells. First, we used miR-27b inhibitor to knock down endogenous miR-27b in GES-1 cells. Q-RT detected that the level of miR-27b was significantly down-regulated by about 70% after the transfection with miR-27b inhibitor (Supplementary Figure 2B). Next, we explored whether miR-27b inhibitor could augment the proliferation of GES-1 cells. Notably, the miR-27b inhibitor dramatically increased the ability of GES-1 cells in colony formation in a serial passage MTT assay and a soft agar assay (Fig. 4A and 4B). Similar data were obtained by Transwell analysis. The miR-27b

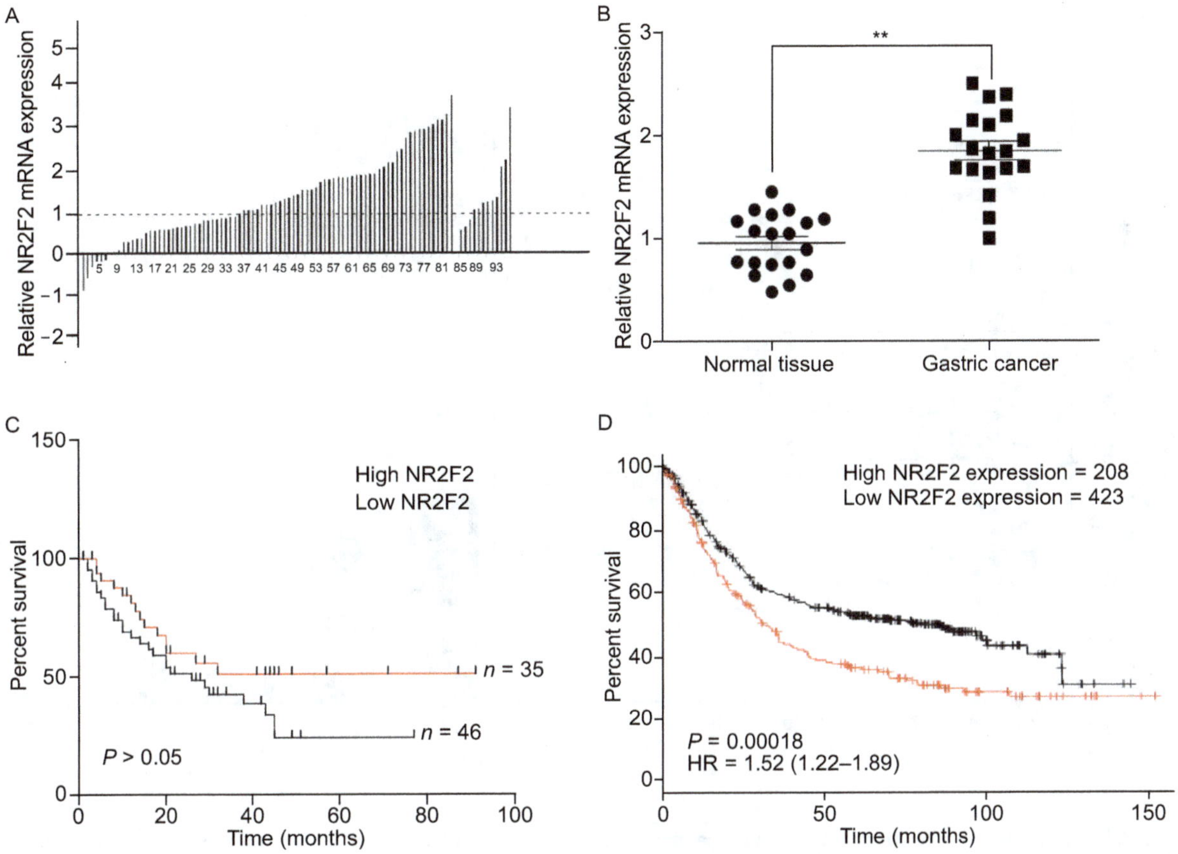

Figure 1. NR2F2 is up-regulated in human gastric cancer tissues and High NR2F2 level have poor clinical outcome. (A) The expression of NR2F2 in human gastric cancer tissue samples from oncomine database. (B) qRT-PCR analysis of NR2F2 expression in human gastric cancer tissue samples and their matched normal gastric tissues from 19 gastric cancer patients. (C) The High expression of NR2F2 in human gastric cancer tissues have trends to poor clinical outcome by oncomine database. (D) The High expression of NR2F2 in human gastric cancer tissues have poor clinical outcome by Kalpan meier-Poitter. Scale bars, 100 μm. *$P < 0.05$, **$P < 0.01$, ***$P < 0.001$.

increased the migration and invasion of GES-1 cells (Fig. 4C) and the mRNA expression of MMP2, MMP9 and CyclinD1, c-Myc. The protein levels of Cyclin D1 and c-Myc were dramatically upregulated (Fig. 4D and 4E).

Gastric tumor growth and metastasis inhibited by miR-27b *in vivo*

To determine whether miR-27b can inhibit the growth and metastasis of gastric cancer cells in vivo, we generated luciferase-labeled MGC-803/miR-27b cells and their control counterparts, and then injected them into the orthotopic sites or spleens of nude mice. Compared with the mice injected with MGC-803/ctrl cells, the mice injected with a combination of MGC-803 with miR-27b cells displayed significantly smaller and lighter tumors four weeks later (Fig. 5A). Moreover, bioluminescence imaging of mice with a 4-week injection of MGC-803/miR-27b or MGC-803/ctrl cells showed that the metastasis of MGC-803 liver cells was significantly impaired by the ectopic overexpression of miR-27b (Fig. 5B).

NR2F2 directly targeted by miR-27b

To reveal the underlying mechanism in which miR-27b inhibits the growth and metastasis of gastric cancer cells, we used silico algorithms Targetscan, miRanda and some other methods to predict the target genes of miR-27b. We found NR2F2 with greater than 30% decreased expression upon ectopic miR-27b overexpression in MGC-803 cells and miR-27b inhibitor increased the level of NR2F2 expression (Fig. 6A). We constructed luciferase reporter vectors containing wide-type or mutant 3′UTRs of NR2F2 (Fig.6B). Luciferase activity assays revealed that miR-27b suppressed the expression of luciferase containing 3′UTRs of NR2F2, compared with controls on 293T cells and MGC-803 cells (Fig. 6C). We found possible binding sites of miR-27b in 3′UTRs of NR2F2, and obliterated these sites by Quick-Change PCR (Zheng et al., 2004). As is shown in Fig. 6D, the mutation of binding sites in 3′UTRs of NR2F2 reversed the downregulation on luciferase activity induced by miR-27b, and abrogated the suppressing effect of miR-27b

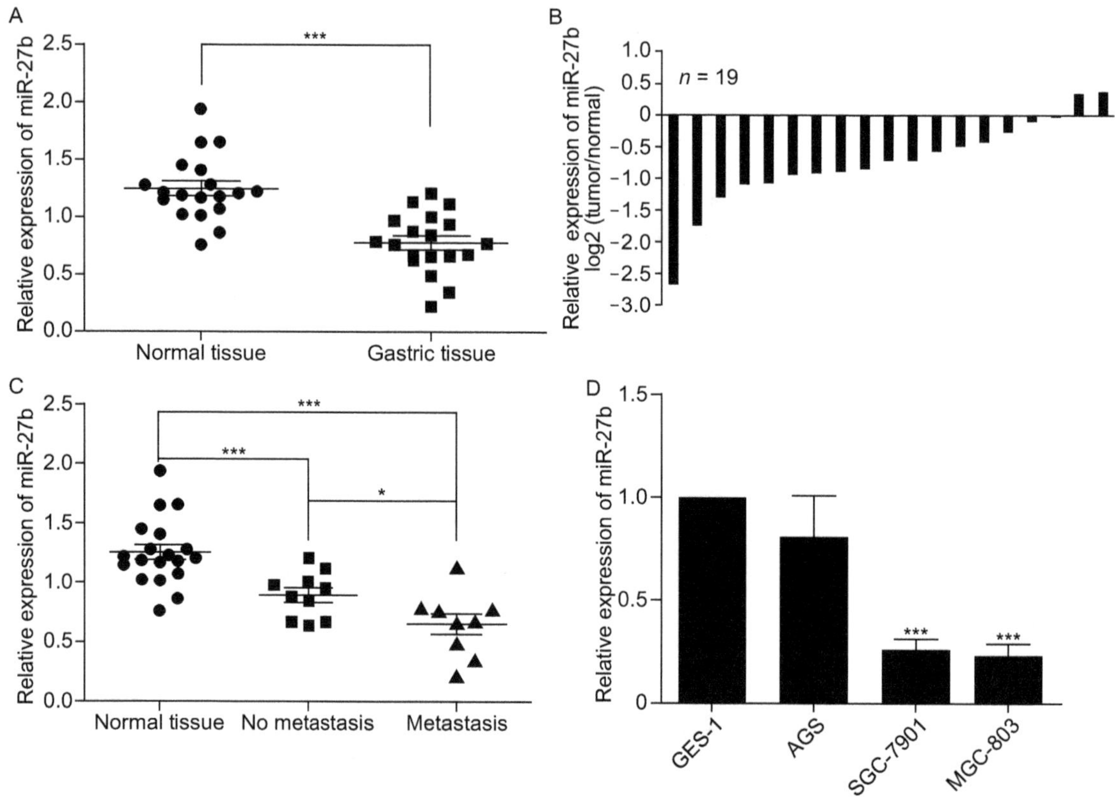

Figure 2. miR-27b is downregulated in human gastric cancer tissues and metastatic gastric cancer cell lines. (A) The expression of miR-27b in human gastric cancer tissue samples and matched normal tissues from 19 gastric cancer patients was analyzed by real-time PCR. (B) miR-27b were expressed as log2 fold change to show the relative expression in every paired sample. (C) Correlation between the lymph node metastasis status and the level of miR-27b in gastric cancers (n = 19). (D) The expression of miR-27b in gastric cancer cell lines with different metastatic potentials was analyzed by real-time PCR. Scale bars, 100 μm. **P* < 0.05, ***P* < 0.01, ****P* < 0.001.

overexpression. These results indicate that NR2F2 is the direct target of miR-27b in gastric cancer cells.

DISCUSSION

Tumorigenesis and metastasis are complex neoplastic processes triggered by a body of transcriptive factors. The abnormal expression of miRNA is involved in tumor proliferation and invasion by inhibiting their target genes (Zhang and Ma, 2012; Zhang et al., 2014b). Therefore, the identification of specific miRNAs and their targets engaged in tumorigenesis and metastasis can provide clues for diagnosing, treating and preventing the cancer. The mechanism that miR-27b participates in gastric cancer proliferation is clear (Tao et al., 2015), but the mechanism in metastasis remains elusive.. In this report, we demonstrated that miR-27b took part in the proliferation and metastasis of gastric cancer. MiR-27b exhibited a decreased expression level in gastric cancer tissues. Moreover, in patients with gastric cancer, a correlation was clarified between lowered miR-27b expression and increased lymphatic metastasis. MiR-27b inhibited proliferation, migration, and invasion of gastric cancer cells *in vitro* and suppressed tumor growth and its metastasis to the liver in vivo. In conclusion, miR-27b is a tumor-suppressing gene in gastric cancer metastasis.

Nuclear receptor subfamily 2 (NR2F2,as known as COUP-TFII) is found in many cancers, including breast cancer (Zhang et al., 2014a), ovarian cancer (Hawkins et al., 2013) and colorectal cancer (Zhou et al., 2016) . Oncomine database and Kaplan meier-piotter defined that NR2F2 was an oncogene in gastric cancer. With Targetscan, miRanda, mirwalk, and Pictar databases, we found highly conserved miR-27b binding sites in the 3′UTRs of NR2F2. Moreover, the miR-27b level was inversely associated with the expression of NR2F2 in gastric cancer tissues, indicating that miR-27b can inhibit gastric cancerous proliferation and metastasis, at least in part, by downregulating the levels of NR2F2.

In all, this experiment provides evidence that miR-27b is a suppressor gene in gastric cancer proliferation and metastasis. MiR-27b can inhibit the proliferation and metastasis of gastric cancer cells by suppressing NR2F2. Furthermore, this experiment provides a potential diagnostic and prognostic marker for gastric cancer metastasis to liver.

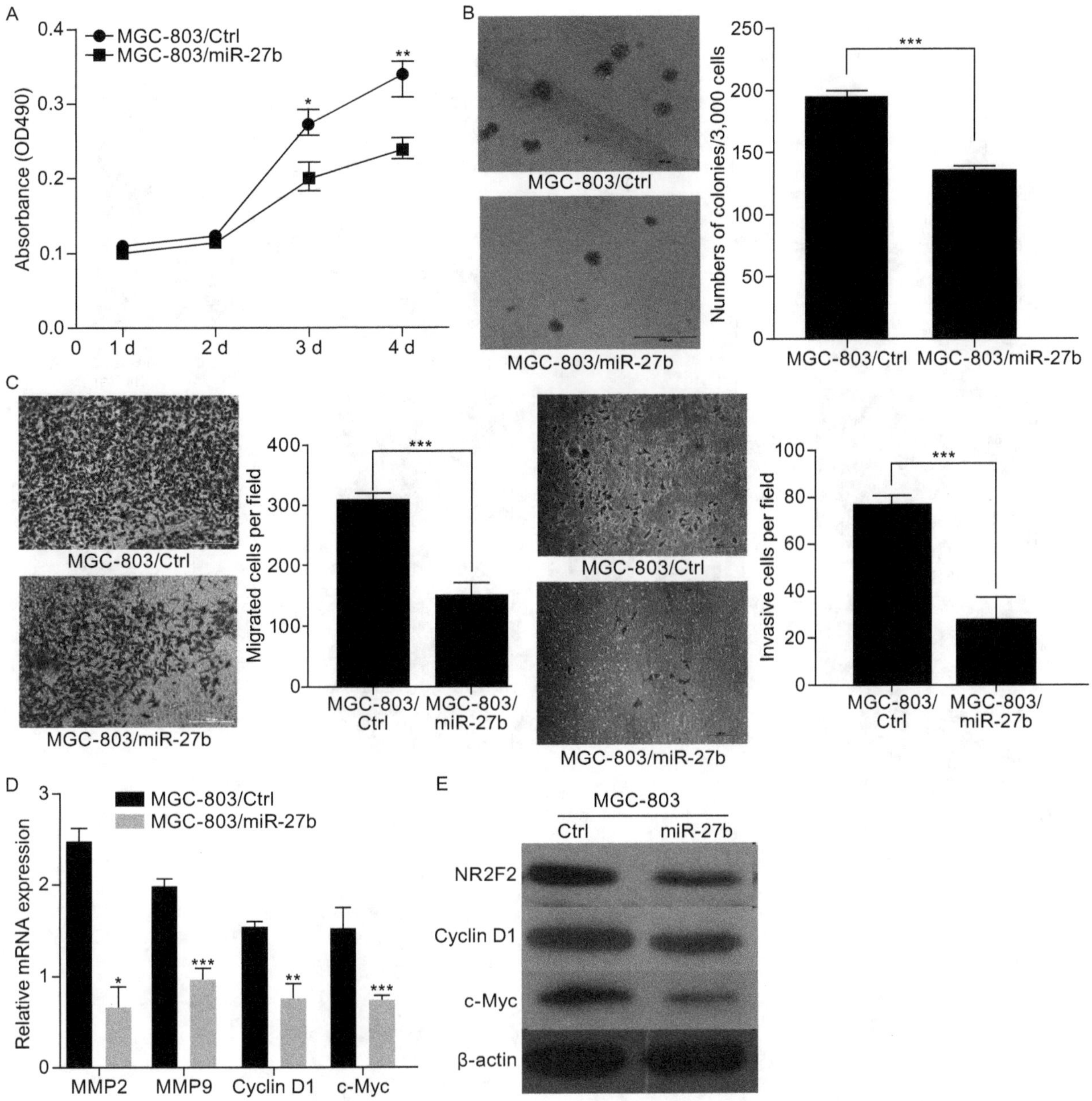

Figure 3. miR-27b inhibits gastric cancer cell growth, migration, and invasion *in vitro*. (A) The effect on cell proliferation of miR-27b overexpression in MGC-803 cells was determined by the MTT assay. (B) Representative images show the colony formation of MGC-803/miR-27b and their control cells (left panel). Average colonies in each well for each group were counted from three independent experiments (right panel). (C) The effects of miR-27b overexpression in MGC-803 cells on cell migration and invasion were analyzed by Transwell migration and Matrigel-coated Transwell invasion analyses. (D) Q-RT analysis of proliferation and metastasis genes. (E) Western blot analysis of proliferation and metastasis genes. $^{*}P < 0.05$, $^{**}P < 0.01$, $^{***}P < 0.001$.

MATERIALS AND METHODS

Cell lines and clinical samples

GES-1, MGC-803, BGC-823, SGC-7901 and 293T cells were saved in our lab (from Institute of Health Sciences, Shanghai Institute for Biological Sciences, Shanghai). 293T cells were cultured in DMEM supplemented with 10% fetal bovine serum. GES-1, MGC-803, BGC-823 and SGC-7901 cells were maintained in RPMI 1640 media supplemented with 10% fetal bovine serum.

Nineteen paired gastric cancer and normal adjacent tissues were collected from The Affiliated Drum Tower Hospital of Medical School of Nanjing University, Jiangsu, China. All patients were diagnosed pathologically according to the criteria of the American Joint Committee on Cancer by two professional pathologists independently.

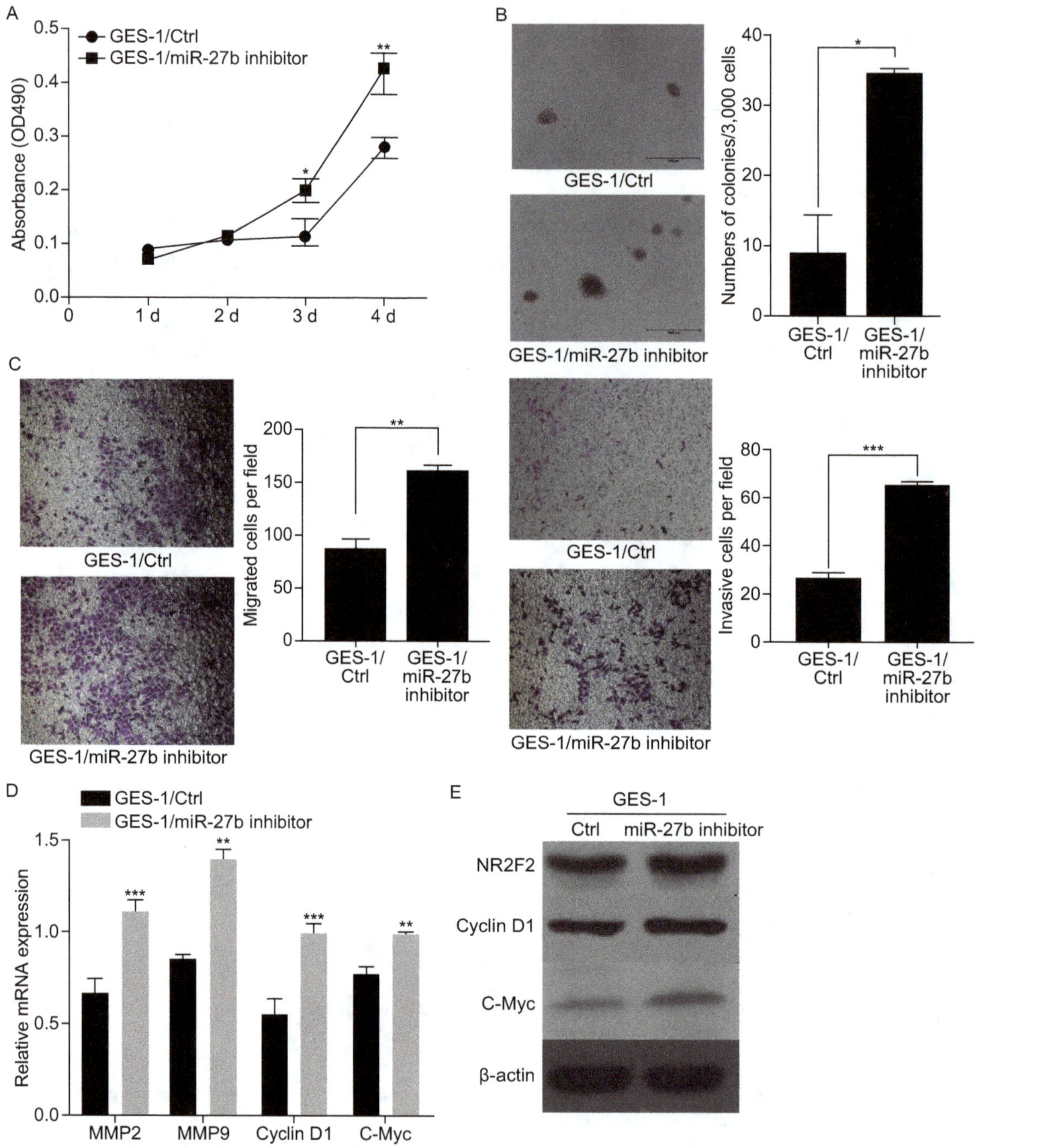

Figure 4. miR-27b inhibitor promotes gastric cancer cell growth, migration, and invasion *in vitro*. (A) The effect on cell proliferation of miR-27b inhibitor in GES-1 cells was determined by the MTT assay. (B) Representative images show the colony formation of GES-1/miR-27b inhibitor and their control cells (left panel). Average colonies in each well for each group were counted from three independent experiments (right panel). (C) The effects of miR-27b inhibitor in GES-1 cells on cell migration and invasion were analyzed by Transwell migration and Matrigel-coated Transwell invasion analyses. (D) Q-RT analysis proliferation and metastasis genes. (E) Western blot analysis of proliferation and metastasis genes. $^{*}P < 0.05$, $^{**}P < 0.01$, $^{***}P < 0.001$.

Specimens were obtained with informed consent and the study was performed in accordance with the approved guidelines by the Medical Ethics Committee of the Affiliated Drum Tower Hospital of Nanjing University (Nanjing, China).

Plasmid construction and generation of stable cell lines

An hsa-miR-27b-containing flank region was amplified from human genomic DNA and inserted into pCDH-CMV-EF1-GFP+puro (System Biosciences). The entire lengths of the 3'UTRs of NR2F2 were

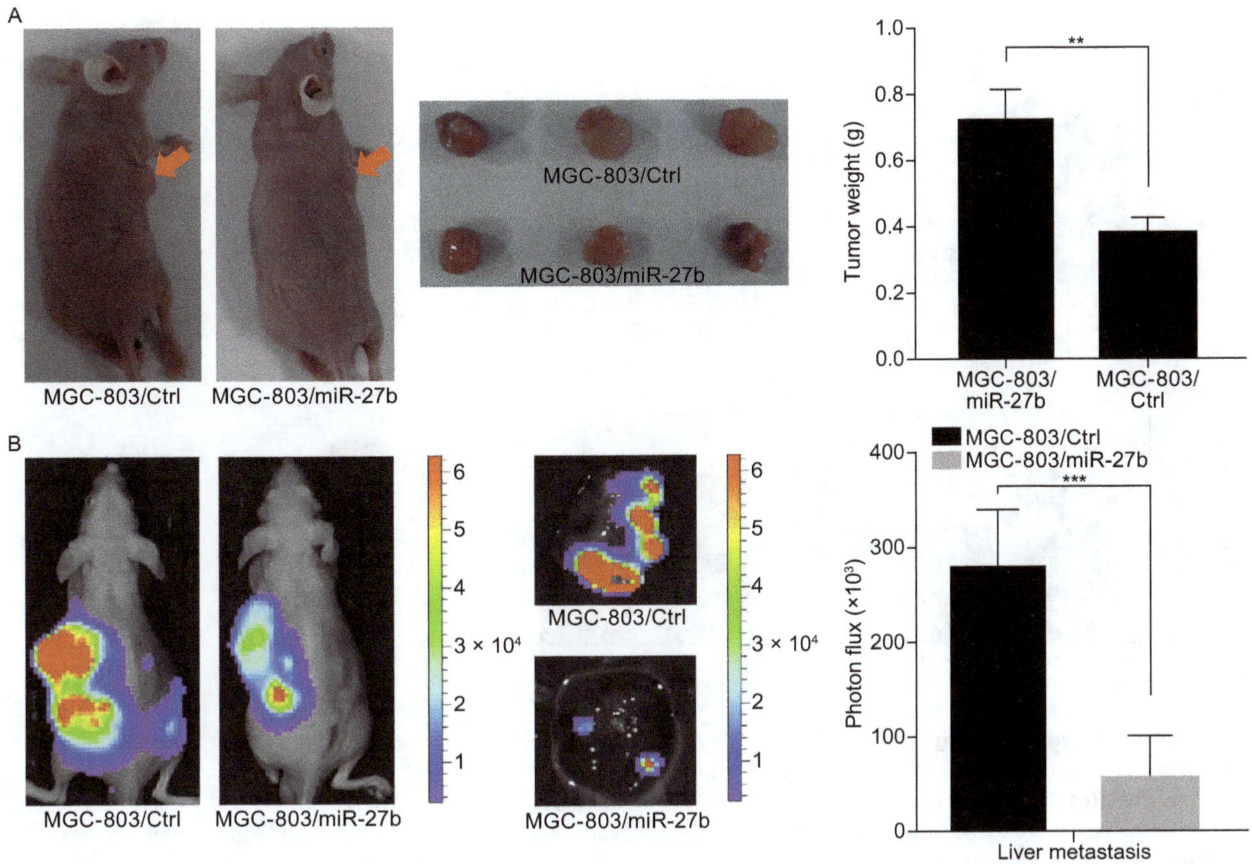

Figure 5. miR-27b suppresses tumor growth and liver metastasis of gastric cancer cells *in vivo*. (A) Representative images of primary tumors in the nude mice orthotopically injected with MGC-803/miR-27b or MGC-803/ctrl cells in the leg (left). Quantification of tumors in each group is shown (right). (B) Representative bioluminescence images of mice injected with MGC-803/miR-27b or MGC-803/ctrl cells into spleen to show liver metastases (left). Quantification of liver metastases was analyzed by bioluminescence measurement (right). Scale bars, 100 μm. $^*P < 0.05$, $^{**}P < 0.01$, $^{***}P < 0.001$.

cloned into the pMIR-REPORT miRNA Expression Reporter Vector (Ambion). To generate a miR-27b-expressing stable cell line, a lentivirus-mediated packaging system containing four plasmids, pCDH-miR-27b or control plasmid, pMDL, REV, and VSVG, was used. To knock down miR-27b in GES-1 cells, miR-27b inhibitor was used GenePharma Colone Primer sequences were as follows: miR-27b Fw-GC TCTAGA TTGCCAGGGATTACCACGCAA; Rv-CG GGATCC CTAGCATTCCCAGCAGGAGACAG NR2F2; 3′UTR Fw-CG ACGCGT AAGAAGGGGGAGTGAAACAGAG; Rv-CCC AAGCTT AGCAAGTTGTTCTGACCGACA.

Western blot

Western blot was performed as described previously (Lin et al., 2015). Cell lysates were separated on SDS-polyacrylamide gels and immunoblot analysis was performed with primary antibodies against NR2F2 (Abcam, Cambridge, MA, USA), β-actin (Millipore, Billerica, MA, USA), c-Myc (Cell Signaling Technology, Danvers, MA) and Cyclin D1 (Sigma-Aldrich).

Cell growth assay

For the MTT assay, 1000 cells were seeded and transfected in a 96-well plate. After transfection for 1, 2, 3, 4 days, and the absorbance at 490 nm was measured. For soft agar assay, 3000 cells were seeded and transfected in a 6-well plate. After transfection for 14 days, and the size and number of soft agar colonies were measured. Experiments were repeated three times.

Oncomine and TCGA database analysis

The Oncomine, TCGA and Kalpan meier-plotter database was searched for NR2F2 gene. The data sets containing expression data for NR2F2 were filtered to display up-regulation in gastric cancer versus normal tissue with $P < 0.05$ and have poor clinical outcome.

Real-time quantitative PCR

Total RNA was extracted from the cultured cells or frozen tissues using TRIzol reagent (Invitrogen, Carlsbad, California, USA). The mRNA was

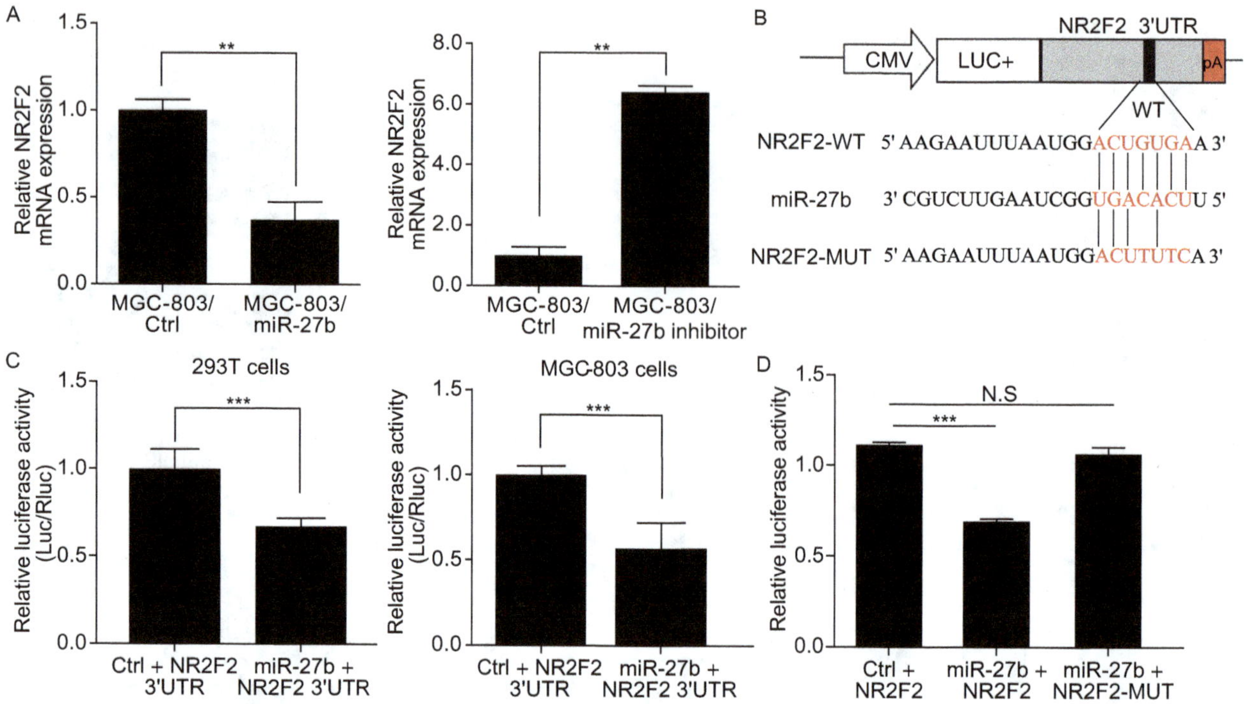

Figure 6. NR2F2 are direct targets of miR-27b. (A) The mRNA levels of NR2F2 in MGC-803/miR-27b and MGC-803/ctrl cells were analyzed by real-time PCR. The mRNA levels of NR2F2 in GES-1/miR-27b inhibitor and GES-1/ctrl cells were further analyzed by real-time PCR. (B) Constructed luciferase reporter vectors containing wild-type or mutant 3′UTRs of NR2F2. (C) The effects of miR-27b overexpression on the activity of the 3′UTRs of target genes in 293T cells and MGC-803 cells were analyzed by the dual luciferase reporter assay. (D) The effects of miR-27b expression on the activity of wild-type and mutant 3′UTRs of NR2F2 were analyzed by the dual luciferase reporter assay. N.S.: no significance, **P* < 0.05, ***P* < 0.01, ****P* < 0.001.

reverse-transcribed by an RT-PCR kit (Invitrogen, Carlsbad, California, USA) according to the manufacturer's instructions. Quantitative PCR was then performed with primers for miR-27b, NR2F2 using SYBR Green PCR Master Mix (Invitrogen, Carlsbad, California, USA) in a real-time PCR System (Applied Biosystems, Carlsbad, California, USA) following a standard quantitative PCR procedure. Primer sequences used are shown in Table S1.

Relative quantification was performed by normalization to the amount of GAPDH or U6.

Cell motility and invasion assay

Migration and invasion assays were performed as described previously (Liu et al., 2014). All experiments were performed at least three times in triplicate.

miRNA reporter luciferase assay

For luciferase reporter assays, 293T cells and MGC-803 cells were seeded into a 24-well plate and co-transfected with 3′UTR-luciferase and either miR-27b or control plasmids. Cells were harvested after two days and assayed using the Dual-Glo Luciferase Assay System (Promega) to determine the relative luciferase activity. The luciferase activity was measured by a luciferin enzyme detection assay kit (Promega) and normalized to Renilla luciferase activity. Each treatment was performed in triplicate in three independent experiments.

Animal studies

All of the experiments using animals were performed in accordance with a protocol approved by the Animal Care and Use Committee of Nanjing University. For tumor growth assays, 5×10^6 cells were subcutaneously injected into the lower back regions of 6-week-old male nude mice for four weeks (n = 3 per group). For orthotropic metastasis, 6-week-old male nude mice were anesthetized and their spleens were exteriorized by laparotomy, and then 5×10^5 cells were injected into the spleens for four weeks (n = 3 per group). Tumor growth and hepatic metastases at day 28 were monitored using the live animal Lumina II system (Xenogen IVIS system).

Statistical analysis

All data were expressed as the mean ± SD. Statistical analysis was performed with Student's *t*-test. A *P* value less than 0.05 was considered statistically significant.

ACKNOWLEDGEMENTS

This work supported by the Fundamental Research Funds for the Central Universities (No. 021414380015, 021414380155).

COMPLIANCE WITH ETHICS GUIDELINES

Qingzhao Feng, Xionglin Wu, Fuchao Li, Beibei Ning, Xiaofeng Lu, Yin Zhang, Ying Pan and Wenxian Guan declare that they have no conflict of interest.

All procedures followed were in accordance with the ethical standards of the responsible committee on human experimentation (institutional and national) and with the Helsinki Declaration of 1975, as revised in 2000 (5). Informed consent was obtained from all patients for being included in the study. All institutional and national guidelines for the care and use of laboratory animals were followed.

REFERENCES

Aleckovic M, Kang Y (2015) Regulation of cancer metastasis by cell-free miRNAs. Biochim Biophys Acta 1855:24–42

Hawkins SM, Loomans HA, Wan YW, Ghosh-Choudhury T, Coffey D, Xiao W, Liu Z, Sangi-Haghpeykar H, Anderson ML (2013) Expression and functional pathway analysis of nuclear receptor NR2F2 in ovarian cancer. J Clin Endocrinol Metab 98:E1152–E1162

Huang CY, Chang AC, Chen HT, Wang SW, Lo YS, Tang CH (2016) Adiponectin promotes VEGF-C-dependent lymphangiogenesis by inhibiting miR-27b through CaMKII/AMPK/p38 signaling pathway in human chondrosarcoma cells. Clin Sci 130(17):1523–1533

Kang W, Tong JH, Chan AW, Zhao J, Dong Y, Wang S, Yang W, Sin FM, Ng SS, Yu J et al (2014) Yin Yang 1 contributes to gastric carcinogenesis and its nuclear expression correlates with shorter survival in patients with early stage gastric adenocarcinoma. J Trans Med 12:80

Kieback DG, Levi T, Kohlberger P, Fiedrich U, Press MF, Rosenthal HE, Mobus VJ, Runnebaum IB, Tong XW, Tsai MJ (1996) Chicken ovalbumin upstream promoter–transcription factor (COUP-TF) expression in human endometrial cancer cell lines. Anticancer Res 16:3371–3376

Lee JJ, Drakaki A, Iliopoulos D, Struhl K (2012) MiR-27b targets PPARgamma to inhibit growth, tumor progression and the inflammatory response in neuroblastoma cells. Oncogene 31:3818–3825

Lin Y, Liu AY, Fan C, Zheng H, Li Y, Zhang C, Wu S, Yu D, Huang Z, Liu F et al (2015) MicroRNA-33b inhibits breast cancer metastasis by targeting HMGA2, SALL4 and Twist1. Sci Rep 5:9995

Liu AY, Cai Y, Mao Y, Lin Y, Zheng H, Wu T, Huang Y, Fang X, Lin S, Feng Q et al (2014) Twist2 promotes self-renewal of liver cancer stem-like cells by regulating CD24. Carcinogenesis 35:537–545

Ma L, Teruya-Feldstein J, Weinberg RA (2007) Tumour invasion and metastasis initiated by microRNA-10b in breast cancer. Nature 449:682–688

Milne AN, Carneiro F, O'Morain C, Offerhaus GJ (2009) Nature meets nurture: molecular genetics of gastric cancer. Human Genet 126:615–628

Prahalad P, Dakshanamurthy S, Ressom H, Byers SW (2010) Retinoic acid mediates regulation of network formation by COUP-TFII and VE-cadherin expression by TGFbeta receptor kinase in breast cancer cells. PloS ONE 5:e10023

Qin J, Chen X, Yu-Lee LY, Tsai MJ, Tsai SY (2010) Nuclear receptor COUP-TFII controls pancreatic islet tumor angiogenesis by regulating vascular endothelial growth factor/vascular endothelial growth factor receptor-2 signaling. Cancer Res 70:8812–8821

Qin J, Wu SP, Creighton CJ, Dai F, Xie X, Cheng CM, Frolov A, Ayala G, Lin X, Feng XH et al (2013) COUP-TFII inhibits TGF-beta-induced growth barrier to promote prostate tumorigenesis. Nature 493:236–240

Ruan K, Fang X, Ouyang G (2009) MicroRNAs: novel regulators in the hallmarks of human cancer. Cancer Lett 285:116–126

Shin SW, Kwon HC, Rho MS, Choi HJ, Kwak JY, Park JI (2009) Clinical significance of chicken ovalbumin upstream promoter-transcription factor II expression in human colorectal cancer. Oncol Rep 21:101–106

Tao J, Zhi X, Zhang X, Fu M, Huang H, Fan Y, Guan W, Zou C (2015) miR-27b-3p suppresses cell proliferation through targeting receptor tyrosine kinase like orphan receptor 1 in gastric cancer. J Exp Clin Cancer Res 34:139

Tsai SY, Tsai MJ (1997) Chick ovalbumin upstream promoter-transcription factors (COUP-TFs): coming of age. Endocr Rev 18:229–240

Yan Y, Wang LF, Wang RF (2015) Role of cancer-associated fibroblasts in invasion and metastasis of gastric cancer. World J Gastroenterol 21:9717–9726

Yao J, Deng B, Zheng L, Dou L, Guo Y, Guo K (2016) miR-27b is upregulated in cervical carcinogenesis and promotes cell growth and invasion by regulating CDH11 and epithelial-mesenchymal transition. Oncol Rep 35:1645–1651

Zhang J, Ma L (2012) MicroRNA control of epithelial-mesenchymal transition and metastasis. Cancer Metastasis Rev 31:653–662

Zhang C, Han Y, Huang H, Qu L, Shou C (2014a) High NR2F2 transcript level is associated with increased survival and its expression inhibits TGF-beta-dependent epithelial-mesenchymal transition in breast cancer. Breast Cancer Res Treat 147:265–281

Zhang Y, Yang P, Wang XF (2014b) Microenvironmental regulation of cancer metastasis by miRNAs. Trends Cell Biol 24:153–160

Zheng L, Baumann U, Reymond JL (2004) An efficient one-step site-directed and site-saturation mutagenesis protocol. Nucleic Acids Res 32:e115

Zhou B, Song J, Han T, Huang M, Jiang H, Qiao H, Shi J, Wang Y (2016) MiR-382 inhibits cell growth and invasion by targeting NR2F2 in colorectal cancer. Mol Carcinog. doi:10.1002/mc.22466

13

MicroRNAs recruit eIF4E2 to repress translation of target mRNAs

Shaohong Chen[1,2], **Guangxia Gao**[1,2]✉

[1] CAS Key Laboratory of Infection and Immunity, CAS Center for Excellence in Biomacromolecules, Institute of Biophysics, Chinese Academy of Sciences, Beijing 100101, China
[2] University of Chinese Academy of Sciences, Beijing 100101, China
✉ Correspondence: gaogx@moon.ibp.ac.cn (G. Gao)

ABSTRACT

MicroRNAs (miRNAs) recruit the RNA-induced silencing complex (RISC) to repress the translation of target mRNAs. While the 5′ 7-methylguanosine cap of target mRNAs has been well known to be important for miRNA repression, the underlying mechanism is not clear. Here we show that TNRC6A interacts with eIF4E2, a homologue of eIF4E that can bind to the cap but cannot interact with eIF4G to initiate translation, to inhibit the translation of target mRNAs. Downregulation of eIF4E2 relieved miRNA repression of reporter expression. Moreover, eIF4E2 downregulation increased the protein levels of endogenous IMP1, PTEN and PDCD4, whose expression are repressed by endogenous miRNAs. We further provide evidence showing that miRNA enhances eIF4E2 association with the target mRNA. We propose that miRNAs recruit eIF4E2 to compete with eIF4E to repress mRNA translation.

KEYWORDS microRNAs, translation repression, 5′ cap, eIF4E2, TNRC6A

INTRODUCTION

MicroRNAs (miRNAs) are 21–23 nucleotides noncoding RNAs that regulate the expression of messenger RNAs. miRNA-mediated gene silencing plays important roles in a variety of biological processes (Flynt and Lai, 2008; Jonas and Izaurralde, 2015). miRNAs in complex with Argonaute proteins (Ago1–4) bind to target mRNAs through nucleotide pairing and recruit one of the TNRC6 proteins (TNRC6A-6C) to form the RNA-induced silencing complex (RISC) (Pfaff and Meister, 2013). RISC can promote target mRNA translation repression, degradation or both (Fabian et al., 2010).

There is increasing evidence indicating that translation initiation is a major target of miRNA repression (Pillai et al., 2005; Meister, 2007; Mathonnet et al., 2007; Fukaya et al., 2014; Humphreys et al., 2005; Thermann and Hentze, 2007; Fukaya and Tomari, 2012; Wang et al., 2008; Ricci et al., 2013; Gu et al., 2014). Cellular mRNAs have a 7-methylguanosine (m7GpppN) cap structure at the 5′ end, which promotes translation and mRNA stability (Varani, 1997). The cap is recognized by the translation initiation factor eIF4E, which recruits other translation initiation factors, including eIF4G, the RNA helicase eIF4A, eIF2, eIF3 and the 40S small ribosomal subunit to initiate translation (Gingras et al., 1999). eIF4A and its homologue eIF4A2 are important for unwinding structures in the 5′ untranslated regions (5′UTRs) (Lu, et al., 2014). It was reported that RISC-induced dissociation of eIF4A or eIF4A2 caused miRNA repression (Lu et al., 2014; Meijer et al., 2013; Fukao et al., 2014). These results partially explain how miRNAs repress translation initiation.

It has been well documented that the 5′ cap is important for miRNA repression. In contrast to the cap-containing mRNA reporters, internal ribosome entry site (IRES)-initiated mRNA reporters are refractory to miRNA silencing (Pillai et al., 2005; Mathonnet et al., 2007; Humphreys et al., 2005; Walters et al., 2010). IRES can bypass the cap to recruit ribosomes to the mRNA internally to initiate translation (Jackson et al., 2010). Addition of a cap to the IRES mRNA reporters conferred sensitivity to miRNA silencing (Walters et al., 2010). On the other hand, some other work showed that IRES-dependent translation can also be repressed by miRNA (Humphreys et al., 2005; Petersen et al., 2006). As is a proper explanation in Valencia-Sanchez et al.'s 2006 review, miRNAs also work through other elements to repress target mRNA, such as poly(A) tail and 5′ cap structure and these miRNA-repression related elements might be missing in some reporters.

That cap optimizes miRNA silencing strongly suggests that a cap-binding protein should be involved in miRNA repression (Kiriakidou et al., 2007). It was reported that human Ago2 bound to the cap through two conserved phenylalanine residues (F450 and F505) and thereby prevented the recruitment of eIF4E in miRNA repression (Kiriakidou et al., 2007). However, this conclusion was challenged by later studies. Those two residues were shown to be important for Ago interaction with GW182 rather than binding to the cap in both human and flies (Eulalio et al., 2008). Further structure analysis showed that in human Ago2 one of the aromatic side chains F450 is buried in the hydrophobic core (Kinch and Grishin, 2009) and MID domain of human Ago2 cannot bind cap analogues significantly (Frank et al., 2011), although Drosophila Ago1 and Neurospora Argonaute contains a 5′ cap binding site (Boland et al., 2010; Djuranovic et al., 2010). Moreover, fly Ago-RISC associated with target mRNA in a manner independent of GW182 and repressed translation without affecting eIF4E recognition of the cap (Fukaya et al., 2014), further arguing against the idea that Ago binding to the cap is required for translation repression.

eIF4E2 is a homologue of eIF4E (Rom et al., 1998). The binding affinity of eIF4E2 for the cap is about 100-fold lower than eIF4E, and the abundance in cells is about 10-fold lower (Zuberek et al., 2007). In addition, unlike eIF4E, eIF4E2 does not interact with eIF4G to initiate translation (Rom et al., 1998). When recruited to the cap of an mRNA by transacting factors, eIF4E2 can repress the translation of the target mRNA. For example, Drosophila Bicoid recruits eIF4E2 to suppress caudal translation (Cho et al., 2005), and TTP recruits eIF4E2 to repress the translation of AU-rich element-containing mRNAs (Tao and Gao, 2015; Fu et al., 2016).

eIF4E2 was reported to interact with 4E-T (Kubacka et al., 2013), which participates in translation repression and mRNA decay in TTP or microRNA mediated silencing (Kamenska et al., 2014; Kamenska et al., 2016; Nishimura et al., 2015). A very recent research shows that eIF4E2 effects miRNA mediated translation silencing by competing with eIF4E for binding 4E-T (Chapat, 2017). In the present study, we confirmed that eIF4E2 is required for miRNA mediated translation repression. In addition, we provide evidence indicating that TNRC6A, the core component of RISC, can directly recruit eIF4E2 to target mRNA to repress translation.

RESULTS

The 5′ cap of target mRNA is required for optimal miRNA-mediated gene silencing

To determine the contributions of the cap and structured 5′ UTR to miRNA silencing, reporter mRNAs expressing firefly luciferase (fLuc) were generated by *in vitro* transcription (Fig. 1A). In the 3′UTR of these reporters, there were eight tandem Let7a-responsive elements (LREs), which are responsive to endogenous let-7a in HeLa cells. Reporters containing mutant responsive elements (REm) were constructed to serve as negative controls (Fig. 1A). In the 5′UTR there was a stem-loop (SL), which mimics a secondary structure, or repetitive CAA sequences, which cannot form secondary structures. The translation of the reporters was driven either by the cap or by EMCV IRES. EMCV IRES-driven translation does not need eIF4E but requires the rest translation initiation factors (Jackson et al., 2010). A reporter expressing renilla luciferase (RL) driven by EMCV IRES was used to serve as a control for transfection efficiency and

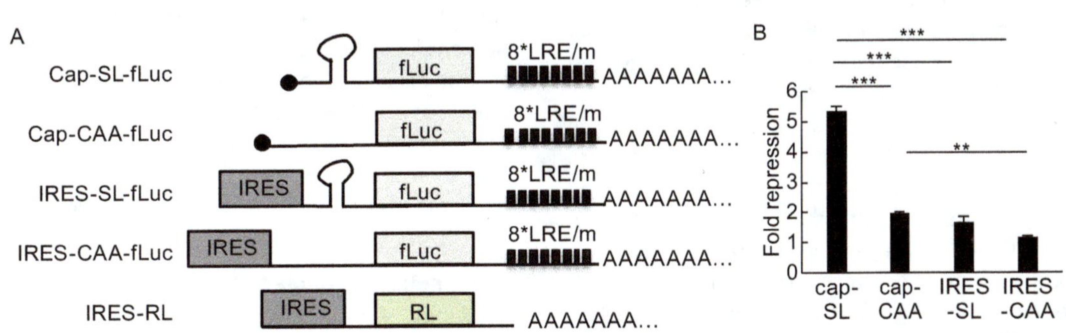

Figure 1. The cap structure of target mRNA is required for optimal miRNA-mediated gene silencing. (A) Schematic representation of the reporter mRNAs. Eight tandem Let-7a response elements (LRE) or mutants (REm) were cloned downstream of the coding sequence of firefly luciferase (fLuc). A stem-loop (SL) structured fragment or a fragment containing $(CAA)_{18}$ was inserted into the 5′UTR of the reporters indicated. (B) Reporter mRNAs were transfected into HeLa cells. A control reporter expressing renilla luciferase (RL) driven by EMCV IRES was included. At 24 h postinfection, luciferase activities were measured and firefly luciferase activity was normalized with renilla luciferase activity. Data presented are means ± SD of three independent experiments. The *P* value is determined by two-tailed Student's *t* test. ns, nonsignificant. $^{**}P < 0.01$; $^{***}P < 0.005$.

sample handling (Fig. 1A). The reporters were transfected into HeLa cells and luciferase activities were measured. The response of a reporter to let-7a silencing was indicated by fold repression, which was calculated as the luciferase activity expressed from the REm reporter divided by that expressed from the LRE reporter. That equivalent amounts of the mRNA reporters were transfected into the cells was confirmed by RT-qPCR (Fig. S1). Data showed that let-7a repressed the expression of the reporter containing both the cap and the stem-loop (Cap-SL-fLuc) by about 5-fold (Fig. 1B). The reporter containing the stem-loop and not the cap (IRES-SL-fLuc) was repressed by about 2-fold (Fig. 1B). The reporter containing the cap but not the stem-loop (Cap-CAA-fLuc) was still responsive, with the repression about 2-fold (Fig. 1B). When both the cap and the stem-loop were removed, the reporter (IRES-CAA-fLuc) was barely responsive (Fig. 1B). These results confirmed that both the cap and a structured 5′UTR contribute to miRNA repression.

TNRC6A interacts with eIF4E2

We next explored the interactions between RISC and cap-binding proteins eIF4E (referred to as eIF4E1 hereafter) and its homologue eIF4E2. TNRC6 proteins, key players in miRNA repression, were tested for their interactions with eIF4E1 and eIF4E2 by coimmunoprecipitation assays. Data showed that immunoprecipitation of both TNRC6A and 6B coprecipitated eIF4E2 (Fig. 2A). In contrast, immunoprecipitation of TNRC6A or TNRC6B failed to coprecipitate eIF4E1 (Fig. 2A). The interaction between TNRC6A/6B and 4E2 is consistent with the BioID results in the recent paper by Chapat et al. (2017). We tried to detect the interaction between endogenous TNRC6A and eIF4E2 but failed (data not shown). This could be accounted for by technical difficulties such as low abundance of the endogenous eIF4E2, degradation of TNRC6A, and lack of good antibodies. We then analyzed the interaction between endogenous TNRC6A and overexpressed eIF4E2. Data showed that immunoprecipitation of endogenous TNRC6A coprecipitated Flag-tagged eIF4E2 but not eIF4E1 (Fig. 2B).

To map the domain(s) of TNRC6A that is involved in its interaction with eIF4E2, TNRC6A truncation mutants were constructed and tested for their interactions with Flag-tagged eIF4E2 (Fig. 2C). Data showed that the C-terminal domain of TNRC6A (TNRC6A-C) displayed considerable interaction with eIF4E2 while the other domains failed to do so although the middle domain (TNRC6A-M) had some weak interaction (Fig. 2C). Further truncations of TNRC6A-C abolished its interaction with eIF4E2 (Fig. 2C). To test whether the interaction between TNRC6A-C and eIF4E2 is direct, they were bacterially expressed, partially purified and analyzed by the pull-down assays. Indeed, the recombinant GST-tagged TNRC6A-C pulled down eIF4E2 but not eIF4E1 (Fig. 2D). These results indicate that eIF4E2 directly interacts with the C-terminal domain of TNRC6A.

Downregulation of eIF4E2 impairs miRNA silencing

To explore the role of eIF4E2 in miRNA-mediated translational repression, we analyzed the effects of eIF4E2 downregulation on miRNA silencing of two reporters, the Let7a-responsive LRE reporter and the miR-19-responsive 19RE reporter (Fig. 3A). Two shRNAs targeting different sites of eIF4E2 were constructed and confirmed for their ability to downregulate the expression of endogenous eIF4E2 (Fig. 3B). In the absence of the shRNA targeting eIF4E2, ectopic expression of let-7a inhibited the LRE reporter expression by about 8-fold (Fig. 3C). In comparison, downregulation of eIF4E2 significantly reduced the repression (Fig. 3C). The relatively high expression level of miR-19 in HEK293 cells allowed us to use the endogenous miR-19 to inhibit 19RE reporter expression. A reporter containing REm (Fig. 3A) was used as a negative control. MiR-19 repression of the 19RE reporter expression was defined as the luciferase activity expressed from the REm reporter divided by that expressed from the 19RE reporter. In the absence of the shRNA targeting eIF4E2, the repression was about 12-fold (Fig. 3D). In comparison, downregulation of eIF4E2 reduced the repression to about 6-fold (Fig. 3D). These results indicate that downregulation of eIF4E2 impaired miRNA repression of reporter expression.

Downregulation of eIF4E2 relieves miRNA repression of target mRNA translation

We next used the polysome profiling assay to confirm that eIF4E2 participates in miRNA mediated translational repression. Overexpression of miRNA let-7a or downregulation of eIF4E2 did not affect the pattern of global mRNA translation (Fig. 4A). Let-7a overexpression reduced the percentage of fLuc-LRE reporter mRNA in the polysome fractions, which indicates that miRNA-mediated repression targets the translational process (Pillai et al., 2005). Downregulation of eIF4E2 relieved the repression, while the control RL mRNA distribution was not affected (Fig. 4B). Notably, let-7a overexpression reduced the total mRNA level of fLuc-LRE but not the control reporter RL, while downregulation of eIF4E2 did not affect the total mRNA level of either reporter (Fig. 4C), indicating that eIF4E2 mainly plays a role in the process of miRNA mediated translational repression but not the process of RNA decay.

Downregulation of eIF4E2 increases the translational efficiency of endogenous IMP1

To further demonstrate the role of eIF4E2 in miRNA-mediated translational repression, we analyzed the effects of eIF4E2 downregulation on the expression of endogenous IMP1 in HeLa cells wherein let-7a is highly expressed. In the 3′UTR of IMP1 coding mRNA, there are six putative let-7 binding sites, which render the mRNA sensitive to the endogenous let-7 (Boyerinas et al., 2008). The IMP1 protein

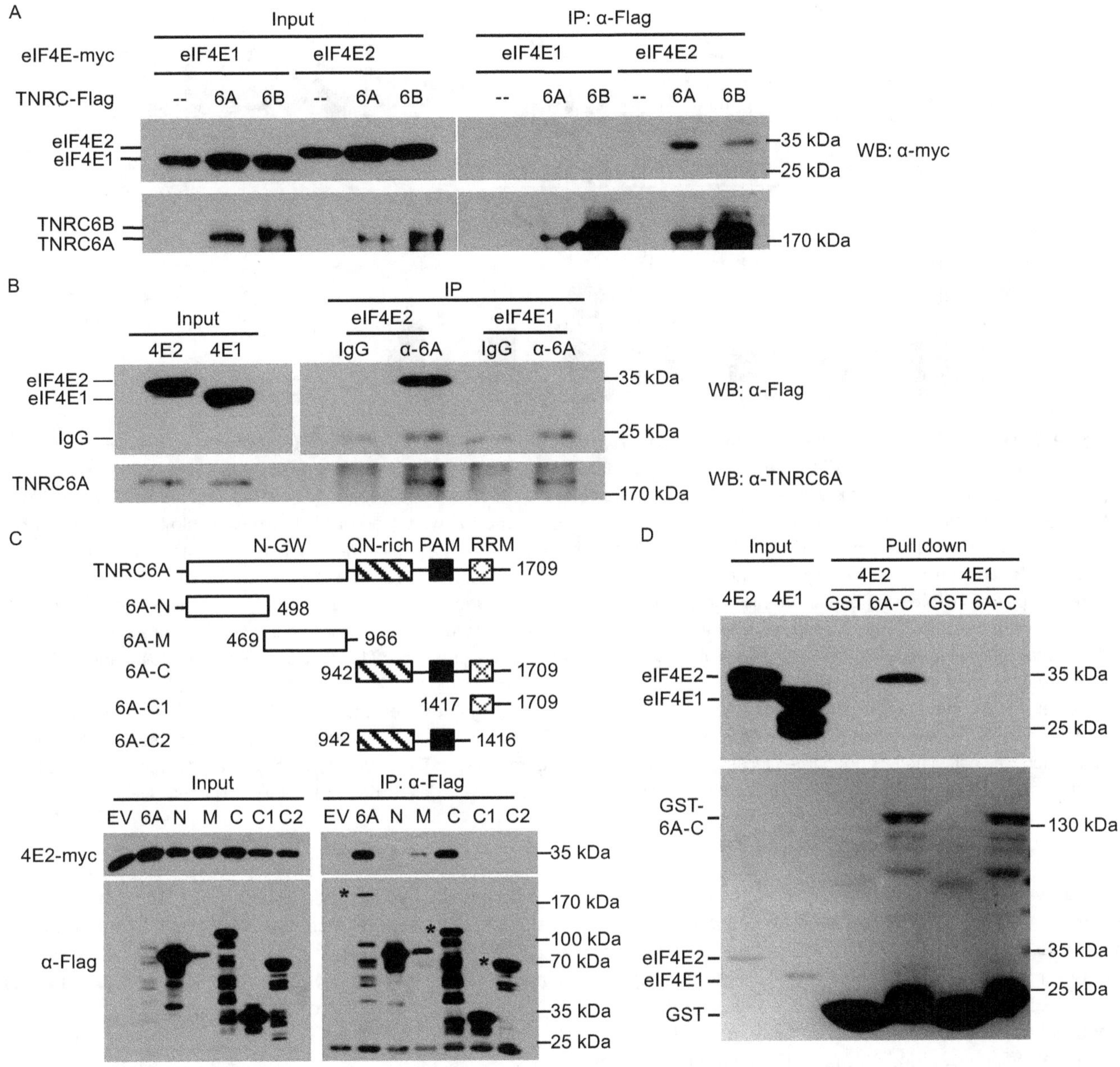

Figure 2. TNRC6A interacts with eIF4E2. (A) Plasmids expressing proteins indicated were transfected into HEK293T cells. At 48 h posttransfection, cells were lysed and the lysates were immunoprecipitated (IP) in the presence of RNase A. The precipitates were resolved on SDS-PAGE followed by Western blotting. (B) A plasmid expressing Flag-tagged eIF4E1 or eIF4E2 was transfected into HEK293T cells. At 48 h posttransfection, cells were lysed and the lysates were immunoprecipitated with the anti-TNRC6A antibody or control IgG in the presence of RNase A. The precipitates were resolved on SDS-PAGE followed by Western blotting. (C) Upper: schematic representation of TNRC6A truncation mutants. Lower: A plasmid expressing the TNRC6A mutant indicated and a plasmid expressing myc-tagged eIF4E2 were transiently transfected into HEK293T cells. At 48 h posttransfection, cells were lysed and the lysates were immunoprecipitated with anti-Flag antibody in the presence of RNase A followed by Western blotting. (D) Bacterially expressed Flag-tagged eIF4E2 or eIF4E1 was incubated with Glutathione Sepharose 4B bound fusion protein of GST and the C-terminal domain of TNRC6A (6A-C). The precipitates were washed and resolved on SDS-PAGE followed by commassie brilliant blue staining (lower) and Western blotting (upper).

levels were assayed in the absence or presence of the siRNAs targeting eIF4E2. Considering that the expression level of TNRC6C is relatively low in HeLa cells compared to TNRC6A and TNRC6B (Yao et al., 2011), only TNRC6A and TNRC6B were downregulated using a mixture of siRNAs (TNi) targeting each to serve as a positive control. To

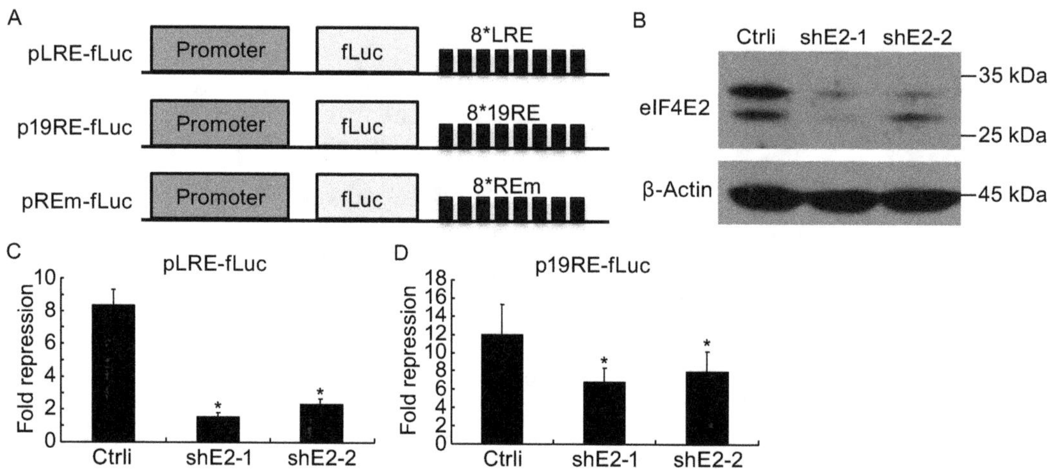

Figure 3. Downregulation of eIF4E2 impairs miRNA repression of reporter expression. (A) Schematic representation of reporter plasmids expressing firefly luciferase (fLuc). Eight tandem Let-7a response elements (LRE), miR-19 response elements (19RE), or the mutant response elements (REm) were cloned downstream of the coding sequence of firefly luciferase. (B) Plasmids expressing shRNAs targeting two different sites of eIF4E2 (shE2-1 and shE2-2) were individually transfected into HEK293 cells. At 48 h postinfection, cells were harvested and the lysates were resolved by SDS-PAGE followed by Western blotting. (C) The pLRE-fLuc reporter and a plasmid expressing a control shRNA or the shRNA targeting eIF4E2 were transfected into HEK293 cells with or without a plasmid expressing Let7a. A renilla luciferase-expressing control reporter was included. At 48 h postinfection, luciferase activities were measured. Firefly luciferase activity was normalized with renilla luciferase activity. Fold repression was calculated as the normalized luciferase activity in the absence of Let7a divided by that in the presence of Let7a. Data presented are means ± SD of three independent experiments. (D) p19RE-fLuc or pREm-fLuc was transfected into HEK293 cells together with a plasmid expressing a control shRNA or the shRNA targeting eIF4E2. A renilla luciferase-expressing reporter was included to serve as a control. At 48 h postinfection, luciferase activities were measured. Firefly luciferase activity was normalized with renilla luciferase activity. Fold repression was calculated as normalized luciferase activity expressed from pREm-fLuc divided by that expressed from p19RE-fLuc. Data presented are means ± SD of three independent experiments. The P value is determined by two-tailed Student's t test. *$P < 0.05$.

determine whether eIF4E2 downregulation had any effect on IMP1 mRNA, the mRNA levels were measured. Relative translational efficiency was calculated as the protein level divided by the mRNA level. Data showed that downregulation of TNRC6A and TNRC6B increased IMP1 protein level by about 3-fold, through increasing both the mRNA level and translational efficiency (Fig. 5A and 5B). Downregulation of eIF4E2 with the two siRNAs increased IMP1 protein levels by about 6-fold and 4-fold, respectively, with little effect on the mRNA levels (Fig. 5A and 5B). In contrast, downregulation of eIF4E1 had little effect on either protein or mRNA levels (Fig. 5A and 5B). Noticeably, downregulation of TNRC6A and TNRC6B increased IMP1 protein levels by a lower magnitude compared with eIF4E2 (Fig. 5A and 5B). If RISC recruits eIF4E2 through TNRC6A to repress translation, one would expect that TNRC6A downregulation should have a more dramatic effect than eIF4E2 downregulation. We speculated that this could be accounted for by that the siRNAs targeting TNRC6A and TNRC6B were not as effective as the ones targeting eIF4E2 (Fig. 5A). Alternatively, eIF4E2 might also work through a yet identified factor to repress translation. In addition to the endogenous IMP1, downregulation of eIF4E2 increased the protein levels of endogenous PDCD4 and PTEN (Fig. S2), whose expressions are repressed by endogenous miR-21 (Asangani et al., 2008; Meng et al., 2007; Qi et al., 2009). Collectively, these results indicate that eIF4E2 is required for miRNA-mediated translational repression of endogenous proteins.

miRNA silencing enhances eIF4E2 association with target mRNA

The affinity of eIF4E2 for the cap is about 100-fold lower than eIF4E1 (Zuberek et al., 2007), and the abundance of eIF4E2 in mammalian cells is about 10-fold lower than eIF4E1 (Kubacka et al., 2013). However, eIF4E2 can be brought to the cap by a transacting factor(s) to increase its local concentrations in the proximity of the cap of target mRNA. Based on the above results, we reasoned that miRNA on a target mRNA would increase eIF4E2 binding to the mRNA. To test this hypothesis, we analyzed the association of eIF4E2 with reporter mRNAs with or without miRNA silencing. The LRE reporter was transiently expressed in HEK293 cells, together with ectopic expression of let-7a, and Flag-tagged eIF4E1, and myc-tagged eIF4E2. The REm reporter was used as a negative control. Tagged eIF4E2 or eIF4E1

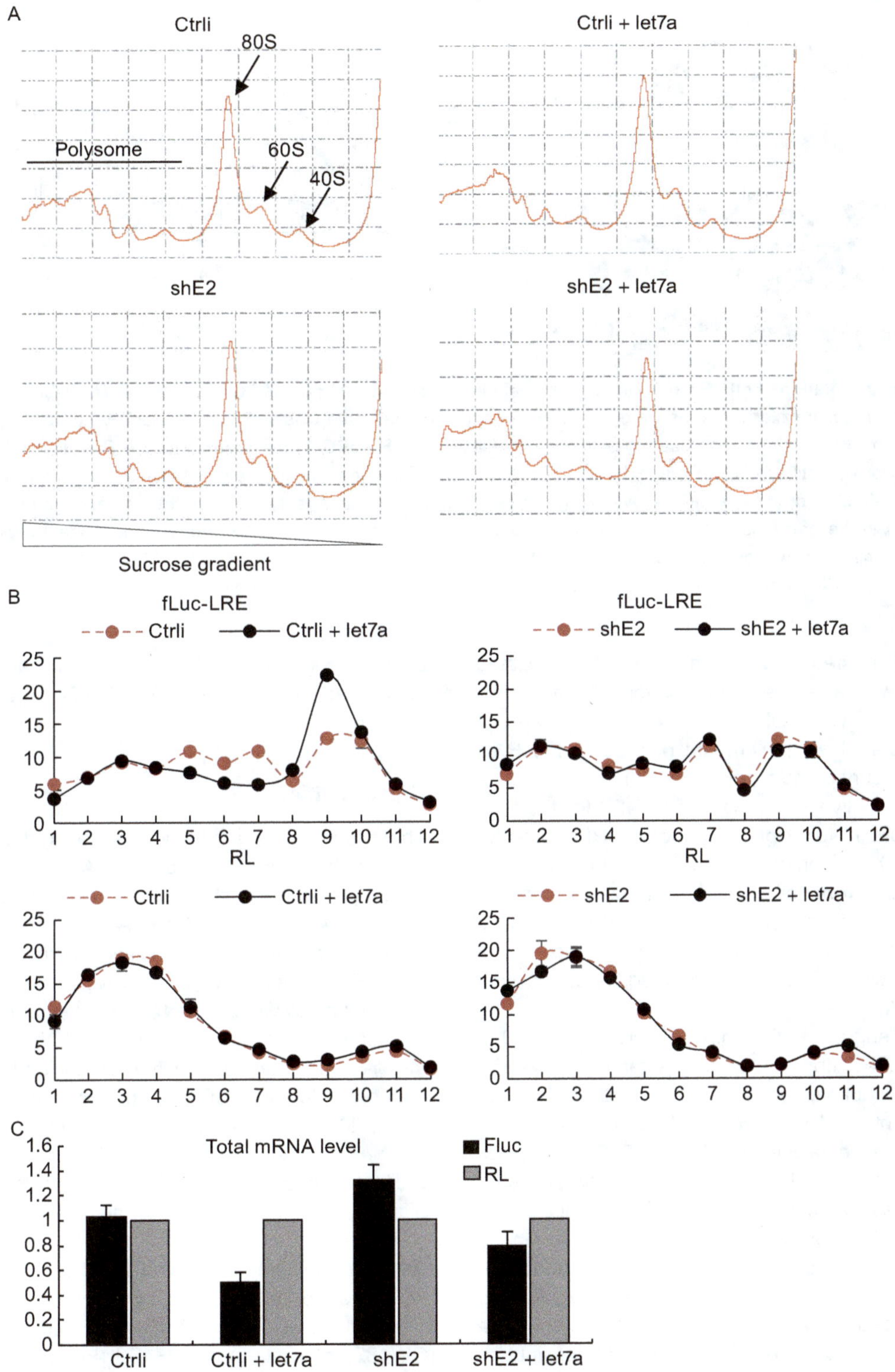

Figure 4. Downregulation of eIF4E2 relieves miRNA repression of target mRNA translation. (A) A control shRNA (Ctrli) or an shRNA targeting eIF4E2 (shE2) were transfected into HEK293 cells with pLRE-fLuc reporter and the control reporter phRL-CMV, together with a plasmid expressing let-7a or a control miRNA. At 48 h posttransfection, the cells were lysed with RNC buffer containing cycloheximide. The lysate was clarified and separated by velocity sedimentation in a sucrose continuous gradient. Fractions were collected and polysome profile was plotted by 254-nm absorption value. (B) fLuc and RL reporter mRNA levels in each fraction were quantified by RT-qPCR. Y-axis represents the percentage of mRNA in each fraction out of the total mRNA in all the fractions. (C) The total mRNA input before ultra-centrifugation and fractioning.

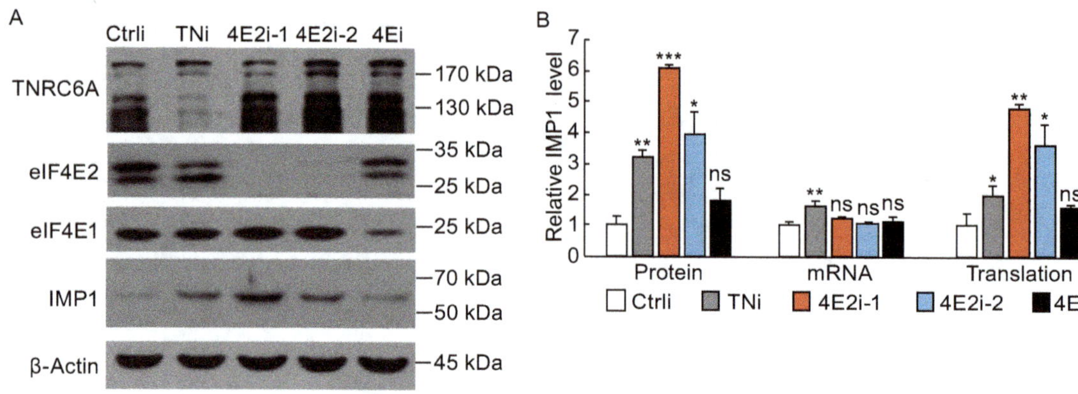

Figure 5. Downregulation of eIF4E2 increases the protein levels of endogenous IMP1. HeLa cells were transfected with siRNAs indicated. At 48 h posttransfection, cells were lysed. (A) A fraction of the lysate was subjected to SDS-PAGE followed by Western blotting. (B) The rest cell lysate was used to extract RNA, followed by RT-qPCR measurement of the RNA levels. Relative IMP1 protein levels were quantified with the Image J software and normalized with the β-actin levels. Translational efficiency was calculated as relative protein level divided by mRNA level. Fold repression was calculated as the value in the presence of the control siRNA divided by that in the presence of the targeting siRNA. Data presented are means ± SD of three independent experiments. The *P* value is determined by two-tailed Student's *t* test. ns, nonsignificant. $^*P < 0.05$; $^{**}P < 0.01$; $^{***}P < 0.005$. Ctrli, control siRNA; TNi, siRNAs targeting TNRC6A and TNRC6B; 4E2i, siRNA targeting eIF4E2; 4Ei, siRNA targeting eIF4E1.

was immunoprecipitated and the amount of associated reporter mRNA was measured. Association of the protein with reporter mRNA was indicated by relative enrichment, which was calculated as the amount of precipitated RNA divided by the amount of the reporter RNA in the input. Data showed that the relative enrichment of eIF4E2 for the LRE reporter was about 10-fold higher than for the REm reporter (Fig. 6A and 6B). In contrast, the relative enrichment of eIF4E1 for the LRE reporter was modestly but significantly lower than the REm reporter (Fig. 6A and 6B). These results indicated that miRNA silencing increased the association of the target mRNA with eIF4E2 but decreased the association with eIF4E1.

These results suggested that miRNA recruited eIF4E2 to compete with eIF4E1 for binding to the target mRNA. To substantiate this notion, we analyzed whether the cap binding activity of eIF4E2 is required for its function in miRNA silencing. The residue W124 of eIF4E2 has been reported to be critical for its cap binding activity, and substitution of this residue with alanine (W124A) abolished its cap binding activity (Rom et al., 1998). The endogenous eIF4E2 was downregulated in HeLa cells. In the meanwhile, the cells were transfected with a rescue plasmid expressing either the wild-type or the mutant eIF4E2, which cannot be targeted by the siRNA. As reported previously, the mutant eIF4E2 lost the cap binding activity (Fig. 6C). Downregulation of endogenous eIF4E2 in HeLa cells reduced miRNA repression of the LRE-fLuc reporter and ectopic expression of the wild-type eIF4E2 restored the repression to some extent (Fig. 6D). In contrast, ectopic expression of the mutant eIF4E2-W124A had little effect (Fig. 6D). Downregulation of the endogenous eIF4E2 and ectopic expression of the wild-type or mutant eIF4E2 were confirmed by Western blotting (Fig. 6E). These results indicated that the cap binding affinity is required for eIF4E2 to function in miRNA silencing.

DISCUSSION

There is increasing evidence indicating that translation initiation is the major target of miRNA repression. It has been recently well established that miRNAs induce the dissociation of eIF4A from target mRNA, resulting in the block of the assembly of translational initiation complex (Fukaya et al., 2014; Fukao et al., 2014). However, this is unlikely the only mechanism by which miRNAs repress translation initiation. It has been extensively reported that the 5′ cap of target mRNA is important for miRNA repression (Pillai et al., 2005; Mathonnet et al., 2007; Humphreys et al. 2005; Walters et al., 2010). In this report, we compared the responses to miRNA repression of reporters with or without a structured 5′UTR. Those with a structured 5′UTR are presumably affected by eIF4A, which is an RNA helicase. We also compared reporters with or without a cap structure. The results collectively confirmed that both the cap and 5′UTR are targets of miRNA repression (Fig. 1).

We show that human TNRC6A and 6B, homologues of fly GW182, interacted with eIF4E2 (Fig. 2). We proposed that TNRC6 proteins recruited eIF4E2 to target mRNA to block translation initiation through competing with eIF4E1. This notion was supported by the observation that miRNA repression increased the association of target mRNA with eIF4E2 but decreased the association of target mRNA with eIF4E1 (Fig. 6). This notion was further supported by the observation that the cap-binding activity of eIF4E2 was required for optimal miRNA repression (Fig. 6). In order to

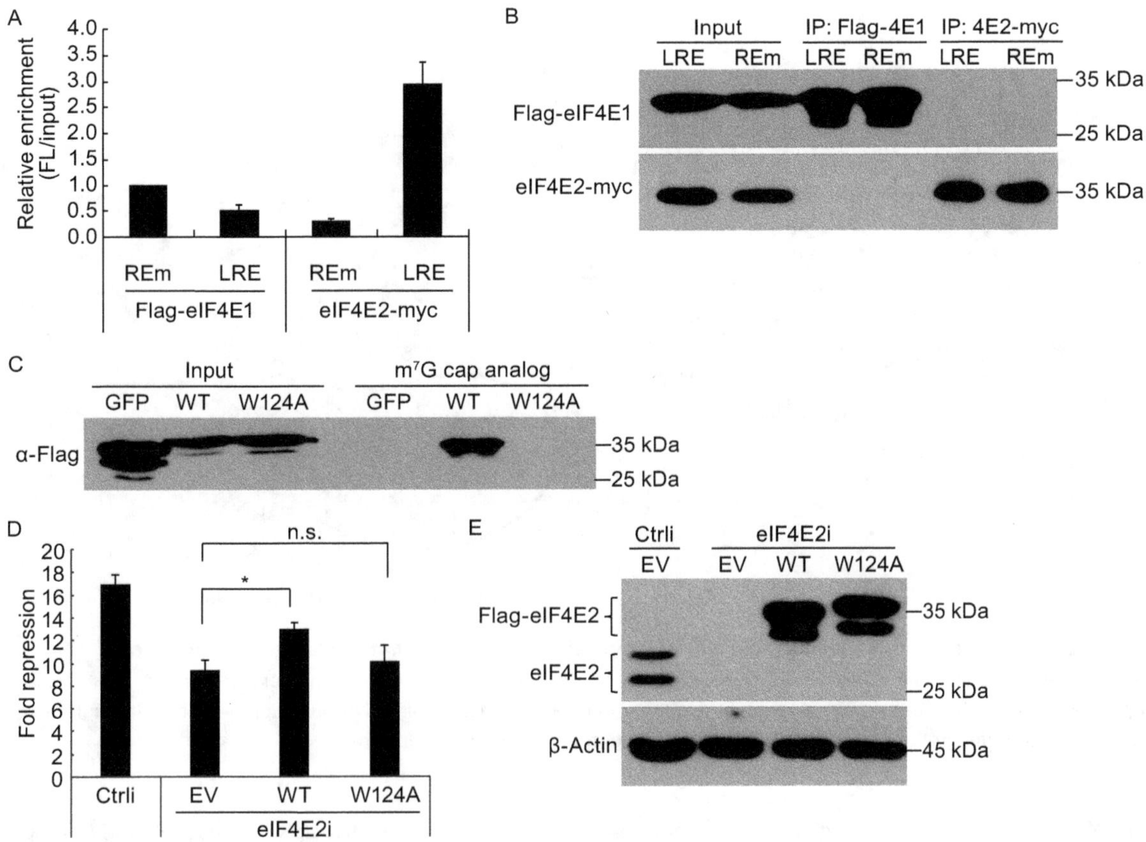

Figure 6. MiRNA silencing increases eIF4E2 association with target mRNA. (A and B) HEK293 cells were transfected with plasmids expressing Flag-tagged eIF4E1, myc-tagged eIF4E2, Let-7a, and reporter pLRE-fLuc or pREm-fLuc. The reporter phRL-CMV was included to serve as a control for transfection efficiency and sample handling. At 48 h postinfection, cells were lysed. (A) The cell lysate was immunoprecipitated with the anti-Flag or anti-myc affinity gel. Reporter mRNA levels were measured by RT-qPCR. The fLuc mRNA levels were normalized with the RL mRNA levels. Relative enrichment was calculated as the amount of the reporter mRNA in the precipitates divided by that in the input cell lysate. The relative enrichment of fLuc-REm mRNA by eIF4E1 was set as 1. Data presented are means ± SD of two independent experiments. (B) The cell lysate and the remaining precipitates were resolved on SDS-PAGE followed by Western blotting. (C) Flag-tagged GFP, eIF4E2 or its mutant eIF4E2-W124A was transiently expressed in HEK293T cells and pulled down with beads conjugated to the cap analog M7-GTP. The cell lysates and precipitates were analyzed by Western blotting. (D and E) HeLa cells were transfected with plasmids expressing the shRNA targeting eIF4E2, pLRE-fLuc or pREm-fLuc reporter and a renilla luciferase-expressing control reporter, and a rescue plasmid expressing the wild-type or mutant eIF4E2. At 48 h postinfection, cells were lysed. (D) Luciferase activities were measured. Firefly luciferase activity was normalized with the renilla luciferase activity. Fold repression was calculated as the normalized luciferase activity expressed from pREm-fLuc divided by that from pLRE-fLuc. Data presented are means ± SD of three independent experiments. The P value is determined by two-tailed Student's t test. ns, nonsignificant. *$P < 0.05$. (E) A fraction of the cell lysate was subjected to SDS-PAGE followed by Western blotting.

distinguish eIF4E2 contribution to translational repression from mRNA decay in miRNA silencing, we measured reporters' mRNA levels and analyzed the polysome profiling (Fig. 4). The results demonstrate that eIF4E2 mainly represses reporter expression on the translational level, without significantly affecting the encoding mRNA levels.

The role of eIF4E2 in miRNA repression was demonstrated by the observation that downregulation of eIF4E2 reduced miRNA repression of two different reporters (Fig. 3). We further showed that eIF4E2 downregulation increased the protein levels of endogenous IMP1, PTEN and PDCD4, whose expression are regulated by endogenous miRNAs including let-7 and miR-21 (Figs. 5 and S2).

During the course of this work, Chapat et al. reported that eIF4E2 participates in miRNA repression of target mRNA translation. They showed that miRNA recruited eIF4E2 through the CNOT1-RCK-4E-T axis. CNOT1, scaffold subunit of CCR4-NOT, recruits RCK and then 4E-T, which interacts with eIF4E2 (Chapat, 2017) (Fig. 7). Here we confirmed their result that eIF4E2 is required for miRNA

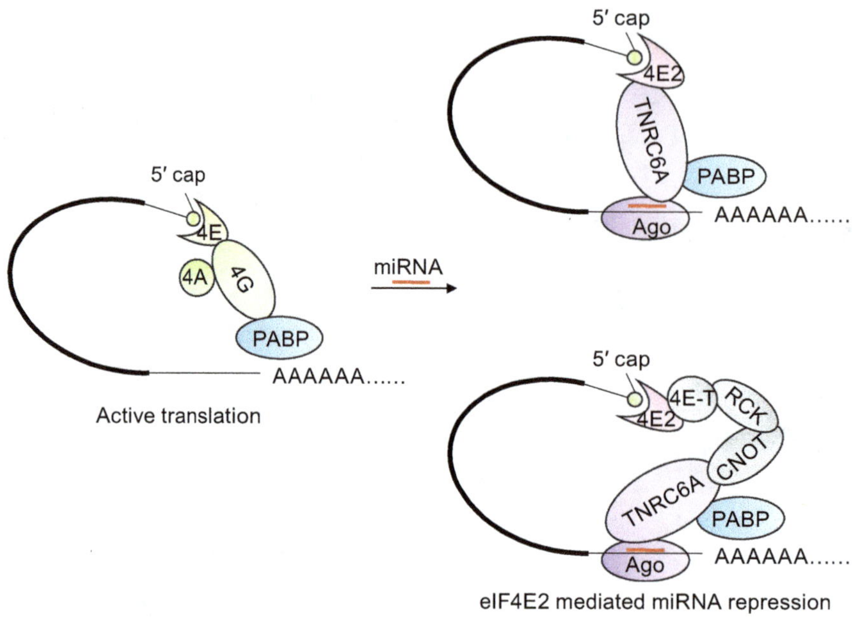

Figure 7. Working models for eIF4E2-mediated miRNA repression of target mRNA translation. In the absence of miRNA, target mRNA is translated actively as a 5′ cap-eIF4E-eIF4G-PABP-3′ poly(A) closed loop (left). In the presence of miRNA, RISC is loaded on the target mRNA, and eIF4E2 is recruited to the 5′ cap through direct interaction with TNRC6A (upper-right) or through the CNOT-RCK-4E-T axis (bottom-right) to compete with eIF4E, resulting in translational repression.

repression of target mRNA translation. However, the underlying mechanism reported here is different from theirs. Here we show that eIF4E2 is recruited by miRNA through its direct interaction with TNRC6 proteins. Notably, these two mechanisms are complementary to each. Which mechanism plays a major role awaits further investigation.

MATERIALS AND METHODS

DNA constructs and siRNAs

To generate reporters pLRE-fLuc, p19RE-fLuc and pREm-fLuc, the coding sequence of renilla luciferase of phRL-CMV (Promega) was first replaced with the coding sequence of firefly luciferase (fLuc). The CMV promoter in the vector was replaced with the PGK promoter. An intron sequence of 137 bp was inserted into the 5′UTR of the reporters such that DNA contamination was excluded in RT-qPCR (Tao and Gao, 2015). Four copies of 2*LRE, 2*19RE or 2*REm were cloned into the 3′UTR of the reporters using restriction sites *Xba*I and *Not*I. Fragments of 2*LRE, 2*19RE, and 2*REm were generated by annealing paired oligonucleotides. The sequences of the oligonucleotides are listed in Table S1.

To generate mRNA reporters cap-CAA-fLuc and cap-SL-fLuc, the coding sequence and 3′UTR of pLRE-fLuc/pREm-fLuc were PCR-amplified using forward primers in which the T7 promoter and the 5′UTR (SL or CAA) sequences were built, and cloned into pMD18-T (Takara). The primer sequences are listed in Table S1.

To generate mRNA reporters EMCV-CAA-fLuc, IRES-SL-fLuc and IRES-RL, the EMCV IRES fragment PCR-amplified from pLet7-EMCV (Meijer et al., 2013) was first cloned before the coding sequence of pLRE-fLuc/pREm-fLuc/phRL-CMV using restriction sites *Sac*I and *Sac*II. CAA or SL sequences were generated by annealing paired oligonucleotides and inserted between the IRES and the firefly luciferase coding sequence using restriction sites *Sac*II and *Nco*I. The cassette covering IRES-CAA/SL-fLuc-LRE/REm was cloned into pBluescript-SK2 (-) (Stratagene) using restriction sites *Xho*I and *Bam*HI. Sequences of the primers and oligonucleotides are listed in Table S1.

The shRNAs targeting eIF4E2 and a control shRNA have been described previously (Tao and Gao, 2015). To downregulate TNRC6A and TNRC6B, a mixture of siRNAs targeting each of them were used. The sequences of the siRNAs targeting eIF4E2, eIF4E1, TNRC6A, and TNRC6B are listed in Table S1

In vitro transcription

To generate mRNA reporters, the plasmids were linearized with *Bam*HI followed by chloroform extraction. The mRNAs were generated using the *in vitro* transcription system (P1300, Promega). Ribo m7G cap analog (P1711, Promega) was added to generate capped RNA transcripts. Then the produced RNA transcripts were added a poly(A) tail using the Poly(A) Tailing Kit (AM1350, Ambion).

Cell culture and transfection

HeLa, HEK293, and HEK293T cells were maintained in Dulbecco's modified Eagle's medium (Invitrogen) supplemented with 10% fetal bovine serum (Gibco). DNA was transfected using Lipofectamine 2000 (Invitrogen) for HeLa cells and using Neofectin DNA transfection reagent (SciLight) for HEK293 and HEK293T cells. siRNA and mRNA reporters were transfected using Lipofectamine 2000.

Antibodies

All the antibodies used in this report were commercially purchased: eIF4E1 (Santa Cruz, A-10, sc-271480), eIF4E2 (CST, D54C2), TNRC6A (Novus Biologicals, NBP1-28751), β-actin (GSGB-BIO, TA-09), IMP1 (Santa Cruz, sc-21026), PDCD4 (CST, D29C6), PTEN (Santa Cruz, A2B1, sc-7974), myc-specific mouse monoclonal antibody 9E10 (Santa Cruz Biotechnology, catalogue No. SC-40), Flag-specific mouse monoclonal antibody M2 (Sigma-Aldrich, catalogue No. F3165), Anti-Flag M2 affinity gel (Sigma), Anti-c-Myc Agarose (Sigma), M7-GTP-Sepharose (GE Healthcare Bioscience).

Immunoprecipitation assay

Cells in a 60-mm dish were lysed in 500 μL Co-IP buffer (30 mmol/L Hepes, pH 7.5, 100mmol/L NaCl, 0.5% NP40), with protease inhibitor cocktail (Roche) and RNase A (sigma). The lysate was clarified at 12,000 rpm for 10 min at 4°C. 50 μL cell lysate was solved with SDS loading buffer and used as the input (10%), and the left was incubated with antibodies and protein G beads for 2–4 h at 4°C. Then the beads were washed 5 times by TBST buffer (20 mmol/L Tris-HCl, pH 7.5, 500 mmol/L NaCl, 0.1% Tween 20), and analyzed by Western Blotting.

Pull-down assay

TNRC6A-C and Flag-tagged eIF4E/4E2 were cloned into the prokaryotic expression vector pGEX-5x-3-linker (modified from pGEX-5x-3, GE). GST-6A-C and GST-Flag-4E/4E2 were expressed in *Escherichia coli* (BL21 (DE3), Transgene) and purified following the handbook for the glutathione S-transferase (GST) fusion protein system (Amersham Biosciences). GST was removed using PreScission protease to produce Flag-4E/4E2, which was then incubated with Glutathione Sepharose 4B bound fusion protein of GST and the C-terminal domain of TNRC6A (6A-C). The precipitates were washed by TBST and resolved on SDS-PAGE followed by commassie brilliant blue staining and Western blotting.

RNA extraction and quantitative PCR

Cytoplasmic RNA was extracted from clarified cell lysate using TRIzol (Invitrogen) following the manufacturer's instructions without DNase treating. RNA concentration was quantified using Nanodrop (Nanodrop Technologies). 1~2 μg RNA was reverse transcribed using random primers (Takara, 3802) and MLV reverse transcriptase. No RT (minus reverse transcriptase) was included as a negative control to exclude the proper effect from plasmids DNA or genomic DNA. Quantitative RT-PCR reactions were performed using specific primers and RealMaster Mix (SYBR Green) (TIANGEN Biotech) in Corbett 6200/6600/65H0 (Corbett research) with the following cycling condition: 95°C, 10 min; 40× (95°C,10 s; 60°C, 15 s; 72°C, 20 s). The fold change on mRNA levels was calculated relative to the control and normalized to GAPDH mRNA level. Data are means of duplicate measurements by two times of 2 independent experiments. The qPCR primers for pLRE-fLuc, pREm-fLuc, and phRL-CMV have been described previously (Tao and Gao, 2015). Other primer sequences are listed in Table S1.

Polysome profiling assay

Cells were lysed with RNC buffer (50 mmol/L Hepes, pH 7.4; 100 mmol/L KAc; 5 mmol/L $MgCl_2$; 0.1% TritonX-100; 1000 U/mL RNase Inhibitor and 100 μg/mL Cycloheximide were added before using). 0.8 mL cell lysate supernatant were loaded on 12 mL 10%–50% sucrose continuous gradient followed by 36,000 rpm ultra-centrifugation for 3 h and 20 min. Then the samples were through a continuous 254-nm absorbance detector and 12 fractions, 1 mL each, were collected for each sample. 300 μL of each fraction was used for RNA extraction with 1 mL Trizol and 0.1 μg NS1 mRNA as the external reference. The fLuc-LRE and RL mRNA levels were measured by RT-qPCR normalized by NS1 level.

RNA binding assay

Cells were lysed with cell lysis buffer (20 mmol/L HEPES KOH (pH 7.3), 150 mmol/L KCl, 10 mmol/L $MgCl_2$ and 1% NP40) (Heiman et al., 2014) supplemented with the protease inhibitor cocktail (Roche), 1 mmol/L DTT, 1000 U/mL RNase inhibitor, and 30 mmol/L $NaBH_3CN$ (Sonenberg and Shatkin, 1977) for 30 min on a roller at 4°C. The lysate was clarified by centrifugation at 12,000 rpm for 15 min at 4°C. The clarified lysate was incubated with Anti-c-Myc Agarose (Sigma) and Anti-Flag M2 affinity gel (Sigma) separately for 2 h on a roller at 4°C. Immunoprecipitates were washed three times with TBST. Ten percents of the immunoprecipitate was resuspended in SDS loading buffer for protein detection by Western blotting. The remaining immunoprecipitate was resuspended in TRIzol (Invitrogen) for RNA extraction.

ACKNOWLEDGEMENTS

This work was supported by a grant to Guangxia Gao from Chinese Academy of Sciences under Grant KJZD-EWL10-01.

ABBREVIATIONS

LREs, Let7a-responsive elements; miRNAs, microRNAs; REm, mutant responsive elements; RISC, RNA-induced silencing complex; RL, renilla luciferase; SL, stem-loop

COMPLIANCE WITH ETHICS GUIDELINES

The authors declare that they have no conflict of interest. This article does not contain any studies with human or animal subjects performed by any of the authors.

REFERENCES

Asangani IA, Rasheed SA, Nikolova DA, Leupold JH, Colburn NH, Post S, Allgayer H (2008) MicroRNA-21 (miR-21) post-transcriptionally downregulates tumor suppressor Pdcd4 and stimulates invasion, intravasation and metastasis in colorectal cancer. Oncogene 27(15):2128–2136

Boland A, Tritschler F, Heimstadt S, Izaurralde E, Weichenrieder O (2010) Crystal structure and ligand binding of the MID domain of a eukaryotic Argonaute protein. Embo Reports 11(7):522–527

Boyerinas B, Park SM, Shomron N, Hedegaard MM, Vinther J, Andersen JS, Feig C, Xu J, Burge CB, Peter ME (2008) Identification of let-7-regulated oncofetal genes. Cancer Res 68 (8):2587–2591

Chapat C, Jafarnejad SM, Matta-Camacho E, Hesketh GG, Gelbart IA, Attig J, Gkogkas CG, Alain T, Stern-Ginossar N, Fabian MR et al (2017) Cap-binding protein 4EHP effects translation silencing by microRNAs. Proc Natl Acad Sci 114:5425–5430

Cho PF, Poulin F, Cho-Park YA, Cho-Park IB, Chicoine JD, Lasko P, Sonenberg N (2005) A new paradigm for translational control: inhibition via 5 '-3 ' mRNA tethering by Bicoid and the eIF4E cognate 4EHP. Cell 121(3):411–423

Djuranovic S, Zinchenko MK, Hur JK, Nahvi A, Brunelle JL, Rogers EJ, Green R (2010) Allosteric regulation of Argonaute proteins by miRNAs. Nat Struct Mol Biol 17(2):144–150

Eulalio A, Huntzinger E, Izaurralde E (2008) GW182 interaction with Argonaute is essential for miRNA-mediated translational repression and mRNA decay. Nat Struct Mol Biol 15(4):346–353

Fabian MR, Sonenberg N, Filipowicz W (2010) Regulation of mRNA translation and stability by microRNAs. Annu Rev Biochem 79:351–379

Flynt AS, Lai EC (2008) Biological principles of microRNA-mediated regulation: shared themes amid diversity. Nat Rev Genet 9 (11):831–842

Frank F, Fabian MR, Stepinski J, Jemielity J, Darzynkiewicz E, Sonenberg N, Nagar B (2011) Structural analysis of 5 '-mRNA-cap interactions with the human AGO2 MID domain. Embo Reports 12(5):415–420

Fu R, Olsen MT, Webb K, Bennett EJ, Lykke-Andersen J (2016) Recruitment of the 4EHP-GYF2 cap-binding complex to tetraproline motifs of tristetraprolin promotes repression and degradation of mRNAs with AU-rich elements. RNA 22(3):373–382

Fukao A, Mishima Y, Takizawa N, Oka S, Imataka H, Pelletier J, Sonenberg N, Thoma C, Fujiwara T (2014) MicroRNAs trigger dissociation of eIF4AI and eIF4AII from target mRNAs in humans. Mol Cell 56(1):79–89

Fukaya T, Tomari Y (2012) MicroRNAs mediate gene silencing via multiple different pathways in drosophila. Mol Cell 48(6):825–836

Fukaya T, Iwakawa HO, Tomari Y (2014) MicroRNAs block assembly of eIF4F translation initiation complex in Drosophila. Mol Cell 56(1):67–78

Gingras AC, Raught B, Sonenberg N (1999) eIF4 initiation factors: effectors of mRNA recruitment to ribosomes and regulators of translation. Annu Rev Biochem 68:913–963

Gu W, Xu Y, Xie X, Wang T, Ko JH, Zhou T (2014) The role of RNA structure at 5' untranslated region in microRNA-mediated gene regulation. RNA 20(9):1369–1375

Heiman M, Kulicke R, Fenster RJ, Greengard P, Heintz N (2014) Cell type-specific mRNA purification by translating ribosome affinity purification (TRAP). Nat Protoc 9(6):1282–1291

Humphreys DT, Westman BJ, Martin DI, Preiss T (2005) MicroRNAs control translation initiation by inhibiting eukaryotic initiation factor 4E/cap and poly(A) tail function. Proc Natl Acad Sci U S A 102 (47):16961–16966

Jackson RJ, Hellen CU, Pestova TV (2010) The mechanism of eukaryotic translation initiation and principles of its regulation. Nat Rev Mol Cell Biol 11(2):113–127

Jonas S, Izaurralde E (2015) NON-CODING RNA Towards a molecular understanding of microRNA-mediated gene silencing. Nat Rev Genet 16(7):421–433

Kamenska A, Lu WT, Kubacka D, Broomhead H, Minshall N, Bushell M, Standart N (2014) Human 4E-T represses translation of bound mRNAs and enhances microRNA-mediated silencing. Nucleic Acids Res 42(5):3298–3313

Kamenska A, Simpson C, Vindry C, Broomhead H, Benard M, Ernoult-Lange M, Lee BP, Harries LW, Weil D, Standart N (2016) The DDX6-4E-T interaction mediates translational repression and P-body assembly. Nucleic Acids Res 44(13):6318–6334

Kinch LN, Grishin NV (2009) The human Ago2 MC region does not contain an eIF4E-like mRNA cap binding motif. Biol Direct 4:2

Kiriakidou M, Tan GS, Lamprinaki S, De Planell-Saguer M, Nelson PT, Mourelatos Z (2007) An mRNA m(7)G cap binding-like motif within human Ago2 represses translation. Cell 129 (6):1141–1151

Kubacka D, Kamenska A, Broomhead H, Minshall N, Darzynkiewicz E, Standart N (2013) Investigating the consequences of eIF4E2 (4EHP) interaction with 4E-transporter on its cellular distribution in HeLa cells. PLoS ONE 8(8):e72761

Lu WT, Wilczynska A, Smith E, Bushell M (2014) The diverse roles of the eIF4A family: you are the company you keep. Biochem Soc Trans 42:166–172

Mathonnet G, Fabian MR, Svitkin YV, Parsyan A, Huck L, Murata T, Biffo S, Merrick WC, Darzynkiewicz E, Pillai RS et al (2007) MicroRNA inhibition of translation initiation in vitro by targeting the cap-binding complex eIF4F. Science 317(5845):1764–1767

Meijer HA, Kong YW, Lu WT, Wilczynska A, Spriggs RV, Robinson SW, Godfrey JD, Willis AE, Bushell M (2013) Translational repression and eIF4A2 activity are critical for microRNA-mediated gene regulation. Science 340(6128):82–85

Meister G (2007) miRNAs get an early start on translational silencing. Cell 131(1):25–28

Meng FY, Henson R, Wehbe-Janek H, Ghoshal K, Jacob ST, Patel T (2007) MicroRNA-21 regulates expression of the PTEN tumor suppressor gene in human hepatocellular cancer. Gastroenterology 133(2):647–658

Nishimura T, Padamsi Z, Fakim H, Milette S, Dunham WH, Gingras AC, Fabian MR (2015) The eIF4E-Binding Protein 4E-T Is a Component of the mRNA Decay Machinery that Bridges the 5 ' and 3 ' Termini of Target mRNAs. Cell Rep 11(9):1425–1436

Petersen CP, Bordeleau ME, Pelletier J, Sharp PA (2006) Short RNAs repress translation after initiation in mammalian cells. Mol Cell 21(4):533–542

Pfaff J, Meister G (2013) Argonaute and GW182 proteins: an effective alliance in gene silencing. Biochem Soc Trans 41 (4):855–860

Pillai RS, Bhattacharyya SN, Artus CG, Zoller T, Cougot N, Basyuk E, Bertrand E, Filipowicz W (2005) Inhibition of translational initiation by Let-7 microRNA in human cells. Science 309 (5740):1573–1576

Qi L, Bart J, Tan LP, Platteel I, Sluis T, Huitema S, Harms G, Fu L, Hollema H, Berg A (2009) Expression of miR-21 and its targets (PTEN, PDCD4, TM1) in flat epithelial atypia of the breast in relation to ductal carcinoma in situ and invasive carcinoma. BMC Cancer 9:163

Ricci EP, Limousin T, Soto-Rifo R, Rubilar PS, Decimo D, Ohlmann T (2013) miRNA repression of translation in vitro takes place during 43S ribosomal scanning. Nucleic Acids Res 41(1):586–598

Rom E, Kim HC, Gingras AC, Marcotrigiano J, Favre D, Olsen H, Burley SK, Sonenberg N (1998) Cloning and characterization of 4EHP, a novel mammalian eIF4E-related cap-binding protein. J Biol Chem 273(21):13104–13109

Sonenberg N, Shatkin AJ (1977) Reovirus messenger-Rna can be covalently crosslinked via 5' cap to proteins in initiation-complexes. Proc Natl Acad Sci USA 74(10):4288–4292

Tao X, Gao G (2015) Tristetraprolin Recruits Eukaryotic Initiation Factor 4E2 To Repress Translation of AU-Rich Element-Containing mRNAs. Mol Cell Biol 35(22):3921–3932

Thermann R, Hentze MW (2007) Drosophila miR2 induces pseudo-polysomes and inhibits translation initiation. Nature 447 (7146):875–878

Valencia-Sanchez MA, Liu J, Hannon GJ, Parker R (2006) Control of translation and mRNA degradation by miRNAs and siRNAs. Genes Dev 20(5):515–524

Varani G (1997) A cap for all occasions. Structure 5(7):855–858

Walters RW, Bradrick SS, Gromeier M (2010) Poly(A)-binding protein modulates mRNA susceptibility to cap-dependent miRNA-mediated repression. Rna 16(1):239–250

Wang BB, Yanaz A, Novina CD (2008) MicroRNA-repressed mRNAs contain 40S but not 60S components. Proc NatL Acad Sci USA 105(14):5343–5348

Yao B, Li SQ, Jung HM, Lian SL, Abadal GX, Han F, Fritzler MJ, Chan EKL (2011) Divergent GW182 functional domains in the regulation of translational silencing. Nucleic Acids Res 39(7):2534–2547

Zuberek J, Kubacka D, Jablonowska A, Jemielity J, Stepinski J, Sonenberg N, Darzynkiewicz E (2007) Weak binding affinity of human 4EHP for mRNA cap analogs. RNA 13(5):691–697

14

Loss of IκB kinase β promotes myofibroblast transformation and senescence through activation of the ROS-TGFβ autocrine loop

Liang Chen[1], Zhimin Peng[1], Qinghang Meng[1], Maureen Mongan[1], Jingcai Wang[1], Maureen Sartor[1], Jing Chen[1], Liang Niu[1], Mario Medvedovic[1], Winston Kao[2], Ying Xia[1,2]✉

[1] Department of Environmental Health and Center of Environmental Genetics, University of Cincinnati Medical Center, Cincinnati, OH 45267, USA
[2] Department of Ophthalmology, University of Cincinnati Medical Center, Cincinnati, OH 45267, USA
✉ Correspondence: ying.xia@uc.edu (Y. Xia)

ABSTRACT

Using forward and reverse genetics and global gene expression analyses, we explored the crosstalk between the IκB kinase β (IKKβ) and the transforming growth factor β (TGFβ) signaling pathways. We show that *in vitro* ablation of *Ikkβ* in fibroblasts led to progressive ROS accumulation and TGFβ activation, and ultimately accelerated cell migration, fibroblast-myofibroblast transformation and senescence. Mechanistically, the basal IKKβ activity was required for anti-oxidant gene expression and redox homeostasis. Lacking this activity, IKKβ-null cells showed ROS accumulation and activation of stress-sensitive transcription factor AP-1/c-Jun. AP-1/c-Jun activation led to up-regulation of the *Tgfβ2* promoter, which in turn further potentiated intracellular ROS through the induction of NADPH oxidase (NOX). These data suggest that by blocking the autocrine amplification of a ROS-TGFβ loop IKKβ plays a crucial role in the prevention of fibroblast-myofibroblast transformation and senescence.

KEYWORDS IκB kinase β (IKKβ), nuclear factor κB (NF-κB), transforming growth factors β (TGFβ), reactive oxygen species (ROS), myofibroblast, senescence

INTRODUCTION

The IκB kinase β (IKKβ) is a key catalytic subunit of the IKK complex, involved in inflammatory responses. It is robustly activated by cytokines, bacterial and viral products and metabolic stresses. IKKβ activation leads to phosphorylation of inhibitor of κB (IκB), and subsequently, translocation of the nuclear factor κB (NF-κB) to nucleus. The nuclear NF-κB binds to the κB elements in gene promoters and enhancers to either activate or repress gene expression (Perkins, 2007). By regulating genes coding for cytokines, chemokines, enzymes and molecules with microbicidal activity, the IKK-NF-κB cascade offers important protection against stress and danger signals (Vallabhapurapu and Karin, 2009). Persistent and unrestrained activation of the cascade, on the other hand, leads to chronic inflammation that may be the underlying cause of detrimental and life-threatening diseases, such as rheumatoid arthritis, atherosclerosis and cancer (Luo et al., 2005; Kim et al., 2006; Chariot, 2009). For this reason, inhibition of IKK signaling is widely considered as a promising strategy for treating many illnesses; the challenge however is to fully recognize, and develop means to offset, the potential harmful consequences of pathway inactivation (Baldwin, Jr., 2001; Li et al., 2002; Bacher and Schmitz, 2004; Courtois and Gilmore, 2006; Karin, 2006).

IKKβ maintains low static activity in the absence of external stimuli. This is associated with slow IκB degradation and equilibrium NF-κB activity (O'Dea et al., 2007). The basal activity is important for redox homeostasis, thus IKKβ inactivation renders cells or tissues vulnerable to oxidative damage (Gerondakis et al., 2006). For example, when *Ikkβ* is knocked out in hepatocytes, the livers of the knockout mice have normal development, but exhibit elevated levels of reactive oxygen species (ROS). In addition, IKKβ-defective livers are susceptible to injuries by carcinogens, concanavalin A and bacterial infection (Lavon et al., 2000; Maeda et al., 2005). When IKKβ is knocked out in fibroblasts, the null cells have elevated ROS levels and are sensitive to damage by stress and injury (Maeda et al., 2005; Chen et al., 2006; Giorgio et al., 2007; May and Madge, 2007; Sen and Roy, 2010). These observations suggest that IKKβ may be involved in a plethora of physiological processes through the regulation of redox homeostasis (Karin, 2008; Pasparakis, 2009).

In the present work, we investigated the role of IKKβ through global gene expression analyses and identified a crosstalk interaction between IKKβ and TGFβ signaling. We showed that loss of IKKβ in fibroblasts led to TGFβ activation, which in turn modulated cell motility, myofibroblast transformation and senescence. These results suggest that IKKβ can act as a repressor of the TGFβ pathway.

RESULTS

IKKβ represses TGFβ signaling

To explore the roles of IKK and NF-κB signaling in fibroblasts, we examined global gene expression in wild type and cells lacking IKKα, IKKβ or the p65 subunit of NF-κB. Comparison of differentially expressed genes between wild type and knockout cells, we found that genes up-regulated in the wild type cells were enriched for the terpenoid backbone biosynthesis pathway, whereas genes down-regulated in the wild type cells were enriched for the focal adhesion and vascular smooth muscle contraction pathways (Table 1).

We further examined differential gene expression between IKKβ-competent (*Ikkβ*$^{-/-}$/Ad-IKKβ) and -deficient (*Ikkβ*$^{-/-}$/Ad-β-Gal and *Ikkβ*$^{-/-}$) cells using the same strategy. Genes up-regulated in the IKKβ-competent cells were, as expected, enriched for pathways involved in immunity and inflammation, such as antigen processing and presentation, rheumatoid arthritis, and B cell receptor signaling pathway and allograft rejection, but intriguingly, genes down-regulated in the IKKβ-competent cells were enriched for focal adhesion, ECM-receptor interaction and, and the TGFβ signaling pathways (Table 2).

We validated the array data focusing on IKKβ-repressed genes of the TGFβ pathway. Compared to the wild type, the *Ikkβ*$^{-/-}$ cells had elevated *Tgfβ2* and *Tgfβ3* mRNA transcripts (Fig. 1A), corresponding to higher gene promoter activities (Fig. 1B). They also exhibited increased SMAD transcriptional activity (Fig. 1C) and phosphorylation (Fig. 1D), as well as increased expression of a number of SMAD target genes, such as *Smad6*, *Ctgf* and *Acta2* (Figs. 1E and S1). In addition, we observed the expression of myofibroblast marker α smooth muscle actin (α-SMA), the product of *Acta2*, in IKKβ-null but not wild type cells (Fig. 1D). Adenoviral-mediated expression of IKKβ, but not of GFP used as control, in the null cells repressed *Tgfβ* expression and promoter activity, decreased SMAD activity and target gene expression, similar to the effects of Ad-SMAD7 and reached the levels same as that in the wild type cells (Fig. 1A–C and 1E). These results indicate that loss of IKKβ leads to the activation of TGFβ expression and signaling.

TGFβ upregulation leads to migration and myofibroblast transformation of IKKβ-null cells

TGFβ plays a pivotal role in cell proliferation, differentiation, wound healing and extracellular matrix production, and it induces growth arrest and myofibroblast transformation in fibroblasts (Datto et al., 1999; Phan, 2002). Chen, et al. have reported that the IKKβ-deficient cells grow slower, but migrate faster (Chen et al., 2006). We confirmed these observations (Fig. 2A and 2B), and furthermore, we showed that the migration rate of the null cells was significantly reduced by expression of IKKβ and inhibitory SMAD7, and by treatment with SB505124, a TGFβ receptor inhibitor (Fig. 2B and 2C).

To assess if promoted migration was due to TGFβ secretion, we collected conditioned medium from wild type and *Ikkβ*$^{-/-}$ cultures and examined its effects on migration of the wild type cells. The wild type-conditioned medium had no effect, but the *Ikkβ*$^{-/-}$-conditioned medium accelerated migration by 50% (Fig. 2D). Additionally, the migration

Table 1. Biological pathways affected by the IKK-NF-κB cascade*

	WT vs. Ikkα$^{-/-}$	WT vs. Ikkβ$^{-/-}$	WT vs. p65$^{-/-}$
Up-regulated genes enriched pathways			
Terpenoid backbone biosynthesis	0.003663	1.53×10^{-5}	0.014653
Down-regulated genes enriched pathways			
Focal adhesion	1.78×10^{-8}	0.039390	3.14×10^{-10}
Vascular smooth muscle contraction	9.00×10^{-7}	0.000589	1.66×10^{-5}

* Each entry is the False Discovery Rate (FDR) adjusted *P*-values for the pathway in the corresponding row in the comparison in the corresponding column. The *P*-values were calculated by R package CLEAN using the KEGG pathway database.

Table 2. The IKKβ-regulated biological pathways*

	Ad-IKKβ vs. uninfected	Ad-IKKβ vs. Ad-β-Gal
Up-regulated genes enriched pathways		
Antigen processing and presentation	3.74×10^{-6}	4.02×10^{-6}
Leishmaniasis	3.69×10^{-5}	3.29×10^{-12}
Phagosome	4.00×10^{-5}	0.002085
Rheumatoid arthritis		1.51×10^{-9}
B cell receptor signaling pathway	0.000402	0.000641
Graft-versus-host disease	0.002208	0.002457
Allograft rejection	0.003035	0.004726
Type I diabetes mellitus	0.009985	0.022062
Autoimmune thyroid disease	0.013477	0.025376
Down-regulated genes enriched pathways		
Focal adhesion	0.000511	6.21×10^{-6}
TGFβ signaling pathway	0.009496	0.028866
ECM-receptor interaction	4.73×10^{-8}	4.70×10^{-7}
Protein digestion and absorption	1.18×10^{-7}	0.001664
Amoebiasis	5.40×10^{-6}	0.000540

* Each entry is the False Discovery Rate (FDR) adjusted *P*-values for the pathway in the corresponding row in the comparison in the corresponding column. The *P*-values were calculated by R package CLEAN using the KEGG pathway database.

stimulatory activity was abolished by TGFβ neutralizing antibodies, supporting the notion that TGFβ secreted by the IKKβ-null cells contributed to the stimulation of fibroblast migration.

Progressive ROS accumulation and TGFβ activation following IKKβ ablation

Infection of $Ikk\beta^{F/F}$ embryonic fibroblasts with Ad-Cre could ablate the *Ikkβ* gene *in vitro*. Using this approach, we generated the $Ikk\beta^{F/F}$/Ad-Cre cells, in which IKKβ expression, NF-κB activity, and NF-κB target gene expression were abolished or significantly reduced (Figs. 3A and S2A–S2E). Infection of the $Ikk\beta^{F/F}$/Ad-Cre cells with Ad-IKKβ, but not Ad-GFP, restored NF-κB activity and target gene expression (Fig. S2D and S2E).

The $Ikk\beta^{F/F}$/Ad-Cre cells lacked IKKβ, but surprisingly, they did not have detectable α-SMA expression immediately following Ad-Cre infection (Fig. 3A). These cells instead displayed a gradual increase in the expression of α-SMA, TGFβ2 and SMAD-target genes (Fig. 3A and 3B), and they exhibited faster migration only after 90 days of Ad-Cre infection (Fig. 3C). Simultaneous ablation of IKKβ and TGFβ receptor 2 reduced α-SMA upregulation and TGFβ1-induced migration (Figs. 3D, 3E and S3). The data derived from the *in vitro* gene ablation system suggest that IKKβ loss leads to a gradual activation of TGFβ signaling and progressive myofibroblast conversion.

Loss of IKKβ leads to activation of the ROS-TGFβ-NOX cascade

Consistent with the notion that IKKβ represses reactive oxygen species (ROS) (Tanaka et al., 1999; Maeda et al., 2005), we showed that the H_2O_2 level, measured by 2′,7′-Dichlorodihydrofluorescein diacetate (CM-H_2DCFDA) labeling, was high in $Ikk\beta^{-/-}$ but low in wild type cells (Fig. S4A). In addition, the expression of the oxidative stress-inducible biomarker gene *Heme oxygenase 1* (*Ho-1*) was more abundant in $Ikk\beta^{-/-}$ than wild type cells (Fig. 4A). The IKKβ- and p65-deficient cells have similar gene expression signatures and faster migration phenotype (Table 1 and Fig. S5), and like the IKKβ-null cells, the $p65^{-/-}$ cells also had increased *Ho-1* expression (Fig. 4B). Furthermore, we detected in the $p65^{-/-}$ cells decreased expression of superoxide dismutase 2 (*Sod2*), encoding for a crucial redox scavenger. Correspondingly, compared to the IKKβ-competent, i.e. wild type and $Ikk\beta^{-/-}$/Ad-IKKβ, cells, the IKKβ-deficient $Ikk\beta^{-/-}$cells had decreased level of RNA pol II recruitment to the *Sod2* promoter and reduced p65 bound at the gene enhancer (Fig. 4C).

To evaluate if SOD2 reduction contributed to TGFβ activation, we expressed SOD2 in IKKβ-null cells and observed

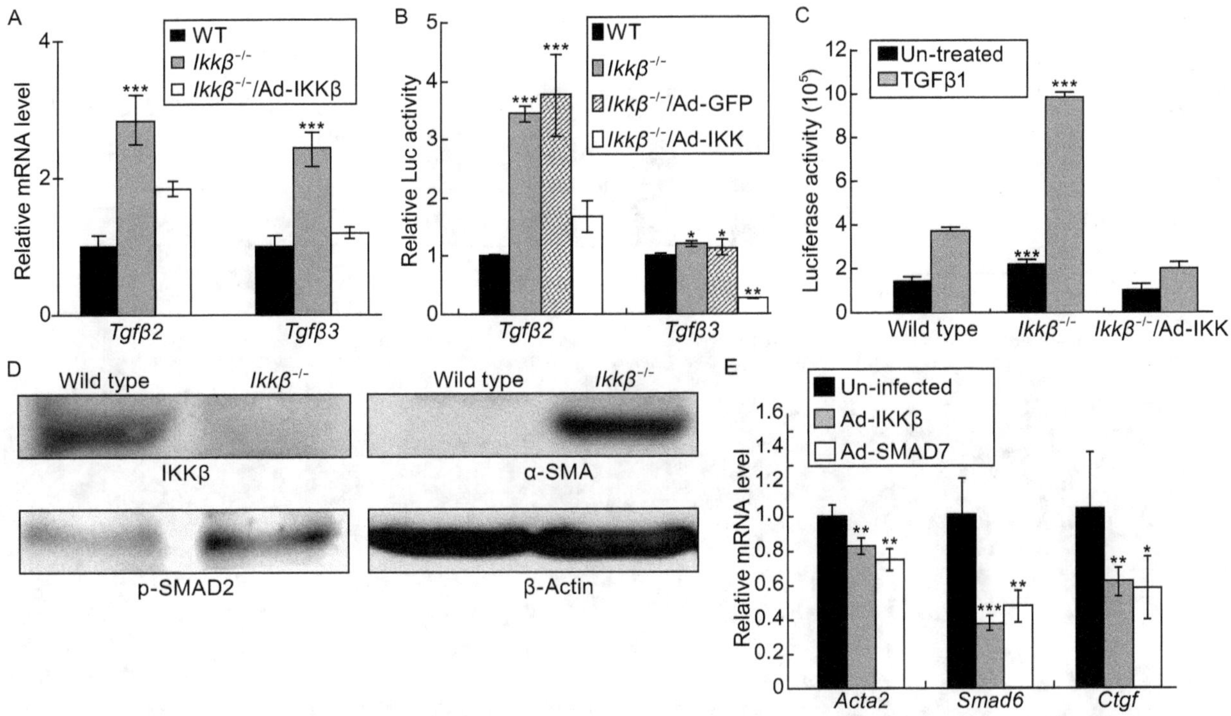

Figure 1. Loss of IKKβ upregulates TGFβ expression and activity. The IKKβ-competent, i.e. wild type and *Ikkβ*$^{-/-}$/Ad-IKKβ, and IKKβ-deficient, i.e. *Ikkβ*$^{-/-}$ and *Ikkβ*$^{-/-}$ /Ad-GFP, fibroblasts were examined for *Tgfβ2* and *3* (A) mRNA expression and (B) promoter activity, and for (C) basal (un-treated) and TGFβ1-induced SMAD activity (SBE-luc) and (D) SMAD phosphorylation, and IKKβ, α-SMA and β-actin expression. (E) The *Ikkβ*$^{-/-}$ cells, either uninfected or infected with Ad-IKKβ and Ad-SMAD7, were examined for the expression of SMAD-target genes, i.e. *Acta2*, *Smad6* and *Ctgf*. Results represent the mean values ± SD from at least three independent experiments. **P* < 0.05, ***P* < 0.01 and ****P* < 0.001 were considered significantly different from the wild type or control samples.

a significant decrease of *Tgfβ2* promoter and SMAD activity (Fig. 4D). We further showed that SOD2 expression caused down-regulation of ROS-sensitive AP-1 activity, raising the possibility that the TGFβ signaling was actually modulated by cellular redox status (Bataller et al., 2003; Fleckenstein et al., 2007; Roy et al., 2011). To test the possibility, we treated the wild type cells with pro-oxidant L-Buthionine sulphoximine (BSO), and the *Ikkβ*$^{-/-}$ cells with anti-oxidant N-acetyl cysteine (NAC). As predicted by the hypothesis, BSO increased *Tgfβ2* gene expression and promoter activity and NAC significantly attenuated them, and reduced SMAD activity and migration of the IKKβ-deficient cells (Fig. 4E–G). These data suggest that SOD2 reduction and ROS accumulation contribute to TGFβ activation in the IKKβ-null cells.

TGFβ, on the other hand, has been shown to activate NADPH oxidases (NOX), which could further augment ROS (Hecker et al., 2009; Bondi et al., 2010). By monitoring intracellular glutathione (GSH), the most abundant redox scavenger, we observed that treating cells with TGFβ1 caused GSH depletion, whereas treating cells with TGFβ inhibitors restored GSH in IKKβ-null and TGFβ1 treated wild type cells (Anderson, 1998) (Fig. 5A). Furthermore, the IKKβ-null cells had high levels of NOX1 and NOX4 expression and NOX inhibitors abolished TGFβ-induced GSH depletion in these cells (Armstrong et al., 2002; Bedard and Krause, 2007) (Fig. 5B and 5C).

Taken together, the above data suggest a scenario that reduction of redox scavengers in IKKβ-null cells could lead to ROS accumulation; the oxidative stresses in turn might activate the TGFβ-NOX cascade to further augment ROS. IKKβ ablation therefore leads to the activation of an autocrine cycle of ROS amplification. Consistent with the conclusion, we found that Ad-Cre infection of the *Ikkβ*$^{F/F}$ cells led to a gradual ROS increase. While 36% *Ikkβ*$^{F/F}$/Ad-Cre cells displayed high H_2O_2 level at 30 days of Ad-Cre infection, the number increased to almost 50% at 90 days after infection (Fig. S4B and S4C). Similarly, the HO-1 expression increased gradually after Ad-Cre infection of *Ikkβ*$^{F/F}$ cells, and by 180 days, it reached the levels similar to that in *Ikkβ*$^{-/-}$ cells and twice that in wild type or Ad-IKKβ-infected *Ikkβ*$^{F/F}$/Ad-Cre cells (Fig. 5D).

AP-1 is involved in ROS-induced TGFβ expression

To identify the molecular link between ROS and TGFβ, we scanned the *Tgfβ2* promoter for transcription factor binding

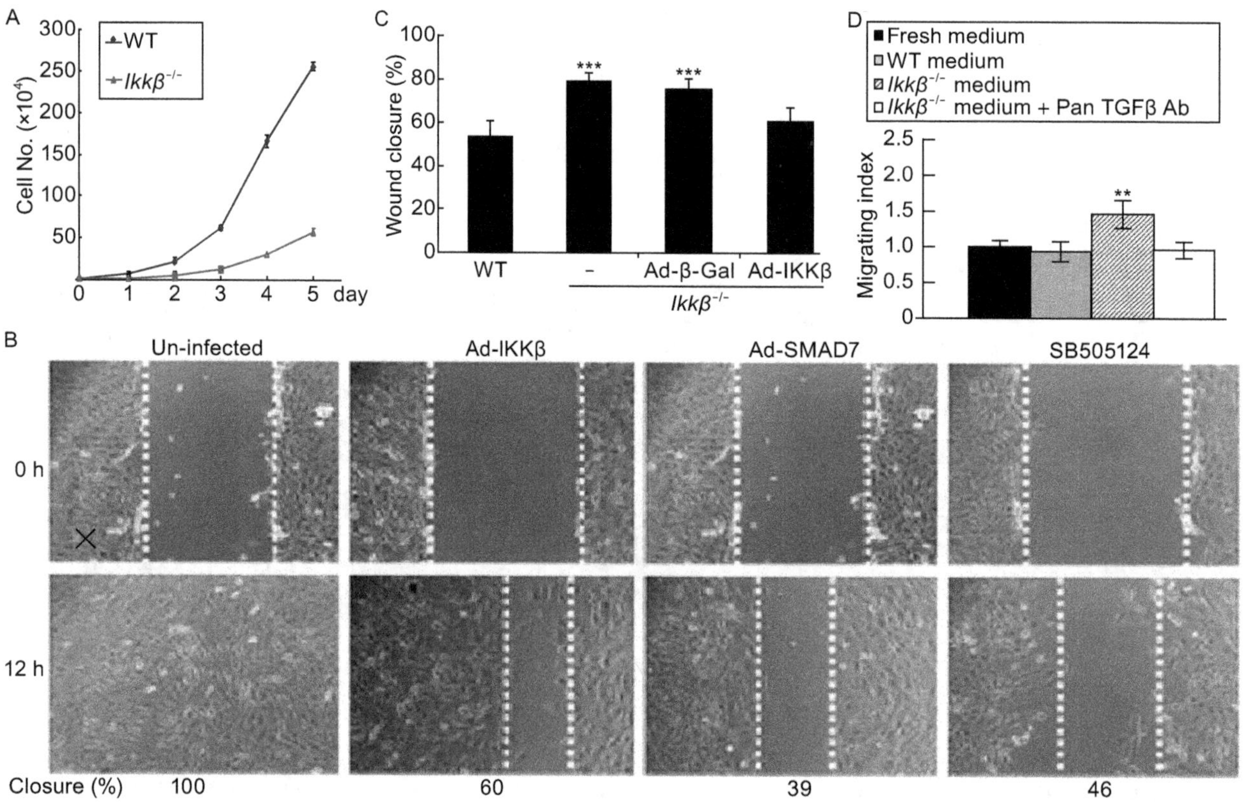

Figure 2. IKKβ loss induces TGFβ expression and cell migration. (A) The wild type and *Ikkβ*$^{-/-}$ fibroblasts were examined for growth rate. The *in vitro* wound healing assay was performed on wild type and *Ikkβ*$^{-/-}$ fibroblasts either uninfected or infected with Ad-GFP or Ad-IKKβ, or pre-treated with SB505124 for 24 h. (B) The cells were photographed at 0 and 12 h after the scratch wound, and (C) the speed of healing was calculated. (D) The *in vitro* wound healing assay was performed on wild type cells in normal growth medium, condition-medium collected from wild type cells or condition-medium collected from *Ikkβ*$^{-/-}$ cells with or without anti-pan-TGFβ. The healing rate was calculated at 12 h after injury. Results represent the mean values ± SD from at least three independent experiments. ***P* < 0.01 and ****P* < 0.001 were considered significantly different from the wild type or control samples.

sites and found two potential AP-1-cJun binding sites (Fig. S6A). AP-1 is a stress responsive transcription factor; we tested its activation with a luciferase reporter bearing an AP-1 binding site and found that luciferase expression was induced by IKKβ ablation, but repressed by IKKβ expression and NAC treatment (Fig. 6A). In addition, AP-1 binding to the *Tgfβ2* promoter, as measured by chromatin immunoprecipitation, was increased, associated with the transcriptionally active H3K4me3 modification on the *Tgfβ2* promoter, in IKKβ-deficient cells (Fig. S6B and S6C). Both AP-1 binding and H3K4me3 were potentiated by IKKβ ablation and BSO treatment, but reduced by IKKβ over-expression and NAC treatment (Fig. 6B and 6C).

To validate the role of AP-1-c-Jun, we expressed a dominant negative mutant c-Jun (bdm-c-Jun) in the *Ikk*β$^{-/-}$ cells and found that its expression repressed *Tgf*β*2* promoter activity and gene expression (Fig. 6D and 6E). We further used c-Jun-competent (*c-Jun*$^{F/F}$/Ad-GFP) and -deficient (*c-Jun*$^{F/F}$/Ad-Cre) cells and showed that while c-Jun ablation did not affect HO-1 induction, it abolished *Tgf*β*2* induction under the oxidative stress conditions created by BSO treatment (Fig. 6F). Collectively, our data suggest that the ROS may act upstream to activate AP-1/c-Jun, which in turn can induce *Tgf*β promoter and gene expression in the *Ikk*β-null cells.

Loss of IKKβ leads to senescence

Chronic oxidative stress can induce, stabilize and amplify senescence, leading ultimately to the detrimental effects of aging (Passos et al., 2010; Nelson et al., 2012). To assess if IKKβ ablation could lead to senescence, we examined the expression of senescence-associated β-Galactosidase (SA-β-Gal) (Dimri et al., 1995). SA-β-Gal activity was low in *Ikkβ*$^{F/F}$ cells, but gradually increased following Ad-Cre infection; by 180 days after infection the activity reached approximately 50% of the level in *Ikkβ*$^{-/-}$cells (Fig. 7A). In Ad-Cre infected *Ikkβ*$^{F/F}$ cells, there was also a progressive increase of the cell cycle regulator cyclin-dependent kinase inhibitor 1A (p21) (Cdkn1a), the extracellular matrix component Fibronectin (*Fn1*), and γH2AX, a histone modification associated with

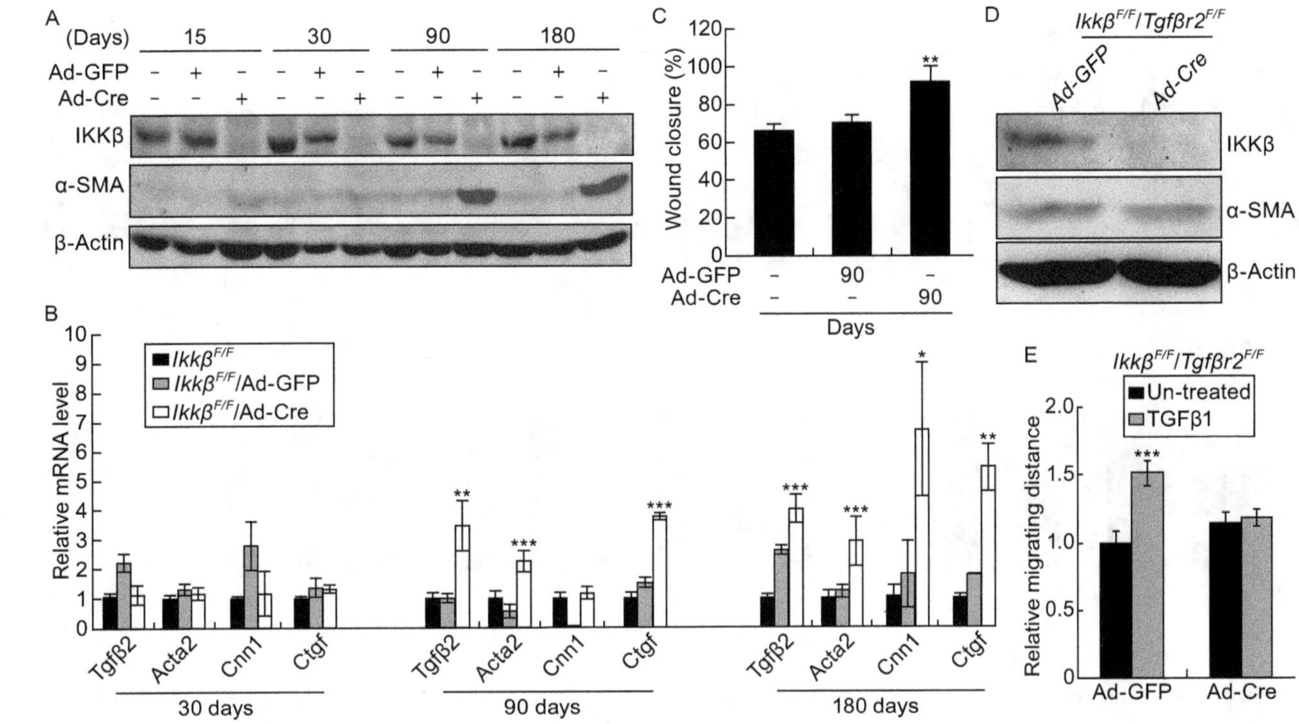

Figure 3. IKKβ ablation leads to progressive activation of TGFβ and cell migration. The *Ikkβ*$^{F/F}$ fibroblasts were infected with Ad-GFP or Ad-Cre and the cells were maintained in culture for various days as indicated. The cells were examined for (A) IKKβ, α-SMA and β-actin protein, (B) mRNA for *Tgfβ2* and SMAD-target genes, and (C) rate of wound healing. The *Ikkβ*$^{F/F}$/*Tgfbr2*$^{F/F}$ fibroblasts were infected with Ad-GFP or Ad-Cre. At 3 months after infection, the cells were examined for (D) IKKβ, α-SMA and β-actin protein and (E) rate of *in vitro* wound healing in the presence or absence of TGFβ1. Results represent the mean values ± SD from at least three independent experiments. *$P < 0.05$, **$P < 0.01$ and ***$P < 0.001$ were considered significantly different from the wild type or control samples.

DNA double strand damage (Dumont et al., 2000; Debacq-Chainiaux et al., 2008; Weyemi et al., 2011) (Fig. 7B and 7C). Furthermore, there was a slight but gradual increase of telomere shortening, suggesting that IKKβ loss may lead to irreversible DNA damage and a senescent phenotype (Balaban et al., 2005; Giorgio et al., 2007) (Fig. 7D). Hence, by repressing the ROS-AP-1-TGFβ axis IKKβ may prevent senescence in fibroblasts (Fig. 7E).

DISCUSSION

The global gene expression signatures provide an initial clue that loss of IKKβ or key components of the NF-κB pathways may lead to activation of TGFβ signaling in fibroblasts. Following this lead, we have identified a molecular link between the IKKβ and TGFβ pathways. We show that the IKKβ-NF-κB cascade sustains the expression of anti-oxidant genes and that inactivation of this cascade impedes the scavenge capacity and results in ROS accumulation. Elevated ROS in turn triggers the feed-forward activation of the ROS-AP-1-TGFβ-NOX loop that leads ultimately to increased motility, fibroblast-myofibroblast transformation, and senescence (Fig. 7E).

The antagonistic relationship between IKK and TGFβ signaling has been reported in other experimental settings. For example, in osteoclasts and head and neck cancers, TGFβ is found acting through the TGFβ activated kinase 1 to activate IKK and NF-κB, whereas NF-κB up-regulates SMAD7 to inhibit TGFβ signaling (Gingery et al., 2008; Freudlsperger et al., 2013). The IKK-NF-κB pathway is also found to modulate transcription factors/cofactors and attenuate SMAD activity (Dennler et al., 2000; Nagarajan et al., 2000; Bitzer et al., 2000; Verrecchia et al., 2001). Here we describe a unique mechanism where the crosstalk of IKK and TGFβ is mediated by ROS. Specifically, the elevated ROS in IKKβ-null cells induce c-Jun binding and activation of the *Tgfβ* promoter.

There are at least two sources for the ROS in IKKβ-null cells. First, IKKβ ablation results in insufficient ROS removal due to down-regulation of antioxidant genes, in agreement with previous reports (Chen et al., 2003; Sakon et al., 2003; Peng et al., 2007; Peng et al., 2010). Second, IKKβ ablation

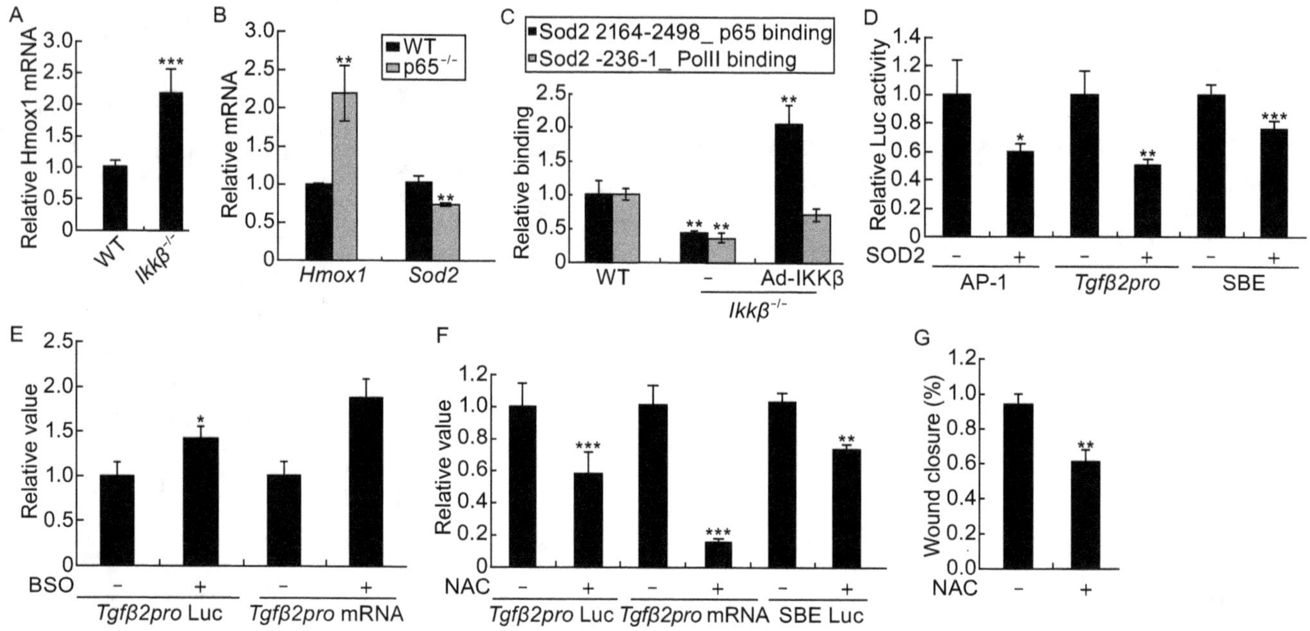

Figure 4. ROS accumulation leads to TGFβ activation in IKKβ-null cells. (A) Wild type and *Ikkβ*$^{-/-}$ cells and (B) wild type and *p65*$^{-/-}$ cells were examined for mRNA of oxidative stress marker *Hmox-1* and/or redox scavenger gene *Sod2*. (C) The wild type and *Ikkβ*$^{-/-}$ cells with or without Ad-IKKβ infection were examined for p65 binding of the *Sod2* enhancer and RNA Pol II occupancy of the *Sod2* promoter. (D) The *Ikkβ*$^{-/-}$ cells were transfected with luciferase reporters for AP-1, SMAD and *Tgfβ2* promoter, together with either an empty vector or SOD2 expression plasmids. The luciferase activities in SOD2 transfected cells were compared to those in empty vector transfected cells, designated as 1. (E) The wild type cells were either un-transfected or transfected with *Tgfβ2* promoter reporter, and treated with pro-oxidant BSO. The *Tgfβ2* mRNA expression and promoter activity were examined. The *Ikkβ*$^{-/-}$ cells were either un-transfected or treanfected with the luciferase reporter plasmids for *Tgfβ2* promoter or SMAD (SEB-luc). The cells were treated with anti-oxidant NAC, and were examined for (F) *Tgfβ2* mRNA expression and luciferase activity, and (G) migration by the *in vitro* wound healing assays. Results represent the mean values ± SD from at least three independent experiments. $^{*}P < 0.05$, $^{**}P < 0.01$ and $^{***}P < 0.001$ were considered significantly different from the wild type or control samples.

causes increased ROS production as the result of TGFβ induced NOX4 expression and NADH activity. Interestingly, the TGFβ-NOX axis itself is also activated by ROS, and thus, this axis and ROS may form an autocrine loop to amplify each other. Such feed-forward signal amplification is likely to be responsible for the progressive ROS accumulation and TGFβ activation in fibroblasts following IKKβ ablation. When the TGFβ signals reach a threshold level, it is able to induce cell migration and myofibroblast transformation; when the chronic ROS reach a threshold level, they may contribute to premature senescence, as it also happens in cells deficient in GSH (Chen et al., 2009).

As the IKK-NF-κB cascade is a major player of the inflammatory response, its inhibition is a promising strategy for treating a vast number of diseases associated with inflammation (McIntyre et al., 2003; Ruocco et al., 2005; Polzer et al., 2008). In particular, this cascade is considered a molecular link between inflammation and cancer; therefore, targeting the cascade has become an attractive rationale in cancer therapy (Vallabhapurapu and Karin, 2009; DiDonato et al., 2012). The caveat is that such treatment may have adverse effects due to disruption of the cascade's pleiotropic physiological functions (DiDonato et al., 2012). Our data in fibroblasts echo this concern and suggest that complete, irreversible and long-term inhibition of IKKβ may lead to chronic oxidative stress, and increase the risks for fibrogenesis and senescence.

MATERIALS AND METHODS

Viruses, plasmids, reagents and antibodies

The adenoviral expression vectors for IKKβ, SMAD7, β-GAL, GFP and GFP-Cre were from Drs. Yi Zheng at the Cincinnati Children's Hospital, Yinling Hu at the National Cancer Institute, and Chia-yang Liu at Indiana University. The reporter plasmids, NF-κB-luc, SBE-Luc, AP-1-Luc, and the *Tgfβ1*, *Tgfβ2* and *Tgfβ3* promoter-luc were obtained from Drs. Edward B. Leof at Mayo Clinic and Alvaro Puga at the University of Cincinnati (Tojima et al., 2000). Expression vector for SOD2 was from Dr. Shanglin Shi at the University of Kentucky and Bdm-c-Jun was described before (Geh et al., 2011). TGFβ1 was from PeproTech, NAC, BSO and DPI were from Sigma-Aldrich, and SB505124 was from EMD Millipore. The following

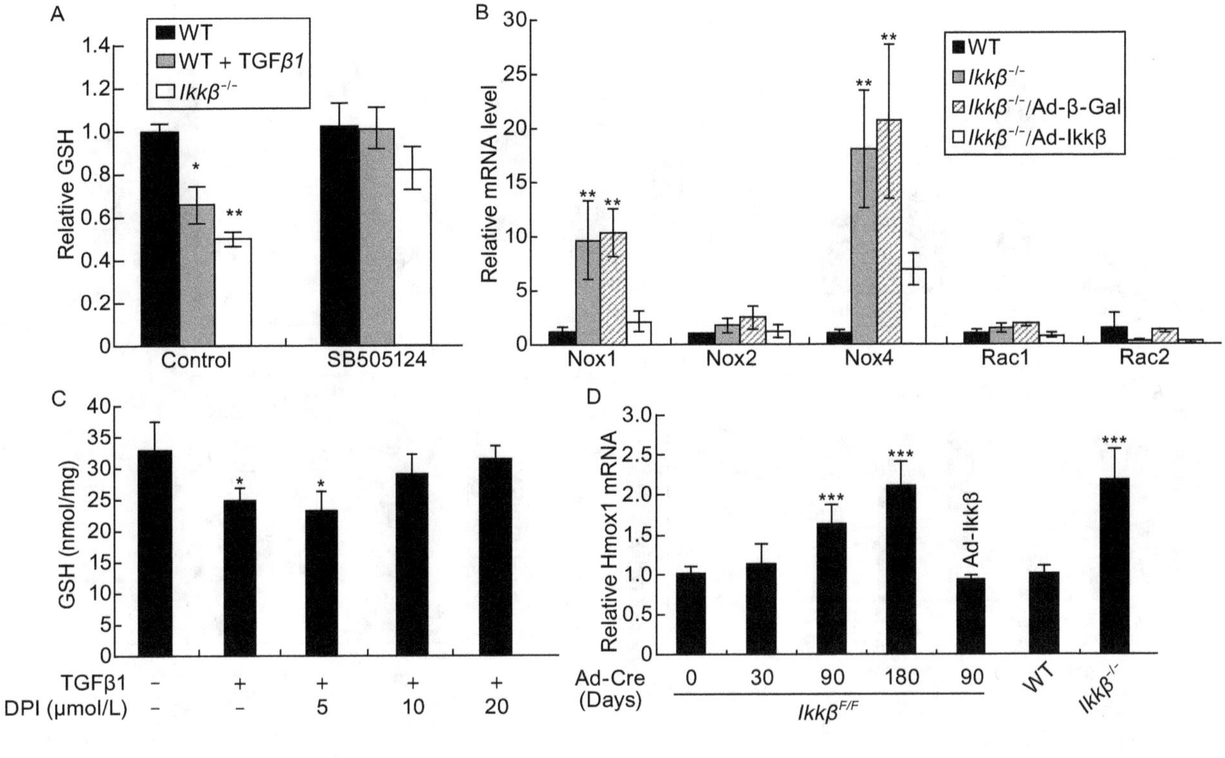

Figure 5. Activation of ROS-TGFβ-NOX amplification loop in the IKKβ-null cells. The intracellular GSH in wild type, TGFβ1 treated wild type and $Ikk\beta^{-/-}$ cells, (A) in the presence or absence of the TGFβ receptor inhibitor SB505124, or (C) in the presence or absence of the NOX inhibitor DPI. (B) The mRNA of genes coding for components of the NADH complexes in IKKβ-competent, i.e. wild type and $Ikk\beta^{-/-}$/Ad-IKKβ, and IKKβ-deficient, i.e. $Ikk\beta^{-/-}$ and $Ikk\beta^{-/-}$/Ad-β-Gal, cells. (D) The mRNA for oxidative stress marker *Hmox-1* in wild type, $Ikk\beta^{-/-}$ and in $Ikk\beta^{F/F}$ fibroblasts infected with Ad-Cre for 0 to 180 days with or without Ad-IKKβ. Results represent the mean values ± SD from at least three independent experiments. *$P < 0.05$, **$P < 0.01$ and ***$P < 0.001$ were considered significantly different from the wild type or control samples.

antibodies were used in the study: anti- IKKα, -IKKβ, -IkBα, and -p-SMAD2 (Ser-465, 467) from Cell Signaling, anti-pan TGFβ from R&D Systems, anti-α-SMA from Abcam, anti-β-actin from Sigma-Aldrich, anti-γH2AX from Novus Biologicals, anti-PolII, -H3, -H3K27Me3, H3K9Me2, H3K9Ac and H3K4Me3 from EMD Millipore, and anti-p65, -c-Jun, and IgG from Santa Cruz Biotechnologies.

Mouse fibroblasts, cell culture, transfection, infection and luciferase assays

The wild type, fibroblasts deficient in IKKβ, IKKα and p65 were gifts from Drs. Karin and Zandi, and were maintained under culture conditions as described (Chen et al., 2006). The $Ikk\beta^{F/F}$, $Ikk\beta^{F/F}$/$Tgfbr2^{F/F}$ and $c\text{-}Jun^{F/F}$ fibroblasts were prepared using E13.5 embryos following standard 3T3 protocol (Aaronson and Todaro, 1968). The cells were cultured in DMEM supplemented with 10% FBS, 50 U/mL penicillin, 50 mg/mL streptomycin for less than 10 passages before used for experiments or adenoviral infection. Some of adenoviral infected cells were allowed to grow for 6 months with approximately 50 passages. Adenoviruses were used at 100–500 PFU to infect 70% confluent cells as described before (Peng et al., 2010). Cells were transfected using the lipofectamin plus method and Firefly and Renilla luciferase activities were measured 24 to 48 h after transfection following the manufacture's protocols (Thermo Fisher Scientific).

Western blotting, ROS measurements, SA-β-Gal activity and *in vitro* wound healing assays

The SA-β-Gal activities were measured at the PH 6.0 using Beta-Glo Assay system (Promega), Western blotting, measurement of ROS and GSH, and the *in vitro* wound healing assays were done as previously described (Zhang et al., 2003; Peng et al., 2010). The conditional medium used for *in vitro* wound healing assays was derived from fresh medium overlaid on wild type or $Ikk\beta^{-/-}$ cells for 24 h.

RNA isolation, reverse transcription and gene expression profiling

RNA was extracted, labeled and hybridized to Affymetrix Mouse Genome 430 2.0 Arrays using standard protocol (Medvedovic et al., 2009). Data was processed by performing background correction, quantile normalization, and calculation of expression set summaries using the Robust Multichip Average (RMA) protocol (Irizarry et al., 2003) as implemented in the Bioconductor affy package. Differentially expressed genes between two groups were identified by two-group comparison using intensity-based empirical Bayes method

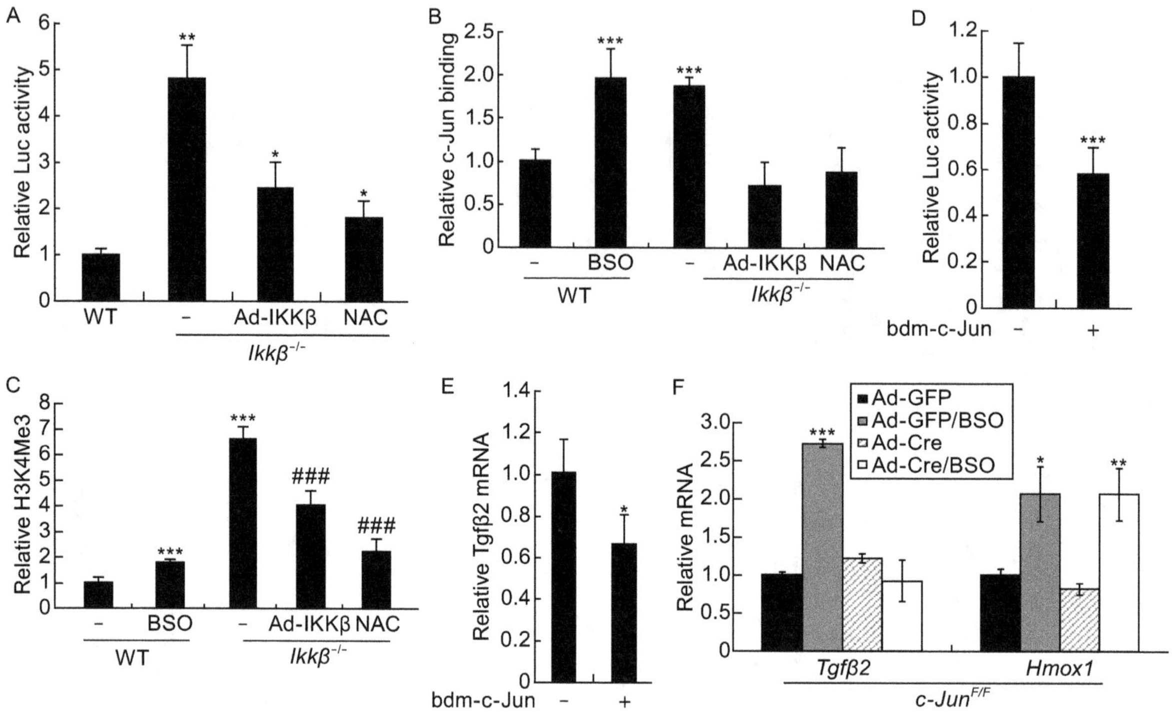

Figure 6. c-Jun regulates TGFβ expression in IKKβ-null cells. The wild type cells with or without BSO treatment, and the $Ikk\beta^{-/-}$ cells with or without Ad-IKKβ infection or NAC treatment were examined for (A) luciferase activity following AP-1-luc plasmid transfection, and (B and C) ChIP assays for (B) c-Jun binding of the *Tgfβ2* enhancer and (C) H3K4me3 modification of the *Tgfβ2* promoters. The $Ikk\beta^{-/-}$ cells were transfected with a dominant negative c-Jun (bdm-c-Jun) expression plasmids, and (D) together with *Tgfb2-luc* and examined for the luciferase activities, and (E) examined for the *Tgfb2* mRNA. (F) The mRNA for *Tgfb2* and *Hmox-1* was examined in $c\text{-}Jun^{F/F}$ cells infected with Ad-Cre or Ad-GFP and treated with BSO. Results represent the mean values ± SD from at least three independent experiments. $^{*}P < 0.05$, $^{**}P < 0.01$ and $^{***}P < 0.001$ were considered significantly different from the un-treated wild type samples; ###$P < 0.001$ was significantly different from the un-treated $Ikk\beta^{-/-}$ samples.

(IBMT) (Sartor et al., 2006). Pathway enrichment analysis was performed using the LRpath methodology (Sartor et al., 2009) implemented in the CLEAN package (Freudenberg et al., 2009).

Quantitative RT-PCR (qRT-PCR), chromatin-immunopreciptation (ChIP) and telomere measurement

qRT-PCR was performed using a DNA Engine Opticon2 Real-Time PCR Detection System (MJ Research) and SYBR Green qPCR MasterMix (Applied Biosystems) and primers for the genes of interest as listed in Table S1. All experiments were performed at least in triplicates. The relative differences in qRT-PCR among samples were determined by the ΔCT value as described previously (Schnekenburger et al., 2007). Hence, the ΔCT value for each sample was calculated by subtracting cycle threshold (CT) value (obtained from the means of replicates) of the input DNA (or *Gapdh* signal) from that of each sample in order to normalize ChIP assay (or to normalize gene expression) results. The ΔΔCT value was calculated by subtracting control ΔCT values from the corresponding experimental ΔCT values. The resulting values were converted to fold changes over control by raising 2 to the power of −ΔCT values.

ChIP was performed following the protocol described previously (Schnekenburger et al., 2007). Briefly, cells were fixed for 10 min with 1% formaldehyde, followed by addition of 0.125 mol/L glycine for 5 min to stop cross-linking. Cells were washed with ice-cold PBS and harvested in cell lysis buffer (5 mmol/L PIPES [pH 8.0], 85 mmol/L KCl, 0.5% NP-40, and protease inhibitor cocktail [Roche]) for 10 min on ice. The nuclei were pelleted, resuspended in nucleus lysis buffer (50 mmol/L Tris-HCl [pH 8.1], 10 mmol/L EDTA, 1% SDS, and protease inhibitor cocktail), and incubated on ice for 10 min. Chromatin was sheared to a size range of 0.3 to 0.8 kb by sonication. After centrifugation to remove cell debris, chromatin was pre-cleared for 1 h at 4°C with a 50% gel slurry of protein A-agarose beads saturated with salmon sperm DNA (Upstate), and then diluted three times in IP dilution buffer (16.7 mmol/L Tris-Cl [pH 8.1], 167 mmol/L NaCl, 1.2 mmol/L EDTA, 1.1% Triton X-100, 0.01% sodium dodecyl sulfate) with 10% of the supernatants used as input. The diluted chromatin was incubated with antibodies specific for the proteins of interest for 2 h at 4°C, followed by addition of a 50% gel slurry of protein A-agarose and incubation overnight (Upstate). The agarose beads were pelleted and washed twice with 1× dialysis buffer (50 mmol/L Tris-HCl [pH 8.0], 2 mmol/L EDTA, 0.2% Sarkosyl) and four times with IP wash buffer (100 mmol/L Tris-HCl [pH 9.0], 500 mmol/L LiCl, 1% NP-40, 1% deoxycholic acid). Precipitated chromatin complexes were removed from the beads by incubation with elution buffer (50 mmol/L NaHCO3, 1% SDS) with mild vortexing. This step was repeated, and the eluates were combined.

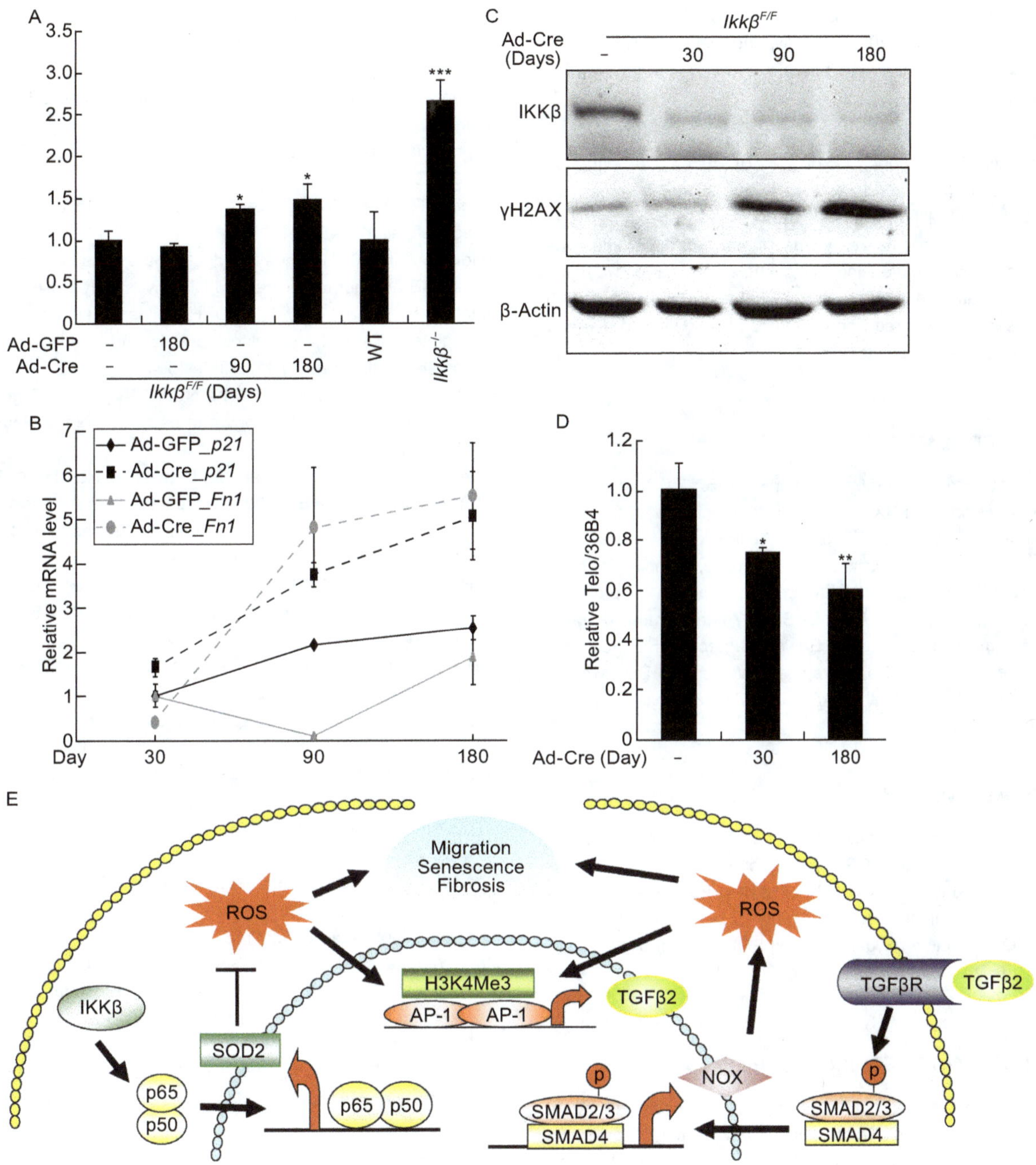

Figure 7. Phenotype of the IKKβ-deficient cells. The wild type, $Ikk\beta^{-/-}$, $Ikk\beta^{F/F}$/Ad-GFP and $Ikk\beta^{F/F}$/Ad-Cre cells were examined for (A) SA-β-gal activity, (B) expression of senescent markers, *p21* and *Fn*, (C) expression of IKKβ, β-actin and γH2AX, a marker for DNA damage, and (D) the telomere length. Results represent the mean values ± SD from at least three independent experiments. * $P < 0.05$, ** $P < 0.01$ and *** $P < 0.001$ were considered significantly different from the wild type or control samples. (E) A proposed model depicting the role of IKKβ in the regulation of the ROS-TGFβ autocrine amplification loop. Specifically, IKKβ acts through p65 to regulate expression of anti-oxidant genes, such as SOD2. Loss of IKKβ decreases SOD2 expression and dampens the scavenger capacity, resulting in ROS accumulation and AP-1/c-Jun activation. The AP-1/c-Jun regulates TGFβ expression thereby activating the TGFβ-NOX axis to further potentiate ROS accumulation. The amplification of the ROS-TGFβ-NOX axis eventually leads to increased cell migration, myofibroblast transformation and senescence.

Cross-linking was reversed by adding NaCl to a final concentration of 0.3 mol/L and incubating overnight at 65°C in the presence of RNase A. Samples were then digested with proteinase K at 45°C for 1.5 h. DNA was purified by chromatography on QIAquick columns (QIAGEN) and eluted in double-distilled water for further qPCR analysis.

The telomere length was measured using Q-PCR as described (Callicott and Womack, 2006). Briefly, the genomic DNA were extracted using a QIAmp DNA micro Kit (Qiagen, Valencia, CA, USA) and quantified.

PCR reactions were performed on the ABI Prism 7700 Sequence Detection System (Applied Biosystems), using telomeric primers for the reference control gene (mouse 36B4 single copy gene). The telomere signal was normalized to the signal from the single-copy gene to generate a relative telomere to single copy gene (T/S) ratio indicative of relative telomere length. Equal amounts of DNA (300 pg) were used for each reaction with several repeats and average telomere length was calculated.

ACKNOWLEDGEMENTS

This work is supported in part by NIH grants EY15227 (YX) and Albert J. Ryan Fellowship (LC); YX, LN and MM are members of Center of Environmental Genetics (P30-ES06096). We thank Drs. Michael Karin for *Ikka*$^{-/-}$ and *Ikkβ*$^{-/-}$ fibroblasts and *Ikkβ*$^{F/F}$ mice, Ebrahim Zandi for *p65*$^{-/-}$ fibroblasts, Yinling Hu, Yi Zheng, Chia-yang Liu for providing adenoviral vectors, Dr. Alvaro Puga for SMAD7 adenovirus and *Tgfβ1*, *Tgfβ2* and *Tgfβ3* promoter-luc, Dr. Edward B. Leof for SBE-luc plasmid, Dr. Xianglin Shi for SOD2 expression vector and Drs. Michael Karin and Harold Moses for the *Ikkβ*$^{F/F}$ and *Tgfbr2*$^{F/F}$ mice.

ABBREVIATIONS

α-SMA, α smooth muscle actin; AP-1, activating protein-1; BSO, L-Buthionine sulphoximine; GSH, glutathione; HO-1, heme oxygenase; IKKβ, IκB kinase β; NAC, N-acetyl cysteine, NF-κB, nuclear factor Kb; NOX, nicotinamide adenine dinucleotide phosphate oxidase; ROS, reactive oxygen species; SOD2, superoxide dismutase 2; TGFβ, transforming growth factor β.

COMPLIANCE WITH ETHICS GUIDELINES

Liang Chen, Zhimin Peng, Qinghang Meng, Maureen Mongan, Jingcai Wang, Maureen Sartor, Jing Chen, Liang Niu, Mario Medvedovic, Winston Kao and Ying Xia declare that they have no conflict of interest. All institutional and national guidelines for the care and use of laboratory animals were followed.

REFERENCES

Aaronson SA, Todaro GJ (1968) Development of 3T3-like lines from Balb-c mouse embryo cultures: transformation susceptibility to SV40. J Cell Physiol 72:141–148

Anderson ME (1998) Glutathione: an overview of biosynthesis and modulation. Chem Biol Interact 111–112:1–14

Armstrong JS, Steinauer KK, Hornung B, Irish JM, Lecane P, Birrell GW, Peehl DM, Knox SJ (2002) Role of glutathione depletion and reactive oxygen species generation in apoptotic signaling in a human B lymphoma cell line. Cell Death Differ 9:252–263

Bacher S, Schmitz ML (2004) The NF-kappaB pathway as a potential target for autoimmune disease therapy. Curr Pharm Des 10:2827–2837

Balaban RS, Nemoto S, Finkel T (2005) Mitochondria, oxidants, and aging. Cell 120:483–495

Baldwin AS Jr (2001) Series introduction: the transcription factor NF-kappaB and human disease. J Clin Investig 107:3–6

Bataller R et al (2003) NADPH oxidase signal transduces angiotensin II in hepatic stellate cells and is critical in hepatic fibrosis. J Clin Investig 112:1383–1394

Bedard K, Krause KH (2007) The NOX family of ROS-generating NADPH oxidases: physiology and pathophysiology. Physiol Rev 87:245–313

Bitzer M, von Gersdorff G, Liang D, Dominguez-Rosales A, Beg AA, Rojkind M, Bottinger EP (2000) A mechanism of suppression of TGF-beta/SMAD signaling by NF-kappa B/RelA. Genes Dev 14:187–197

Bondi CD, Manickam N, Lee DY, Block K, Gorin Y, Abboud HE, Barnes JL (2010) NAD(P)H oxidase mediates TGF-beta1-induced activation of kidney myofibroblasts. J Am Soc Nephrol 21:93–102

Callicott RJ, Womack JE (2006) Real-time PCR assay for measurement of mouse telomeres. Comp Med 56:17–22

Chariot A (2009) The NF-kappaB-independent functions of IKK subunits in immunity and cancer. Trends Cell Biol 19:404–413

Chen F, Castranova V, Li Z, Karin M, Shi X (2003) Inhibitor of nuclear factor kappaB kinase deficiency enhances oxidative stress and prolongs c-Jun NH2-terminal kinase activation induced by arsenic. Cancer Res 63:7689–7693

Chen F, Lu Y, Castranova V, Li Z, Karin M (2006) Loss of Ikkbeta promotes migration and proliferation of mouse embryo fibroblast cells. J Biol Chem 281:37142–37149

Chen Y, Johansson E, Fan Y, Shertzer HG, Vasiliou V, Nebert DW, Dalton TP (2009) Early onset senescence occurs when fibroblasts lack the glutamate–cysteine ligase modifier subunit. Free Radic Biol Med 47:410–418

Courtois G, Gilmore TD (2006) Mutations in the NF-kappaB signaling pathway: implications for human disease. Oncogene 25:6831–6843

Datto MB, Frederick JP, Pan L, Borton AJ, Zhuang Y, Wang XF (1999) Targeted disruption of Smad3 reveals an essential role in transforming growth factor beta-mediated signal transduction. Mol Cell Biol 19:2495–2504

Debacq-Chainiaux F, Pascal T, Boilan E, Bastin C, Bauwens E, Toussaint O (2008) Screening of senescence-associated genes with specific DNA array reveals the role of IGFBP-3 in premature senescence of human diploid fibroblasts. Free Radic Biol Med 44:1817–1832

Dennler S, Prunier C, Ferrand N, Gauthier JM, Atfi A (2000) c-Jun inhibits transforming growth factor beta-mediated transcription by repressing Smad3 transcriptional activity. J Biol Chem 275:28858–28865

DiDonato JA, Mercurio F, Karin M (2012) NF-kappaB and the link between inflammation and cancer. Immunol Rev 246:379–400

Dimri GP, Lee X, Basile G, Acosta M, Scott G, Roskelley C, Medrano EE, Linskens M, Rubelj I, Pereira-Smith O (1995) A biomarker that identifies senescent human cells in culture and in aging skin in vivo. Proc Natl Acad Sci USA 92:9363–9367

Dumont P, Burton M, Chen QM, Gonos ES, Frippiat C, Mazarati JB, Eliaers F, Remacle J, Toussaint O (2000) Induction of replicative senescence biomarkers by sublethal oxidative stresses in normal human fibroblast. Free Radic Biol Med 28:361–373

Fleckenstein K, Zgonjanin L, Chen L, Rabbani Z, Jackson IL, Thrasher B, Kirkpatrick J, Foster WM, Vujaskovic Z (2007) Temporal onset of hypoxia and oxidative stress after pulmonary irradiation. Int J Radiat Oncol Biol Phys 68:196–204

Freudenberg JM, Joshi VK, Hu Z, Medvedovic M (2009) CLEAN: clustering enrichment analysis. BMC Bioinform 10:234

Freudlsperger C, Bian Y, Contag WS, Burnett J, Coupar J, Yang X, Chen Z, Van WC (2013) TGF-beta and NF-kappaB signal pathway cross-talk is mediated through TAK1 and SMAD7 in a subset of head and neck cancers. Oncogene 32:1549–1559

Geh E, Meng Q, Mongan M, Wang J, Takatori A, Zheng Y, Puga A, Lang RA, Xia Y (2011) Mitogen-activated protein kinase kinase kinase 1 (MAP3K1) integrates developmental signals for eyelid closure. Proc Natl Acad Sci USA 108:17349–17354

Gerondakis S, Grumont R, Gugasyan R, Wong L, Isomura I, Ho W, Banerjee A (2006) Unravelling the complexities of the NF-kappaB signalling pathway using mouse knockout and transgenic models. Oncogene 25:6781–6799

Gingery A, Bradley EW, Pederson L, Ruan M, Horwood NJ, Oursler MJ (2008) TGF-beta coordinately activates TAK1/MEK/AKT/NFkB and SMAD pathways to promote osteoclast survival. Exp Cell Res 314:2725–2738

Giorgio M, Trinei M, Migliaccio E, Pelicci PG (2007) Hydrogen peroxide: a metabolic by-product or a common mediator of ageing signals? Nat Rev Mol Cell Biol 8:722–728

Hecker L, Vittal R, Jones T, Jagirdar R, Luckhardt TR, Horowitz JC, Pennathur S, Martinez FJ, Thannickal VJ (2009) NADPH oxidase-4 mediates myofibroblast activation and fibrogenic responses to lung injury. Nat Med 15:1077–1081

Irizarry RA, Hobbs B, Collin F, Beazer-Barclay YD, Antonellis KJ, Scherf U, Speed TP (2003) Exploration, normalization, and summaries of high density oligonucleotide array probe level data. Biostatistics 4:249–264

Karin M (2006) Nuclear factor-kappaB in cancer development and progression. Nature 441:431–436

Karin M (2008) The IkappaB kinase: a bridge between inflammation and cancer. Cell Res 18:334–342

Kim HJ, Hawke N, Baldwin AS (2006) NF-kappaB and IKK as therapeutic targets in cancer 3. Cell Death Differ 13:738–747

Lavon I et al (2000) High susceptibility to bacterial infection, but no liver dysfunction, in mice compromised for hepatocyte NF-kappaB activation. Nat Med 6:573–577

Li X, Massa PE, Hanidu A, Peet GW, Aro P, Savitt A, Mische S, Li J, Marcu KB (2002) IKKalpha, IKKbeta, and NEMO/IKKgamma are each required for the NF-kappa B-mediated inflammatory response program. J Biol Chem 277:45129–45140

Luo JL, Kamata H, Karin M (2005) IKK/NF-kappaB signaling: balancing life and death–a new approach to cancer therapy. J Clin Investig 115:2625–2632

Maeda S, Kamata H, Luo JL, Leffert H, Karin M (2005) IKKbeta couples hepatocyte death to cytokine-driven compensatory proliferation that promotes chemical hepatocarcinogenesis 1. Cell 121:977–990

May MJ, Madge LA (2007) Caspase inhibition sensitizes inhibitor of NF-kappaB kinase beta-deficient fibroblasts to caspase-independent cell death via the generation of reactive oxygen species. J Biol Chem 282:16105–16116

McIntyre KW, Shuster DJ, Gillooly KM, Dambach DM, Pattoli MA, Lu P, Zhou XD, Qiu Y, Zusi FC, Burke JR (2003) A highly selective inhibitor of I kappa B kinase, BMS-345541, blocks both joint inflammation and destruction in collagen-induced arthritis in mice. Arthritis Rheum 48:2652–2659

Medvedovic M et al (2009) Influence of fatty acid diets on gene expression in rat mammary epithelial cells. Physiol Genom 38 (1):80–88

Nagarajan RP, Chen F, Li W, Vig E, Harrington MA, Nakshatri H, Chen Y (2000) Repression of transforming-growth-factor-beta-mediated transcription by nuclear factor kappaB. Biochem J 348 (Pt 3):591–596

Nelson G, Wordsworth J, Wang C, Jurk D, Lawless C, Martin-Ruiz C, von Zglinicki T (2012) A senescent cell bystander effect: senescence-induced senescence. Aging Cell 11:345–349

O'Dea EL, Barken D, Peralta RQ, Tran KT, Werner SL, Kearns JD, Levchenko A, Hoffmann A (2007) A homeostatic model of IkappaB metabolism to control constitutive NF-kappaB activity. Mol Syst Biol 3:111

Pasparakis M (2009) Regulation of tissue homeostasis by NF-kappaB signalling: implications for inflammatory diseases. Nat Rev Immunol 9:778–788

Passos JF et al (2010) Feedback between p21 and reactive oxygen production is necessary for cell senescence. Mol Syst Biol 6:347

Peng Z, Peng L, Fan Y, Zandi E, Shertzer HG, Xia Y (2007) A critical role for IkappaB kinase beta in metallothionein-1 expression and protection against arsenic toxicity. J Biol Chem 282:21487–21496

Peng Z, Geh E, Chen L, Meng Q, Fan Y, Sartor M, Shertzer HG, Liu ZG, Puga A, Xia Y (2010) IkB kinase b regulates redox homeostasis by controlling the constitutive levels of glutathione. Mol Pharmacol 77(5):784–792

Perkins ND (2007) Integrating cell-signalling pathways with NF-kappaB and IKK function. Nat Rev Mol Cell Biol 8:49–62

Phan SH (2002) The myofibroblast in pulmonary fibrosis. Chest 122:286S–289S

Polzer K, Baeten D, Soleiman A, Distler J, Gerlag DM, Tak PP, Schett G, Zwerina J (2008) Tumour necrosis factor blockade increases lymphangiogenesis in murine and human arthritic joints. Ann Rheum Dis 67:1610–1616

Roy S, Dontamalla SK, Mondru AK, Sannigrahi S, Veerareddy PR (2011) Downregulation of apoptosis and modulation of TGF-beta1 by sodium selenate prevents streptozotocin-induced diabetic rat renal impairment. Biol Trace Elem Res 139:55–71

Ruocco MG, Maeda S, Park JM, Lawrence T, Hsu LC, Cao Y, Schett G, Wagner EF, Karin M (2005) IkB kinase IKKb, but not IKKa, is a critical mediator of osteoclast survival and is required for inflammation-induced bone loss. J Exp Med 201:1677–1687

Sakon S et al (2003) NF-kappaB inhibits TNF-induced accumulation of ROS that mediate prolonged MAPK activation and necrotic cell death. EMBO J 22:3898–3909

Sartor MA, Tomlinson CR, Wesselkamper SC, Sivaganesan S, Leikauf GD, Medvedovic M (2006) Intensity-based hierarchical Bayes method improves testing for differentially expressed genes in microarray experiments. BMC Bioinform 7:538

Sartor MA, Leikauf GD, Medvedovic M (2009) LRpath: a logistic regression approach for identifying enriched biological groups in gene expression data. Bioinformatics 25:211–217

Schnekenburger M, Peng L, Puga A (2007) HDAC1 bound to the Cyp1a1 promoter blocks histone acetylation associated with Ah receptor-mediated trans-activation. Biochim Biophys Acta 1769:569–578

Sen CK, Roy S (2010) Oxygenation state as a driver of myofibroblast differentiation and wound contraction: hypoxia impairs wound closure. J Investig Dermatol 130:2701–2703

Tanaka M, Fuentes ME, Yamaguchi K, Durnin MH, Dalrymple SA, Hardy KL, Goeddel DV (1999) Embryonic lethality, liver degeneration, and impaired NF-kappa B activation in IKK-beta-deficient mice. Immunity 10:421–429

Tojima Y et al (2000) NAK is an IkappaB kinase-activating kinase. Nature 404:778–782

Vallabhapurapu S, Karin M (2009) Regulation and function of NF-kappaB transcription factors in the immune system. Annu Rev Immunol 27:693–733

Verrecchia F, Vindevoghel L, Lechleider RJ, Uitto J, Roberts AB, Mauviel A (2001) Smad3/AP-1 interactions control transcriptional responses to TGF-beta in a promoter-specific manner. Oncogene 20:3332–3340

Weyemi U (2011) ROS-generating NADPH oxidase NOX4 is a critical mediator in oncogenic H-Ras-induced DNA damage and subsequent senescence. Oncogene 31(9):1117–1129

Zhang L, Wang W, Hayashi Y, Jester JV, Birk DE, Gao M, Liu CY, Kao WW, Karin M, Xia Y (2003) A role for MEK kinase 1 in TGF-beta/activin-induced epithelium movement and embryonic eyelid closure. EMBO J 22:4443–4454

15

NSC-640358 acts as RXRα ligand to promote TNFα-mediated apoptosis of cancer cell

Fan Chen[1,2], Jiebo Chen[1], Jiacheng Lin[1], Anton V. Cheltsov[3], Lin Xu[1], Ya Chen[4], Zhiping Zeng[1] Liqun Chen[1], Mingfeng Huang[1], Mengjie Hu[1], Xiaohong Ye[1], Yuqi Zhou[1], Guanghui Wang[1], Ying Su[1,4], Long Zhang[5], Fangfang Zhou[6], Xiao-kun Zhang[1,4]✉, Hu Zhou[1]✉

[1] School of Pharmaceutical Sciences, Xiamen University, Xiamen 361102, China
[2] School of Biological Science and Biotechnology, Minnan Normal University, Zhangzhou 363000, China
[3] Q-MOL LLC, San Diego, CA 92105, USA
[4] Cancer Center, Sanford-Burnham Medical Research Institute, 10901 N. Torrey Pines Road, La Jolla, CA 92037, USA
[5] Life Science Institute, Zhejiang University, Hangzhou 310058, China
[6] Institutes of Biology and Medical Sciences, Soochow University, Suzhou 215123, China
✉ Correspondence: xkzhang@xmu.edu.cn (X. Zhang), huzhou@xmu.edu.cn (H. Zhou)

ABSTRACT

Retinoid X receptor α (RXRα) and its N-terminally truncated version tRXRα play important roles in tumorigenesis, while some RXRα ligands possess potent anticancer activities by targeting and modulating the tumorigenic effects of RXRα and tRXRα. Here we describe NSC-640358 (N-6), a thiazolyl-pyrazole derived compound, acts as a selective RXRα ligand to promote TNFα-mediated apoptosis of cancer cell. N-6 binds to RXRα and inhibits the transactivation of RXRα homodimer and RXRα/TR3 heterodimer. Using mutational analysis and computational study, we determine that Arg316 in RXRα, essential for 9-*cis*-retinoic acid binding and activating RXRα transactivation, is not required for antagonist effects of N-6, whereas Trp305 and Phe313 are crucial for N-6 binding to RXRα by forming extra π–π stacking interactions with N-6, indicating a distinct RXRα binding mode of N-6. N-6 inhibits TR3-stimulated transactivation of Gal4-DBD-RXRα-LBD by binding to the ligand binding pocket of RXRα-LBD, suggesting a strategy to regulate TR3 activity indirectly by using small molecules to target its interacting partner RXRα. For its physiological activities, we show that N-6 strongly inhibits tumor necrosis factor α (TNFα)-induced AKT activation and stimulates TNFα-mediated apoptosis in cancer cells in an RXRα/tRXRα dependent manner. The inhibition of TNFα-induced tRXRα/p85α complex formation by N-6 implies that N-6 targets tRXRα to inhibit TNFα-induced AKT activation and to induce cancer cell apoptosis. Together, our data illustrate a new RXRα ligand with a unique RXRα binding mode and the abilities to regulate TR3 activity indirectly and to induce TNFα-mediated cancer cell apoptosis by targeting RXRα/tRXRα.

KEYWORDS NSC-640358, ligand, RXRα, tRXRα, TNFα, apoptosis

INTRODUCTION

Retinoid X receptor α (RXRα), a unique member of the nuclear receptor superfamily, plays a pivotal role in diverse biological and physiological processes, including cell growth, differentiation, apoptosis, and homeostasis (Szanto et al., 2004; Evans and Mangelsdorf, 2014). RXRα has the typical structure of nuclear receptor, including an N-terminal region, a central DNA-binding domain (DBD), and a C-terminal ligand-binding domain (LBD) (de Lera et al., 2007). The unique property of RXRα among nuclear receptors lies in its ability to form heterodimers with many other nuclear receptors such as retinoic acid receptors (RARs) and peroxisome proliferator-activated receptors (PPARs) (Lefebvre et al., 2010; Evans and Mangelsdorf, 2014). Like other nuclear receptors, RXRα has important genomic functions mainly through its binding to responding DNA elements to regulate target gene transcription (Tang and Gudas, 2011; Evans and

Mangelsdorf, 2014). Recently, accumulating evidence indicate that RXRα also resides in the cytoplasm to exhibit its non-genomic actions (Katagiri et al., 2000; Cao et al., 2004; Ghose et al., 2004; Zimmerman et al., 2006). For example, under certain stimuli, RXRα migrates from nucleus to cytoplasm with nerve growth factor IB (NGFIB, also known as Nur77 and TR3) to induce cell apoptosis by binding to Bcl-2 and converting Bcl-2 from an anti-apoptotic to a pro-apoptotic molecule (Li et al., 2000a; Cao et al., 2004; Lin et al., 2004).

Dysfunctions of RXRα are implicated in the initiation and development of a variety of diseases including cancer (Shulman and Mangelsdorf, 2005; Tang and Gudas, 2011). For examples, targeted disruption of *RXRα* gene leads to skin abnormalities and prostatic preneoplastic lesions (Li et al., 2000b; Huang et al., 2002), while abnormal phosphorylation of RXRα is associated with the progresses of colon cancer and liver cancer (Matsushima-Nishiwaki et al., 2001). The expression level of RXRα is often found reduced in cancer cells and tissues, implying the association of the less RXRα expression and carcinogenesis (Jiang et al., 1999; Lotan et al., 2000; Takiyama et al., 2004). A number of studies have shown that proteolytic cleavage of RXRα is one of the mechanisms for lower expression of RXRα in cancer cells (Nagaya et al., 1998; Nomura et al., 1999; Casas et al., 2003). Recently, we showed that calpain II cleaves RXRα to produce a truncated RXRα–tRXRα in cancer cells (Gao et al., 2013). Different from full-length RXRα, tRXRα is able to reside in the cytoplasm and interact with p85α, the subunit of phophoinositide 3-kinase (PI3K), which leads to the enhanced TNFα-induced AKT activation (Zhou et al., 2010; Wang et al., 2013). tRXRα-mediated activation of the TNFα/PI3K/AKT pathway significantly promotes cancer cell growth both *in vitro* and *in vivo* (Zhou et al., 2010; Wang et al., 2013), providing a potential approach to inhibit cancer cell growth by targeting tRXRα with small molecules to inhibit TNFα/PI3K/AKT survival pathway.

The functions of nuclear receptors are tightly and delicately regulated by their cognate ligands (Gronemeyer et al., 2004). A number of natural and synthetic compounds have been identified as RXRα selective ligands (Altucci et al., 2007; de Lera et al., 2007; Dawson and Xia, 2012). In general, the chemical structures of RXRα ligands, such as 9-*cis*-retinoic acid (9-*cis*-RA) and LG100754, contain an acidic moiety and a large hydrophobic moiety. The acidic moiety is able to form salt bridges with Arg316 residue in RXRα ligand binding pocket (LBP), which contributes immensely to ligand-receptor interaction (Altucci et al., 2007; Dawson and Xia, 2012). Comparison of RXRα apo (unliganded) and holo (liganded) crystal structures reveals a ligand-induced conformational change of RXRα, which significantly affects the interactions of RXRα with other proteins such as co-activators and co-repressors (de Lera et al., 2007). Although binding to the same LBP, different ligands induce distinct functions of RXRα, which is due to the distinct conformational changes of RXRα induced by different ligands (Altucci et al., 2007; de Lera et al., 2007; Perez et al., 2012). TR3 is an orphan nuclear receptor which can form heterodimer with RXRα (Moll et al., 2006). Crystal studies have shown that the LBD of TR3 lacks a typical hydrophobic pocket for accommodating ligands (Wang et al., 2003). However, TR3 is able to recruit co-activator for transactivation independent of ligand binding (Wansa et al., 2002), reflecting its active unliganded conformation resembling other agonist-activated nuclear receptors.

In this study, we show that NSC-640358 (N-6) is a selective RXRα antagonist with a unique RXRα binding mode. Through binding to RXRα, N-6 inhibits RXRα transactivation and RXRα-mediated transcriptional activity of TR3. Moreover, N-6 inhibits tumor necrosis factor α (TNFα)-induced AKT activation and enhances TNFα-mediated cancer cell apoptosis in an RXRα/tRXRα dependent manner.

RESULTS

N-6 binds to RXRα

We used Q-MOL molecular modeling package to perform a virtual screening of a compound library for RXRα ligands, and found that NSC-640358 ((NZ)-N-[1-[2-(3,5-diphenylpyrazol-1-yl)-4-methyl-1,3-thiazol-5-yl]ethylidene]droxylamine, N-6), a thiazolyl-pyrazole derived compound, was a potential RXRα ligand (Fig. 1A). The interaction of N-6 and RXRα was first examined by ligand competition binding assay. As a positive control, unlabeled 9-*cis*-RA displaced [^{3}H]-labeled 9-*cis*-RA for binding to RXRα-LBD protein in a dose-dependent manner (Fig. 1B). Similarly, N-6 dose-dependently competed with [^{3}H]-labeled 9-*cis*-RA for binding to RXRα-LBD protein with an IC_{50} at 11.3 μmol/L (Fig. 1B), suggesting the direct binding of N-6 to RXRα. Then we performed surface plasmon resonance (SPR) technology-based experiment to confirm their physical binding. As shown in Fig. 1C, N-6 dose-dependently binds to RXRα-LBD with a K_d value of 15.7 μmol/L. Furthermore, our fluorescence ligand binding assay showed that RXRα-LBD protein potently enhanced the fluorescence intensity of N-6 (Fig. 1D). Together, these results demonstrate that N-6 is able to bind to RXRα directly.

N-6 selectively inhibits RXRα transactivation

To examine whether N-6 regulated RXRα transactivation, a reporter gene containing RXRα homodimer-responsive elements, $(TREpal)_2$-*tk*-CAT (Dawson et al., 2001), was transfected with a RXRα expression vector into CV-1 cells that lack detectable levels of endogenous RXRα (Zhang et al., 1992a, b). Treatment of cells with 9-*cis*-RA potently induced the transactivation of RXRα homodimer, which was dose-dependently inhibited by N-6 (Fig. 2A). In contrast, N-6 failed to inhibit 9-*cis*-RA-induced transactivation of RARγ homodimer or the ligand-independent transactivation of Nur77 (Fig. 2B and 2C) (Kolluri et al., 2003). Thus, N-6 selectively

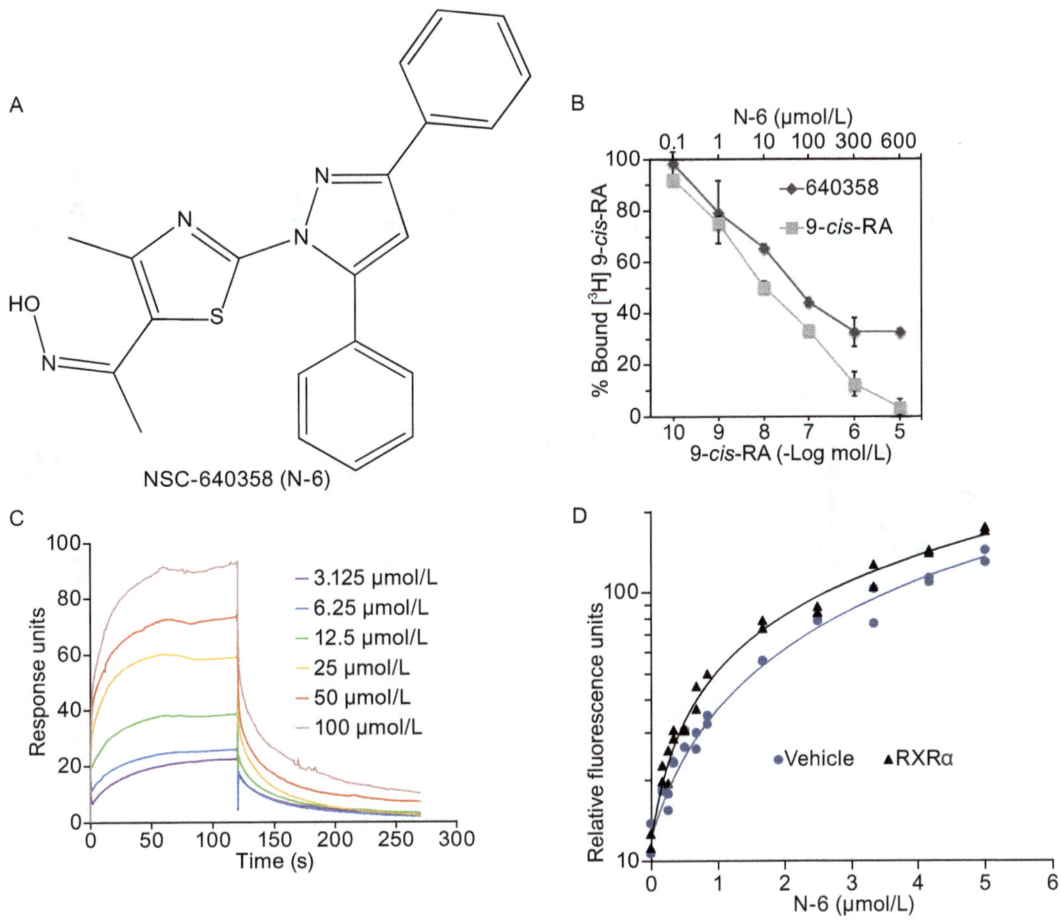

Figure 1. N-6 binds to RXRα. (A) Structure of NSC-640358 (N-6). (B) RXRα-LBD protein was incubated with $[^3H]$9-*cis*-RA in the presence or absence of unlabeled 9-*cis*-RA or N-6. Bound $[^3H]$9-*cis*-RA was quantitated by liquid scintillation counting. (C) Gradient concentrations of N-6 were injected through flow cells immobilized with RXRα-LBD. The kinetic profiles are shown and the dissociation constant (K_d) of the N-6/RXRα-LBD complex was calculated to be 15.755×10^{-6} mol/L. (D) RXRα-LBD protein enhanced fluorescent intensity of N-6 (ex 278 nm, em 338 nm, cutoff 325 nm, delay 50 μs, integration 450 μs). One of three similar experiments is shown.

inhibits RXRα transactivation, which was further confirmed by our mammalian one-hybrid technology-based assay. N-6 strongly inhibited 9-*cis*-RA-induced Gal4-DBD-RXRα-LBD transactivation but not the other chimeric nuclear receptors, such as RARγ, TR3, GR, LXRα and PPARγ (Fig. 2D–I). We next examined the effects of N-6 on the transcriptional activity of RXRα/TR3 heterodimer. To this end, CV-1 cells were transfected with RXRα and TR3 expression vectors together with βRARE-*tk*-CAT reporter. 9-*cis*-RA-induced transactivation of RXRα/TR3 heterodimer was strongly inhibited by N-6 (Fig. 2J). It is known that the binding of 9-*cis*-RA induces RXRα conformational change, leading to co-activator recruitment. We then examined whether N-6 inhibited 9-*cis*-RA-induced co-activator recruitment. As shown in Fig. 2K, N-6 dose-dependently inhibited 9-*cis*-RA-induced interaction of co-activator and RXRα-LBD with an IC_{50} at 32.9 μmol/L. Taken together, these findings indicate that N-6 is a selective antagonist of RXRα.

N-6 regulates TR3 activity by binding to RXRα

In the absence of RXRα agonists, the fusion protein Gal4-DBD-RXRα-LBD only exhibited basal transactivation. However, transfection of TR3 remarkably enhanced its transactivation (Fig. 3A). The salt bridges formed by Glu453/456 in helix 12 and Arg302 in helix 4 are essential for maintaining co-activator binding site in RXRα (Egea et al., 2000; Egea et al., 2002). The fusion protein Gal4-DBD-RXRα-LBD/E453/456A, with Ala substituted with Glu, completely lost transactivation ability in response to 9-*cis*-RA and CD3254 (Fig. 3B). However, TR3 was still able to stimulate the reporter gene activity together with Gal4-DBD-RXRα-LBD/

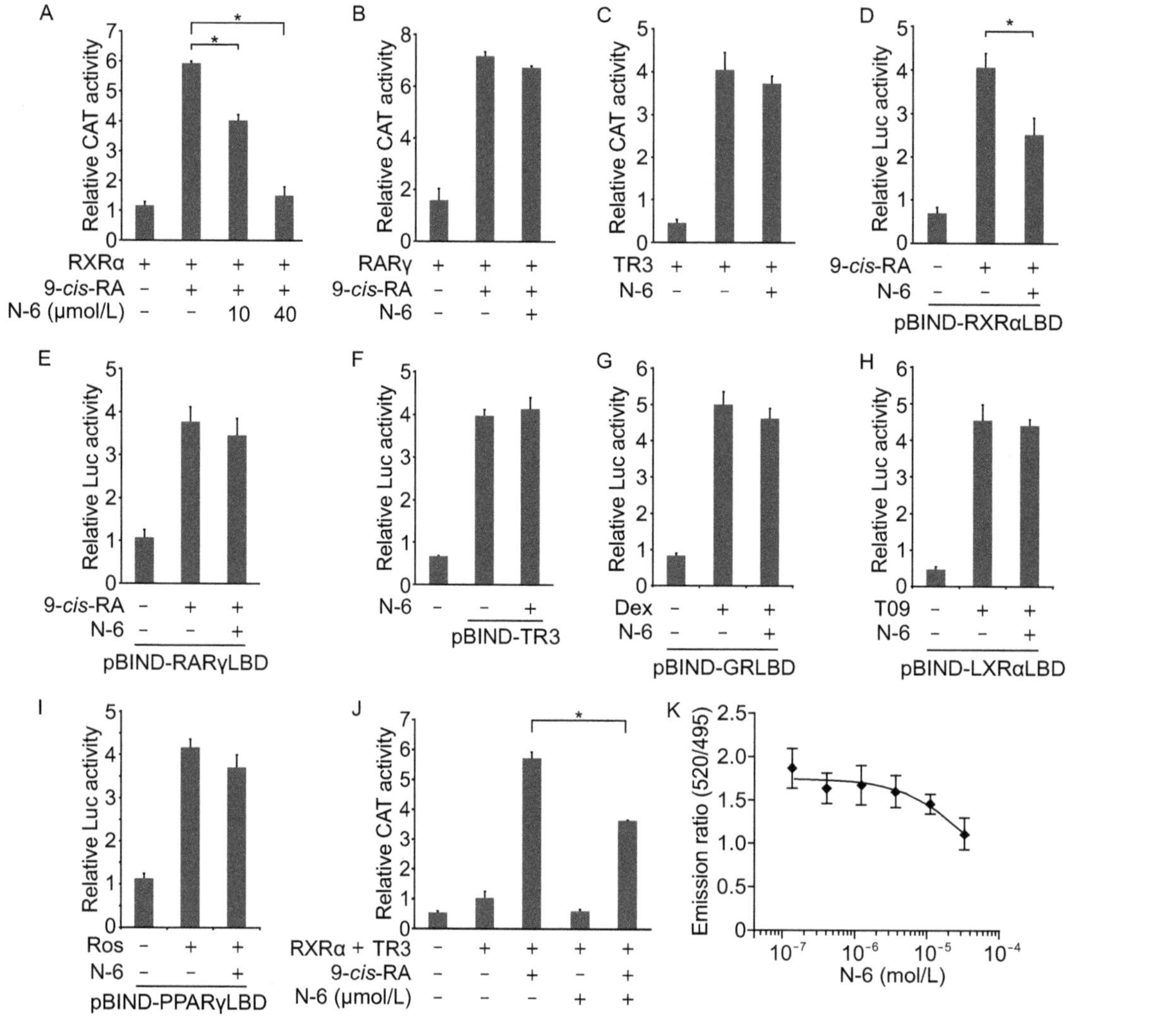

Figure 2. N-6 is a selective antagonist of RXRα. (A–C) CV-1 cells were cotransfected with $(TREpal)_2$-*tk*-CAT and pCMV-Myc-RXRα (A), $(TREpal)_2$-*tk*-CAT and pCMV-Myc-RARγ (B), or NurRE-*tk*-CAT and pCMV-Myc-Nur77 (C). Cells were treated with N-6 (10 μmol/L) and 9-*cis*-RA (10^{-7} mol/L) for 18 h. CAT activities were measured and normalized to β-galactosidase activities (**P* < 0.05). (D–I) MCF-7 cells were cotransfected with pG5-Gaussia-Dura and pBIND-RXRαLBD (D), pBIND-RARγLBD (E), pBIND-TR3 (F), pBIND-GRLBD (G), pBIND-LXRαLBD (H), or pBIND-PPARγLBD (I). Cells were treated with N-6 (10 μmol/L) in the presence or absence of 0.1 μmol/L 9-*cis*-RA (D and E), 1 μmol/L Dexamethasone (Dex) (G), 1 μmol/L T0901317 (T09) (H), or 1 μmol/L Rosiglitazone (Ros) (I) for 18 h. Reporter activities were measured and normalized (**P* < 0.05). (J) CV-1 cells transfected with βRARE-*tk*-CAT, RXRα and TR3 expression vectors were treated with N-6 (10 μmol/L) in the presence or absence of 9-*cis*-RA (10^{-7} mol/L) for 18 h. CAT activities were measured and normalized to β-galactosidase activities (**P* < 0.05). (K) GST-RXRα-LBD was incubated with different concentrations of N-6 in the presence of 9-*cis*-RA (10^{-8} mol/L) for 2 h. FRET signals were measured and normalized. The IC_{50} of N-6 was calculated to be 3.29×10^{-5} mol/L. One of three similar experiments is shown. Data shown are mean ± SD.

E453/456A (Fig. 3C). Sequence analysis indicated there were no cognate TR3 binding sites in the G5 reporter vector. Thus, the ability of TR3 to induce the transactivation was due to the interaction of Gal4-DBD-RXRα-LBD with TR3 and TR3 ligand-independent transactivation. No matter what above fusion proteins were used, TR3-induced transactivation was potently inhibited by N-6 (Fig. 3D). However, in the case of Gal4-DBD-RXRα-LBD/C432Y with disrupted RXRα-LBP due to the substitution of Cys with Tyr, N-6 completely lost its activity (Fig. 3D), suggesting the necessity of the binding of N-6 to RXRα-LBD. Similar results were observed when RXRα antagonist UVI3003 was examined (Fig. 3D). Thus, N-6 and UVI3003 bind to RXRα to indirectly inhibit TR3 transactivation. One possible mechanism by which N-6 inhibited TR3 transactivation was the disruption of the interactions of TR3 with the fusion proteins, which was analyzed by our co-immunoprecipitation experiments. As shown in Fig. 3E, Myc-TR3 interacted with the fusion

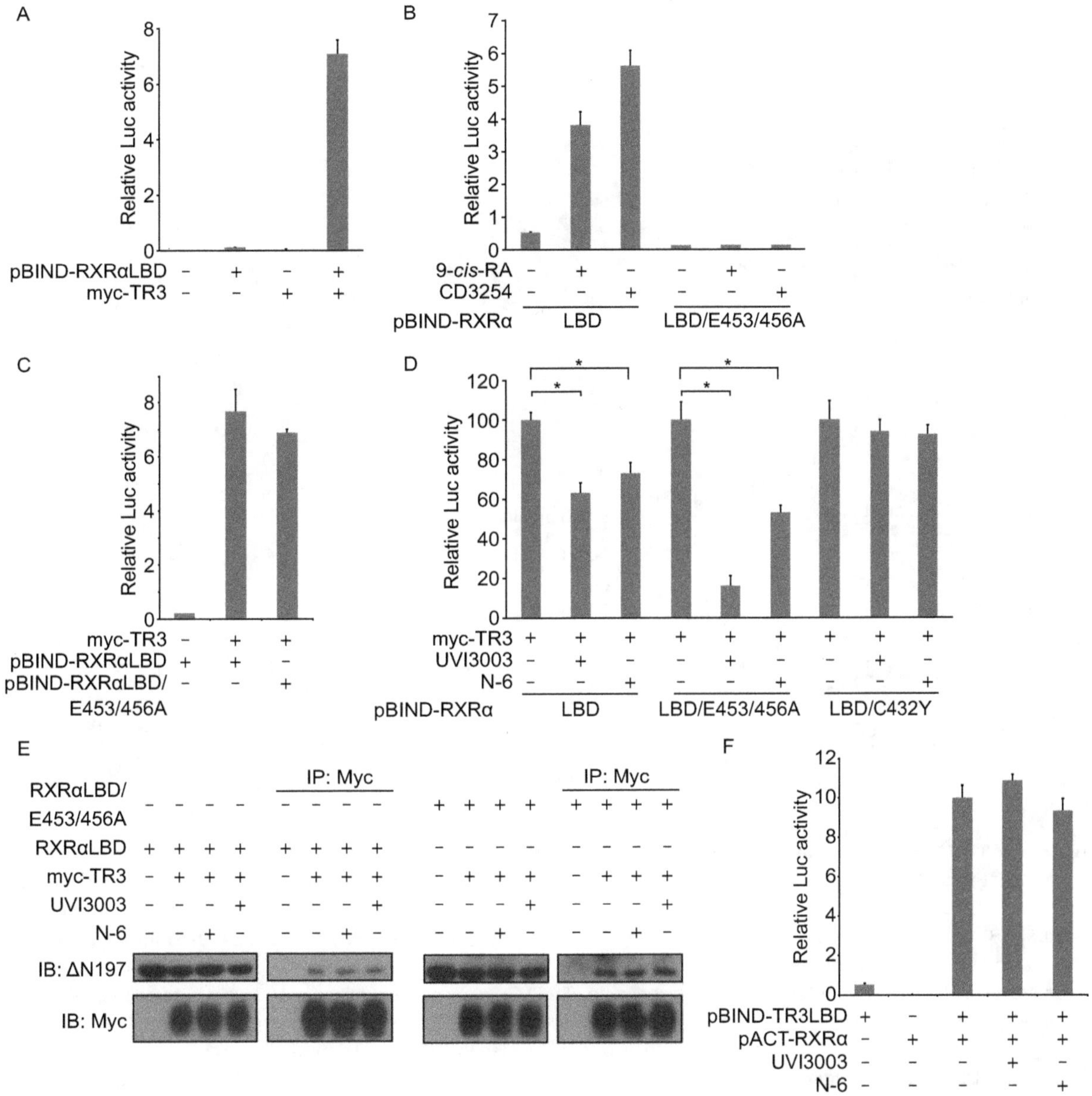

Figure 3. N-6 inhibits TR3 transcriptional activity by binding to RXRα. (A–D, and F) MCF-7 cells cotransfected with pG5-Gaussia-Dura reporter vector and the indicated expression vectors were treated with or without N-6 (10 μmol/L), UVI3003 (1 μmol/L), and 9-*cis*-RA (10^{-7} mol/L) for 18 h. Reporter activities were measured and normalized. Data shown are mean ± SD (**P* < 0.05). (E) HEK293T cells cotransfected with pCMV-Myc-TR3 and pBIND-RXRα-LBD expression vectors were treated with UVI3003 (1 μmol/L) or N-6 (10 μmol/L) for 18 h. Cell lysate were analyzed for heterodimerization of Nur77 and Gal4-DBD-RXRα-LBD by co-immunoprecipitation with anti-myc antibody. One of three similar experiments is shown.

proteins, as evidenced by their co-precipitation. Neither N-6 nor UVI3003 affected their interactions (Fig. 3E). GAL4-DBD-TR3-LBD had low, if there was, transactivation activity, which was due to the loss of the N-terminal TR3 (Wansa et al., 2002). When both GAL4-DBD-TR3-LBD and p16-ACT-RXRα fusion proteins existed, a dramatic induction of the reporter gene activity was observed, which reflected a typical mammalian two-hybrid assay to show the interaction of RXRα and TR3-LBD. In this context, neither N-6 nor UVI3003 was able to inhibit the induced transactivation (Fig. 3F), which also demonstrated the inability of N-6 and UVI3003 for blocking the interaction of TR3 and RXRα. Therefore, it is not through the disruption of the interaction of TR3 and RXRα for N-6 to inhibit TR3 transactivation.

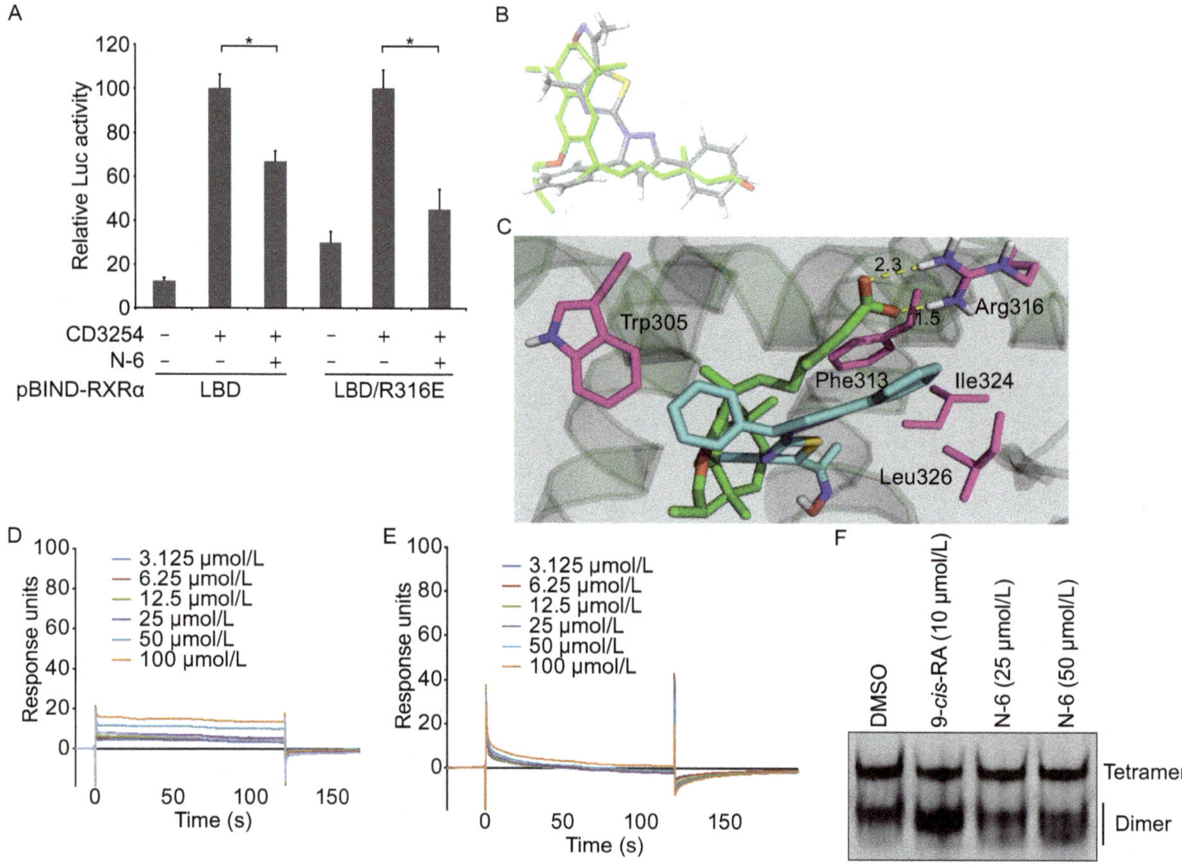

Figure 4. Arg316 is not required for N-6 binding to RXRα. (A) MCF-7 cells cotransfected with pG5-Gaussia-Dura reporter vector and pBIND-RXRα-LBD or pBIND-RXRα-LBD/R316E expression vectors were treated with or without N-6 (10 μmol/L) in the presence or absence of CD3254 (10^{-7} mol/L) for 18 h. Reporter activities were measured and normalized. Data shown are mean ± SD (*$P < 0.05$). (B) Comparison of the docked conformation of N-6 (gray) with the crystal structure of LG100754 (green). (C) N-6 was docked into the LBP of the co-crystal structure of LG100754 and RXRα-LBD (PDB 3A9E). Salt bridges are shown as dotted yellow lines, and residues interacting with N-6 are shown in magenta. (D–E) Gradient concentrations of N-6 were injected through flow cells immobilized with RXRα-LBD/Trp305Ala (D) and RXRα-LBD/Phe313Ala (E), respectively. The kinetic profiles are shown and the dissociation constants (K_d) of the N-6/RXRα-LBD complex were calculated to be 1.0×10^{-3} mol/L (D) and more than 1.0×10^{-2} mol/L (E). (F) RXRα-LBD proteins (2 mg/mL) was incubated with DMSO, 10 μmol/L 9-*cis*-RA, 25 μmol/L N-6 or 50 μmol/L N-6 for 3 h, and proteins were separated by 8% nondenaturing PAGE followed by Commassie Blue staining.

N-6 binds to RXRα in a distinct mode

Unlike many natural and synthetic RXRα ligands, N-6 lacks a carboxylate moiety known to form salt bridges with Arg316 in RXRα-LBP. To examine the requirement of Arg316 for N-6, it was replaced with Glutamic acid, and the resulting mutant Gal4-DBD-RXRαLBD/R316E was evaluated for its transactivation regulated by N-6. As shown in Fig. 4A, CD3254 was able to stimulate the transactivation of the mutant. Similar to RXRα-LBD, CD3254-induced transactivation of RXRα-LBD/R316E was potently inhibited by N-6 (Fig. 4A), implying that Arg316 was not required for N-6 binding to RXRα. We then used computer-aided and docking-based techniques to analyze the binding mode of N-6. In light of the antagonist activity of N-6 and a large degree of structural overlapping between N-6 and LG100754 (Fig. 4B), an antagonist of RARα/RXRα heterodimer (Sato et al., 2010), it was reasonable that the model of RXRα under its antagonist conformation should be used for the docking of N-6. In fact, our docking study showed that N-6 was well accommodated into the LBP of RXRα with an antagonist conformation (Fig. 4C). Unlike LG100754, N-6 did not form ionic bonds (2.3 Å and 1.5 Å for LG100754) with Arg316. However, N-6 possesses two aromatic rings that could establish extra π-π stacking interactions with Phe313 and Trp305 (Fig. 4C). The essential role of Phe313 and Trp305 in N-6 binding was confirmed by our SPR assays, showing that N-6 had much lower binding affinity to two RXRα-LBD point mutants with Ala substitution of Phe313 or Trp305 (Fig. 4D and 4E). Furthermore, our

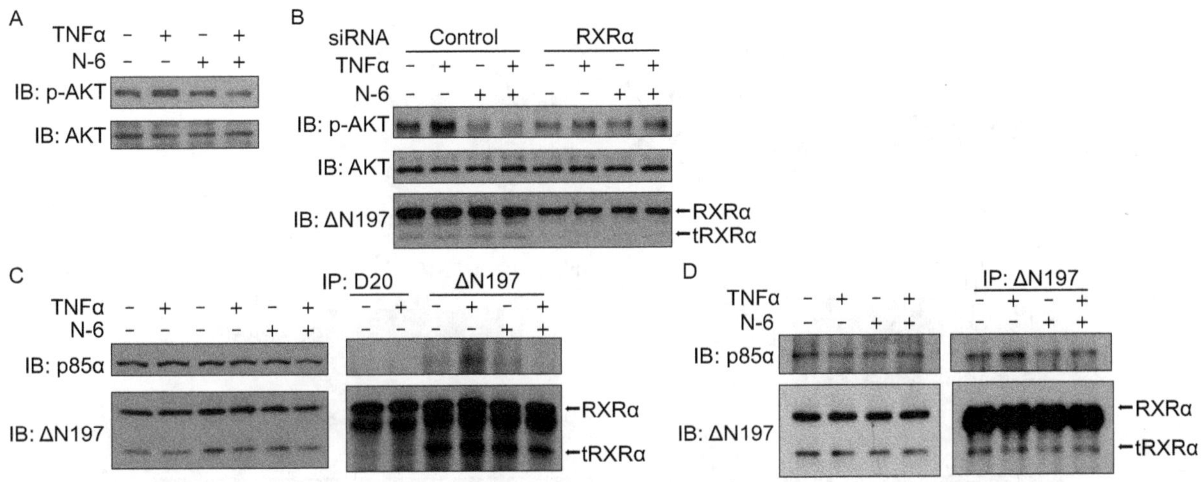

Figure 5. N-6 inhibits TNFα-induced AKT activation in a tRXRα-dependent manner. (A) HCT116 cells were pretreated with N-6 (10 μmol/L) for 2 h in serum free medium before being exposed to TNFα (10 ng/mL) for an additional 30 min. Lysates prepared were analyzed by Western blotting for AKT activation. (B) A549 cells transfected with RXRα siRNA or control siRNA for 48 h were treated with N-6 (10 μmol/L) for 2 h in serum free medium before being exposed to TNFα (10 ng/mL) for an additional 30 min. Lysates prepared were analyzed by Western blotting for AKT activation. (C and D) H292 cells (C) or A549 cells (D) pretreated with N-6 (10 μmol/L) for 2 h in serum free medium before being exposed to TNFα (10 ng/mL) for an additional 30 min were analyzed for p85α/tRXRα interaction by co-immunoprecipitation assay using anti-RXRα antibodies of D20 or ΔN197. Immunoprecipitates were analyzed by Western blotting for the presence of p85α and tRXRα. One of three similar experiments is shown.

docking study showed that two hydrophobic residues Ile324 and Leu326 could produce additional hydrophobic interactions with N-6 (Fig. 4C). Ligand binding often leads to the conformational change of RXRα protein, which was investigated for N-6 by our native gel electrophoresis assay. Consistent with previous reports, 9-*cis*-RA strongly induced the homodimeric formation of RXRα-LBD protein. However, N-6 dose-dependently induced the conformational changes of RXRα-LBD dimer, indicated by the pattern changes of the dimer bands (Fig. 4F). Taken together, our data indicate that N-6 has a distinct binding model comparing with classic RXRα ligands.

N-6 inhibits TNFα-induced AKT activation in an RXRα/tRXRα dependent manner

It has been reported that RXRα antagonist K-80003 inhibits TNFα-induced AKT activation in a tRXRα-dependent pathway (Zhou et al., 2010). Here we examined whether N-6 had a similar function. Figure 5A showed that N-6 significantly inhibited TNFα-induced AKT activation in HCT116 cells, revealed by its suppression of the expression of phosphor-AKT but not total AKT (Fig. 5A). Similar results were obtained in A549 cells (Fig. 5B). We then determined the requirement of RXRα/tRXRα for the inhibition of AKT activation by N-6. Transfection of RXRα siRNA in A549 cells, which reduced both the full-length RXRα and tRXRα expression, abolished the inhibitory effect of N-6 on TNFα-induced AKT activation (Fig. 5B). Thus, the expression of RXRα/tRXRα is essential for AKT inhibition by N-6. We have reported that K-80003 inhibited AKT activation by blocking the TNFα-induced interaction of tRXRα and p85α (Zhou et al., 2010). We then investigated whether N-6 had the similar action. Two anti-RXRα antibodies, ΔN197 and D20, were used in the co-immunoprecipitation assay, of which ΔN197 but not D20 could recognize both RXRα and tRXRα (Zhou et al., 2010). Consistent with our previous reports (Zhou et al., 2010), TNFα strongly induced the interaction of tRXRα but not RXRα with p85α showed by our co-immunoprecipitation experiment (Fig. 5C). However, when cells were treated with N-6 together with TNFα, the interaction of tRXRα and p85α promoted by TNFα was completely blocked (Fig. 5C and 5D). Thus, the inhibitory effect of N-6 on TNFα-induced AKT activation might rely on its disruption of the interaction of tRXRα with p85α.

N-6 induces RXRα/tRXRα-dependent apoptosis of cancer cell

The ability of N-6 to inhibit AKT activation prompted us to examine the effect of N-6 on the survival of cancer cells by colonogenic survival assay. Treatment of cells with N-6 significantly inhibited colony formation of HeLa cells, which was dramatically enhanced by TNFα (Fig. 6A). The inhibition of TNFα-induced AKT activation often leads to the activation of TNFα-mediated apoptosis (Zhou et al., 2010; Wang et al., 2013). Indeed, the combination of TNFα and N-6 synergistically induced PARP cleavage in HCT116 cells (Fig. 6B).

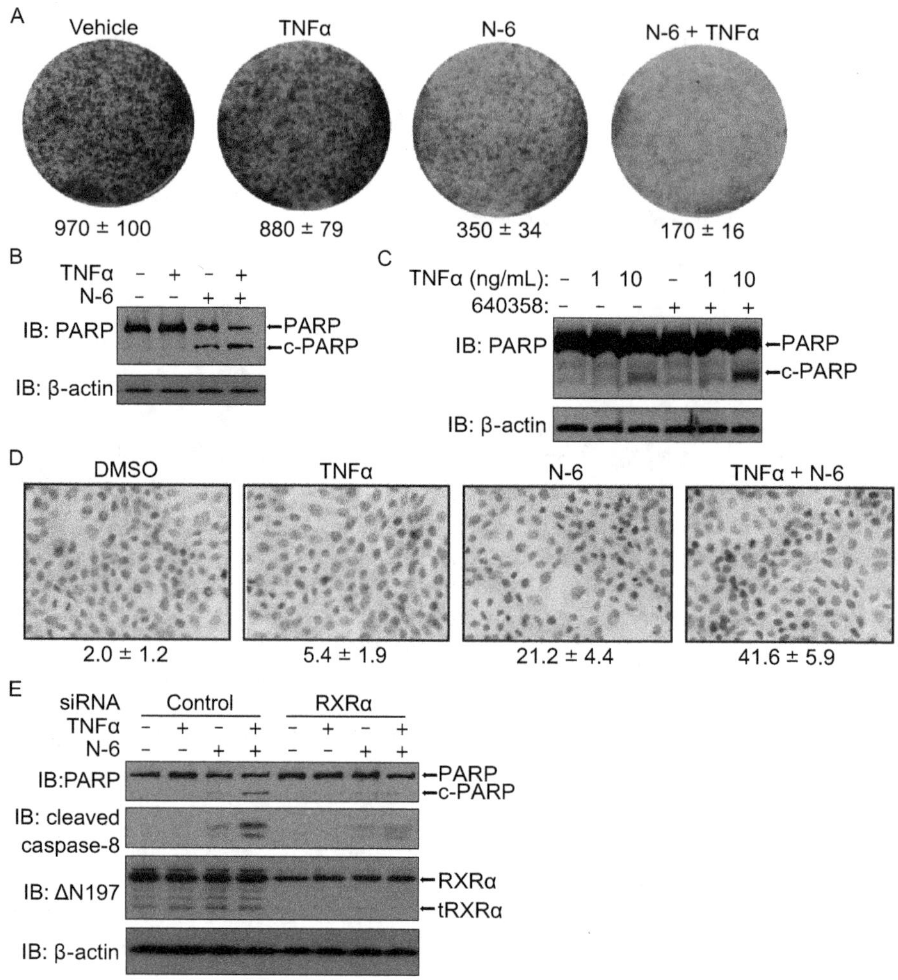

Figure 6. Combination of TNFα and N-6 induces cancer cell apoptosis in an RXRα/tRXRα-dependent manner. (A) HeLa cells grown in 6-well plates were treated with or without N-6 (5 μmol/L) and TNFα (10 ng/mL) for 3 days. Colonies were stained with 0.1% crystal violet and counted. (B and C) HCT116 and MCF-7 cells were treated with or without N-6 (10 μmol/L) and TNFα (10 ng/ mL) for 15 h in serum free medium. Cell lysates prepared were analyzed by Western blotting for PARP cleavage. (D) MCF-7 cells grown in 24-well plates were treated with or without N-6 (10 μmol/L) and/or TNFα (10 μg/mL) for 3 days. Apoptotic cells were detected by TUNEL staining and counted. (E) HCT116 cells transfected with RXRα siRNA or control siRNA for 48 h were treated with or without N-6 (10 μmol/L) and TNFα (10 ng/mL) for 15 h. Cell lysates prepared were analyzed by Western blotting for PARP and caspase-8 cleavage. One of three similar experiments is shown.

The synergistic induction of PARP cleavage was TNFα-concentration dependent in MCF-7 cells (Fig. 6C). Furthermore, our TUNEL staining assay confirmed the synergistic pro-apoptosis of the combination (Fig. 6D). We then determined the role of RXRα/tRXRα in apoptosis induction of the combination using knocking down approach. Compared to control siRNA, RXRα siRNA impaired the ability of the combination of N-6 and TNFα for the induction of PARP cleavage (Fig. 6E). Accompanying with PARP cleavage, the combined treatment also induced caspase-8 activation, which was also inhibited by the suppression of RXRα/tRXRα expression (Fig. 6E). Thus, RXRα/tRXRα is involved in the apoptotic induction of the combination of N-6 and TNFα.

DISCUSSION

RXRα is a promising drug target for cancer therapy, as it has an inner LBP to accommodate "drug-like" small molecule modulators (Altucci et al., 2007; de Lera et al., 2007). In addition, dysfunctions or abnormal expressions of RXRα are implicated in tumor development (Tang and Gudas, 2011; Thomas et al., 2012). Recently, we reported that RXRα antagonist K-80003 exhibits striking anti-cancer effects by inhibiting the cytoplasmic activation of PI3K/AKT by tRXRα (Zhou et al., 2010). Here, we present a new RXRα ligand with a distinct structure and an RXRα/tRXRα-dependent induction of cancer cell apoptosis with TNFα.

The structure of N-6 is very different from known RXRα ligands such as 9-*cis*-RA and LG100754. However, it bound to RXRα with a unique mode and acted as an RXRα antagonist (Figs. 1, 2 and 4). The holo-crystal structure of 9-*cis*-RA and RXRα-LBD indicates that the salt bridge formed via Arg316 at the end of the L-shaped LBP and the carboxylate group of 9-*cis*-RA is crucial for 9-*cis*-RA binding to RXRα (Egea et al., 2000; Egea et al., 2002; Sato et al., 2010). N-6 lacks such a carboxylate moiety, and therefore it should not require Arg316 for binding, which was verified by our mammalian one-hybrid assay showing that the mutation of Arg316 did not affect the inhibition of RXRα transactivation by N-6 (Fig. 4A). Recently, several compounds without carboxylate moiety such as CF31, Danthron, Magnolol, and Bigelovin were reported to bind to RXRα (Zhang et al., 2011a; Zhang et al., 2011b; Zhang et al., 2011c; Wang et al., 2013). Therefore, salt bridges are not necessary for some ligands binding to RXRα. Instead, in the case of N-6, the π-π stacking interactions and hydrophobic interactions contribute to the binding of N-6 to RXRα (Fig. 4C–E). N-6, CF31, and Danthron are RXRα antagonists while Magnolol and Bigelovin are RXRα agonists. Both as antagonist of RXRα, Danthron and N-6 might use different antagonistic mechanisms. In contrast with stabilization of RXRα tetramer by Danthron (Zhang et al., 2011c), N-6 might induce the conformational changes of RXRα dimer to inhibit its homodimer transactivation (Figs. 2A and 4F). The different mechanisms should be due to the distinct spatial structures and binding modes of Danthron and N-6 that lead to different conformational changes of RXRα. It will be of great interests to study how distinct RXRα conformational changes induced by these compounds without acidic group define their regulation properties and mechanisms of RXRα activities.

Crystal structure of TR3 indicates that it lacks a normal LBP (Wang et al., 2003). Although many compounds have been identified to bind to TR3 (Lee et al., 2011), the controversy of the cognate ligands of TR3 still exists. Thus, it might be difficult to design and apply strategy to regulate TR3 physiopathological functions by small molecules directly targeting TR3. An alternative approach is to use small molecules to regulate TR3 functions indirectly by targeting TR3 interacting proteins. Our study provides an example to show the feasibility of this approach. In the absence of binding to its cognate DNA elements, the transactivation of TR3 associated with RXRα was able to be inhibited by N-6 through binding to RXRα (Fig. 3D). The conformational change of RXRα induced by N-6 binding may enhance RXRα recruiting co-repressors that would prevent co-activator recruitment of TR3 due to steric hindrance, or may affect the activating conformation of TR3, especially its AB domain, in the RXRα/TR3 heterodimer. The activities of TR3 modulated by N-6 may not be limited to its transcriptional activities. Besides regulating gene transcription, TR3 also interacts with many proteins to exert its non-genomic functions in the absence of DNA binding (Lin et al., 2004; Moll et al., 2006). Therefore, our study also provides a potential strategy to use small molecules to target RXRα to regulate TR3 non-genomic functions.

The physiological functions of ligands often depend on their abilities of inducing conformational changes of their cognate nuclear receptors followed by the changes of interactions of nuclear receptors with other proteins. The conformational change of tRXRα induced by N-6 disrupted the interaction of tRXRα and p85α, leading to the inhibition of TNFα-induced AKT activation (Fig. 5). However, in the case of the interaction of RXRα and TR3, N-6 did not disrupt the interaction but definitely changed the interaction manner, leading to the inhibition of TR3 transactivation (Fig. 3). Therefore, there are two modes for N-6 to regulate RXRα interactions with other proteins. One is to regulate the interaction strength, and the other one is to regulate the interaction manner. The regulation mode of RXRα interaction by N-6 should be determined by the interacting domains of both RXRα and its interacting partners such as TR3 as well as other interacting proteins in the RXRα-formed complex such as co-activators and co-repressors in the RXRα/TR3 complex.

TNFα is a double-edged sword in tumor development (Balkwill, 2009; Waters et al., 2013). It stimulates caspase8-dependent apoptosis, and/or promotes survival mainly through PI3K/AKT and NFκB pathways (Mocellin and Nitti, 2008; Balkwill, 2009; Waters et al., 2013). In most cases, TNFα acts as a promoter rather than a killer in tumor cells and tissues because of its abnormally elevated survival functions (Mocellin and Nitti, 2008; Balkwill, 2009; Waters et al., 2013). To take the advantage of TNFα's apoptotic functions in cancer treatment, its survival signals in cancer cells need to be blocked. Our previous discovery provides a potential target—tRXRα for small molecules targeting to inhibit TNFα survival signals. Our current study presents several lines of evidence that N-6 could target tRXRα to convert TNFα signal from survival to death. First, N-6 strongly inhibited TNFα-induced AKT activation in cancer cells (Fig. 5A and 5B). Second, N-6 and TNFα synergistically induced caspase-8 activation and cancer cell apoptosis (Fig. 6B–E). Third, the functions of N-6 on AKT inhibition and apoptosis-induction with TNFα were dependent on RXRα/tRXRα expression (Figs. 5B and 6E). Forth, N-6 inhibited the interactions of tRXRα and p85α (Fig. 5C and 5D).

Taken together, our results identify a new RXRα antagonist with abilities to modulate TR3 functions indirectly by binding to RXRα and to convert TNFα signal to induce cancer cell apoptosis by targeting tRXRα.

MATERIALS AND METHODS

Reagents

NSC-640358 was provided by the NCI/DTP Open Chemical Repository and the DTP website address is http://dtp.cancer.gov. Fermentas TurboFect *in vitro* transfection reagent, DharmaFECT 1 transfection reagent, Gaussia-Dura Luciferase Glow Assay Kit, goat anti-rabbit and anti-mouse secondary antibody conjugated to horseradish peroxidase from Thermo Fisher Scientific, Inc. (Waltham, MA, USA); CD3254, UVI3003, anti-AKT1 (C-20, sc-1618), anti-c-Myc (9E10, sc-40), anti-RXR (ΔN197, sc-774), anti-RXR (D20, sc-553) from Santa Cruz Biotechnology (Santa Cruz, CA, USA); anti-caspase-8 (1C12, #9746), anti-phospho-AKT (Ser473) (D9E, 4060) from Cell Signaling Technology (Boston, MA, USA); anti-poly (ADP-ribose) polymerase (PARP, 556494) from BD Biosciences (San Diego, CA, USA); 9-*cis*-Retinoic Acid, anti-β-actin antibody from Sigma (St. Louis, MO, USA); anti-PI3 kinase (p85α, 04-403), polyvinylidene difluoride membranes from Millipore (Billerica, MA, USA); TR-FRET Retinoic X Receptor alpha coactivator assay kit (PV4797) from invitrogen (Carlsbad, CA, USA); rhTNF-α (Tumor Necrosis Factor-α, Human, Recombinant) from Promega (Beijing, China); enhanced chemiluminescence reagents from GE Healthcare (Buckinghamshire, UK) and a cocktail of proteinase inhibitors from Roche (Meylan, France); $[^3H]$9-*cis*-Retinoic Acid from PerkinElmer (Boston, MA, USA) were used in this study.

Cell culture and transient transfection

MCF-7 breast cancer cells, HCT116 colon cancer cells, HeLa cervix cancer cells, H292 lung cancer cells, A549 lung cancer cells, HEK293T cells, and CV-1 cells were obtained from American Type Culture Collection (ATCC) and cultured in Dulbecco's modified Eagle's medium containing 10% fetal bovine serum (FBS). All cell lines used were passaged in our lab for fewer than 4 months after resuscitation, and were used at the fifth through tenth passage in culture for this study. Transient transfection was performed using TurboFect according to the instructions of the manufacturer in regular growth medium.

Fluorescence ligand binding assay

A dilution series of the compound N-6 in potassium phosphate buffer (10 mmol/L, pH 7.50) was made on a 96-well polypropylene plate, which was subsequently transferred onto a 96-well, half-area UV-transparent quartz plate in duplicate, 50 μL/well. The stock solution of the protein RXRα ligand binding domain was diluted into potassium phosphate buffer to reach 0.60 μmol/L, which was immediately transferred onto the quartz plate, 10 μL/well. The fluorescent intensities of the compound were acquired on a FlexStation 3 Microplate Reader (Molecular Devices, Sunnyvale, CA). In order to minimize the interference from the intrinsic fluorescence of Trp residues, time-resolved fluorescence mode was used. Data analyses were performed with GraphPad Prism 5.0d.

Mammalian one-hybrid assays

MCF-7 cells seeded in 48-well plates were transiently cotransfected with the luciferase reporter plasmid pG5-Gaussia (40 ng), pBind-receptors (40 ng), and β-Gal plasmid (2 ng). Twenty-four hours after transfection, the medium was replaced by medium containing a respective ligand and/or a testing compound. After 18 h, the medium was collected and assayed by using Gaussia-Dura Luciferase Glow Assay Kit (Pierce). Cells were washed, lysed and assayed for β-Gal activity. Transfection efficiency was normalized to β-Gal activity.

Mammalian two-hybrid assays

MCF-7 cells seeded in 48-well plates were transiently cotransfected with the luciferase reporter plasmid pG5-Gaussia (40 ng), pBind-TR3-LBD (40 ng), pACT-RXRα and β-Gal plasmid (2 ng). Twenty-four hours after transfection, the medium was replaced by medium containing a respective ligand and/or a testing compound. After 18 h, the medium was collected and assayed by using Gaussia-Dura Luciferase Glow Assay Kit (Pierce). Cells were washed, lysed, and assayed for β-Gal activity. Transfection efficiency was normalized to β-Gal activity.

CAT assays for NR response elements

CV-1 cells seeded in 24-well plates were transiently cotransfected with one of the luciferase reporter plasmids pBLCAT2-TREpal (100 ng)/pBLCAT2-βRARE (100 ng)/pBLCAT2-NurRE (100 ng), one or two NR expressing plasmids pCMV-Myc-RXRα (20 ng)/pCMV-Myc-RARγ (20 ng)/pCMV-Myc-TR3 (20 ng), and pCMV-Myc-β-Gal (5 ng) as internal reference. Twenty-four hours after transfection, the medium was replaced by medium containing a respective ligand and/or a testing compound. After 18 h, cells were harvested and assayed for CAT and β-Gal activity. Transfection efficiency was normalized to β-Gal activity.

TR-FRET RXRα coactivator assays

Invitrogen's LanthsScreen time-resolved fluorescence resonance energy transfer (TR-FRET) assay was used. Retinoic X Receptor alpha coactivator assay was conducted according to manufacture's protocol. The TR-FRET ratio was calculated by dividing the emission signal at 520 nm by the emission signal at 495 nm.

SPR measurements

The binding kinetics between RXRα-LBD and N-6 was analyzed at 25°C on a BIAcore T200 machine with CM5 chips (GE Healthcare). PBSP was used for all measurements. A blank channel was used as negative control. About 10,000 response units of RXRα-LBD were immobilized on the chip. When the data collection was finished in each cycle, the sensor surface was regenerated with Glycine-HCl 2.5. A series of concentrations of N-6 ranging from 3.125 to 100 μmol/L were applied for experiment. Sensograms were fit globally with BIAcore T200 analysis using 1:1 Langumuir binding mode.

siRNA and transfections

siRNAs against RXRα (SASI-HS01-00097639) and control siRNA (SIC001) used were from Sigma. Cells were transfected with 5 μL aliquot of 20 μmol/L siRNA/well in six-well plates using

DharmaFECT 1 transfection reagent, according to the manufacturer's recommendations.

Western blotting

Cell lysates were prepared by lysing cells with lysis buffer containing 25 mmol/L Tris, 150 mmol/L NaCl, 1 mmol/L EDTA, 1% NP-40, 5% glycerol, pH 7.4 with a cocktail of proteinase inhibitors on ice for 30 min. Equal amounts of the lysates were electrophoresed on an SDS-PAGE gel (8% or 12%) and transferred onto polyvinylidene difluoride membranes, which were then blocked with 5% nonfat milk in TBST [50 mmol/L Tris-HCl (pH 7.4), 150 mmol/L NaCl, and 0.1% Tween 20] for 1 h, incubated with various primary antibodies overnight at 4°C and incubated with secondary antibodies for 1 h at room temperature. Immunoreactive products were detected by chemiluminescence with an enhanced chemiluminescence system.

Co-immunoprecipitation (Co-IP) assays

For Co-IP assay, cells grown in 10-cm dishes were transfected with various plasmids for 24 h. After transfection and treatments, cells were harvested. Cell lysates were incubated with the appropriate antibody for 2 h, and subsequently incubated with protein A-Sepharose beads for 2 h. The protein-antibody complexes recovered on beads were subjected to Western blotting using appropriate antibodies after separation by SDS-polyacrylamide gel electrophoresis.

Ligand-binding competition assays

The His-tagged human RXRα-LBD (223–462) was incubated in tubes with unlabeled 9-*cis*-RA or different concentrations of compounds in 200 μL binding buffer [0.15 mol/L KCl, 10 mmol/L Tris·HCl (pH 7.4), 8% glycerol, 0.5% CHAPS] at 4°C for 1 h. [^{3}H]-9-*cis*-RA was added to the tubes to a final concentration of 7.5 nmol/L and final volume of 300 μL then incubated overnight at 4°C. The RXRα-LBD was then captured by nickel-coated beads. Bound [^{3}H]-9-*cis*-RA was quantitated by liquid scintillation counting.

Colony formation assay

HeLa cells were seeded in 6-well plate (500 cells/well) for 3 days, and were treated with N-6 (5 μmol/L) and TNFα (10 ng/mL) alone or in combination in 1% serum medium for 3 days. Cells were fixed with 4% paraformaldehyde. Colonies were stained with 0.1% crystal violet and counted.

Tunel staining

MCF-7 cells were fixed in 4% paraformaldehyde and permeabilized with 1% Triton X-100 in phosphate-buffered saline. Cells were blocked using 3% H_2O_2 (80 mL methanol + 10 mL ddH_2O + 10 mL 30% H_2O_2) for 10 min. TUNEL reaction mixture was obtained by adding terminal deoxynucleotidyl transferase and Biotin-11-dUTP. Each sample was then incubated with TUNEL reaction mixture in a humidified chamber at 37°C in dark for 60 min, and then labeled with Streptavidin-HRP. DAB staining was used for coloration. Nucleus was stained by hematoxylin.

Virtual ligand screening

Virtual ligand screening (VLS) of approximate 200,000 compound library of the Developmental Therapeutics Program (DTP) NCI/NIH (http://dtp.nci.nih.gov) was performed using Q-MOL molecular modeling package (Q-MOL L.L.C., San Diego, CA, USA; www.q-mol.com) (Shiryaev et al., 2011; Remacle et al., 2012; Shiryaev et al., 2012). The Optimized Potential for Liquid Simulations (OPLS) all atom force field (Jorgensen et al., 1996) is uniformly utilized within the Q-MOL program. The ligand docking simulations were conducted using the RXRα crystal structure coordinates from 3FUG PDB. The docking site was prepared as an area encompassing both the ligand binding pocket and the peptide cofactor binding site. The protein molecule preparation included adding of hydrogen atoms and the assignment of the standard OPLS atom types. To increase the speed of calculations and to incorporate implicitly the flexibility of RXR, the protein molecule was treated as a set of grid-based potentials accounting for the relevant protein-ligand interactions. The ligands were docked into the grid-based potentials using the Monte Carlo simulation in the internal coordinate space as implemented in the Q-MOL program. The preparation of each ligand for docking simulation initially included an automatic OPLS atom type assignment and conversion of the 2D sketch-like models (input as the MDL MOL format) into the 3D molecular models. The full-atom ligand structure was then minimized using the Q-MOL small molecule minimization protocol. The protocol combines minimization in both internal and Cartesian coordinates to properly optimize the rotable bonds of a small molecule. In the course of VLS, the compounds were minimally filtered by applying lower molecular mass cut-off of 220 Da, a polyphenols sub-structure filter, and a filter detecting chlorine atoms attached to aliphatic carbons. Because of the stochastic nature of the Q-MOL docking protocol, each ligand was docked at least three times. The best energy conformers with the lowest binding energy were then selected. To differentiate between the true and false binders, Q-MOL VLS uses a proprietary protein-ligand binding energy evaluation function. This function is based on the re-parameterized OPLS force field, and, in addition to the protein-ligand interactions, it accounts for the internal energy change of the docked ligand. Upon completion of VLS, the hits were ranked by relative binding energy, and the best (lowest) energy hits were then selected. The selected hits were visually inspected and the highly symmetric or heavily halogenated ligands and the high molecular weight compounds were discarded and not analyzed further. Out of 100 predicted hits, only 11 were selected and ordered from NCI DTP for *in vitro* testing.

Modeling of protein-ligand complex

The structure of RXRα was retrieved from crystal complex LG100754-RXRα (PDB entry: 3a9e). Glide was used to study the interaction of ligand-RXRα complex (LG100754 and N-6) in the suite of Schrodinger with default docking parameter settings. The interaction image was produced by the Pymol software.

Statistical analysis

Data were expressed as mean ± SD. Each assay was repeated in triplicate in three independent experiments. Statistical significance of differences between groups was analyzed by using Student's *t* test. $P < 0.05$ was considered significant.

ACKNOWLEDGEMENTS

This work was supported by the National Natural Science Foundation of China (Grant Nos. NSFC-31271453, NSFC-31471318, NSFC-91129302, NSFC-91429306, U1405229, NSFC-31471315, and NSFC-81370097); the Fundamental Research Funds for the Central Universities (2013121038); Funding from Xiamen Oceanic Administration (14PYY051SF04); Zhejiang Provincial Natural Science Foundation of China (R14C070002); the National Institutes of Health (CA179379); and the California Breast Cancer Research Program (20IB-0138).

ABBREVIATIONS

9-*cis*-RA, 9-*cis*-retinoic acid; Dex, dexamethasone; GR, glucocorticoid receptor; IP, immunoprecipitate; LXR, liver X receptor; N-6, NSC-640358; NR, nuclear receptor; PARP, poly (ADP-ribose) polymerase; PPAR, peroxisome proliferator-activated receptor; RARγ, retinoic acid receptor γ; RXRα, retinoid X receptor α; Ros, rosiglitazone; siRNA, small interference RNA; SPR, surface plasmon resonance; T09, T0901317; TNFα, tumor necrosis factor α; tRXRα, N-terminally truncated RXRα; TUNEL, terminal deoxynucleotidyltransferase dUTP nick end labeling.

COMPLIANCE WITH ETHICS GUIDELINES

Fan Chen, Jiebo Chen, Jiacheng Lin, Anton V Cheltsov, Lin Xu, Ya Chen, Zhiping Zeng, Liqun Chen, Mingfeng Huang, Mengjie Hu, Xiaohong Ye, Yuqi Zhou, Guanghui Wang, Ying Su, Long Zhang, Fangfang Zhou, Xiao-kun Zhang, and Hu Zhou declare that they have no conflict of interest.

This article does not contain any studies with human or animal subjects performed by the any of the authors.

REFERENCES

Altucci L, Leibowitz MD, Ogilvie KM, de Lera AR, Gronemeyer H (2007) RAR and RXR modulation in cancer and metabolic disease. Nat Rev Drug Discov 6:793–810

Balkwill F (2009) Tumour necrosis factor and cancer. Nat Rev Cancer 9:361–371

Cao X, Liu W, Lin F, Li H, Kolluri SK, Lin B, Han YH, Dawson MI, Zhang XK (2004) Retinoid X receptor regulates Nur77/TR3-dependent apoptosis [corrected] by modulating its nuclear export and mitochondrial targeting. Mol Cell Biol 24:9705–9725

Casas F, Daury L, Grandemange S, Busson M, Seyer P, Hatier R, Carazo A, Cabello G, Wrutniak-Cabello C (2003) Endocrine regulation of mitochondrial activity: involvement of truncated RXRalpha and c-Erb Aalpha1 proteins. FASEB J 17:426–436

Dawson MI, Xia Z (2012) The retinoid X receptors and their ligands. Biochim Biophys Acta 1821:21–56

Dawson MI, Hobbs PD, Peterson VJ, Leid M, Lange CW, Feng KC, Chen G, Gu J, Li H, Kolluri SK et al (2001) Apoptosis induction in cancer cells by a novel analogue of 6-[3-(1-adamantyl)-4-hydroxyphenyl]-2-naphthalenecarboxylic acid lacking retinoid receptor transcriptional activation activity. Cancer Res 61:4723–4730

de Lera AR, Bourguet W, Altucci L, Gronemeyer H (2007) Design of selective nuclear receptor modulators: RAR and RXR as a case study. Nat Rev Drug Discov 6:811–820

Egea PF, Mitschler A, Rochel N, Ruff M, Chambon P, Moras D (2000) Crystal structure of the human RXRalpha ligand-binding domain bound to its natural ligand: 9-*cis* retinoic acid. EMBO J 19:2592–2601

Egea PF, Mitschler A, Moras D (2002) Molecular recognition of agonist ligands by RXRs. Mol Endocrinol 16:987–997

Evans RM, Mangelsdorf DJ (2014) Nuclear receptors, RXR, and the big bang. Cell 157:255–266

Gao W, Liu J, Hu M, Huang M, Cai S, Zeng Z, Lin B, Cao X, Chen J, Zeng JZ et al (2013) Regulation of proteolytic cleavage of retinoid X receptor-alpha by GSK-3beta. Carcinogenesis 34:1208–1215

Ghose R, Zimmerman TL, Thevananther S, Karpen SJ (2004) Endotoxin leads to rapid subcellular re-localization of hepatic RXRalpha: a novel mechanism for reduced hepatic gene expression in inflammation. Nucl Recept 2:4

Gronemeyer H, Gustafsson JA, Laudet V (2004) Principles for modulation of the nuclear receptor superfamily. Nat Rev Drug Discov 3:950–964

Huang J, Powell WC, Khodavirdi AC, Wu J, Makita T, Cardiff RD, Cohen MB, Sucov HM, Roy-Burman P (2002) Prostatic intraepithelial neoplasia in mice with conditional disruption of the retinoid X receptor alpha allele in the prostate epithelium. Cancer Res 62:4812–4819

Jiang SY, Shen SR, Shyu RY, Yu JC, Harn HJ, Yeh MY, Lee MM, Chang YC (1999) Expression of nuclear retinoid receptors in normal, premalignant and malignant gastric tissues determined by in situ hybridization. Br J Cancer 80:206–214

Jorgensen WL, Maxwell DS, TiradoRives J (1996) Development and testing of the OPLS all-atom force field on conformational energetics and properties of organic liquids. J Am Chem Soc 118:11225–11236

Katagiri Y, Takeda K, Yu ZX, Ferrans VJ, Ozato K, Guroff G (2000) Modulation of retinoid signalling through NGF-induced nuclear export of NGFI-B. Nat Cell Biol 2:435–440

Kolluri SK, Bruey-Sedano N, Cao X, Lin B, Lin F, Han YH, Dawson MI, Zhang XK (2003) Mitogenic effect of orphan receptor TR3 and its regulation by MEKK1 in lung cancer cells. Mol Cell Biol 23:8651–8667

Lee SO, Li X, Khan S, Safe S (2011) Targeting NR4A1 (TR3) in cancer cells and tumors. Expert Opin Ther Targets 15:195–206

Lefebvre P, Benomar Y, Staels B (2010) Retinoid X receptors: common heterodimerization partners with distinct functions. Trends Endocrinol Metab 21:676–683

Li H, Kolluri SK, Gu J, Dawson MI, Cao X, Hobbs PD, Lin B, Chen G, Lu J, Lin F et al (2000a) Cytochrome c release and apoptosis induced by mitochondrial targeting of nuclear orphan receptor TR3. Science 289:1159–1164

Li M, Indra AK, Warot X, Brocard J, Messaddeq N, Kato S, Metzger D, Chambon P (2000b) Skin abnormalities generated by temporally controlled RXRalpha mutations in mouse epidermis. Nature 407:633–636

Lin B, Kolluri SK, Lin F, Liu W, Han YH, Cao X, Dawson MI, Reed JC, Zhang XK (2004) Conversion of Bcl-2 from protector to killer by interaction with nuclear orphan receptor Nur77/TR3. Cell 116:527–540

Lotan Y, Xu XC, Shalev M, Lotan R, Williams R, Wheeler TM, Thompson TC, Kadmon D (2000) Differential expression of nuclear retinoid receptors in normal and malignant prostates. J Clin Oncol 18:116–121

Matsushima-Nishiwaki R, Okuno M, Adachi S, Sano T, Akita K, Moriwaki H, Friedman SL, Kojima S (2001) Phosphorylation of retinoid X receptor alpha at serine 260 impairs its metabolism and function in human hepatocellular carcinoma. Cancer Res 61:7675–7682

Mocellin S, Nitti D (2008) TNF and cancer: the two sides of the coin. Front Biosci 13:2774–2783

Moll UM, Marchenko N, Zhang XK (2006) p53 and Nur77/TR3: transcription factors that directly target mitochondria for cell death induction. Oncogene 25:4725–4743

Nagaya T, Murata Y, Yamaguchi S, Nomura Y, Ohmori S, Fujieda M, Katunuma N, Yen PM, Chin WW, Seo H (1998) Intracellular proteolytic cleavage of 9-*cis*-retinoic acid receptor alpha by cathepsin L-type protease is a potential mechanism for modulating thyroid hormone action. J Biol Chem 273:33166–33173

Nomura Y, Nagaya T, Yamaguchi S, Katunuma N, Seo H (1999) Cleavage of RXRalpha by a lysosomal enzyme, cathepsin L-type protease. Biochem Biophys Res Commun 254:388–394

Perez E, Bourguet W, Gronemeyer H, de Lera AR (2012) Modulation of RXR function through ligand design. Biochim Biophys Acta 1821:57–69

Remacle AG, Golubkov VS, Shiryaev SA, Dahl R, Stebbins JL, Chernov AV, Cheltsov AV, Pellecchia M, Strongin AY (2012) Novel MT1-MMP small-molecule inhibitors based on insights into hemopexin domain function in tumor growth. Cancer Res 72:2339–2349

Sato Y, Ramalanjaona N, Huet T, Potier N, Osz J, Antony P, Peluso-Iltis C, Poussin-Courmontagne P, Ennifar E, Mely Y et al (2010) The "Phantom Effect" of the Rexinoid LG100754: structural and functional insights. PLoS One 5:e15119

Shiryaev SA, Cheltsov AV, Gawlik K, Ratnikov BI, Strongin AY (2011) Virtual ligand screening of the National Cancer Institute (NCI) compound library leads to the allosteric inhibitory scaffolds of the West Nile Virus NS3 proteinase. Assay Drug Dev Technol 9:69–78

Shiryaev SA, Cheltsov AV, Strongin AY (2012) Probing of exosites leads to novel inhibitor scaffolds of HCV NS3/4A proteinase. PLoS One 7:e40029

Shulman AI, Mangelsdorf DJ (2005) Retinoid X receptor heterodimers in the metabolic syndrome. N Engl J Med 353:604–615

Szanto A, Narkar V, Shen Q, Uray IP, Davies PJ, Nagy L (2004) Retinoid X receptors: X-ploring their (patho)physiological functions. Cell Death Differ 11(Suppl 2):S126–S143

Takiyama Y, Miyokawa N, Sugawara A, Kato S, Ito K, Sato K, Oikawa K, Kobayashi H, Kimura S, Tateno M (2004) Decreased expression of retinoid X receptor isoforms in human thyroid carcinomas. J Clin Endocrinol Metab 89:5851–5861

Tang XH, Gudas LJ (2011) Retinoids, retinoic acid receptors, and cancer. Annu Rev Pathol 6:345–364

Thomas M, Sukhai MA, Kamel-Reid S (2012) An emerging role for retinoid X receptor alpha in malignant hematopoiesis. Leuk Res 36:1075–1081

Wang Z, Benoit G, Liu J, Prasad S, Aarnisalo P, Liu X, Xu H, Walker NP, Perlmann T (2003) Structure and function of Nurr1 identifies a class of ligand-independent nuclear receptors. Nature 423:555–560

Wang GH, Jiang FQ, Duan YH, Zeng ZP, Chen F, Dai Y, Chen JB, Liu JX, Liu J, Zhou H et al (2013) Targeting truncated retinoid X receptor-alpha by CF31 induces TNF-alpha-dependent apoptosis. Cancer Res 73:307–318

Wansa KD, Harris JM, Muscat GE (2002) The activation function-1 domain of Nur77/NR4A1 mediates trans-activation, cell specificity, and coactivator recruitment. J Biol Chem 277:33001–33011

Waters JP, Pober JS, Bradley JR (2013) Tumour necrosis factor and cancer. J Pathol 230:241–248

Zhang XK, Hoffmann B, Tran PB, Graupner G, Pfahl M (1992a) Retinoid X receptor is an auxiliary protein for thyroid hormone and retinoic acid receptors. Nature 355:441–446

Zhang XK, Lehmann J, Hoffmann B, Dawson MI, Cameron J, Graupner G, Hermann T, Tran P, Pfahl M (1992b) Homodimer formation of retinoid X receptor induced by 9-*cis* retinoic acid. Nature 358:587–591

Zhang H, Li L, Chen L, Hu L, Jiang H, Shen X (2011a) Structure basis of bigelovin as a selective RXR agonist with a distinct binding mode. J Mol Biol 407:13–20

Zhang H, Xu X, Chen L, Chen J, Hu L, Jiang H, Shen X (2011b) Molecular determinants of magnolol targeting both RXRalpha and PPARgamma. PLoS One 6:e28253

Zhang H, Zhou R, Li L, Chen J, Chen L, Li C, Ding H, Yu L, Hu L, Jiang H et al (2011c) Danthron functions as a retinoic X receptor antagonist by stabilizing tetramers of the receptor. J Biol Chem 286:1868–1875

Zhou H, Liu W, Su Y, Wei Z, Liu J, Kolluri SK, Wu H, Cao Y, Chen J, Wu Y et al (2010) NSAID sulindac and its analog bind RXRalpha and inhibit RXRalpha-dependent AKT signaling. Cancer Cell 17:560–573

Zimmerman TL, Thevananther S, Ghose R, Burns AR, Karpen SJ (2006) Nuclear export of retinoid X receptor alpha in response to interleukin-1beta-mediated cell signaling: roles for JNK and SER260. J Biol Chem 281:15434–15440

16

The superior fault tolerance of artificial neural network training with a fault/noise injection-based genetic algorithm

Feng Su[1,2,3], **Peijiang Yuan**[1], **Yangzhen Wang**[2,3], **Chen Zhang**[2,3]✉

[1] Robotics Institute, Beihang University, Beijing 100191, China
[2] State Key Laboratory of Membrane Biology, School of Life Sciences, Beijing 100871, China
[3] PKU-IDG/McGovern Institute for Brain Research, Peking University, Beijing 100871, China
✉ Correspondence: ch.zhang@pku.edu.cn (C. Zhang)
Received June 20, 2016 Accepted July 12, 2016

ABSTRACT

Artificial neural networks (ANNs) are powerful computational tools that are designed to replicate the human brain and adopted to solve a variety of problems in many different fields. Fault tolerance (FT), an important property of ANNs, ensures their reliability when significant portions of a network are lost. In this paper, a fault/noise injection-based (FIB) genetic algorithm (GA) is proposed to construct fault-tolerant ANNs. The FT performance of an FIB-GA was compared with that of a common genetic algorithm, the back-propagation algorithm, and the modification of weights algorithm. The FIB-GA showed a slower fitting speed when solving the exclusive OR (XOR) problem and the overlapping classification problem, but it significantly reduced the errors in cases of single or multiple faults in ANN weights or nodes. Further analysis revealed that the fit weights showed no correlation with the fitting errors in the ANNs constructed with the FIB-GA, suggesting a relatively even distribution of the various fitting parameters. In contrast, the output weights in the training of ANNs implemented with the use the other three algorithms demonstrated a positive correlation with the errors. Our findings therefore indicate that a combination of the fault/noise injection-based method and a GA is capable of introducing FT to ANNs and imply that the distributed ANNs demonstrate superior FT performance.

KEYWORDS artificial neural networks, fault tolerance, genetic algorithm

INTRODUCTION

The brain is composed of biological neural networks (BNNs) that contain billions of interconnecting neurons with the ability to perform computations. Artificial neural networks (ANNs), mathematical models that mimic BNNs, are typically built as structured node groups with activation functions and connection weights that are adjusted based on the applied learning rules (Hampson, 1991, 1994; Basheer and Hajmeer, 2000; Krogh, 2008). Because of their powerful computational and learning abilities, ANNs are being used increasingly in various fields, including computation, engineering, machine learning, clinical medicine, and cognitive science (Presnell and Cohen, 1993; Baxt, 1995; Dybowski and Gant, 1995; Forsstrom and Dalton, 1995; Kamimura et al., 1996; Almeida, 2002; Lisboa, 2002; Rajan and Tolley, 2005; Lisboa and Taktak, 2006; Patel and Goyal, 2007; Hu et al., 2013; Street et al., 2013; Azimi et al., 2015).

Fault tolerance (FT), an important feature of BNNs, ensures the fidelity and reality of a system's input-output relationship. The FT of BNNs is thought to rely on extensive parallel interconnections, distributed information storage and processing, and self-learning and self-organizing characteristics. For instance, Alzheimer's patients lose a significant number of neurons (sometimes equaling half the normal brain mass) but still maintain certain brain functions (Fayed et al., 2012; Li et al., 2012; Weiner et al., 2015; Pini et al.,

2016). Moreover, structural measurements of various areas of the brain have revealed that brain volume has no direct correlation with cognitive decline in patients (Braskie and Thompson, 2014). Fault tolerance is also an important consideration in the construction of ANNs, especially in highly variable or "fail-safe" systems (Protzel et al., 1993; Phatak and Koren, 1995b). A fault-tolerant ANN is a special ANN system designed to work normally, or at least to a certain degree of normalcy, even if some of its components are unexpectedly damaged. Recently, FT performance has become more important, partly due to the fact that the soft errors caused by transient faults are an unavoidable concern in very large-scale integration (VLSI) technology, whose dimension is approaching the nanoscale (Mahdiani et al., 2012).

To construct a fault-tolerant ANN, neurons (nodes) are replicated in the hidden layer (Emmerson and Damper, 1993; Medler and Dawson, 1994; Phatak and Koren, 1995a; Tchernev et al., 2005). In this way, FT is introduced to the ANNs at the expense of increased complexity. For instance, ANNs with thousands of artificial neurons and up to a million interconnections in the hidden layer are required to solve complex problems, such as mapping landslide susceptibility (Arnone et al., 2014), modeling pectus excavatum corrective prostheses (Rodrigues et al., 2014), and reconstructing traffic networks (Jiang et al., 2014). However, this increase in network complexity makes the hardware implementation of ANNs relatively difficult and inefficient. Other methods that have been proposed to build fault-tolerant ANNs include an adjustment of the distributing weight values (Cavalieri and Mirabella, 1999a) using an empirical equation to deduce the mean prediction error (Sum and Leung, 2008) and adopting two objective functions (i.e., one that deals with open-weight fault and another that deals with open node fault (Mak et al., 2011)) during the training process to improve network FT. However, to our knowledge, no studies have investigated in detail whether a genetic algorithm might enhance the FT of ANNs.

A genetic algorithm (GA) is a heuristic algorithm used to search for a non-random optimal solution to a problem by mimicking the evolutionary process of natural selection (Holland, 1975; Goldberg, 1989). The GA process is iterative and includes initialization, selection, and genetic operation. The genetic operation usually consists of inheritance, mutation, and crossover. In each iteration, which is also called a generation, a fitness function is used to evaluate the fitness of individuals to find the best solution. Thus, a GA requires only a solvable function, which makes it suitable for complex and non-linear problems. Genetic algorithms have been applied to solve a variety of problems, especially when the basic functions are not discontinuous or non-differentiable (Forrest, 1993; Maddox, 1995; Willett, 1995; Pedersen and Moult, 1996; Meurice et al., 1998; Weber, 1998; Liu and Wang, 2001; Rothlauf et al., 2002; Jamshidi, 2003; Leardi, 2007; Wu, 2007; Gerlee et al., 2011; Manning et al., 2013; Pena-Malavera et al., 2014).

Thus, this study proposes an approach that combines an FIB learning algorithm with a GA to build fault-tolerant ANNs and to demonstrate this method's superior FT performance in comparison with that of a general GA (GE-GA) and two classic existing algorithms, the back-propagation (BP) algorithm and the modification of weight (MW) methods, in solving an exclusive OR (XOR) problem or an overlapping classification problem.

RESULTS

Training ANNs to solve an XOR problem with a GA

An XOR problem was used to train the ANN with either a GE-GA or an FIB-GA. Figure 1A illustrates the architecture of the ANN, and Fig. S1 illustrates the artificial neuron model. Two classic algorithms were included in the comparisons: the back-propagation (BP) and the modification of weights (MW). The BP algorithm is a traditional learning method based on a gradient descent, and the MW algorithm modifies the weight during the learning phase if the absolute value of the weight exceeds a certain threshold. For each experiment, the training proceeded until the terminating condition (i.e., error less than 0.001 or number of iterations reaching 1,000) was satisfied. Figure 1B illustrates changes in the error (or minimum error in training with a GE-GA or FIB-GA) in one iteration versus the number of iterations. Errors with all four fitting methods declined with increasing iterations. The BP and MW methods reached the terminating condition much faster (BP: 3.0 ± 0.0, MW: 3.8 ± 0.8; GA: 610.6 ± 274.8; FT: 905.6 ± 148.4 iterations) compared with the other two methods. Furthermore, the $Error_{BP}$, $Error_{MW}$ and $Error_{GE-GA}$ decayed approximately in an exponential manner (τ = 5.6196 ± 0.8967, fitting function: $f(x) = a \cdot e^{-\tau x}$, 4.1171 ± 0.6494, and 0.0186 ± 0.0050 $iteration^{-1}$, respectively); however, the $Error_{FIB-GA}$ decayed much more slowly and in an irregular manner (Fig. 1B), suggesting low efficiency in the parameter optimization process. As there are four elements in the output vector ($c = [c_1\ c_2\ c_3\ c_4]$) and the calculated error in one iteration comprises the average from all four individual elements, we also examined, in each training period, the error of each element that is given by

$$Error - c_i = c_i^{calculated} - c_i^{actual} \quad (i = 1, 2, 3, 4).$$

Figure 1C illustrates the fluctuations in the error of each element during 20 independent trainings. The average fluctuations during the training of ANNs with BP, MW, GE-GA, and FIB-GA were BP: 0.0001 ± 0.0000, MW: 0.0001 ± 0.0000, GE-GA: 0.0007 ± 0.0009, and FIB-GA: 0.0024 ± 0.0025, respectively. Statistical analysis revealed that the FIB-GA method showed the biggest fluctuation when compared with the other three methods (Table S1). Taken together, all four methods demonstrated the capability of training the ANN successfully, although at different speeds.

Typically, in an FT ANN, the impact of each node is distributed as evenly as possible so as to avoid *dominant*

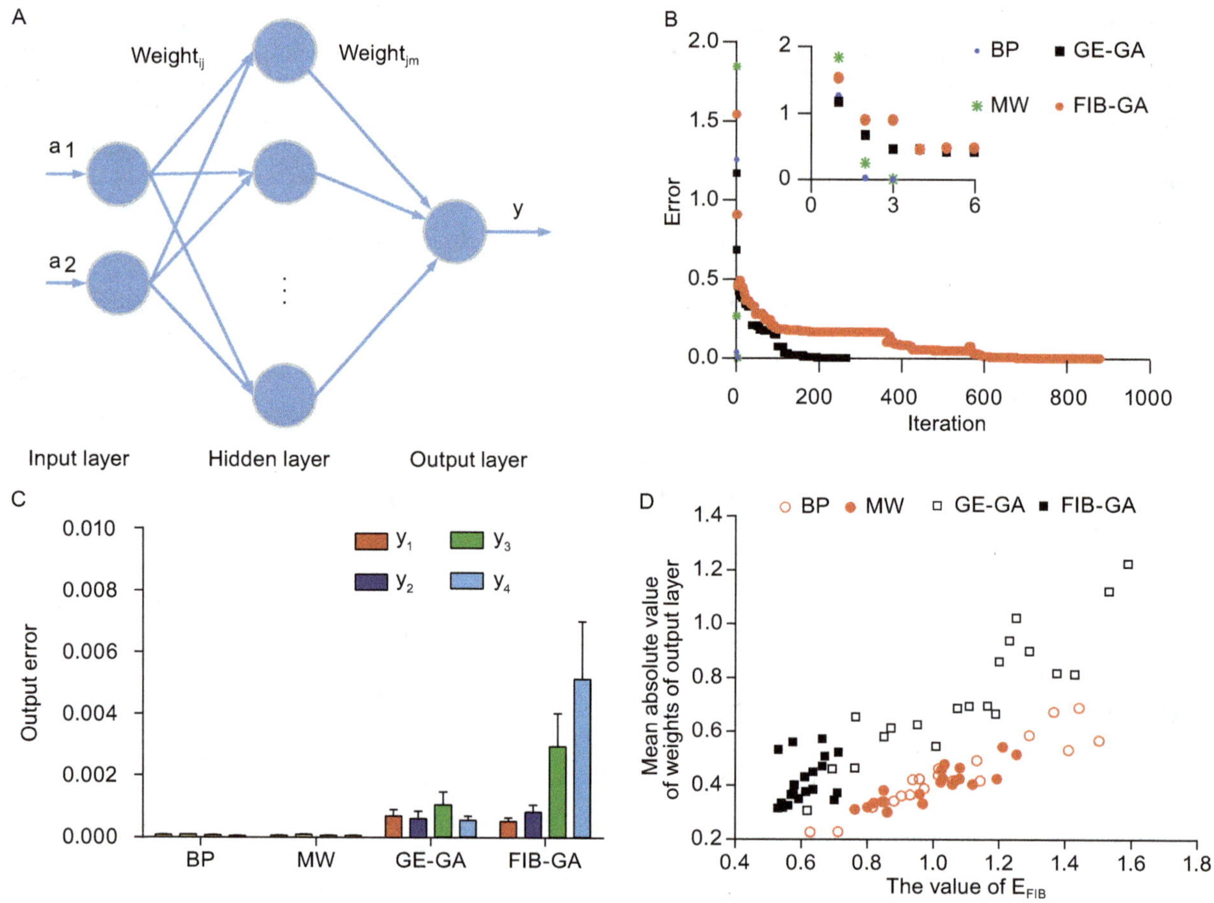

Figure 1. The use of BP, MW, GE-GA, and FIB-GA in training ANNs to solve an XOR problem. (A) The topology of the artificial neural networks. (B) The plot of the fitting errors versus the number of iterations. The inset shows the changes of the error within the initial six iterations. (C) The summary graph comparing the error of each element in the output vectors between BP ANN, MW ANN, GE-GA ANN, and FIB-GA ANN. (D) The plot of the weights for the output neuron versus the fitting errors in ANN training with the BP, MW, GE-GA, and FIB-GA.

Table 1. Correlation between each category of parameters and average errors

ANN	BP ANN				MW ANN			
Parameter	$\sqsupseteq^{\mathcal{H}}_{\mathcal{MA}}$	$\lfloor^{\mathcal{H}}_{\mathcal{MA}}$	$\sqsupseteq^{\mathcal{O}}_{\mathcal{MA}}$	$\lfloor^{\mathcal{O}}$	$\sqsupseteq^{\mathcal{H}}_{\mathcal{MA}}$	$\lfloor^{\mathcal{H}}_{\mathcal{MA}}$	$\sqsupseteq^{\mathcal{O}}_{\mathcal{MA}}$	$\lfloor^{\mathcal{O}}$
r	−0.659	0.158	0.936	−0.152	0.051	0.018	0.857	0.064
p	0.002	0.507	0	0.521	0.831	0.940	0	0.788
ANN	GE-GA ANN				FIB-GA ANN			
Parameter	$\sqsupseteq^{\mathcal{H}}_{\mathcal{MA}}$	$\lfloor^{\mathcal{H}}_{\mathcal{MA}}$	$\sqsupseteq^{\mathcal{O}}_{\mathcal{MA}}$	$\lfloor^{\mathcal{O}}$	$\sqsupseteq^{\mathcal{H}}_{\mathcal{MA}}$	$\lfloor^{\mathcal{H}}_{\mathcal{MA}}$	$\sqsupseteq^{\mathcal{O}}_{\mathcal{MA}}$	$\lfloor^{\mathcal{O}}$
r	−0.789	−0.070	0.901	0.291	−0.260	0.524	0.339	−0.258
p	0	0.768	0	0.214	0.268	0.018	0.144	0.272

nodes. Thus, we evaluated the correlation index between different ANN parameters and the fitting errors. All 25 parameters were grouped into four categories: 12 weights ($weight_{ij}$) and six biases ($bias_j$) for neurons in the hidden layer and six weights ($weight_{jm}$) and one bias ($bias_m$) for the output neuron. Table 1 summarizes the correlation efficiency and significance between each category of the parameters and errors. The weights for the neurons in the hidden-layer

ANN training using the BP and the GE-GA strongly correlated negatively with the fitting error; however, those in the other two ANNs did not. The bias for the output neuron has no significant correlation with the fitting errors in all four algorithms. All the output neuron weights in the ANN training with the GE-GA, BP, and MW strongly correlated with the errors (Fig. 1D); however, those in the FIB-GA ANN training did not. These results showed that in the ANN training with the FIB-GA, no parameter set correlated with the fitting error, a finding that implies there is no dominant parameter in ANNs trained via the use of an FIB-GA.

The FT performance of ANNs in solving an XOR problem with a single fault

Fault tolerance is the property that allows an ANN or BNN to operate properly in the event one or more components are lost. We began by comparing the errors among the ANNs generated by the BP, MW, GE-GA, and FIB-GA methods in which one randomly selected network parameter was changed to 0 (void). The plot of the errors versus the faulty parameters in 20 independent experiments clearly shows that the ANNs constructed using the FIB-GA contains the least number of errors (Fig. 2A). The averaged errors from 20 independent experiments are as follows: BP: 0.2623 ± 0.0614, MW: 0.2507 ± 0.0355, GE-GA: 0.2746 ± 0.0698, and FIB-GA: 0.1527 ± 0.0150 (statistical test in Table S2). Assuming the error equals or exceeds 0.4 as a fault output, the error rates show a similar trend (Fig. 2B): BP: 24.80 ± 9.68%, MW: 21.60 ± 7.94%, GE-GA: 23.80 ± 10.66%, and FIB-GA: 8.60 ± 5.24% (statistical test in Table S2). Next, the FT performances of the four ANNs were compared when one neuron (rather than one parameter) in the hidden layer completely lost its responsiveness. The performances of all four ANNs were reduced when compared with the fully functional ANNs, while the FIB-GA ANNs showed the fewest errors (Fig. 2C): BP: 0.3900 ± 0.1041, MW: 0.3645 ± 0.0567, GE-GA: 0.5167 ± 0.1413, and FIB-GA: 0.2936 ± 0.0410; (statistical test in Table S3) and the lowest error rates, with a 0.4 threshold (Fig. 2D): BP: 45.00 ± 19.57%, MW: 40.83 ± 13.76%, GE-GA: 54.17 ± 20.86%, and FIB-GA: 27.50 ± 13.55% (statistical test in Table S3).

As the output matrix is composed of four elements in the XOR problem, the errors of the individual elements were compared. The distributions of Error $_{-}c_i$ among the four ANNs were plotted while voiding one parameter or one neuron in the hidden layer (Fig. S2). Among the four algorithms, ANN training with the FIB-GA consistently showed the least number of errors (Fig. S2A): BP: 0.2623 ± 0.0614, MW: 0.2507 ± 0.0355, GE-GA: 0.2746 ± 0.0698, and FIB-GA: 0.1527 ± 0.0150 (statistical test in Table S4) and the lowest error rate (Fig. S2B): BP: 25.40 ± 7.94%, MW: 25.90 ± 5.39%, GE-GA: 25.75 ± 7.35%, and FIB-GA: 14.85 ± 3.30% (statistical test in Table S4). When one neuron in the hidden layer was voided randomly, the average error and the error rate with the 0.4 threshold showed a similar trend (Fig. S2C): average error: BP: 0.3900 ± 0.1041, MW: 0.3645 ± 0.0567, GE-GA: 0.5167 ± 0.1413, and FIB-GA: 0.2936 ± 0.0410 (statistical test in Table S5). Figure S2D shows the error rate with a 0.4 threshold: BP: 43.75 ± 17.34%, MW: 42.50 ± 12.72%, GE-GA: 48.13 ± 16.36%, and FIB-GA: 31.25 ± 6.55% (statistical test in Table S5). Together, these results clearly show that the FIB-GA ANN has superior FT when one parameter or one neuron in the network is lost.

The FT performance of ANNs in solving an XOR problem with multiple faults

In both ANN and BNN, errors typically happen at multiple sites but are not restricted to one element. Thus, the performances of ANNs were compared to solve an XOR problem in which two to four parameters are disabled simultaneously. As each ANN has 25 parameters, there are 300 (C_{25}^{2}), 2,300 (C_{25}^{3}), and 12,650 (C_{25}^{4}) combinations when two, three, and four parameters are all set to 0, respectively. Figure 3A illustrates the distribution of error; the summarized data clearly show that the FIB-GA-trained ANN still performed best under multiple-fault conditions (Table 2; statistical test in Tables S6–7). Next, we examined the errors occurring when two to six neurons in the hidden layer are voided. Under these circumstances, the ANN trained using the GE-GA showed the largest number of errors, while the ANN trained using the FIB-GA demonstrated the best performance (Fig. 3B). The error rates with a 0.4 threshold displayed the same order in the fitting performance (Fig. 3C). Not surprisingly, the performance of the ANNs trained using the FIB-GA was significantly better than the other three ANNs; however, the performance of the FIB-GA-trained ANNs weakened as the number of voided neurons increased (Table 3 and Fig. S3; statistical test in Tables S8–9). Since the performance of ANNs highly relies on the number of nodes in the hidden layer(Xu and Xu, 2013; Sasakawa et al., 2014). we next investigated whether the number of hidden neurons could affect the FT performance of the four ANNs in solving the XOR problem. In the ANNs with three or nine neurons in the hidden layer, the FIB-GA-trained ANN continued to demonstrate an FT performance that was superior to that of the BP, MW, and GE-GA ANNs (three neurons: Table 4; statistical test in Tables S10–11; nine neurons: Table 5; statistical test in Tables S12–13).

Together, these results demonstrate that the FIB-GA ANN has superior FT in solving XOR problems when multiple parameters or neurons in the network are lost.

The FT performance of ANNs in solving an overlapping classification problem

Next, we examined the FT performance of the four ANNs in solving an overlapping classification problem (Fig. 4A) that is more complicated than an XOR problem. Solving

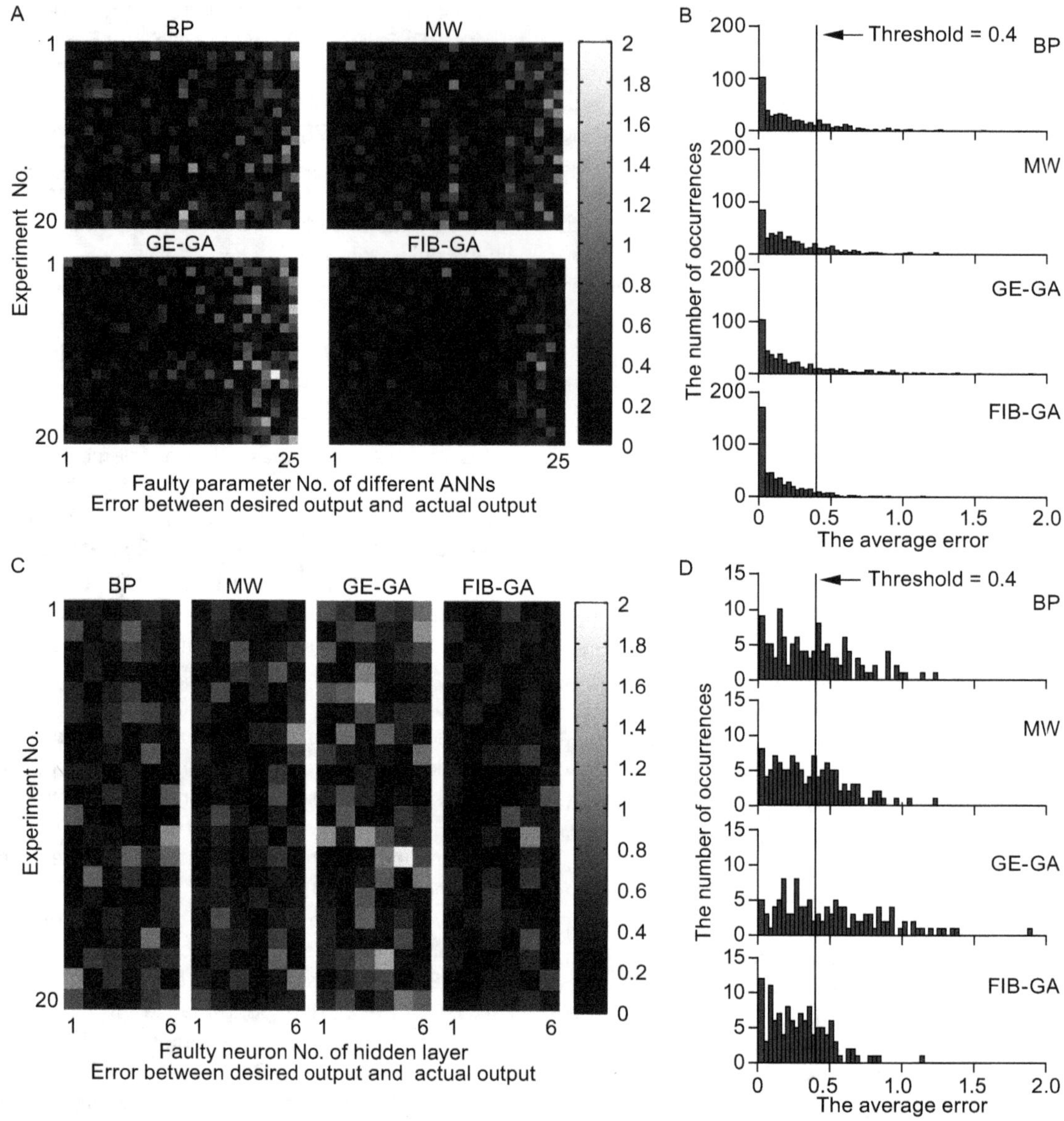

Figure 2. The FT performance of ANNs in solving an XOR problem with a single faulty parameter or neuron in the hidden layer. (A and B) The plot of errors versus the faulty parameter (A) and a histogram of error occurrence (B) in 20 independent experiments using ANNs trained via BP, MW, GE-GA, and FIB-GA methods. (C and D) The plot of errors versus the faulty neurons in the hidden layer (C) and a histogram of error occurrence (D) in 20 independent experiments using ANNs trained via BP, MW, GE-GA, and FIB-GA methods. In panels C and D, a fault output represents a fitting with the error equal to or exceeding a threshold of 0.4 (red lines).

overlapping classification problems using ANNs has been investigated extensively in the pattern recognition and machine-learning areas (Lovell and Bradley, 1996; Tang et al., 2010; Xiong et al., 2010). We adopted an ANN with the same structures used in these previous studies (see Fig. 1A). The terminating condition was set at 1,000 iterations, since none of the four ANNs could satisfy the condition that the fitting error must be equal to or less than 0.001 within 1,000 iterations, partly due to the complexity of the problem. Figure 4B illustrates the changes in the correct rates versus the number of iterations. The correct rates of BP and MW ANNs increased significantly faster compared with those of GE-GA and FIB-GA (fitting function: $f(x) = a \cdot e^{-\tau x}$, square class RCR: BP: 0.4662 ± 0.0006, $\tau = 0.2717$, $R^2 = 0.9880$, MW: 0.4672 ± 0.0010, $\tau = 0.2529$, $R^2 = 0.8855$, GE-GA: 0.4605 ± 0.0166, $\tau = 0.0040$, $R^2 = 0.9068$, FIB-GA: 0.4578 ± 0.0083, $\tau = 0.0030$, $R^2 = 0.8502$, and circle class RCR: BP: 0.4542 ± 0.0006, $\tau = 0.2210$, $R^2 = 0.9230$, MW: 0.4541 ± 0.0115, $\tau = 0.2581$, $R^2 = 0.8996$, GE-GA: 0.4547 ± 0.0115, $\tau = 0.0084$, $R^2 = 0.9014$, FIB-GA: 0.4605 ± 0.0074, $\tau = 0.0111$, $R^2 = 0.7846$ (statistical test in

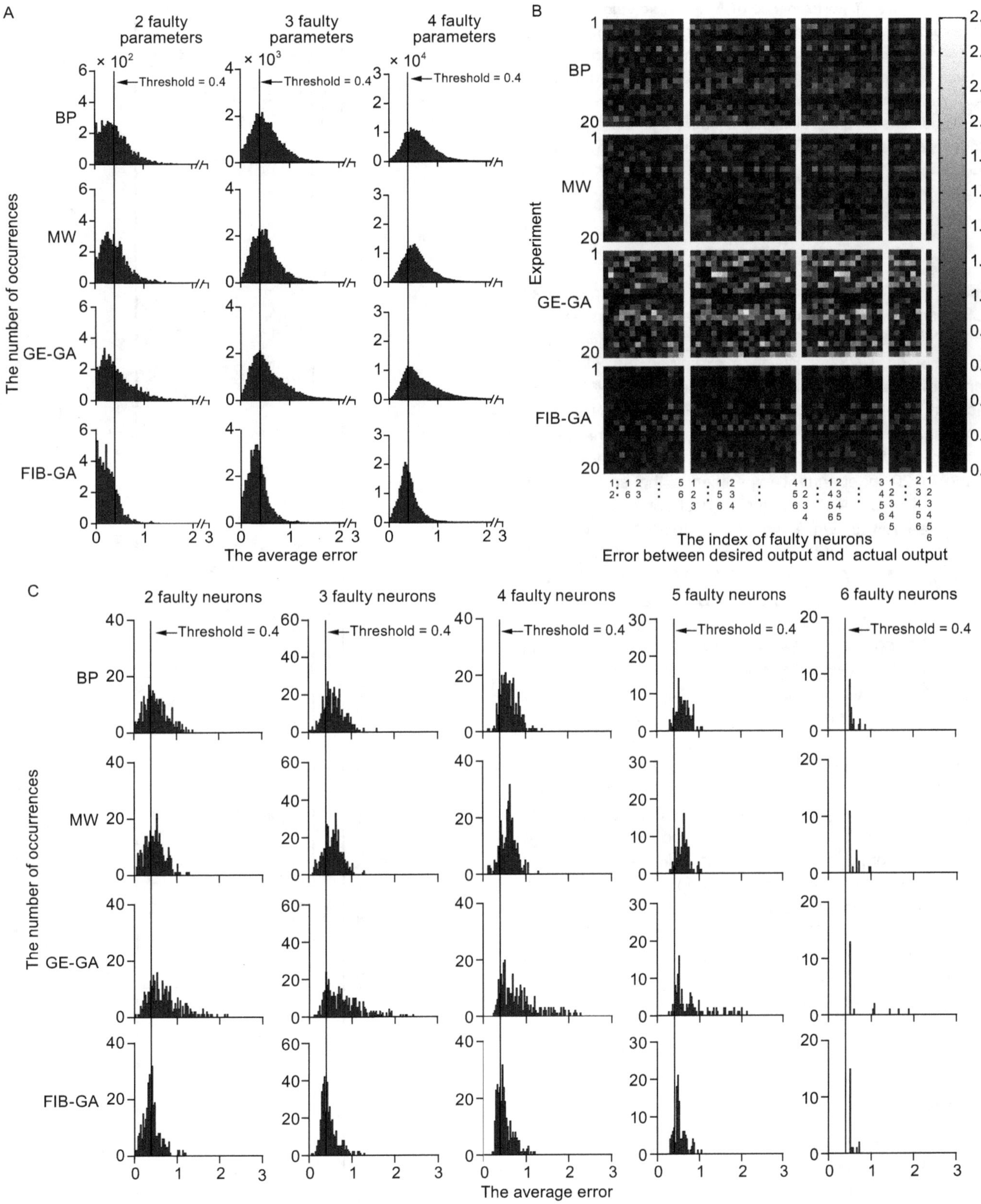
A
2 faulty parameters
3 faulty parameters
4 faulty parameters
× 10^2
× 10^3
× 10^4
Threshold = 0.4
BP
MW
GE-GA
FIB-GA
The number of occurrences
The average error
B
Experiment
The index of faulty neurons
Error between desired output and actual output
C
2 faulty neurons
3 faulty neurons
4 faulty neurons
5 faulty neurons
6 faulty neurons
Threshold = 0.4
BP
MW
GE-GA
FIB-GA
The number of occurrences
The average error

◄ **Figure 3. The FT performance of ANNs in solving an XOR problem with multiple faulty parameters or neurons in the hidden layer.** (A) The histogram of error occurrence in 20 independent experiments using ANNs trained via BP, MW, GE-GA, and FIB-GA methods. (B and C) The plot of errors versus faulty neurons (B) and a histogram of error occurrence (C) in 20 independent experiments using ANNs trained via BP, MW, GE-GA, and FIB-GA methods.

Table S14). The number of iterations that occurred when RCR reached 0.45 are shown for each method as follows: BP 23.2 ± 14.3, MW 15.8 ± 8.3, GE-GA 613.7 ± 341.8, and FIB-GA 583.1 ± 287.7. We then examined the FT performance of these four ANNs in solving an overlapping classification problem and found that the FIB-GA ANN showed significantly fewer errors when one parameter or one neuron was voided. When any of the parameters were voided one at a time in 20 independent experiments, the square class RCRs for FIB-GA ANN was the highest compared with the other three ANNs, while the other three ANNs were not significantly different (square class RCRs: BP: 0.3407 ± 0.0397, MW: 0.3536 ± 0.0317, GE-GA: 0.2863 ± 0.0501, and FIB-GA: 0.4116 ± 0.0247; circle class RCRs: BP: 0.2978 ± 0.0600, MW: 0.3075 ± 0.0301, GE-GA: 0.2151 ± 0.0601, and FIB-GA: 0.4045 ± 0.0258 (statistical test in Table S15) (Fig. 4C). When one neuron in the hidden layer was voided randomly, FIB-GA ANN still significantly outperformed the other three ANNs (square class RCRs: BP: 0.2392 ± 0.1049, MW: 0.2646 ± 0.0743, GE-GA: 0.1334 ± 0.1160, and FIB-GA: 0.4224 ± 0.0586; circle class RCRs: BP: 0.2396 ± 0.1523, MW: 0.2574 ± 0.0911, GE-GA: 0.0073 ± 0.1242, and FIB-GA: 0.4164 ± 0.0640 (statistical test in Table S16) (Fig. 4D). We next randomly voided two to three parameters or neurons in these ANNs and compared their FT performance. As illustrated in Figure 5, voiding two to three parameters or neurons reduced the performance of all four ANNs. The FIB-GA showed the fewest errors and lowest error rates under almost all the fault conditions tested (Tables S17–18). Thus, our data clearly demonstrate that, compared to the ANNs trained using the BP, MW, and GE-GA methods, the FT ability of the ANN trained using the FIB-GA, at a relatively low training speed, is superior.

DISCUSSION

This study compared the FT performances of ANNs constructed with four different algorithms: BP, MW, GE-GA, and FIB-GA. The FIB-GA was constructed via the use of FIB learning algorithms, which has been proven to be a common and efficient method of training fault-tolerant neural networks that includes the addition of noise to the input, weights, or nodes (Leung and Sum, 2008; Ho et al., 2010). Our results clearly show that the FIB learning algorithm is an efficient

Table 2. Average errors and error rates of multiple faulty parameters with six hidden neurons

Error		BP	MW	GE-GA	FIB-GA
Average error	2	0.4295 ± 0.0925	0.4092 ± 0.0538	0.4534 ± 0.1092	0.2603 ± 0.0271
	3	0.5419 ± 0.1086	0.5158 ± 0.0649	0.5752 ± 0.1334	0.3392 ± 0.0373
	4	0.6199 ± 0.1167	0.5903 ± 0.0716	0.6609 ± 0.1490	0.3983 ± 0.0453
Error rate	2	47.03% ± 13.11%	45.03% ± 9.92%	45.45% ± 15.25%	20.05% ± 7.50%
	3	63.52% ± 12.68%	62.87% ± 9.52%	62.07% ± 15.54%	31.93% ± 9.06%
	4	74.61% ± 10.75%	74.31% ± 8.18%	73.43% ± 13.86%	43.29% ± 9.75%

Table 3. Average errors and error rates of multiple faulty neurons with six hidden neurons

Error		BP	MW	GE-GA	FIB-GA
Average error	2	0.5229 ± 0.1127	0.4891 ± 0.0794	0.7041 ± 0.1804	0.4077 ± 0.0657
	3	0.5960 ± 0.1231	0.5593 ± 0.0878	0.7891 ± 0.2036	0.4665 ± 0.0837
	4	0.6227 ± 0.1210	0.5974 ± 0.0969	0.8235 ± 0.2398	0.5009 ± 0.0909
	5	0.6115 ± 0.1174	0.6133 ± 0.1176	0.8096 ± 0.3082	0.5267 ± 0.0905
	6	0.5765 ± 0.1094	0.6037 ± 0.1489	0.7618 ± 0.4398	0.5382 ± 0.0740
Error rate	2	62.67% ± 12.31%	62.00% ± 13.87%	77.67% ± 18.00%	47.33% ± 15.58%
	3	80.00% ± 13.76%	78.00% ± 12.61%	81.75% ± 16.96%	55.25% ± 16.26%
	4	88.33% ± 14.00%	86.67% ± 11.03%	89.00% ± 13.90%	65.00% ± 14.49%
	5	90.83% ± 16.64%	94.17% ± 11.18%	95.00% ± 9.52%	81.67% ± 13.13%
	6	100.00% ± 0.00%	100.00% ± 0.00%	100.00% ± 0.00%	100.00% ± 0.00%

Table 4. Error rates of multiple faulty parameters and neurons with three hidden neurons

Error		BP	MW	GE-GA	FIB-GA
Parameters	1	33.85% ± 14.21%	41.15% ± 16.03%	38.85% ± 13.08%	8.46% ± 6.06%
	2	63.91% ± 12.94%	71.15% ± 13.27%	70.51% ± 13.50%	30.77% ± 7.51%
	3	82.52% ± 8.51%	87.13% ± 7.39%	86.00% ± 9.24%	56.59% ± 4.78%
Neurons	1	71.67% ± 16.31%	75.00% ± 18.34%	86.67% ± 19.94%	18.33% ± 17.01%
	2	95.00% ± 16.31%	95.00% ± 16.31%	88.33% ± 16.31%	46.67% ± 42.44%
	3	100.00% ± 0.00%	100.00% ± 0.00%	100.00% ± 0.00%	100.00% ± 0.00%

Table 5. Error rates of multiple faulty parameters and neurons with nine hidden neurons

Error		BP	MW	GE-GA	FIB-GA
Parameters	1	22.70% ± 9.62%	23.11% ± 8.01%	20.95% ± 11.29%	1.22% ± 1.38%
	2	46.67% ± 12.70%	47.70% ± 12.56%	38.20% ± 15.48%	4.44% ± 1.91%
	3	63.45% ± 13.11%	64.35% ± 12.40%	52.43% ± 16.25%	9.57% ± 2.85%
Neurons	1	41.67% ± 15.24%	45.00% ± 14.18%	48.89% ± 20.52%	1.67% ± 4.07%
	2	60.14% ± 16.05%	65.69% ± 14.20%	66.39% ± 17.40%	10.83% ± 6.62%
	3	71.43% ± 12.67%	76.31% ± 11.55%	75.42% ± 14.32%	20.89% ± 7.74%

method for improving the FT performance of ANNs. The data of this study show that FIB-GA results in a significant improvement in errors between the actual and desired inputs when one or multiple neurons are voided. It is worth noting that, when solving an XOR problem or an overlapping classification problem with the continuous and differentiable basis function, the GE-GA does not offer an advantage over the BP and MW algorithms. This might suggest that FT is not an intrinsic property of GA ANNs. In contrast, compared to the two GAs, the BP and MW algorithms showed much faster training speeds, which implies that efficiency competes with fault tolerance in ANNs. An option for increasing the FT performances of ANNs is to avoid weights or neurons with significant effects on errors. Our analysis showed that the weights between the hidden layer and the output layer in the ANNs trained with the GE-GA, BP, or MW, but not those trained with the FIB-GA, are correlated with the output errors, a finding which clearly supports the notion that robustness is greater in a distributed ANN. This is also consistent with previous attempts to improve the partial FT of ANNs by distributing the absolute value of weights uniformly (Cavalieri and Mirabella, 1999a, b). In addition, Macia and Sole reported that degeneracy, rather than redundancy, is necessary for reliable designs of NAND (NOT AND, a binary operation) gate-forming systems(Macia and Sole, 2009). Considering the fact that the BNNs are also distributed systems yielding a high FT performance, distributed storage and processing seem to be key properties in both ANNs and BNNs. Together, our results propose that a fault/noise injection-based genetic algorithm would serve as an efficient approach for improving the FT in ANNs.

METHODS

The architecture of the ANN

In this study, a three-layer ANN was constructed: an input layer of two neurons, a hidden layer comprised of different numbers of neurons, and an output layer of one neuron. Figure 1A shows the architecture. Each neuron receives multiple weighted inputs and sends one weighted output to the connected neurons. Simply stated, neurons are interconnected through a unidirectional manner (input → hidden → output direction), and there is no connection within a layer (Fig. 1A). The input of the neuron *j* in the hidden layer is given by

$$x_j = \sum_{i=1}^{n} (\text{Input}_{ij} \times \text{weight}_{ij}) + b_j$$

where Input_{ij}, weight_{ij}, and b_j denote the input, weight, and input bias of the postsynaptic neuron *j*, which is connected to the presynaptic neuron *i*, and n denotes the number of the presynaptic neurons connecting to the neuron *j*.

The output of the neuron *j* is given by the following tansig function:

$$y_j = f(x_j) = \frac{1 - e^{-2x_j}}{1 + e^{-2x_j}}$$

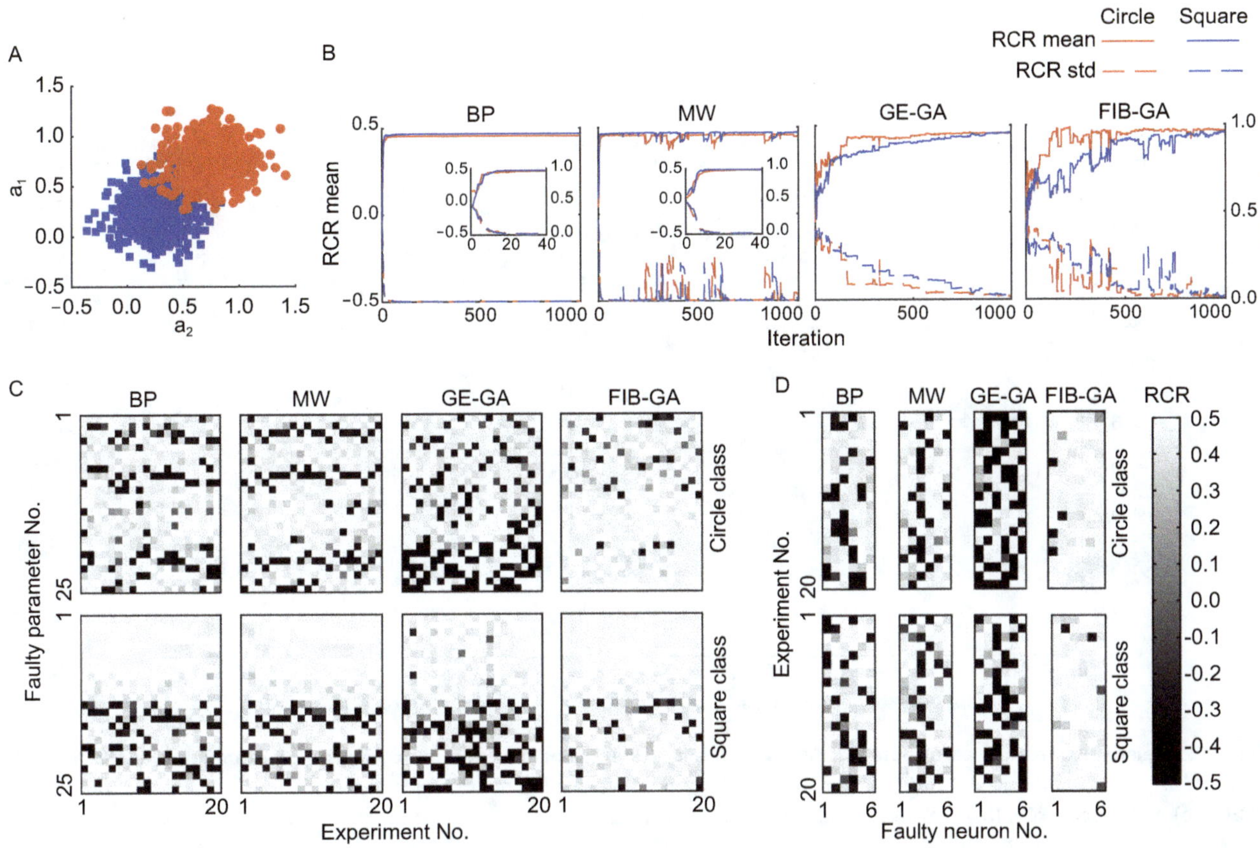

Figure 4. The FT performance of different ANNs with six hidden-layer neurons in solving an overlapping classification problem. (A) Two classes of Gaussian noise sources. (B) The plot of the relative correct rate versus the number of iterations. The inset shows the changes of the error within the initial 40 iterations. (C and D) The plot of errors versus the faulty parameters (C) and the plot of errors versus the faulty neurons in the hidden layer (D) in 20 independent experiments when using ANNs trained via BP, MW, GE-GA, and FIB-GA methods.

Calculation of the ANN to solve the XOR problem

A classic XOR problem was selected for use in training and examining the performance of the ANN. According to the architecture of the ANN used in this study (Fig. 1A), the training data, a_1 and a_2, are defined as follows:

$$a_1 = [1 \quad 0 \quad 1 \quad 0]$$

$$a_2 = [1 \quad 1 \quad 0 \quad 0]$$

Thus, the actual output of the XOR problem with these two inputs is given by

$$y = [y_1 \quad y_2 \quad y_3 \quad y_4] = a_1 \oplus a_2 = (\neg a_1 \wedge a_2) \vee (a_1 \wedge \neg a_2) = [0 \quad 1 \quad 1 \quad 0].$$

For one solution set, the output of the ANN is given by

$$c = [c_1\, c_2\, c_3\, c_4]$$

where $c_p = \sum_{j=1}^{6}(f(x_j) \times \text{weight}_{jm}) + b_m$ and x_j denote the input to the neuron j in the hidden layer from the two presynaptic neurons (a_1 and a_2), which is given by $x_j = \sum_{i=1}^{2}(a_i(p) \times \text{weight}_{ij}) + b_j$.and Weight_{jm} and b_m denote the weight and bias of the neuron m in the output layer, respectively.

Thus, the error for one solution set is given by

$$\text{Error} = \frac{1}{4}\sum_{p=1}^{4} |c_p - y_p|$$

Training of the ANN using a GE-GA

A GE-GA was adopted for training the ANN. The basic idea of a GA is to mimic the process of natural selection and to find the best solution to a problem after several generations. In this study, the upper limit of iteration was set at 1,000, and 20 individuals (i.e., sets of solutions) were used in each generation (i.e., training cycle). The best individual is defined as the one set of solutions having minimum errors. In the first generation, each individual was assigned randomly. In the subsequent generations, the 20 individuals consisted of three parts: two elite individuals (N_{elite}), which are the two individuals carried forward from the previous generation and

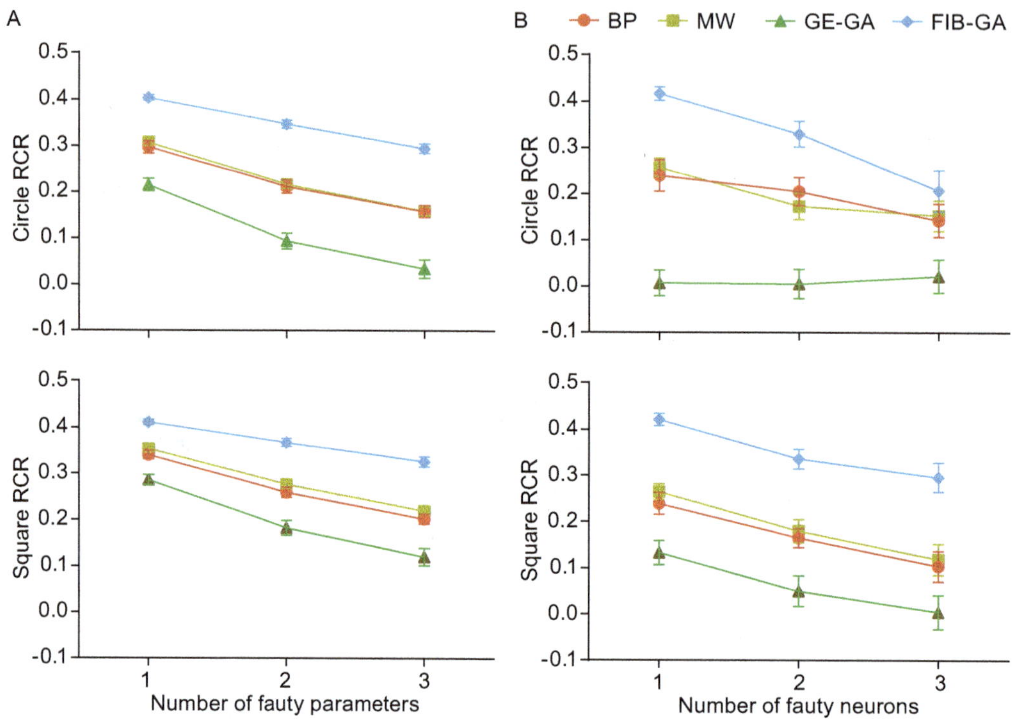

Figure 5. Relative correct rates of different ANNs with six hidden-layer neurons in solving an overlapping classification problem. (A) The plot of circle RCR (top) and square RCR (bottom) versus the number of faulty parameters. (B) The plot of circle RCR (top) and square RCR (bottom) versus the number of faulty neurons.

having the fewest errors, 14 crossover individuals ($N_{crossover}$), which are generated by combining two selected parents, and four mutation individuals ($N_{mutation}$). The selection criteria for parents is based on the scaled position ($Scaled_i$) of each individual within its generation. $Rank_i$ is redefined as the position of $Individual_i$ when sorting all the individuals in one generation by Error in ascending order. Thus, for individual i, the probability to be selected as a parent is calculated as follows:

$$P_i = Scaled_i \Big/ . \sum_{j=1}^{20} \frac{1}{\sqrt{j}} = Scaled_i/7.5953$$

where $Scaled_i = 1/.\sqrt{Rank_i}$, ($Rank_i \neq Rank_j$ if*semicoloni* $\neq$ j).

A line segment was then drawn that consisted of lines whose lengths were proportional to the P_i of each individual. A step size was given by $1/N_{parent}$, where $N_{parent} = 2 \cdot N_{crossover} + N_{mutation} = 2 \times 14 + 4 = 32$ and an initial position is denoted as $Initial_{position}$, where $0 < Initial_{position} < 1/N_{parent}$. A cursor is then placed at $Initial_{position}$ and is moved along in steps of $1/N_{parent}$. For each step, the position on which the cursor lands is selected as a parent. Thus, this algorithm generates 32 parents in one generation (Fig. S4). The crossover process generates a child by crossing two parents ($Parent_1 = [parP1_1, \cdots\cdots, parP1_{25}]$ and $Parent_2 = [parP2_1, \cdots\cdots, parP2_{25}]$) with a randomly generated binary vector $Coef = [Coe_1, \cdots\cdots, Coe_{25}]$, where Coe_i is assigned to 0 or 1 based on rounding a value that is randomly selected in the open interval (0,1). The parameter vector of the $Child = [parC_1, \cdots\cdots, parC_{25}]$ generated by the crossover is given by

$$parC_i = parP1_i \times Coe_i + parP2_i \times |Coe_i - 1|.$$

The mutation process generates a child from one parent ($Parent_1 = [parP1_1, \cdots\cdots, parP1_{25}]$) with a vector $Coef = [Coe_1, \cdots\cdots, Coe_{25}]$, where Coe_i follows a Gaussian distribution centered at 0 (Fig. S4C). The standard deviation of the Gaussian distribution in the first generation is 1, and it shrinks to 0 linearly when reaching the last generation. The parameter vector of the $Child = [parC_1, \cdots\cdots, parC_{25}]$ generated by the mutation is given by

$$parC_i = parP1_i + Coe_i.$$

The goal of training the ANN with a GE-GA is to search for the individual with the minimal $Error_{GE-GA}$ which is given by

$$Error_{GE-GA} = \frac{1}{4} \sum_{p=1}^{4} |c_p - y_p| \quad (1)$$

Training of ANN using with a FIB-GA

In addition to the GE-GA, a FIB-GA was another approach used to train the ANN. In the FIB-GA, faults on the ANN parameters were considered during the training process. Thus, the error for one set of solutions is given by

$$\mathrm{Error}_{\mathrm{FIB-GA}} = \frac{1}{25} \times \sum_{p=1}^{25} \mathrm{Error}_i \tag{2}$$

where Error_i is the error when the ith of the 25 parameters is forced to 0, assuming the corresponding parameter becomes faulty.

Overlapping classification problem

Two classes of Gaussian noise sources were considered (shown in Fig. 4A). The first class is shown as a blue square, with a mean at (a_1 , $semicolona_2$) coordinates of (0.25, 0.25). The second class is shown as a red circle, with a mean at (a_1 , $semicolona_2$) coordinates of (0.75, 0.75). Both classes have a standard deviation of 0.2. The coordinates (a_1 , $semicolona_2$) were used as input in the ANN training, and the output is 0.5 and −0.5 for the circle class and the square class, respectively. Each class had 500 scatters in total, and all the data were shuffled before the training was initiated. An actual output value larger than 0 for a point in the circle class and an actual output value less than 0 for a point in the square class were regarded correct.

For each class with N ($0 \leq N \leq 500$) points classified correctly, the relative correct rate (RCR) is defined as follows:

$$\mathrm{RCR} = \frac{N}{500} - 0.5.$$

ACKNOWLEDGEMENTS

This work was supported by grants from the National Basic Research Program (973 Program) (Nos. 2014CB942804, 2014BAI03B01, and 2012YQ0302604), Beijing Institute of Collaborative Innovation (15I-15-BJ), and the Seeding Grant for Medicine and Life Sciences of Peking University (2014-MB-11).

ABBREVIATIONS

ANNs, artificial neural networks; BP, back-propagation; FIB, fault/noise injection-based; FT, fault tolerance; GA, genetic algorithm; MW, modification of weights; VLSI, very large-scale integration; XOR, exclusive OR.

COMPLIANCE WITH ETHICS GUIDELINES

Feng Su, Peijiang Yuan, Yangzhen Wang and Chen Zhang declare that they have no conflict of interest. This article does not contain any studies with human or animal subjects performed by the any of the authors.

AUTHOR CONTRIBUTIONS

F.S., P.Y., Y.W., and C.Z. carried out the experiments; F.S., P.Y., Y. W., and C.Z. contributed to the planning of the work; F.S., P.Y., Y.W., and C.Z. wrote the paper.

REFERENCES

Almeida JS (2002) Predictive non-linear modeling of complex data by artificial neural networks. Curr Opin Biotechnol 13:72–76

Arnone E, Francipane A, Noto LV, Scarbaci A, La Loggia G (2014) Strategies investigation in using artificial neural network for landslide susceptibility mapping: application to a Sicilian catchment. J Hydroinf 16:502–515

Azimi P, Mohammadi HR, Benzel EC, Shahzadi S, Azhari S, Montazeri A (2015) Artificial neural networks in neurosurgery. J Neurol Neurosurg Psychiatry 86:251–256

Basheer IA, Hajmeer M (2000) Artificial neural networks: fundamentals, computing, design, and application. J Microbiol Methods 43:3–31

Baxt WG (1995) Application of artificial neural networks to clinical medicine. Lancet 346:1135–1138

Braskie MN, Thompson PM (2014) A focus on structural brain imaging in the Alzheimer's disease neuroimaging initiative. Biol Psychiatry 75:527–533

Cavalieri S, Mirabella O (1999a) A novel learning algorithm which improves the partial fault tolerance of multilayer neural networks. Neural Netw 12:91–106

Cavalieri S, Mirabella O (1999b) A novel learning algorithm which improves the partial fault tolerance of multilayer neural networks. Neural Netw 12:91–106

Dybowski R, Gant V (1995) Artificial neural networks in pathology and medical laboratories. Lancet 346:1203–1207

Emmerson MD, Damper RI (1993) Determining and improving the fault-tolerance of multilayer perceptrons in a pattern-recognition application. IEEE Trans Neural Netw 4:788–793

Fayed N, Modrego PJ, Salinas GR, Gazulla J (2012) Magnetic resonance imaging based clinical research in Alzheimer's disease. J Alzheimers Dis 31:S5–18

Forrest S (1993) Genetic algorithms: principles of natural selection applied to computation. Science 261:872–878

Forsstrom JJ, Dalton KJ (1995) Artificial neural networks for decision support in clinical medicine. Ann Med 27:509–517

Gerlee P, Basanta D, Anderson AR (2011) Evolving homeostatic tissue using genetic algorithms. Prog Biophys Mol Biol 106:414–425

Goldberg DE (1989) Genetic algorithms in search, optimization, and machine learning. Addison-Wesley Pub. Co, Reading

Hampson S (1991) Generalization and specialization in artificial neural networks. Prog Neurobiol 37:383–431

Hampson S (1994) Problem solving in artificial neural networks. Prog Neurobiol 42:229–281

Ho KI, Leung CS, Sum J (2010) Convergence and objective functions of some fault/noise-injection-based online learning algorithms for RBF networks. IEEE Trans Neural Netw 21:938–947

Holland JH (1975) Adaptation in natural and artificial systems: an introductory analysis with applications to biology, control, and artificial intelligence. University of Michigan Press, Ann Arbor

Hu X, Cammann H, Meyer HA, Miller K, Jung K, Stephan C (2013) Artificial neural networks and prostate cancer–tools for diagnosis and management. Nat Rev Urol 10:174–182

Jamshidi M (2003) Tools for intelligent control: fuzzy controllers, neural networks and genetic algorithms. Philos Trans R Soc Lond A 361:1781–1808

Jiang DD, Zhao ZY, Xu ZZ, Yao CP, Xu HW (2014) How to reconstruct end-to-end traffic based on time-frequency analysis and artificial neural network. Aeu-Int J Electron Commun 68:915–925

Kamimura R, Konstantinov K, Stephanopoulos G (1996) Knowledge-based systems, artificial neural networks and pattern recognition: applications to biotechnological processes. Curr Opin Biotechnol 7:231–234

Krogh A (2008) What are artificial neural networks? Nat Biotechnol 26:195–197

Leardi R (2007) Genetic algorithms in chemistry. J Chromatogr A 1158:226–233

Leung CS, Sum JP (2008) A fault-tolerant regularizer for RBF networks. IEEE Trans Neural Netw 19:493–507

Li J, Pan P, Huang R, Shang H (2012) A meta-analysis of voxel-based morphometry studies of white matter volume alterations in Alzheimer's disease. Neurosci Biobehav Rev 36:757–763

Lisboa PJ (2002) A review of evidence of health benefit from artificial neural networks in medical intervention. Neural Netw 15:11–39

Lisboa PJ, Taktak AF (2006) The use of artificial neural networks in decision support in cancer: a systematic review. Neural Netw 19:408–415

Liu F, Wang J (2001) Genetic algorithms and its application to spectral analysis. Guang Pu Xue Yu Guang Pu Fen Xi 21:331–335

Lovell BC, Bradley AP (1996) The multiscale classifier. IEEE Trans Pattern Anal Mach Intell 18:124–137

Macia J, Sole RV (2009) Distributed robustness in cellular networks: insights from synthetic evolved circuits. J R Soc Interface 6:393–400

Maddox J (1995) Genetics helping molecular-dynamics. Nature 376:209

Mahdiani HR, Fakhraie SM, Lucas C (2012) Relaxed fault-tolerant hardware implementation of neural networks in the presence of multiple transient errors. IEEE Trans Neural Netw Learn Syst 23:1215–1228

Mak SK, Sum PF, Leung CS (2011) Regularizers for fault tolerant multilayer feedforward networks. Neurocomputing 74:2028–2040

Manning T, Sleator RD, Walsh P (2013) Naturally selecting solutions: the use of genetic algorithms in bioinformatics. Bioengineered 4:266–278

Medler DA, Dawson MR (1994) Training redundant artificial neural networks: imposing biology on technology. Psychol Res 57:54–62

Meurice N, Leherte L, Vercauteren DP (1998) Comparison of benzodiazepine-like compounds using topological analysis and genetic algorithms. SAR QSAR Environ Res 8:195–232

Patel JL, Goyal RK (2007) Applications of artificial neural networks in medical science. Curr Clin Pharmacol 2:217–226

Pedersen JT, Moult J (1996) Genetic algorithms for protein structure prediction. Curr Opin Struct Biol 6:227–231

Pena-Malavera A, Bruno C, Fernandez E, Balzarini M (2014) Comparison of algorithms to infer genetic population structure from unlinked molecular markers. Stat Appl Genet Mol Biol 13:391–402

Phatak DS, Koren I (1995a) Complete and partial fault-tolerance of feedforward neural nets. IEEE Trans Neural Netw 6:446–456

Phatak DS, Koren I (1995b) Complete and partial fault tolerance of feedforward neural nets. IEEE Trans Neural Netw 6:446–456

Pini L, Pievani M, Bocchetta M, Altomare D, Bosco P, Cavedo E, Galluzzi S, Marizzoni M, Frisoni GB (2016) Brain atrophy in Alzheimer's disease and aging. Ageing Res Rev 28:30002

Presnell SR, Cohen FE (1993) Artificial neural networks for pattern recognition in biochemical sequences. Annu Rev Biophys Biomol Struct 22:283–298

Protzel PW, Palumbo DL, Arras MK (1993) Performance and fault-tolerance of neural networks for optimization. IEEE Trans Neural Netw 4:600–614

Rajan P, Tolley DA (2005) Artificial neural networks in urolithiasis. Curr Opin Urol 15:133–137

Rodrigues PL, Rodrigues NF, Pinho ACM, Fonseca JC, Correia-Pinto J, Vilaca JL (2014) Automatic modeling of pectus excavatum corrective prosthesis using artificial neural networks. Med Eng Phys 36:1338–1345

Rothlauf F, Goldberg DE, Heinzl A (2002) Network random keys: a tree representation scheme for genetic and evolutionary algorithms. Evol Comput 10:75–97

Sasakawa T, Sawamoto J, Tsuji H (2014) Neural network to control output of hidden node according to input patterns. Am J Intell Syst 4:196–203

Street ME, Buscema M, Smerieri A, Montanini L, Grossi E (2013) Artificial neural networks, and evolutionary algorithms as a systems biology approach to a data-base on fetal growth restriction. Prog Biophys Mol Biol 113:433–438

Sum J, Leung ACS (2008) Prediction error of a fault tolerant neural network. Neurocomputing 72:653–658

Tang W, Mao KZ, Mak LO, Ng GW (2010) Classification for overlapping classes using optimized overlapping region detection and soft decision. Paper presented at: information fusion

Tchernev EB, Mulvaney RG, Phatak DS (2005) Investigating the fault tolerance of neural networks. Neural Comput 17:1646–1664

Weber L (1998) Applications of genetic algorithms in molecular diversity. Curr Opin Chem Biol 2:381–385

Weiner MW, Veitch DP, Aisen PS, Beckett LA, Cairns NJ, Cedarbaum J, Green RC, Harvey D, Jack CR, Jagust W et al (2015) 2014 update of the Alzheimer's disease neuroimaging initiative: a review of papers published since its inception. Alzheimers Dement 11:001

Willett P (1995) Genetic algorithms in molecular recognition and design. Trends Biotechnol 13:516–521

Wu AH (2007) Use of genetic and nongenetic factors in warfarin dosing algorithms. Pharmacogenomics 8:851–861

Xiong H, Wu J, Liu L (2010) Classification with class overlapping: a systematic study. ICEBI-10

Xu C, Xu C (2013) Optimization analysis of dynamic sample number and hidden layer node number based on BP neural network. Springer, Berlin

17

Mutation analysis of large tumor suppressor genes *LATS1* and *LATS2* supports a tumor suppressor role in human cancer

Tian Yu[1], **John Bachman**[2], **Zhi-Chun Lai**[1,2,3]✉

[1] Intercollege Graduate Degree Program in Molecular, Cellular and Integrative Biosciences, The Pennsylvania State University, University Park, PA 16802, USA
[2] Department of Biology, The Pennsylvania State University, University Park, PA 16802, USA
[3] Department of Biochemistry and Molecular Biology, The Pennsylvania State University, University Park, PA 16802, USA
✉ Correspondence: zcl1@psu.edu (Z.-C. Lai)
Received October 9, 2014 Accepted November 13, 2014

ABSTRACT

In recent years, human cancer genome projects provide unprecedented opportunities for the discovery of cancer genes and signaling pathways that contribute to tumor development. While numerous gene mutations can be identified from each cancer genome, what these mutations mean for cancer is a challenging question to address, especially for those from less understood putative new cancer genes. As a powerful approach, *in silico* bioinformatics analysis could efficiently sort out mutations that are predicted to damage gene function. Such an analysis of human large tumor suppressor genes, *LATS1* and *LATS2*, has been carried out and the results support a role of *hLATS1/l2* as negative growth regulators and tumor suppressors.

KEYWORDS *LATS1* & *LATS2*, hippo signaling, cancer genome, human cancer

INTRODUCTION

Hippo signaling plays a crucial role in animal development and tumorigenesis (Harvey et al., 2013; Yu and Guan, 2013). As a key regulator of this growth-inhibitory pathway, the large tumor suppressor (*Lats*)/*warts* (*wts*) gene encodes a Ser/Thr protein kinase and somatic mutations in human *LATS1* and *LATS2* have been identified in primary tumors (e.g. Murakami et al., 2011; Visser and Yang, 2010). To systematically evaluate how *LATS1/2* genes play a critical role in human cancer, a mutation analysis has been carried out. In the Catalogue of Somatic Mutation in Cancer (COSMIC) database, 101 non-synonymous *LATS1* somatic mutations have been identified from 9183 unique human tumor samples (Fig. 1A). Similarly, there are 80 *LATS2* non-synonymous mutations out of 9516 samples (Fig. 1B). Therefore, an overall mutation rate is 1.10% for *LATS1* and 0.84% for *LATS2*. In the cBioPortal database, *LATS1* and *LATS2* exhibited similar overall mutation rates, 1.83% (135/7390) and 1.50% (111/7390), respectively. The top three highest mutation rates with relative larger total sample size for *LATS1* occurred in stomach adenocarcinoma (5.91%, n = 220), uterine corpus endometrial carcinoma (4%, n = 248), and bladder urothelial carcinoma (3.1%, n = 130). Meanwhile, the highest *LATS2* non-synonymous mutation rate occurred in uterine corpus endometrial carcinoma (5.2%, n = 248), stomach adenocarcinoma (4.1%, n = 220), and lung adenocarcinoma (3.9%, n = 229) (Table S1).

To determine the mutation distribution across different domains, analysis through Fisher's exact test shows that both the kinase domain (P = 0.01075) and proline-rich (P = 0.0312) of LATS1 displayed the highest mutation frequency among all the LATS1 domains. The proline-rich domain had 7 mutations in a 31-amino acid (aa) region (2.2 mutations/10 aa), and the kinase domain had 43 mutations in a 306-aa region (1.41 mutations/10 aa) (Fig. 1A). In LATS2, the kinase domain ($P = 5.66 \times 10^{-5}$) and insertion domain (P = 0.03121) had the highest mutation frequency. LATS2 kinase domain

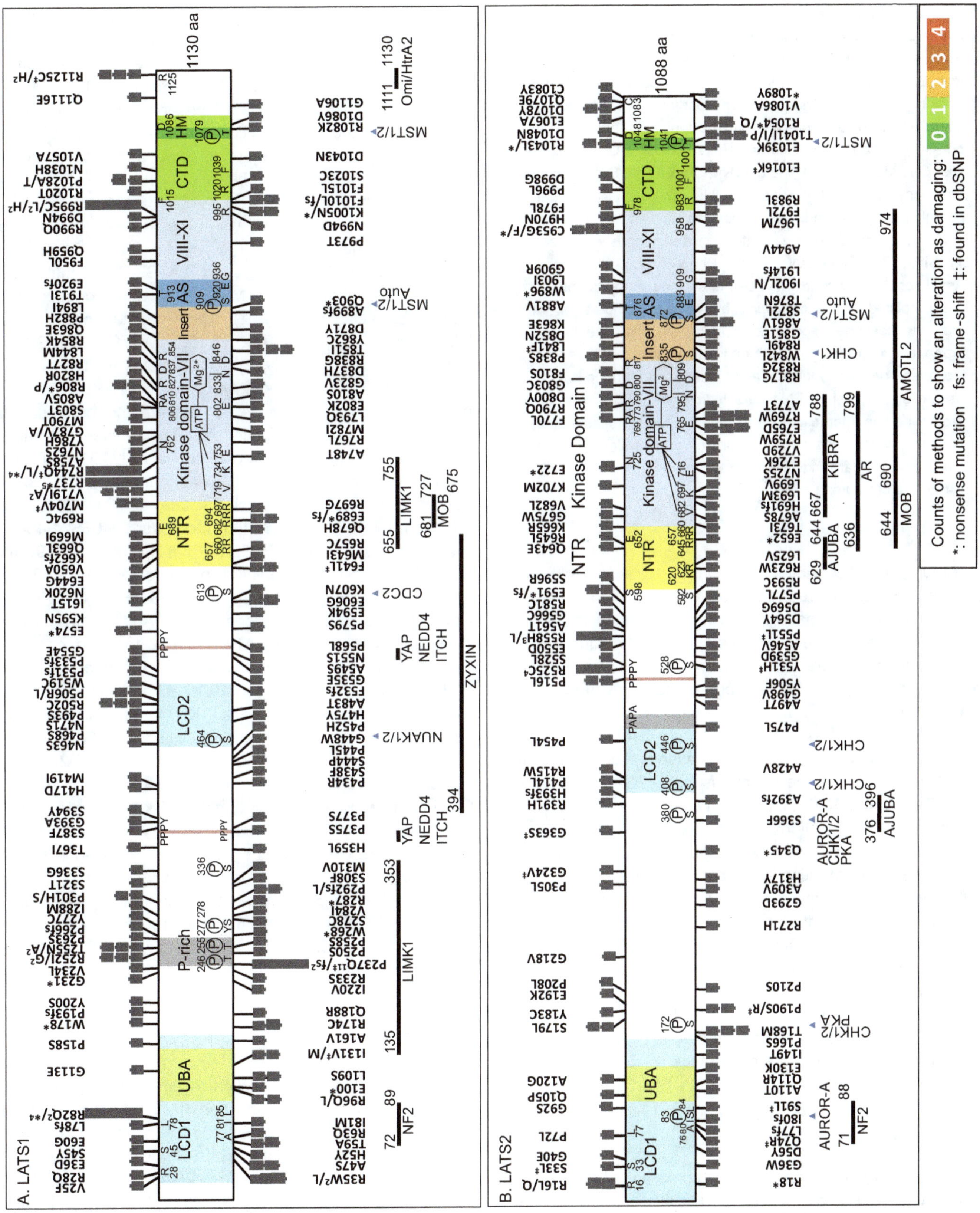

had 44 mutations in a 306-aa region (1.48 mutation/10 aa) and LATS2 insertion domain had 9 in 45 aa (2 mutation/10 aa) (Fig. 1B). These data support that selections have occurred to enrich mutations in functionally significant regions such as the kinase domain to facilitate tumorigenesis.

Figure 1. Human cancer mutations in LATS1 (A) and LATS2 (B) are mapped to their corresponding open reading frames. Human LATS1 and LATS2 mutation data was collected from Catalogue of Somatic Mutations in Cancer (COSMIC) (top portions) and cBioPortal (bottom portions) databases. (A complete list of all mutations can be found in Table S1). All non-synonymous mutations are analyzed, using Uniprot identifier O95835 for LATS1 and Q9NRM7 for LATS2. Synonymous mutations were not included in this analysis. To evaluate potential impact of individual mutations on LATS1/2 structure and function, the following bioinformatics resources were used: 1) SIFT; 2) PROVEAN; 3) PolyPhen-2; and 4) Mutation Assessor. A color code was used to distinguish mutations that are predicted to be damaging by various numbers of the methods described above (zero in dark green, one in light green, two in orange yellow, three in brown, and four in red). "*" indicates nonsense mutation and "‡" for ones found in dbSNP. "fs" is for frame-shift. Blue bars indicate regions involved in protein-protein interactions as indicated. Blue triangles identify phosphorylation sites by corresponding protein kinases. Each small block square indicates one unique mutation sample for *LATS1/2* from human cancer.

Among all the mutations, nonsense and frame-shift mutations clearly disrupt *LATS1/2* gene function. Thirteen and 10 unique nonsense mutations were found in *LATS1* and *LATS2*, respectively. Moreover, 12 and 7 unique frame-shift mutations were detected in *LATS1* and *LATS2*, respectively (Fig. 1). The pattern of equal distribution of *LATS1/2* nonsense or frame-shift mutations is consistent with the idea that *LATS1* and *LATS2* are tumor suppressor genes. The percentage of either nonsense or frame-shift mutations among all the mutations for *LATS1* and *LATS2*, was 17.43% and 10.69%, respectively.

To predict the functional relevance of other non-synonymous mutations in LATS1/2, we performed analyses of evolutionary conservation and protein structure through four different mutation-assessing methods: SIFT, PROVEAN, PolyPhen-2, and Mutation Assessor (Ng and Henikoff, 2003; Adzhubei et al., 2010; Reva et al., 2011; Choi et al., 2012). We found that 73.85% (161/218) of missense mutations from *LATS1* and 77.98% (124/159) from *LATS2* are predicted to be damaging by at least one method (Fig. 1). In regards to mutations that were considered severe as determined by all four methods, LATS1 had a percentage of 25.69% (56/218) and LATS2 had 31.44% (50/159).

LCD1 AND LCD2

LATS1/2 proteins share LATS conserved domain 1 (LCD1) and LATS conserved domain 2 (LCD2), which are conserved in all vertebrate LATS1/2 homologues. LCD1 and LCD2 are critical for LATS1/2 function and regulation. The deletions of either LCD1 or LCD2 in mouse Lats2 abolished its tumor suppressor activity in immortalized mouse cell line (Visser and Yang, 2010). Lats1 LCD1 knockout mice were born with a low birth rate, from which the mouse embryonic fibroblasts displayed chromosomal instability and tumorigenesis (Yabuta et al., 2011). Within LCD1, a short segment called Conserved N-terminal Motif (CNM) (aa 72–89 for LATS1 and 71–88 for LATS2) is important for membrane recruitment and activation of LATS1/2 by Merlin/NF2. The alterations of three highly conserved residues in LATS1/2-A77/76P-I81/80T-L85/84P prevented its interaction with Merlin/NF2, membrane localization and activation (Yin et al., 2013). Therefore, LATS1-I81M, LATS1-R82Q, and LATS2-P72L mutations may fail to interact with Merlin/NF2 and consequently cannot be activated (Fig. 1). The truncated products of LATS1 such as L78fs, R82*, E100*, W178*, and G231* may compete with wild-type LATS proteins for the binding partners of LCD1. Interestingly, LATS2-S83 within CNM is phosphorylated by Aurora-A to regulate the centrosomal localization and mitotic activity of LATS2 (Visser and Yang, 2010). Finally, the conserved residues in LCD1 could be critical for function. Mutations were found at certain conserved sites in LCD1 which include LATS1/2-R28/R16, S45/S33, L78/L77, and I81/I80. Moreover, N463S, P468S, H475Y, A483T, P493S, R502C, P506R/L, and W519C in LCD2 for LATS1 are predicted to be damaging. P468S is close to the phosphorylation site LATS1-S464 by NUAK, which promotes LATS1 degradation. Additionally, CDC2 phosphorylates S613 of LATS1. CDC2 forms a complex with LATS1 in the centrosome and phosphorylation of S613 occurs during mitosis (Visser and Yang, 2010). Mutations near S613, such as K607N and I615T, may interfere with phosphorylation at this site. Four lesser deleterious LATS2 mutations were found in LCD2, which contains the phosphorylation sites S408 and S446 by Chk1/2 in response to UV damage (Okada et al., 2011). Next to LCD2, LATS2-S380 is phosphorylated by Aurora A during mitosis, which is critical for Aurora A-LATS-Aurora B axis to regulate mitotic progression (Yabuta et al., 2011). Moreover, LATS2 S380 is located within an Ajuba-binding region of LATS2 (aa 376–396), which regulates the spindle apparatus formation (Visser and Yang, 2010). Mutations like R391H may affect the interactions and cell cycle control.

PROLINE-RICH REGION

LATS1 has a unique proline-rich region (Fig. 1A). Previous studies detected phosphorylation of T246 and T255 in P-stretch, as well as S336 located downstream (Hornbeck et al., 2012). T255A/N and S336G mutations in LATS1 would prevent phosphorylation of these residues, and the R252I/G mutations nearby may affect these phosphorylation events. Moreover, Y277 and S278 located within the LIMK-binding site of LATS1 were also found to be phosphorylated (Visser and Yang, 2010). While the functional significance of these phosphorylations is unknown, Y277C and S278C mutations clearly prevent phosphorylation of these two residues.

BINDING WITH YAP AND KIBRA

The interaction between the Proline-Proline-x-Tyrosine (PPxY) motif and the hydrophobic pocket of WW domain is critical for the LATS1/2 binding with either their substrate YAP or their activator KIBRA. It has been reported that LATS1-Y559F and Y376A abrogated the binding with YAP (Visser and Yang, 2010), while LATS2-Y518A partially abolish the binding with KIBRA (Xiao et al., 2011). These interactions may be affected by the third proline mutation P375S in the first PPxY motif of LATS1 and the second proline mutation P516L in the only PPxY motif of LATS2.

BINDING WITH MOB, AJUBA AND AMOTL2

The N-terminal regulatory domain (NTR) adjacent to the kinase domain is required for LATS1/2 activation by MOB, AJUBA, and AMOTL2 (Visser and Yang, 2010; Paramasivam et al., 2011; Xiao et al., 2011). Positively charged residues such as LATS1-R657/LATS2-K620, R660/R623, R682/R645, R694/R657, and R697/R660 in LATS1/2 NTR are conserved in the NDR subfamily from yeast to human. Previous studies have shown that LATS1 mutations (e.g. R657A, R660A and R694A) abolish their interaction with the negatively charged surface of MOB1A/B and kinase activity (Hergovich et al., 2006). Therefore, LATS1 mutants (R657C, R694C and R697G) and LATS2 mutants (R623W and R645L), are expected to become inactive in human cancer due to loss of interactions with LATS activators such as MOB. The *Drosophila* Wts-R702 residue is equivalent to LATS1/2-R694/R657 and critical for Mob-binding, kinase activity, and inhibition of tissue growth in development (Ho et al., 2010).

THE KINASE DOMAIN

LATS1 and LATS2 are members of the AGC (named after PKA, PKG, and PKC) protein kinase family. Although crystal structures of some AGC proteins have been deciphered, there is no structure information available for LATS1/2 kinase domains. To estimate mutation-induced structural changes, we performed structure remodeling for LATS1/2 based on two AGC family proteins, ROCK and PKC. Conserved in most eukaryotic protein kinases, the N-terminal catalytic domain of LATS1/2 interacts with the phosphate donor ATP through a crucial network, composited by GxGxxGxV loop (LATS1: 712–719/ LATS2: 675–682), K734/697, E753/716, DxKxxN (828–833/791–795) and DFG motif (846–848/809–811) ("x" represents any aa) (Endicott et al., 2012; Hanks and Hunter, 1995). In addition, previous studies also have verified that mutants LATS1-K734M/LATS2-K697A/mLats2-K655M are all kinase-dead (Zhao et al., 2007; Wei et al., 2007; Visser and Yang, 2010). Mutated within or close to these conserved catalytic elements, LATS1 V719I/A, R744Q/L, A748T, R827T, R837H, L844M and LATS2 G675W, A678S, V682L, L693M, L699V, D800Y, G803C are most likely to disrupt ATP binding and catalysis. LATS1-R995C/L/H and LATS2-G909R also change highly conserved residues. Therefore, these cancer mutants may affect their kinase activity. Moreover, conserved residues in the kinase domain of LATS1/2 are expected to be important for their kinase activity. Mutations occurring in these residues included LATS1-N762/LATS2-N725, R806/R769, A810/A773, R827/R790, D837/D800, R854/R817, and R995/R958 (Fig. 1).

Unlike other AGC family members, LATS1/2 and the NDR subfamily members have an insert between the kinase subdomain VII and the activation segment (AS) in the subdomain VIII. Basic residues in the insert inhibit the activity of AS likely through an auto-inhibitory mechanism (Hergovich et al., 2006). Missense mutations in this region (e.g. LATS1-D871Y) may have an impact on the kinase activity (Fig. 1).

PHOSPHORYLATION AND ACTIVATION BY MST1/2 KINASES

The phospho-S909/S872 and other residues in AS organize interaction between ATP and substrates of LATS1/2. Previous study has shown that LATS1 mutant S909A/D/E cannot be activated by the MST2 kinase. Therefore, LATS2-S872L would abolish the kinase activity, and LATS1-T913I and LATS2-T876N, A881V also likely damage catalytic activity. In the C-terminal of kinase IX and XI subdomains, the LATS2-G909 and LATS1-R995 are highly conserved among eukaryotic kinases (Endicott et al., 2012). LATS1-R995C/L/H and LATS2-G909R would probably damage kinase activity. LATS2-G909R and C953* has been experimentally shown to be defective for kinase activity and YAP regulation (Yu et al., 2013).

In the C-terminal domain (CTD) of AGC family kinase, the conserved NFD (Asn-Phe-Asp) motif interacts with hydrophobic pocket in the N-terminus of kinase domain to facilitate kinase activation (e.g. PKC, Leonard et al., 2011). In LATS1 NFD, N1038H may disrupt the kinase activity. In addition, LATS1-R1020T, P1028A/T, S1023C and LATS2-P996L, D998G may also affect this activation. LATS1/2 activation also requires phosphorylation of T1079/T1041 by MST1/2 in hydrophobic motif (HM), which has a consensus sequence F-x-x-Y/F-T-Y/F-K/R in the NDR protein subfamily (Hergovich et al., 2006). Thus, LATS2-T1041I/P mutations clearly abolish phosphorylation at this site to cause kinase inactivation. LATS1-R1082K, D1086Y and LATS2-E1039K, R1043L, D1048N may disrupt kinase activity as well. Mutations at conserved sites between LATS1/2 such as LATS1-F1015/LATS2-F978, R1020/R983, F1039/F1001, and D1086/D1048 may affect function of the C-terminus.

CONCLUDING REMARKS

Genetic analysis of Lats/Wts family genes using *Drosophila* and mice models has revealed their role as negative growth regulators and tumor suppressors in animal (Visser and Yang, 2010; Harvey et al., 2013; Yu and Guan, 2013). The fact that human LATS1 can functionally replace Wts in *Drosophila* supports that LATS may function as a tumor suppressor in human cells. From human cancer genome projects, an increasing number of mutations in *LATS1/2* are detected. Compared to some well-established cancer genes such as *TP53* and *Rb*, *LATS1* and *LATS2* are not frequently mutated. However, our *in silico* analysis provides supporting evidence that *LATS1/2* mutations drive human tumor development based on the following observations: 1) Cancer mutations in *hLATS1/2* do not appear to be random mutations. Damaging mutations have been accumulated more preferentially in important protein domains such as the kinase domain. Majority of the mutations including nonsense and frame-shift mutations clearly disrupt LATS1/2 function; 2) Some mutations occurred in regions and residues important for LATS1/2 activation. Examples include MST1/2 phosphorylation sites, MOB-binding domain, and the region critical for NF2 interaction; 3) In certain cancer types like stomach adenocarcinoma and uterine corpus endometrial carcinoma, mutation rates can be reasonably high (5.2%–5.9%). On the other hand, mechanisms other than gene mutation could also be effective to alter gene activity. In this regard, *LATS1/2* genes are known to be down-regulated by promoter methylation. Mutations in other Hippo pathway genes such as *MST1/2* are also expected to reduce *LATS1/2* function. While the functional significance of some of these mutations have been experimentally tested *in vivo* (Yu et al., 2013), results reported here further support that *LATS1/2* act normally as tumor suppressors and loss of their functions contributes to human cancer development.

ACKNOWLEDGEMENTS

This work was partly supported by the National Science Foundation. We would like to thank Nicholas Anzalone and Samar Almarzooqi for assistance and other members of the Lai laboratory for discussions.

ABBREVIATIONS

CTD, C-terminal domain; HM, hydrophobic motif; LCD1, LATS conserved domain 1; LCD2, LATS conserved domain 2; NTR, N-terminal regulatory domain; PPxY, proline-proline-x-tyrosine.

COMPLIANCE WITH ETHICS GUIDELINES

Tian Yu, John Bachman, and Zhi-Chun Lai declare that they have no conflict of interest. This article does not contain any studies with human or animal subjects performed by the any of the authors.

REFERENCES

Adzhubei IA, Schmidt S, Peshkin L, Ramensky VE, Gerasimova A, Bork P, Kondrashov AS, Sunyaev SR (2010) A method and server for predicting damaging missense mutations. Nat Methods 7:248–249

Choi Y, Sims GE, Murphy S, Miller JR, Chan AP (2012) Predicting the functional effect of amino acid substitutions and indels. PLoS One 7:e46688

Endicott JA, Noble ME, Johnson LN (2012) The structural basis for control of eukaryotic protein kinases. Annu Rev Biochem 81:587–613

Hanks SK, Hunter T (1995) Protein kinases 6. The eukaryotic protein kinase superfamily: kinase (catalytic) domain structure and classification. FASEB J 9:576–596

Harvey KF, Zhang X, Thomas DM (2013) The Hippo pathway and human cancer. Nat Rev Cancer 13:246–257

Hergovich A, Stegert MR, Schmitz D, Hemmings BA (2006) NDR kinases regulate essential cell processes from yeast to humans. Nat Rev Mol Cell Biol 7:253–264

Ho LL, Wei X, Shimizu T, Lai ZC (2010) Mob as tumor suppressor is activated at the cell membrane to control tissue growth and organ size in *Drosophila*. Dev Biol 337:274–283

Hornbeck PV, Kornhauser JM, Tkachev S, Zhang B, Skrzypek E, Murray B, Latham V, Sullivan M (2012) PhosphoSitePlus: a comprehensive resource for investigating the structure and function of experimentally determined post-translational modifications in man and mouse. Nucleic Acids Res 40:D261–D270

Leonard TA, Różycki B, Saidi LF, Hummer G, Hurley JH (2011) Crystal structure and allosteric activation of protein kinase C βII. Cell 144:55–66

Murakami H, Mizuno T, Taniguchi T, Fujii M, Ishiguro F, Fukui T, Akatsuka S, Horio Y, Hida T, Kondo Y, Toyokuni S, Osada H, Sekido Y (2011) LATS2 is a tumor suppressor gene of malignant mesothelioma. Cancer Res 71:873–883

Ng PC, Henikoff S (2003) SIFT: predicting amino acid changes that affect protein function. Nucleic Acids Res 31:3812–3814

Okada N, Yabuta N, Suzuki H, Aylon Y, Oren M, Nojima H (2011) A novel Chk1/2–Lats2–14-3-3 signaling pathway regulates P-body formation in response to UV damage. J Cell Sci 124:57–67

Paramasivam M, Sarkeshik A, Yates JR, Fernandes MJ, McCollum D (2011) Angiomotin family proteins are novel activators of the LATS2 kinase tumor suppressor. Mol Biol Cell 22:3725–3733

Reva B, Antipin Y, Sander C (2011) Predicting the functional impact of protein mutations: application to cancer genomics. Nucleic Acids Res 39:e118

Visser S, Yang X (2010) LATS tumor suppressor: a new governor of cellular homeostasis. Cell Cycle 9:3892–3903

Wei X, Shimizu T, Lai ZC (2007) Mob as tumor suppressor is activated by Hippo kinase for growth inhibition in Drosophila. EMBO J 26:1772–1781

Xiao L, Chen Y, Ji M, Dong J (2011) KIBRA regulates Hippo signaling activity via interactions with large tumor suppressor kinases. J Biol Chem 286:7788–7796

Yabuta N, Mukai S, Okada N, Aylon Y, Nojima H (2011) The tumor suppressor Lats2 is pivotal in Aurora A and Aurora B signaling during mitosis. Cell Cycle 10:2724–2736

Yin F, Yu J, Zheng Y, Chen Q, Zhang N, Pan D (2013) Spatial organization of Hippo signaling at the plasma membrane mediated by the tumor suppressor Merlin/NF2. Cell 154:1342–1355

Yu FX, Guan KL (2013) The Hippo pathway: regulators and regulations. Genes Dev 27:355–371

Yu T, Bachman J, Lai ZC (2013) Evidence for a tumor suppressor role for the large tumor suppressor genes LATS1 and LATS2 in human cancer. Genetics 195:1193–1196

Zhao B, Wei X, Li W, Udan RS, Yang Q, Kim J, Xie J, Ikenoue T, Yu J, Li L, Zheng P, Ye K, Chinnaiyan A, Halder G, Lai ZC, Guan KL (2007) Inactivation of YAP oncoprotein by the Hippo pathway is involved in cell contact inhibition and tissue growth control. Genes Dev 21:2747–2761

18

MiR-33a suppresses breast cancer cell proliferation and metastasis by targeting ADAM9 and ROS1

Chuankai Zhang[1,2], Yunda Zhang[1,2], Weiji Ding[1,3], Yancheng Lin[2], Zhengjie Huang[1,3]✉, Qi Luo[1,3]✉

[1] Department of Gastrointestinal Surgery, First Affiliated Hospital of Xiamen University, Xiamen 361003, China
[2] State Key Laboratory of Cellular Stress Biology, Innovation Center for Cell Signaling Network, School of Life Sciences, Xiamen University, Xiamen 361102, China
[3] Department of Gastrointestinal Surgery, First Clinical Medical College of Fujian Medical University, Fuzhou 350005, China
✉ Correspondence: huangzhengjie@xmu.edu.cn (Z. Huang), luoqixmzsh@126.com (Q. Luo)

ABSTRACT

MicroRNAs (miRNAs) are small noncoding RNAs that have a pivotal role in the post-transcriptional regulation of gene expression by sequence-specifically targeting multiple mRNAs. Although miR-33a was recently reported to play an important role in lipid homeostasis, atherosclerosis, and hepatic fibrosis, the functions of miR-33a in tumor progression and metastasis are largely unknown. Here, we found that downregulated miR-33a in breast cancer tissues correlates with lymph node metastasis. MiR-33a expression is significantly lower in the highly metastatic breast cancer cell lines than the noncancerous breast epithelial cells and non-metastatic breast cancer cells. Moreover, the overexpression of miR-33a in metastatic breast cancer cells remarkably decreases cell proliferation and invasion *in vitro* and significantly inhibits tumor growth and lung metastasis *in vivo*, whereas its knockdown in non-metastatic breast cancer cells significantly enhances cell proliferation and invasion *in vitro* and promotes tumor growth and lung metastasis *in vivo*. Combining bioinformatics prediction and biochemical analyses, we showed that ADAM9 and ROS1 are direct downstream targets of miR-33a. These findings identified miR-33a as a negative regulator of breast cancer cell proliferation and metastasis.

KEYWORDS miR-33a, breast cancer, proliferation, metastasis

INTRODUCTION

Breast cancer is the leading cause of cancer-related death for women worldwide, and distant metastasis is the most common cause of death in breast cancer patients (Hong and Dong, 2014). Similar to other cancers, the invasion-metastasis cascade of breast cancer consists of multiple steps, including local invasion, entry into the circulation, arrival at distant secondary sites, extravasation and colonization in distant organs (Chaffer and Weinberg, 2011; Wan et al., 2013; Wang and Ouyang, 2012). Considerable progress in breast cancer treatment has been made over the past decades; however, metastatic breast cancer is still difficult to cure. The lack of curative treatment options for metastatic breast cancer patients emphasizes the need to better understand the molecular mechanisms that drive tumor metastasis.

MicroRNAs (miRNAs) are a class of endogenous small noncoding RNAs that are typically 18–22 nucleotides in length. By binding to completely or partially complementary sites in the 3′-untranslated-region (3′UTR) of target mRNAs, miRNAs suppress the protein translation of these transcripts and/or degrade target mRNAs. Currently, miRNAs have been shown to contribute to regulate tumor onset, growth, and progression (Aleckovic and Kang, 2015; Ruan et al., 2009; Volinia et al., 2012). A single miRNA can downregulate the expression of multiple target genes, thereby coordinately inhibiting or promoting tumor metastasis. Therefore, miRNAs may be attractive targets for modulating the invasion-metastasis cascade. MiR-33a is an intronic miRNA located in intronic sequences of the sterol-response-element-binding protein gene SREBF2, and it plays a critical role in the regulation of cholesterol and fatty acid metabolism (Marquart et al., 2010; Najafi-Shoushtari et al., 2010; Rayner et al., 2011; Rayner et al., 2010). MiR-33a is also involved in liver fibrosis aggravation in nonalcoholic steatohepatitis in mice (Li et al., 2014). Interestingly, a recent report demonstrated that higher miR-33a expression correlates with poor prognosis for GBM patients and promotes glioma-initiating cell self-renewal via the activation of the PKA and NOTCH pathways by targeting PDE8A and UVRAG (Wang et al., 2014). However, little is known about the function of miR-33a in breast cancer progression. In this report, we revealed that miR-33a is downregulated in breast cancer tissues. The expression level of miR-33a in breast cancer tissues is inversely correlated with lymph node metastasis. MiR-33a overexpression inhibits breast cancer cell proliferation and invasion *in vitro* and suppresses tumor growth and lung metastasis of breast cancer cells *in vivo*. Conversely, miR-33a knockdown significantly enhances breast cancer cell proliferation and invasion *in vitro* and promotes tumor growth and lung metastasis *in vivo*. Our study demonstrated that miR-33a plays a tumor suppressor role in breast cancer metastasis.

RESULTS

Identification of miR-33a as a tumor suppressor gene in breast cancer

Recent studies demonstrate that miR-33a is highly expressed in osteosarcoma, enhances osteosarcoma cell resistance to cisplatin by targeting Twist1 (Zhou et al., 2014), and promotes glioma-initiating cell self-renewal in glioma (Wang et al., 2014). However, miR-33a acts as a tumor suppressor miRNA in colon cancer by directly downregulating of the oncogene Pim-1 (Ibrahim et al., 2011; Thomas et al., 2012). Currently, the function of miR-33a in breast cancer progression remains unknown. Microarray analysis by Blenkiron et al. showed that miR-33a expression is often lost in human primary breast cancers because of chromosome alterations, indicating that miR-33a may function as a tumor suppressor (Blenkiron et al., 2007). To determine the exact function of miR-33a in breast cancer, we performed real-time PCR and *in situ* hybridization assays to detect the expression level of miR-33a in breast cancer tissues and cell lines. As shown in Fig. 1A, in 23 cases matched breast cancer samples and normal breast tissues, miR-33a expression was significantly decreased in the breast cancer samples compared to the matched normal tissues. *In situ* hybridization assays confirmed that miR-33a was highly expressed in the normal breast tissue, whereas little signal was observed in tumor tissue (Fig. 1B). We further determined the correlation between the miR-33a level and the metastatic status of patients with breast cancer. We found that the expression of miR-33a was negatively associated with lymph node metastasis (Fig. 1C) and the progression of clinical stage (Fig. 1D) in breast cancer patients. The relevance between the miR-33a expression level and prognostic factors of breast cancer is summarized in Fig. 1E. We also observed that miR-33a expression was significantly lower in the highly metastatic breast cancer cell lines MDA-MB-231 and BT-549 than in the noncancerous breast epithelial cell line MCF-10A and non-metastatic breast cancer cell line MCF-7 (Fig. 1F). These results suggest that the miR-33a level is downregulated in breast cancer tissues and breast cancer cell lines and that it is negatively correlated with the metastatic ability of breast cancer cells.

MiR-33a inhibits breast cancer cell growth, migration and invasion *in vitro*

To evaluate the biological function of miR-33a in breast cancer, the highly metastatic breast cancer cell line MDA-MB-231, which has very low endogenous miR-33a expression, was stably transfected with miR-33a by lentiviral infection, whereas endogenous miR-33a was knocked down in non-metastatic MCF-7 breast cancer cells via a lentivirus-based antagomir expression system (Fig. 2A). MTT assays showed that the ectopic overexpression of miR-33a inhibited the proliferation of MDA-MB-231 cells, whereas sh-miR-33a promoted MCF-7 cell proliferation (Fig. 2B). Moreover, colony formation assays revealed that miR-33a-overexpressing MDA-MB-231 cells displayed fewer colonies compared with the control group. Conversely, the knockdown of miR-33a in MCF-7 cells increased colony numbers (Fig. 2C). Because miR-33a expression was inversely correlated with the metastatic abilities of breast cancer cell lines, the effects of miR-33a on the migration and invasion of breast cancer cells were further examined by Transwell assays. As shown in Fig. 2D, miR-33a overexpression in MDA-MB-231 cells dramatically inhibited cell migration and invasion. In contrast, miR-33a knockdown in MCF-7 cells could promote cell migration and invasion (Fig. 2E). These results demonstrate that miR-33a can inhibit the proliferation, migration, and invasion of breast cancer cells *in vitro*.

MiR-33a suppresses the tumor growth and lung metastasis of breast cancer cells *in vivo*

To determine whether miR-33a can inhibit the tumor growth and metastasis of breast cancer cells *in vivo*, we generated luciferase-labeled MDA-MB-231/miR-33a cells, MCF-7/sh-miR-33a cells, and their control counterparts, and then injected them into the orthotopic sites or tail vein of nude mice. Thirty days after injection at orthotopic sites, the mice injected with MDA-MB-231/miR-33a cells displayed significant smaller and lower weight tumors compared with the mice injected with MDA-MB-231/ctrl cells (Fig. 3A), whereas the mice injected with MCF-7/sh-miR-33a cells formed larger tumors with higher weights than mice injected with MCF-7/ctrl cells (Fig. 3B). Moreover, bioluminescence imaging of mice injected with MDA-MB-231/miR-33a or MDA-MB-231/ctrl cells for 30 days showed that the lung metastasis of MDA-MB-231 cells was significantly impaired by the ectopic

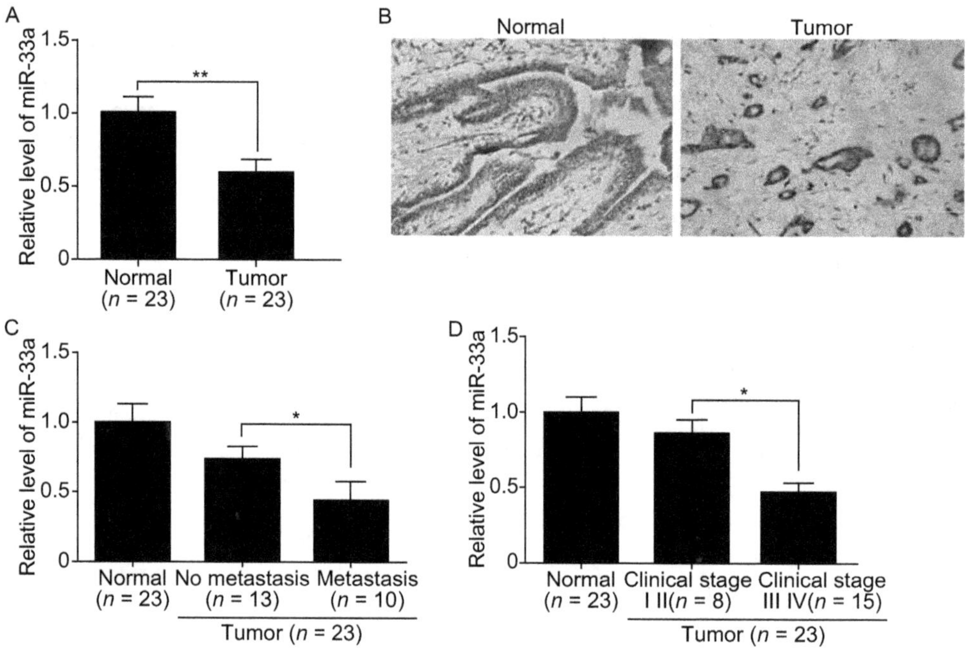

E

Relationship between miR-33a level and prognostic factors of breast cancer

Variable	Number of cases	miR-33a High	miR-33a Low	*P* value
Age				
≥50	20	4	16	0.730
<50	3	1	2	
Clinical stage				
I,II	8	3	5	0.015
III	15	2	13	
Lymph node status				
No metastasis	13	3	10	0.029
Metastasis	10	2	8	
ER status				
Postive	15	3	12	0.280
Negative	8	2	6	
HER-2 status				
Postive	7	2	5	0.130
Negative	16	3	13	

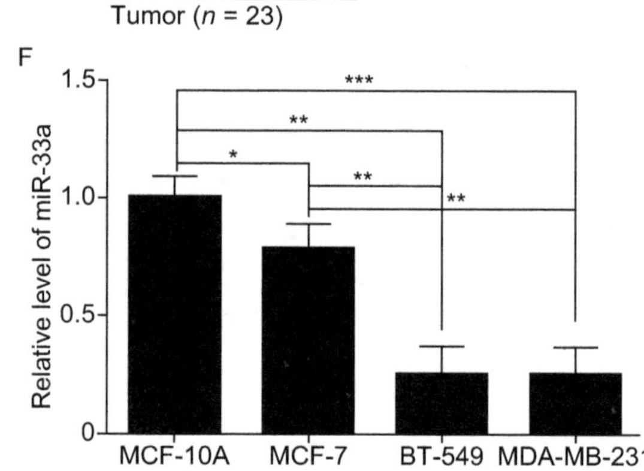

Figure 1. MiR-33a is markedly downregulated in human breast cancer tissues and metastatic breast cancer cell lines. (A) qRT-PCR analysis of miR-33a expression in human breast cancer tissue samples and their matched normal breast tissues from 23 breast cancer patients. (B) *In situ* hybridization analysis of miR-33a expression in human breast cancer tissues and matched normal tissues. (C) Correlation between miR-33a expression and the lymph node metastasis status of breast cancer. (D) Correlation between miR-33a expression and the progression of the clinical stage of breast cancer. (E) Correlation between clinicopathological features and miR-33a expression in 23 breast cancer tissues. (F) qRT-PCR analysis of miR-33a expression in noncancerous human mammary epithelial cells and breast cancer cell lines with different metastatic potential. Scale bars = 100 µm; $*P < 0.05$; $**P < 0.01$; $***P < 0.001$.

overexpression of miR-33a (Fig. 3C). However, mice bearing MCF-7/sh-miR-33a cells after tail vein injection for 42 days displayed large lung metastases compared with the control group (Fig. 3D). These data demonstrate that miR-33a suppresses the tumor growth and lung metastasis of breast cancer cells *in vivo*.

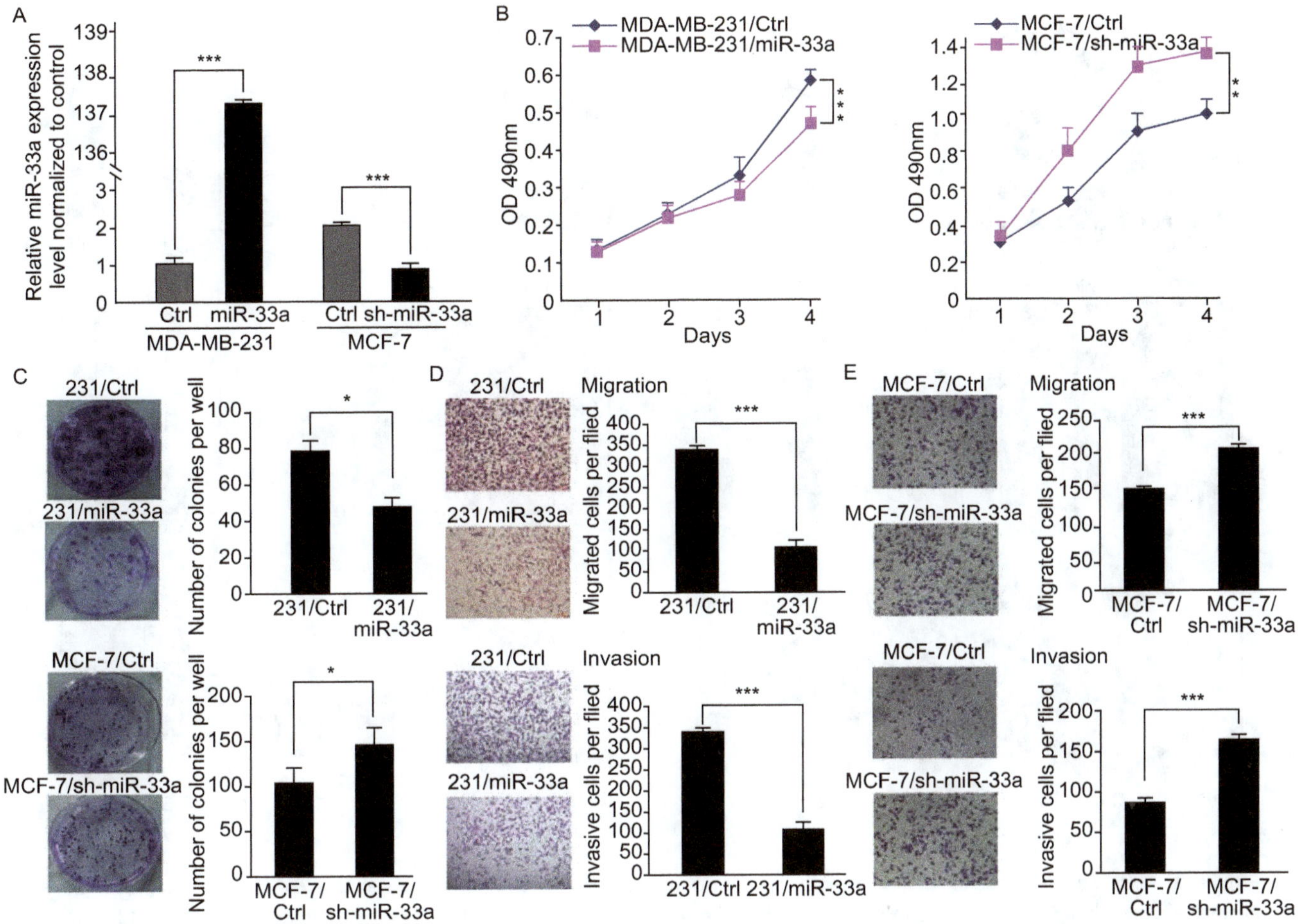

Figure 2. MiR-33a inhibits breast cancer cell growth, migration, and invasion *in vitro*. (A) The miR-33a expression level in MDA-MB-231 cells after ectopic expression of miR-33a and the knockdown efficiency of miR-33a in MCF-7 cells were detected by real-time PCR; (B) The effect on cell proliferation of miR-33a overexpression in MDA-MB-231 cells or knockdown in MCF-7 cells was determined by the MTT assay; (C) Representative images show the colony formation of MDA-MB-231/miR-33a and MCF-7/sh-miR-33a and their control cells (left panel). Average colonies in each well for each group were counted from three independent experiments (right panel); (D) The effects of miR-33a overexpression in MDA-MB-231 cells on cell migration and invasion were analyzed by Transwell migration and Matrigel-coated Transwell invasion analyses; (E) The effects of miR-33a knockdown in MCF-7 cells on cell migration and invasion were analyzed by Transwell migration and Matrigel-coated Transwell invasion analyses. *$P < 0.05$; **$P < 0.01$; ***$P < 0.001$.

ADAM9 and ROS1 are direct targets of miR-33a

To reveal the underlying mechanism by which miR-33a inhibits the tumor growth and lung metastasis of breast cancer cells, we used four *in silico* algorithms (Targetscan, miRanda, mirwalk, and Pictar) to predict the target genes of miR-33a and then used real-time PCR to detect the expression of putative miR-33a targets. We found four candidate targets with greater than 30% decreased expression upon ectopic miR-33a overexpression in MDA-MB-231 cells (Fig. 4A and 4B). To examine whether these four predicted oncogene targets were true targets of miR-33a, we constructed luciferase reporter vectors containing wild-type or mutant 3′UTRs of these candidate target genes. Luciferase activity assays revealed that miR-33a suppressed the expression of luciferase containing the 3′UTRs of ADAM9 and ROS1 compared with controls (Fig. 4C). We also performed Western blot analyses to examine the levels of ADAM9 and ROS1 proteins. As shown in Fig. 4D, the levels of ADAM9 and ROS1 were markedly decreased in MDA-MB-231/miR-33a cells compared with MDA-MB-231/ctrl cells. Conversely, the levels of ADAM9 and ROS1 were increased in MCF-7/sh-miR-33a cells compared with MCF-7/ctrl cells. We found two putative binding sites of miR-33a in the ADAM9 3′UTR and one putative binding site in the ROS1

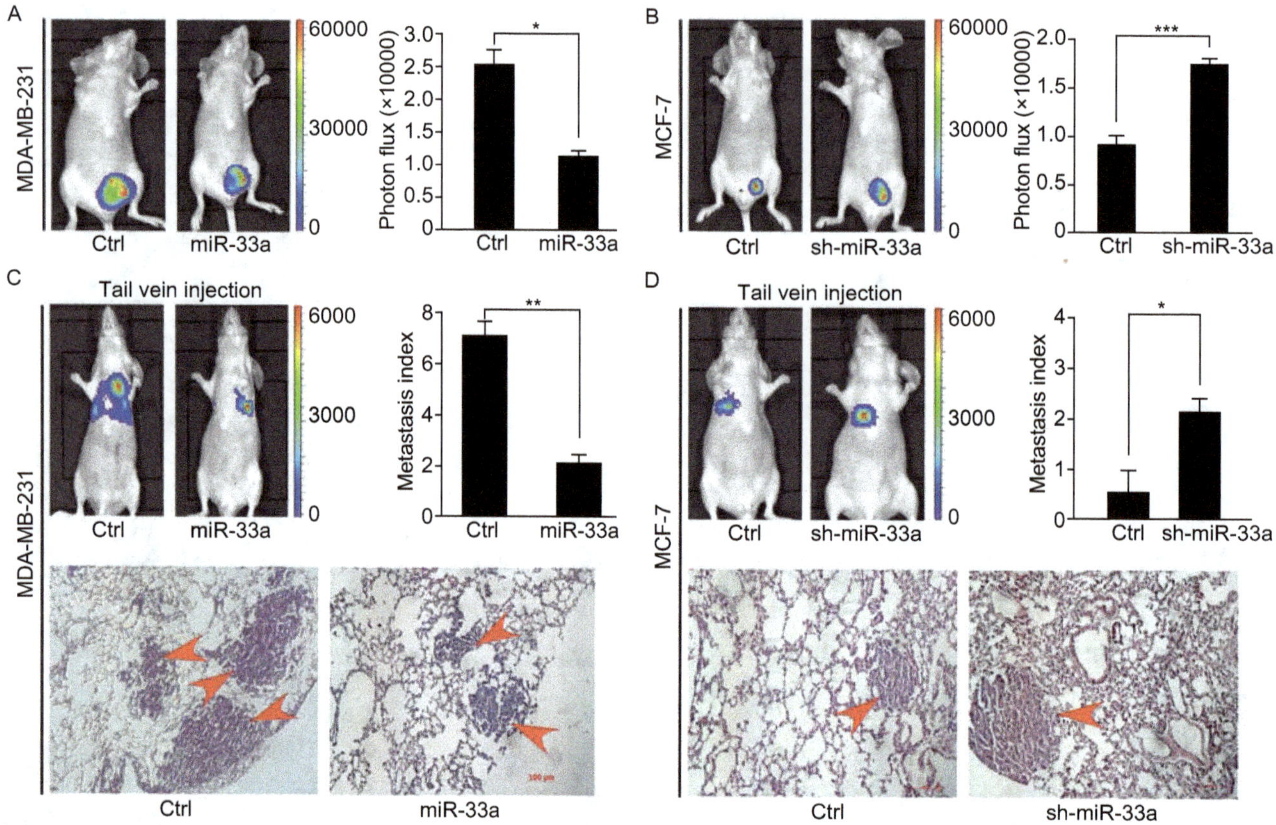

Figure 3. MiR-33a suppresses tumor growth and lung metastasis of breast cancer cells *in vivo*. (A) Representative bioluminescence images of primary tumors in the nude mice orthotopically injected with MDA-MB-231/miR-33a or MDA-MB-231/ctrl cells in the mammary fat pad (left). Quantification of tumors in each group is shown (right); (B) Representative bioluminescence images of primary tumors in the nude mice orthotopically injected with MCF-7/sh-miR-33a or MCF-7/ctrl cells in the mammary fat pad (left). Quantification of tumors in each group is shown (right); (C) Representative bioluminescence images of mice injected with MDA-MB-231/miR-33a or MDA-MB-231/ctrl cells into tail vein to show lung metastases (top-left). Quantification of lung metastases was analyzed by bioluminescence measurement (top-right). Lung metastases in the mice were detected by H&E staining (bottom); (D) Representative bioluminescence images of mice injected with MCF-7/sh-miR-33a or MCF-7/ctrl cells into tail vein to show lung metastases (top-left). Quantification of lung metastases was analyzed by bioluminescence measurement (top-right). Lung metastases in the mice were detected by H&E staining (bottom). Scale bars = 100 μm; $^*P < 0.05$; $^{**}P < 0.01$; $^{***}P < 0.001$.

3′UTR, and we then obliterated these miR-33a binding sites in the 3′UTRs of ADAM9 and ROS1 by QuickChange PCR (Zheng et al., 2004). As shown in Fig. 4E, the mutation of binding site 1, binding site 2, or both sites in the ADAM9 3′UTR reversed the miR-33a-induced downregulation of luciferase activity. Mutation of the binding sites of miR-33a in the ROS1 3′UTR also abrogated the suppressive effect of miR-33a overexpression. Immunohistochemical staining showed that breast cancer tissues with high miR-33a expression have low expression of ADAM9 and ROS1, whereas breast cancer tissues with low miR-33a expression exhibit high expression of ADAM9 and ROS1 (Fig. 4F). Taken together, these results indicate that ADAM9 and ROS1 are direct targets of miR-33a in breast cancer cells.

DISCUSSION

Recent studies have revealed that the aberrant expression of miRNAs is involved in tumor progression; these miRNAs function by inhibiting their target genes, and they play critical roles in the coordination of tumor cell proliferation, invasion, intravasation, survival, extravasation, and/or colonization. Therefore, the identification of specific miRNAs and their targets involved in tumorigenesis and metastasis would provide important clues for identifying new diagnostic and therapeutic targets for cancer prevention and treatment. Interestingly, the microarray analysis by Blenkiron et al. revealed that miR-33a is often lost in human breast cancer (Blenkiron et al., 2007). Here, we found that miR-33a

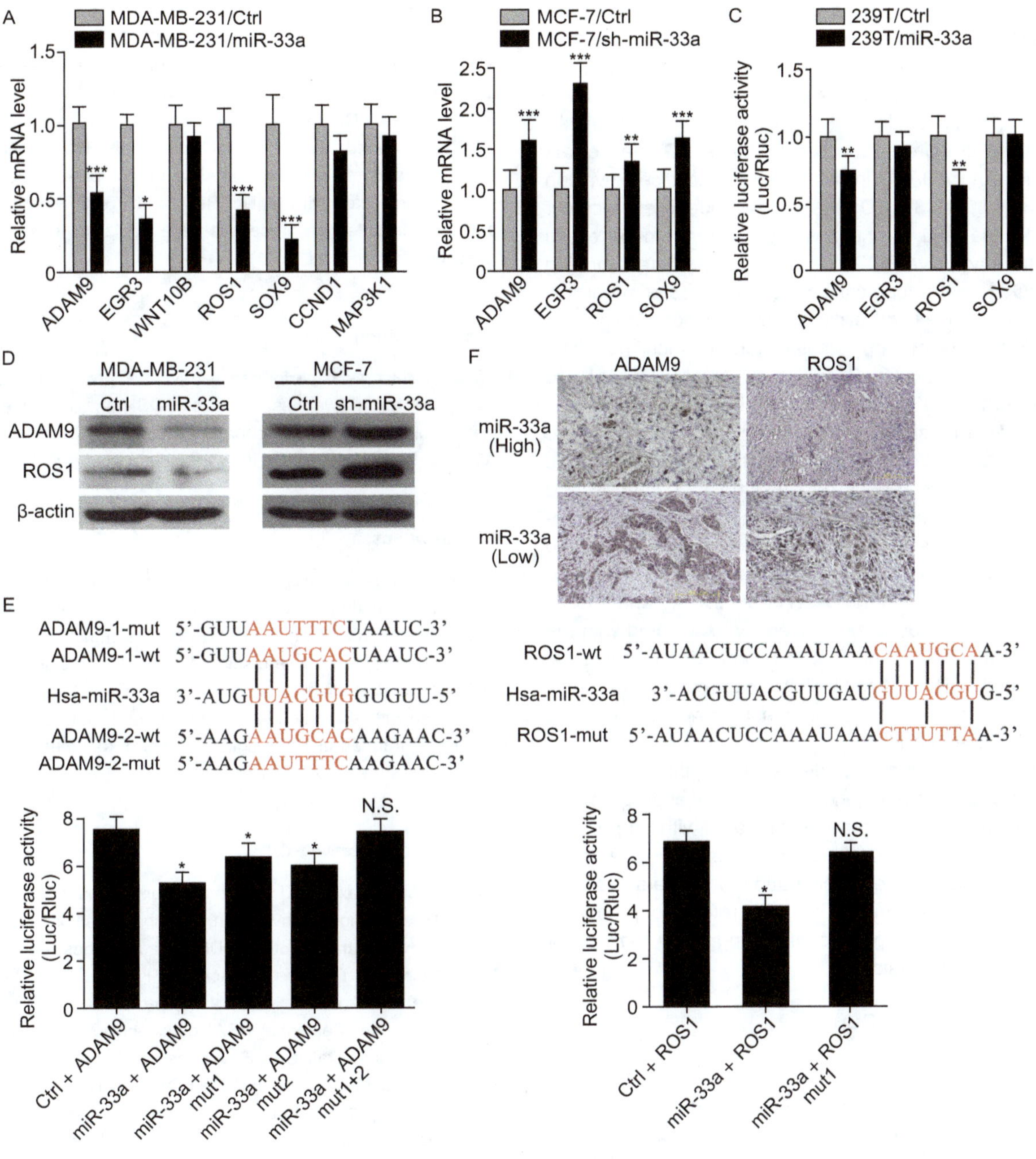

Figure 4. ADAM9 and ROS1 are direct targets of miR-33a. (A) The mRNA levels of predicted target genes of miR-33a in MDA-MB-231/miR-33a and MDA-MB-231/ctrl cells were analyzed by real-time PCR; (B) The mRNA levels of predicted target genes of miR-33a in MCF-7/sh-miR-33a and MCF-7/ctrl cells were further analyzed by real-time PCR; (C) The effects of miR-33a overexpression on the activity of the 3′UTRs of target genes in 293T cells were analyzed by the dual luciferase reporter assay; (D) The effect of miR-33a overexpression in MDA-MB-231 cells or knockdown in MCF-7 cells on the expression of ADAM9 and ROS1 was detected by Western blotting; (E) The effects of miR-33a expression on the activity of wild-type and mutant 3′UTRs of ADAM9 and ROS1 were analyzed by the dual luciferase reporter assay; (F) The levels of ADAM9 and ROS1 were negatively correlated with miR-33a expression in human breast cancer tissues. Scale bars = 200 μm; **P* < 0.05; ***P* < 0.01; ****P* < 0.001; N.S. no significance.

exhibits a decreased expression level in breast cancer tissues compared with matched normal tissues. Moreover, in patients with breast cancer, a correlation is observed between lower miR-33a expression and increased lymph node metastasis. MiR-33a inhibits breast cancer cell proliferation, migration, and invasion *in vitro* and suppresses tumor growth and lung metastasis *in vivo*. Therefore, our study identifies miR-33a as a tumor-suppressive gene in breast cancer metastasis.

Members of the ADAM family are involved in tumor metastasis (Rocks et al., 2008). One member of this family, ADAM9, is overexpressed in breast cancer (O'Shea et al., 2003). Silencing ADAM9 inhibits breast cancer cell invasion (Micocci et al., 2013). ROS1 is a newly identified receptor tyrosine kinase, and its chromosomal rearrangement results in constitutively active ROS1, which has potent transforming ability. ROS1 rearrangement can be used as a predictive marker for response to crizotinib, a tyrosine kinase inhibitor, in patients with advanced non-small cell lung cancer (NSCLC) (Bergethon et al., 2012; Shaw et al., 2014). Recently, a ROS1-rearranged NSCLC patient who had choroidal metastases that did not respond to initial chemotherapy but showed a rapid and complete response to crizotinib was reported (Lu et al., 2015). These reports demonstrate that ADAM9 and ROS1 are oncogenes that may contribute to tumor metastasis. In this report, using Targetscan, miRanda, miRwalk, and Pictar databases, we found that there are highly conserved miR-33a binding sites in the 3′UTRs of ADAM9 and ROS1. Moreover, the miR-33a level is inversely associated with the expression of ADAM9 and ROS1 in breast cancer tissues, indicating that miR-33a may inhibit breast cancer cell proliferation and metastasis, at least in part, by downregulating the levels of ADAM9 and ROS1.

Collectively, our study provides evidence that miR-33a is a novel tumor suppressor gene in breast cancer progression. MiR-33a expression is negatively correlated with lymph node metastasis in patients with breast cancer. MiR-33a might inhibit breast cancer cell proliferation and metastasis by suppressing ADAM9 and ROS1. Therefore, our findings provide a potential diagnostic and prognostic marker for breast cancer metastasis.

MATERIALS AND METHODS

Cell lines and clinical samples

MCF-7, MCF-10A, and 293T cells were obtained from Dr. Kunxin Luo (University of California at Berkeley, Berkeley, CA). MDA-MB-231 cells were provided by Dr. Guohong Hu (Institute of Health Sciences, Shanghai Institute for Biological Sciences, Shanghai). 293T cells were cultured in DMEM supplemented with 10% fetal bovine serum. MCF-7 and MDA-MB-231 cells were maintained in RPMI 1640 media supplemented with 10% fetal bovine serum.

Thirteen matched breast cancer, normal adjacent tissues, and lymph node metastases were collected from the First Affiliated Hospital of Xiamen University, Xiamen, China. Specimens were obtained with informed consent and the study was performed in accordance with the approved guidelines by the Ethics and Scientific Committees of Xiamen University.

Plasmid construction and generation of stable cell lines

An hsa-miR-33a-containing flank region was amplified from human genomic DNA and inserted into pCDH-CMV-EF1-GFP+puro (System Biosciences). The entire lengths of the 3′UTRs of ADAM9 and ROS1 were cloned into the pMIR-REPORT miRNA Expression Reporter Vector (Ambion). To generate a miR-33a-expressing stable cell line, a lentivirus-mediated packaging system containing four plasmids, pCDH-miR-33a or control plasmid, pMDL, REV, and VSVG, was used. To stably knock down miR-33a in MCF-7 cells, pLL3.7-puro containing anti-miR-33a or anti-pre-miR-33a shRNA was co-transfected with pMDL, REV, and VSVG. The transfection and lentiviral infection processes were similar to those previously described (Fang et al., 2011).

Western blotting

Western blotting was performed as described previously (Fang et al., 2011). Cell lysates were separated on SDS–polyacrylamide gels and immunoblot analysis was performed with primary antibodies against ADAM9 (Invitrogen), ROS1 (Cell Signaling), and β-actin (Millipore).

Cell growth assay

For the MTT assay, cells were seeded and transfected in a 96-well plate. After transfection for 1, 2, 3, or 4 days, MTT was added to the cells, and the absorbance at 490 nm was measured. For colony formation assays, cells were seeded in 6-well plates and maintained in RPMI 1640 medium containing 10% FBS for 15 days. Colonies were fixed with 4% PFA, stained with 0.1% crystal violet for 15 min, and photographed using a digital camera. Experiments were repeated three times.

Real-time quantitative PCR

Total RNA was extracted from the cultured cells or frozen tissues using TRIzol reagent (Invitrogen, Carlsbad, California, USA). For miRNA reverse transcription, cDNA was synthesized using TaqMan® MicroRNA Reverse Transcription Kit (ABI) with 100 ng total RNA. The mRNA was reverse-transcribed by an RT-PCR kit (Invitrogen, Carlsbad, California, USA) according to the manufacturer's instructions. Quantitative PCR was then performed with primers for miR-33a, ADAM9, EGR3, ROS1, SOX9, WNT10B, CCND1, and MAP3K1 using SYBR Green PCR Master Mix (Invitrogen, Carlsbad, California, USA) in a real-time PCR System (Applied Biosystems, Carlsbad, California, USA) following a standard quantitative PCR procedure. Primer sequences were as follows: miR-33a, 5′-GGGGG TGCATTGTAGTTG-3′ and 5′-TGCGTGTCGTGGAGTC-3′; ADAM9, 5′-GCTAGTTGGACTGGAGATTTGG-3′ and 5′-TTATTACCACAGGA GGGAGCAC-3′; EGR3, 5′-CTGCCTGACAATCTGTACCC-3′ and 5′-GTAGGTCACGGTCTTGTTGC-3′; ROS1, 5′-ATGGGCTCCTGTATT GGTTG-3′ and 5′-CATCAGTGCATTCTGGGAAA-3′; SOX9, 5′-GAG GAAGTCGGTGAAGAACG-3′ and 5′-GGAGTGCACCTCGCTCAT-3′; WNT10B, 5′-TGGGATGTGTAGCCTTCTCC-3′ and 5′-CCCAGC CAAAAGGAGTATGA-3′; CCND1, 5′-CCCTCGGTGCCTACTTCAA-3′ and 5′-CTCCTCGCACTTCTGTTCCT-3′; MAP3K1, 5′-TGATGTAT GGAGTGTTGGCTG-3′ and 5′-AATGTGAAGGGATCGATGGAG-3′; GAPDH, 5′-GCACCGTCAAGGCTGAGAAC-3′ and 5′-TGGTGAAG ACGCCAGTGGA-3′. Relative quantification was performed by normalization to the amount of GAPDH.

MiRNA reporter luciferase assay

For luciferase reporter assays, 293T cells were seeded into a 24-well plate and co-transfected with 3′UTR-luciferase and either miR-33a or control plasmids. Cells were harvested after two days and assayed using the Dual-Glo Luciferase Assay System (Promega) to determine the relative luciferase activity. The luciferase activity was measured by a luciferin enzyme detection assay kit (Promega) and normalized to *Renilla* luciferase activity. Each treatment was performed in triplicate in three independent experiments.

Cell motility and invasion assay

Migration and invasion assays were performed as described previously (Liu et al., 2014). All experiments were performed at least three times in triplicate.

Animal studies

All experiments using animals were performed in accordance with a protocol approved by the Animal Care and Use Committee of Xiamen University. For orthotopic assays, 5×10^6 MDA-MB-231 or MCF-7 cells in Matrigel (BD, Biosciences) plus growth media were injected into the mammary fat pads of nude mice (n = 3–4 per group). Mice were euthanized 30 days after orthotopic injection, and the growth of subcutaneous tumors was examined by live animal Lumina II system (Xenogen IVIS system). For tail vein metastasis, 1×10^6 MDA-MB-231 cells or 3×10^6 MCF-7 cells were injected into the tail veins of nude mice (n = 3 per group). Mice were euthanized 30 (MDA-MB-231) or 42 (MCF-7) days after tail vein injection, and the lung metastases were examined by live animal Lumina II system (Xenogen IVIS system) and further detected by H&E staining.

Statistical analysis

All of our experiments were performed 3 biological repeats. All data were expressed as the mean ± SD. Statistical analysis was performed with Student's t test. A P value less than 0.05 was considered statistically significant.

ACKNOWLEDGMENTS

This study was supported by grants from the Medical Innovations Topic in Fujian Province (No. 2012-CXB-29), Key Project of Science and Technology Project of Fujian Province (No. 2014D017), Project of Xiamen Scientific and Technological Plan (No. 3502Z20134011), and Science and Technology Project of Natural Science Foundation of Fujian Province (No. 2015J01564).

ABBREVIATIONS

FBS, fetal bovine serum; miR-33a, microRNA-33a; miRNA, microRNA; MTT, 3-(4,5-dimethyl-2-thiazolyl)-2,5-diphenyl-2-H-tetrazolium bromide; NSCLC, non-small cell lung cancer; PFA, paraformaldehyde; 3′UTR, 3′-untranslated-region.

COMPLIANCE WITH ETHICS GUIDELINES

Chuankai Zhang, Yunda Zhang, Weiji Ding, Yancheng Lin, Zhengjie Huang and Qi Luo declare that they have no conflict of interest. All procedures followed were in accordance with the ethical standards of the responsible committee on human experimentation (institutional and national) and with the Helsinki Declaration of 1975, as revised in 2000 (5). Informed consent was obtained from all patients for being included in the study.

REFERENCES

Aleckovic M, Kang Y (2015) Regulation of cancer metastasis by cell-free miRNAs. Biochim Biophys Acta 1855:24–42

Bergethon K, Shaw AT, Ou SH, Katayama R, Lovly CM, McDonald NT, Massion PP, Siwak-Tapp C, Gonzalez A, Fang R et al (2012) ROS1 rearrangements define a unique molecular class of lung cancers. J Clin Oncol 30:863–870

Blenkiron C, Goldstein LD, Thorne NP, Spiteri I, Chin SF, Dunning MJ, Barbosa-Morais NL, Teschendorff AE, Green AR, Ellis IO et al (2007) MicroRNA expression profiling of human breast cancer identifies new markers of tumor subtype. Genome Biol 8: R214

Chaffer CL, Weinberg RA (2011) A perspective on cancer cell metastasis. Science 331:1559–1564

Fang X, Cai Y, Liu J, Wang Z, Wu Q, Zhang Z, Yang CJ, Yuan L, Ouyang G (2011) Twist2 contributes to breast cancer progression by promoting an epithelial-mesenchymal transition and cancer stem-like cell self-renewal. Oncogene 30:4707–4720

Hong W, Dong E (2014) The past, present and future of breast cancer research in China. Cancer Lett 351:1–5

Ibrahim AF, Weirauch U, Thomas M, Grunweller A, Hartmann RK, Aigner A (2011) MicroRNA replacement therapy for miR-145 and miR-33a is efficacious in a model of colon carcinoma. Cancer Res 71:5214–5224

Li ZJ, Ou-Yang PH, Han XP (2014) Profibrotic effect of miR-33a with Akt activation in hepatic stellate cells. Cell Signal 26:141–148

Liu AY, Cai Y, Mao Y, Lin Y, Zheng H, Wu T, Huang Y, Fang X, Lin S, Feng Q et al (2014) Twist2 promotes self-renewal of liver cancer stem-like cells by regulating CD24. Carcinogenesis 35:537–545

Lu S, Azada MC, Ou SH (2015) Choroidal metastasis response to crizotinib in a ROS1-rearranged NSCLC patient. Lung Cancer 87:207–209

Marquart TJ, Allen RM, Ory DS, Baldan A (2010) miR-33 links SREBP-2 induction to repression of sterol transporters. Proc Natl Acad Sci USA 107:12228–12232

Micocci KC, Martin AC, Montenegro Cde F, Durante AC, Pouliot N, Cominetti MR, Selistre-de-Araujo HS (2013) ADAM9 silencing inhibits breast tumor cell invasion in vitro. Biochimie 95:1371–1378

Najafi-Shoushtari SH, Kristo F, Li Y, Shioda T, Cohen DE, Gerszten RE, Naar AM (2010) MicroRNA-33 and the SREBP host genes cooperate to control cholesterol homeostasis. Science 328:1566–1569

O'Shea C, McKie N, Buggy Y, Duggan C, Hill AD, McDermott E, O'Higgins N, Duffy MJ (2003) Expression of ADAM-9 mRNA and protein in human breast cancer. Int J Cancer (Journal international du cancer) 105:754–761

Rayner KJ, Suarez Y, Davalos A, Parathath S, Fitzgerald ML, Tamehiro N, Fisher EA, Moore KJ, Fernandez-Hernando C (2010) MiR-33 contributes to the regulation of cholesterol homeostasis. Science 328:1570–1573

Rayner KJ, Esau CC, Hussain FN, McDaniel AL, Marshall SM, van Gils JM, Ray TD, Sheedy FJ, Goedeke L, Liu X et al (2011) Inhibition of miR-33a/b in non-human primates raises plasma HDL and lowers VLDL triglycerides. Nature 478:404–407

Rocks N, Paulissen G, El Hour M, Quesada F, Crahay C, Gueders M, Foidart JM, Noel A, Cataldo D (2008) Emerging roles of ADAM and ADAMTS metalloproteinases in cancer. Biochimie 90:369–379

Ruan K, Fang X, Ouyang G (2009) MicroRNAs: novel regulators in the hallmarks of human cancer. Cancer Lett 285:116–126

Shaw AT, Ou SH, Bang YJ, Camidge DR, Solomon BJ, Salgia R, Riely GJ, Varella-Garcia M, Shapiro GI, Costa DB et al (2014) Crizotinib in ROS1-rearranged non-small-cell lung cancer. N Engl J Med 371:1963–1971

Thomas M, Lange-Grunweller K, Weirauch U, Gutsch D, Aigner A, Grunweller A, Hartmann RK (2012) The proto-oncogene Pim-1 is a target of miR-33a. Oncogene 31:918–928

Volinia S, Galasso M, Sana ME, Wise TF, Palatini J, Huebner K, Croce CM (2012) Breast cancer signatures for invasiveness and prognosis defined by deep sequencing of microRNA. Proc Natl Acad Sci USA 109:3024–3029

Wan L, Pantel K, Kang Y (2013) Tumor metastasis: moving new biological insights into the clinic. Nat Med 19:1450–1464

Wang Z, Ouyang G (2012) Periostin: a bridge between cancer stem cells and their metastatic niche. Cell Stem Cell 10:111–112

Wang H, Sun T, Hu J, Zhang R, Rao Y, Wang S, Chen R, McLendon RE, Friedman AH, Keir ST et al (2014) miR-33a promotes glioma-initiating cell self-renewal via PKA and NOTCH pathways. J Clin Investig 124:4489–4502

Zheng L, Baumann U, Reymond JL (2004) An efficient one-step site-directed and site-saturation mutagenesis protocol. Nucleic Acids Res 32:e115

Zhou Y, Huang Z, Wu S, Zang X, Liu M, Shi J (2014) miR-33a is up-regulated in chemoresistant osteosarcoma and promotes osteosarcoma cell resistance to cisplatin by down-regulating TWIST. J Exp Clin Cancer Res 33:12

19

Potential coordination role between *O*-GlcNAcylation and epigenetics

Donglu Wu[1], Yong Cai[1,2,3], Jingji Jin[1,2,3]✉

[1] School of Life Sciences, Jilin University, Changchun 130012, China
[2] National Engineering Laboratory for AIDS Vaccine, Jilin University, Changchun 130012, China
[3] Key Laboratory for Molecular Enzymology and Engineering, The Ministry of Education, Jilin University, Changchun 130012, China
✉ Correspondence: jjjin@jlu.edu.cn (J. Jin)

ABSTRACT

Dynamic changes of the post-translational *O*-GlcNAc modification (*O*-GlcNAcylation) are controlled by *O*-linked β-N-acetylglucosamine (*O*-GlcNAc) transferase (OGT) and the glycoside hydrolase *O*-GlcNAcase (OGA) in cells. *O*-GlcNAcylation often occurs on serine (Ser) and threonine (Thr) residues of the specific substrate proteins via the addition of *O*-GlcNAc group by OGT. It has been known that *O*-GlcNAcylation is not only involved in many fundamental cellular processes, but also plays an important role in cancer development through various mechanisms. Recently, accumulating data reveal that *O*-GlcNAcylation at histones or non-histone proteins can lead to the start of the subsequent biological processes, suggesting that *O*-GlcNAcylation as 'protein code' or 'histone code' may provide recognition platforms or executive instructions for subsequent recruitment of proteins to carry out the specific functions. In this review, we summarize the interaction of *O*-GlcNAcylation and epigenetic changes, introduce recent research findings that link crosstalk between *O*-GlcNAcylation and epigenetic changes, and speculate on the potential coordination role of *O*-GlcNAcylation with epigenetic changes in intracellular biological processes.

KEYWORDS *O*-GlcNAcylation, post-translational modification, histone modification, epigenetics

INTRODUCTION

A large number of intracellular proteins often undergo post-translational modifications. As one of the highly dynamic, inducible, and reversible post-translational modification, *O*-GlcNAc on Ser/Thr residues of nuclear, cytoplasmic, and mitochondrial proteins is ubiquitous in eukaryotic cells (Wells and Hart, 2003; Bullen et al., 2014). Dynamic changes of *O*-GlcNAcylation are controlled by OGT and OGA (Gao et al., 2001). The sugar nucleotide UDP-GlcNAc serves as a donor for *O*-GlcNAc addition to nucleocytoplasmic proteins whereas OGA removes the *O*-GlcNAc group from proteins (Mailleux et al., 2016). Numerous studies have demonstrated that *O*-GlcNAcylation is involved in many fundamental cellular processes, including gene transcription (Ha and Lim, 2015), cell signaling (Cassey, 1995), and apoptosis (Butkinaree et al., 2008). More than those, many transcriptional factors, tumor suppressors, oncogenes, and cell receptors have been found to be tightly associated with *O*-GlcNAcylation in tumorigenesis (Bond and Hanover, 2015; Charoensuksai et al., 2015). Recently, accumulating data reveal that *O*-GlcNAcylation is frequently related to epigenetic changes. For example, *O*-GlcNAcylation at serine 112 site of histone H2B (H2BS112) by OGT stimulates H2B at lysine 120 (H2BK120) ubiquitination that further activates gene transcription such as ring finger protein 20 (RNF20) (Fujiki et al., 2011; Nakamura et al., 2011), suggesting the existence of complicated coordinative role between *O*-GlcNAcylation and the other histone post-translational modifications. In this review, we focus on the *O*-GlcNAcylation by OGT, summarize the current understanding of the crosstalk between *O*-GlcNAcylation and epigenetic changes in cells.

INTRACELLULAR *O*-GLCNAC MODIFICATION IS REGULATED BY OGT AND OGA

As is well known, *O*-GlcNAcylation is the process of adding *O*-GlcNAc via an *O*-linkage to Ser or Thr residues of intracellular proteins. Thousands of *O*-GlcNAcylated proteins have been found to implicate in the regulation of core processes including cellular metabolism, growth, and other important function (Copeland et al., 2008; Hanover et al., 2012). In cells, *O*-GlcNAcylation is often referred to as a nutrient sensor. The sensitivity of nutrient is based on the levels of UDP-GlcNAc, the donor substrate for protein *O*-GlcNAcylation, while UDP-GlcNAc is the end product of the hexosamine biosynthetic pathway (HBP). Thus, the levels of UDP-GlcNAc in cells are closely associated with flux through the HBP (Marshall et al., 1991). According to the current reports, OGT and OGA are the only two enzymes to control intracellular *O*-GlcNAcylation.

OGT is highly conserved from *Caenorhabditis elegans* to human. Functional domain studies have been clarified that full length of OGT is composed of two important regions: an N-terminal tetratricopeptide-repeat (TPR) super-helical structure and a C-terminal catalytic domain (Jinek et al., 2004; Kreppel and Hart, 1999). Wherein, TPR domain is involved in substrate recognition through serving as protein: protein docking sites for substrate targeting proteins (Lubas and Hanover, 2000), and C-terminal catalytic domain binds to donor UDP-GlcNAc and is responsible for catalyzing the substrate targeting proteins by adding *O*-GlcNAc group (Lazarus et al., 2011). In cells, three different transcripts encoding nucleocytoplasmic isoform (ncOGT, 116 kDa), the mitochondrial isoform (mOGT, 103 kDa), and the short isoform (sOGT, 75 kDa) are produced by alternative gene splicing. C-terminal of these isoforms is completely the same, and only differs in the number of N-terminal TPRs (Hanover et al., 2003). As shown in Fig. 1, 13.5, 9, and 3 TPRs are contained in ncOGT, mOGT, and sOGT, respectively. The different localizations and the number of TPRs in each isoform suggest that each OGT isoform possesses its own substrate specificity and function (Lazarus et al., 2011; Love et al., 2003; Trapannone et al., 2016; Sacoman et al., 2017; Shin et al., 2011; Riu et al., 2008).

Another important enzyme that affects intracellular *O*-GlcNAcylation is *O*-GlcNAc hydrolase OGA. OGA was initially isolated from crude cellular extract, and it catalyzes hydrolytic cleavage of *O*-GlcNAc from proteins (Gao et al., 2001). Full-length OGA is composed of N-terminal N-acetyl-β-D-glucosamindase domain and C-terminal with acetyltransferase-like (*AT*) domain (Toleman et al., 2004). Alternative gene splicing results in two different lengths of OGA isoforms: one isoform has a 916 amino acid (OGA-L) and predominantly localizes in the cytoplasm, while the other only has a 677 amino acid (OGA-S), and localizes in nuclear and lipid-droplet (Comtesse et al., 2001).

In view of the OGT and OGA are the only enzymes that involved in *O*-GlcNAc post-translational modification in cells, it is not difficult to understand that both enzymes are essential to cell growth, ontogeny, and survival in mammals. As reported, OGT modulates the somatic cell function and embryo viability through *O*-GlcNAc modification of X-chromosome-linked proteins in mice (O'Donnell et al., 2004). Furthermore, global *O*-GlcNAcylation of rat brain protein is relatively high at early development stage, and gradually decreases during the development, suggesting the coordinative role between OGT and OGA in cell growth and differentiation (Liu et al., 2012).

MODE OF OGT IN CELLS

Although the expression level of OGT in the pancreas and brain is higher than other tissues (Kreppel et al., 1997), OGT is ubiquitously expressed in all tissues, showing the essential role of OGT in cells. In line with this view, numerous studies have confirmed the broad presence of *O*-GlcNAc modification involved in many fundamental cellular processes such as cell morphogenesis, cell signaling, apoptosis, and transcription (Lazarus et al., 2006). It is worth noting that, in addition to free form, OGT can also be assembled in complex to participate in intracellular biological processes.

Ten-eleven translocation (TET) family enzymes including TET1, TET2, TET3 catalyze the conversion of 5-methylcytosine (mC) to 5-hydroxymethylation (hmC) (Ito et al., 2011). Recent studies have found that each enzyme of TET family (TETs) can be complexed with OGT to play a cellular function. The complexes of TET2/TET3 with OGT can *O*-GlcNAcylate the host cell factor 1 (HCF1), a shared subunit of the methyltransferase SET1/COMPASS and histone acetyltransferase NSL complexes, therefore impact the histone H3K4 tri-methylation (H3K4me3) via SET1/COMPASS (Deplus et al., 2013). In-depth study demonstrated that the TET3-OGT complex enhances the recruitment of OGT to chromatin through its stabilization (Ito et al., 2014). TET1, being no exception, can interact with OGT, and the TET1-OGT complex regulates the recruitment of OGT to chromatin and TET1 activity in embryonic stem cells (Vella et al., 2013). Thus, we speculate that the TETs-OGT crosstalk not only help OGT to raise in chromatin, but also regulates gene transcription by affecting histone modification and chromatin structure. In addition to TETs proteins, the function of OGT is also connected to other intracellular proteins. For example, a heterotrimeric complex formed by URI, OGT, and PP1γ regulates cellular *O*-GlcNAcylation in response to metabolic stress (Burén et al., 2016). However, in C2C12 skeletal muscle cells, OGT and AMPK (AMP-activated protein kinase) cooperatively regulate nutrient-sensitive intracellular metabolism, growth, and proliferation (Bullen et al., 2014). Furthermore, a stable complex of MLL5 (mixed lineage leukemia 5), OGT, and USP7 (ubiquitin specific protease 7) was verified by Zhang Y's group (Ding et al., 2015), and the coordinative expression between three proteins was observed in primary cervical adenocarcinomas. The above results suggest that OGT participates in intracellular different biological processes via forming different protein complexes.

We previously described a second human MOF (males absent on the first)-containing histone acetyltransferases NSL (non-specific lethal) complex, which comprises 9 subunits and can acetylate histone H4 at lysine 16 (K16), 5 (K5),

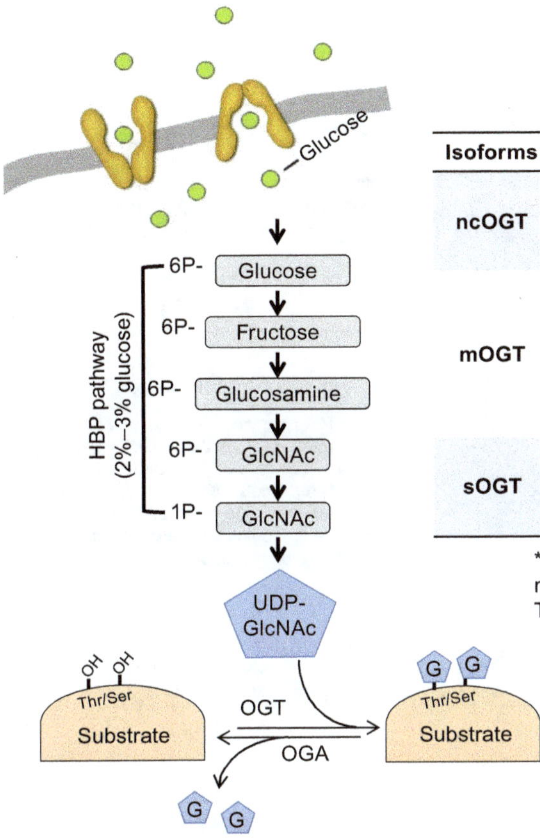

Isoforms	TPRs	Location	Function	Reference
ncOGT	13.5	Nucleus and cytoplasm	*O*-GlcNAcylates the nucleus and cytoplasm proteins, as well as the mitochondrial protein to participate individual pathways. Not toxic to cell.	(Lazarus et al., 2011) (Love et al., 2003) (Trapannone et al., 2016)
mOGT	9	Mitochondria	Rarely localizes with *O*-GlcNAcylated proteins, but preferentially associates with the mitochondrial inner membrane, and functions in maintaining the structure and function of mitochondria. Over expressed of mOGT triggers the apoptosis of cells.	(Lazarus et al., 2011) (Love et al., 2003) (Sacoman et al., 2017) (Shin et al., 2011)
sOGT	3	Nucleus and cytoplasm	can be self-*O*-GlcNAcylated.	(Lazarus et al., 2011) (Love et al., 2003) (Riu et al., 2008)

* ncOGT, nucleocytoplasmic *O*-GlcNAc transferase; mOGT, mitochondria *O*-GlcNAc transferase; sOGT, short *O*-GlcNAc transferase; TPR, tetratrico repeat domain.

Figure 1. The process of *O*-GlcNAc modification and the molecular structure of OGT isoforms. A range of 2%–3% glucose uptake by the cells goes through the HBP pathway, experiences a series of modification and synthesize UDP-GlcNAc, the substrates of *O*-GlcNAc reaction. OGT/OGA are the enzymes in charge of add/remove UDP-GlcNAc at Ser or Thr residues respectively to control the balance of intracellular *O*-GlcNAcylation level. The molecular structure of 3 isoforms of OGT is highly conserved. By alternative splicing, 3 isoforms of OGT including nucleocytoplasmic isoform (ncOGT, 116 kDa), the mitochondrial isoform (mOGT, 103 kDa), and the short isoform (sOGT, 75 kDa) are produced.

and 8 (K8) (Cai et al., 2010). Although the function of OGT in NSL complex is unclear, the coordinative role of OGT and HCF1 has been reported. For instance, *O*-GlcNAcylation by OGT to the specific repeats region of HCF-1 may provide instructions for the HCF-1 proteolysis (Capotosti et al., 2011). In addition, the interaction of OGT/*O*-GlcNAcylation and HCF1 stabilizes PGC-1α, a master regulator of gluconeogenesis. Knockdown both OGT and HCF1 significantly decreases the expression of gluconeogenic genes induced by PGC-1α in fed mice, suggesting the OGT/HCF1 subcomplex-mediated cooperative regulation in gluconeogenesis (Ruan et al., 2012). Also, OGT and HCF1 were characterized as major cellular binding partner of THAP1, a sequence-specific DNA binding factor (Mazars et al., 2010). Besides those stable complexes, OGT can also form a temporary complex with OGA, PP1γ, and Aurora B to regulate the post-translational status of vimentin under the coordination of *O*-GlcNAcylation and phosphorylation (Slawson et al., 2008), indicating the manifold functions of OGT assembled in complexes (Fig. 2).

CROSSTALK BETWEEN *O*-GLCNACYLATION AND EPIGENETIC MECHANISMS

Changes in chromatin structure are mainly caused by N-terminal tails of histone modification and ATP-dependent chromatin remodelers (Gerhold and Gasser, 2014; Längst and Manelyte, 2015; Jin et al., 2005). These complexes regulate the gene expression by epigenetic mechanism in most cellular biological processes (Bannister and Kouzarides, 2011). Recent research evidence shows that *O*-GlcNAcylation by OGT is often implicated in chromatin-mediated processes in a coordinated manner (Fig. 3). According to OGT is an important and indispensable enzyme in cell metabolic pathway, unbalanced *O*-GlcNAc modification on substrate proteins leads to various kinds of diseases, such as diabetes, neurologic disorders, cardiovascular disease, and cancer (Gao et al., 2001). Especially, high *O*-GlcNAc modification has been found in breast cancer (Caldwell et al., 2010; Gu et al., 2010; Krzeslak et al., 2012), prostate cancer (Lynch et al., 2012; Kamigaito et al., 2014;

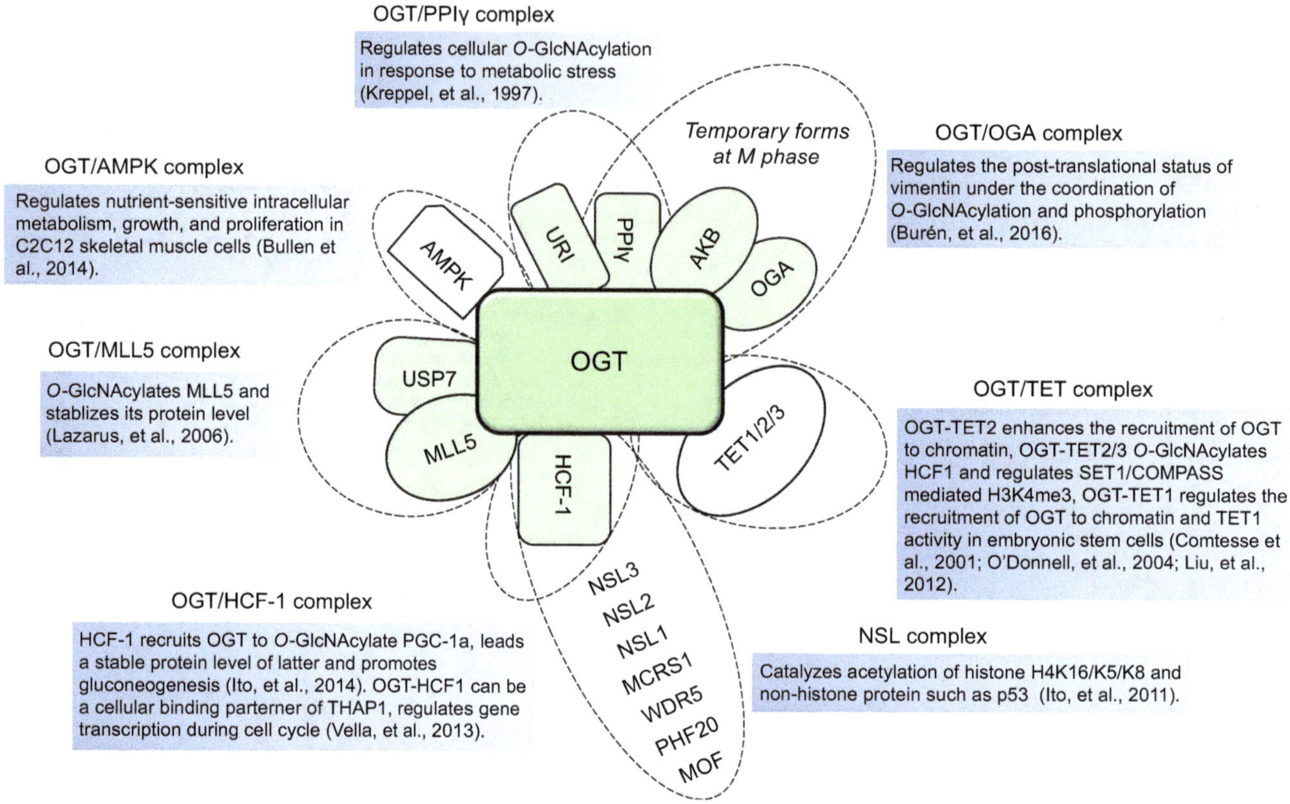

Figure 2. Regulation and function of OGT-mediated protein complexes.

Gu et al., 2014), lung and colon cancer (Mi et al., 2011), esophageal cancer (Qiao et al., 2012), and hepatocellular carcinoma (Zhu et al., 2012). More importantly, numerous tumor-related proteins are also modified and regulated by OGT, such as oncoprotein c-Myc (Chou et al., 1995; Itkonen et al., 2013) and tumor suppressor gene p53 (Yang et al., 2006; de Queiroz et al., 2016). The role of OGT/*O*-GlcNAcylation in some important signal transduction pathways cannot be ignored. In cholangiocarcinoma (CCA) cells, matrix metalloproteinases (MMP)-mediated migration and invasion of CCA cells are regulated by *O*-GlcNAcylation through affecting the nuclear translocation of NFκB (Phoomak et al., 2016).

Coordination of *O*-GlcNAcylation and TET proteins

Given that TETs implicated in controlling genome-wide DNA methylation and cellular differentiation, the impact of TETs proteins in tumorigenesis has been investigated (Shi et al., 2013; Fu et al., 2013). Subsequent research found that TET1 is a fusion partner of mixed-lineage leukemia (MLL)-rearranged acute myeloid leukemia (AML) (Ono et al., 2002; Lorsbach et al., 2003). More importantly, TET genes, especially TET2 are frequently mutated in various cancers, including myelodysplastic syndrome (MDS), cohort of chronic myelomonocytic leukemia (CMML), primary and secondary AML, blastic plasmacytoid dendritic neoplasm (Scourzic et al., 2015; Delhommeau et al., 2009; Langemeijer et al., 2009; Tefferi et al., 2009). However, transcriptional regulation of TETs might be associated with its interplay with histone modifications. Supporting this speculation, OGT-dependent H2BS112 *O*-GlcNAcylation is regulated by TET2 in embryonic stem cells (Chen et al., 2013). While genome-wide analysis confirms that *O*-GlcNAcylation of H2BS112 is widely distributed over chromosomes, and is frequently co-localizes with H2BK120 mono-ubiquitination (Fujiki et al., 2011), suggesting the complexity among TET2, *O*-GlcNAcylation, and H2BK120 ubiquitination.

As we mentioned earlier, TETs proteins function to recruit OGT to chromatin. Thus, it is not difficult to understand that knocking down TETs proteins declines the *O*-GlcNAcylation of chromatin proteins. In mouse embryonic stem (ES) cells, TET1-OGT interaction links to multiple chromatin regulators including Sin3A and NuRD complexes (Shi et al., 2013). Follow-up studies clarify that TET1 with those chromatin regulators together commands the level of local 5hmC to regulate the expression of target genes (Wu et al., 2011; Williams et al., 2011; Yildirim et al., 2011), suggesting the existence of crosstalk between TET1-OGT and TET1-chromatin remodelers in gene transcription. Consistent with this conjecture, knocking down either TET1 or OGT in mouse ES cells leads to reduced genomic targeting of both Ezh2

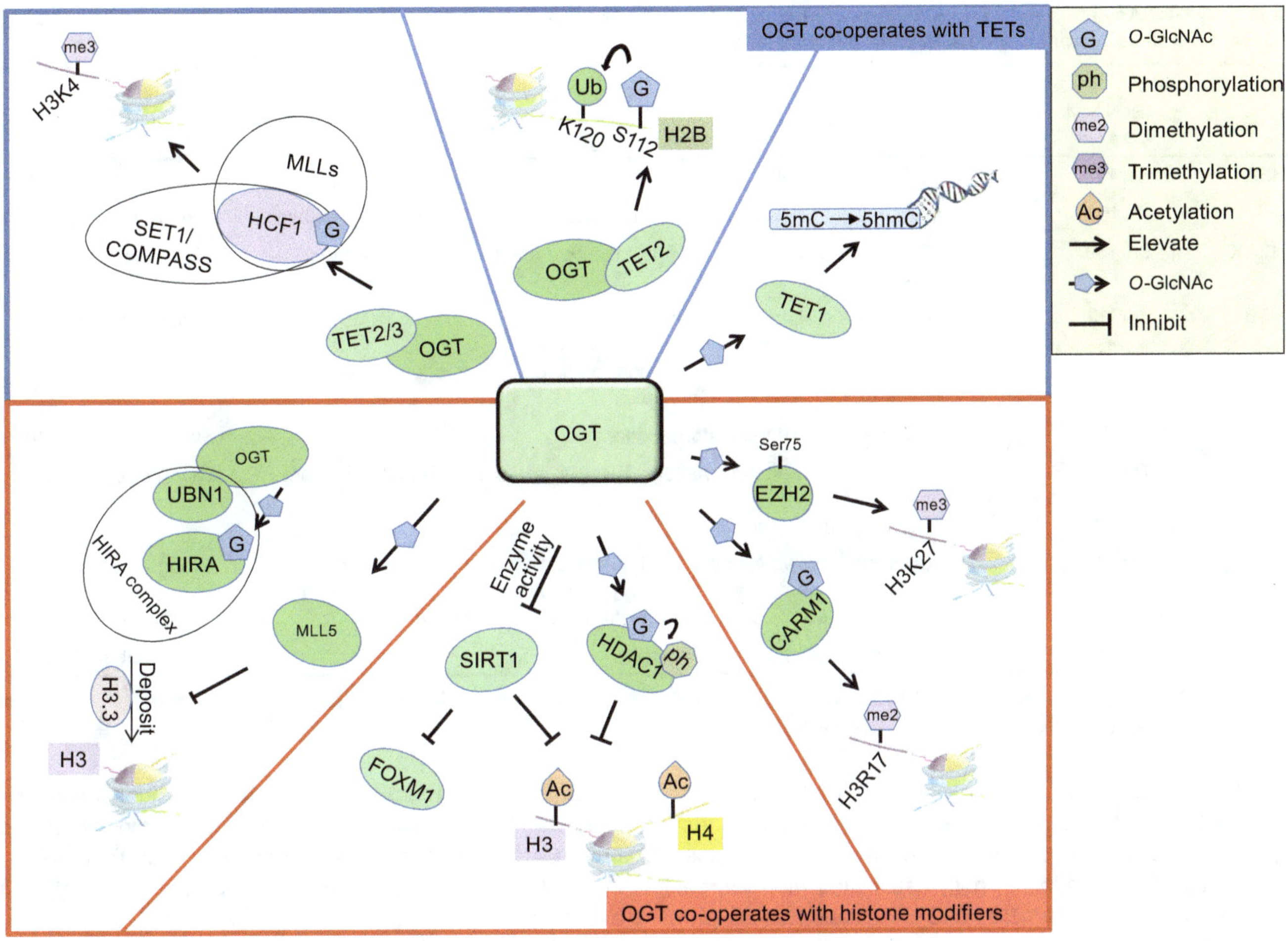

Figure 3. Crosstalk between *O*-GlcNAcylation and chromatin remodelers. Extensive connections of *O*-GlcNAcylation with chromatin modifiers are detected in cells.

(enhancer of Zeste 2 polycomb repressive complex 2 subunit, a subunit of PRC2 complex) and Sin3A, indicating the importance of TET1-OGT coordination in repressing developmental genes (Shi et al., 2013). In *drosophila*, polycomb group gene *super sex comb* (*sxc*) that regulates the bithorax complex is turned out to encode OGT, and play essential role in polycomb repression (Gambetta et al., 2009; Ingham, 1984). Confirming to this conclusion, TET1-OGT complex represses gene transcription through interaction with Sin3A/HDAC1/Sirt1 and PCR2/Ezh2 complexes (Hardivillé and Hart, 2012; Williams et al., 2012). Similar effects are reflected in TET2-OGT or TET3-OGT interaction. They can target OGT to transcriptional start site (TSS), and in collaboration with HCF1 recruit SET1/COPASS complexes, therefore activate gene transcription through histone H3K4me3 at target promoters (Deplus et al., 2013). TETs-OGT/*O*-GlcNAcylation is involved in some cancer development. In 5-fluorouracilresistant colon cancer cells (SNUC5/5-FUR), highly expressed OGT binds to TET1 and recruits to the Nrf2 (nuclear factor erythroid 2-related factor 2) promoter region, suggesting the role of OGT in TET1-mediated Nrf2 expression (Kang et al., 2016). Taken together, TETs proteins in collaboration with OGT play a critical role in regulating chromatin structure and gene transcription.

Coordination of *O*-GlcNAcylation and other modifiers on histones

Intracellular biological processes are extremely complex, and are often regulated by two or more histone modifiers in a co-ordinated manner. *O*-GlcNAcylation-mediated mechanism is no exception, often in collaboration with other histone modifications to regulate the biological processes in cells. Besides all four histones (H2A, H2B, H3, and H4) can be *O*-GlcNAcylated by OGT (Zhang et al., 2011; Sakabe et al., 2010) (Table 1), the accurate modified sites are also identified gradually. Genome-wide studies confirm that *O*-GlcNAcyaltion of histone H2A at Ser40 is dramatically changed during the differentiation in mouse trophoblast stem cells (Hirosawa et al., 2016). However, *O*-GlcNAcylation on histone H2A at Thr101 can relax chromatin structure through destabilizing H2A-H2B dimer (Lercher et al., 2015). On the

Table 1. *O*-GlcNAcylation sites and functions of histone tails. *O*-GlcNAc modification is observed in all four histones (H2A, H2B, H3 and H4) as well as histone variants H3.3 at indicated sites.

Histones	*O*-GlcNAcylated sites	Functions	References
H2A	Ser40	Tightly relates with the differentiation in mouse trophoblast stem cells	(Hirosawa et al., 2016)
	Thr101	Destabilizes H2A-H2B dimer, further relaxes the structure of chromatin	(Lercher et al., 2015)
H2AX	Ser139	Co-localizes with DNA damage foci, may function in DNA damage repair	(Chen and Yu 2016)
H2B	Ser112	Preserves a stable chromatin and represses gene transcription at the early stage of adipocyte differentiation Promotes H2BK120 ubiquitination, participates the regulation of H3K4me3 and gene transcription	(Ronningen et al., 2015) (Deplus et al., 2013)
	Ser36	May be a part of the histone code	(Sakabe et al., 2010)
H3	Thr32	Increases the phosphorylation of Thr32, Ser28, and Ser10, which are the specific mark of mitosis	(Zhang et al., 2011) (Fong et al., 2005)
	Ser10	Competitively reduces the levels of H3S10 phosphorylation, therefore regulates the pathway that H3S10P involved in, such as passing the G_2-M phase check point, regulating the H4K16ac	(Zhang et al., 2011)
H4	Ser47	May be a part of the histone code	(Sakabe et al., 2010)

Ser40/139/112/10/36/47, serine residues 40/139/112/10/36/47; Thr101/32, threonine residue 101/32; H2BK120, H2B lysine 120; H3K4me3, H3 lysine 4 tri-methylation; H4K16ac, H4 lysine 16 acetylation.

other hand, occurrence of *O*-GlcNAcylation on histone variant H2AX at Ser139 is often detected in DNA damage foci (Chen and Yu, 2016). Obviously, *O*-GlcNAcylation on histone H2A at different sites is tightly associated with different intracellular functions. In cells, *O*-GlcNAcylation on histone H2BS112 may preserve a stable chromatin at the early stage of adipocyte differentiation, thus repressing gene transcription in cell fate (Ronningen et al., 2015). And H2BS112-*O*-GlcNAcylation facilitates H2B at K120 ubiquitination, the latter further acts as a platform recruiting the SET1/COMPASS complex binding to histone H3, thereby activates gene transcription through histone H3K4me3 (Deplus et al., 2013).

Interestingly, competitive modification between *O*-GlcNAcylation and phosphorylation on Ser/Thr residues of substrate proteins may be closely related with functional switch. For example, higher *O*-GlcNAcylated H3 at Thr32 is observed during interface than mitosis. Further research demonstrates that *O*-GlcNAcylated Thr32 reduces mitosis-specific phosphorylation of Thr32, Ser28, and Ser10 on H3, suggesting the switching function of *O*-GlcNAcylation-mediated Thr32 in mitosis (Zhang et al., 2011; Fong et al., 2005). Importantly, according to *O*-GlcNAcylation of H3 can competitively reduce the level of H3S10 phosphorylation, but removal of *O*-GlcNAc from H3S10 is required for entering mitosis during the G_2-M transition phase (Zhang et al., 2011), H3S10 has been considered as a molecular checkpoint for entering mitosis (Van Hooser et al., 1998). Further study confirmed that H3S10 phosphorylation provides a binding platform for the phospho-binding 14-3-3 proteins and histone acetyltransferase MOF to trigger acetylation of histone H4 at lysine 16 (H4K16ac), and H3S10 phosphorylation and H4K16ac further coordinatively regulate the binding site for bromodomain protein BRD4 (Zippo et al., 2009). Moreover, in addition to up-regulation of H3S10 phosphorylation in hepatocellular carcinoma and primary lung cancer (Zhu et al., 2016), there is sufficient evidence to prove that H3S10 phosphorylation is responsible for neoplastic cell transformation and oncogene c-fos/c-Jun activation (Choi et al., 2005), suggesting the important coordinative role between *O*-GlcNAcylation and other histone modifications in tumor development.

Coordination of *O*-GlcNAcylation and the other chromatin remodelers

In addition to direct glycosylation of histones, OGT can also affect other histones modifications by crosslinking other chromatin remodelers. In line with this view, overexpression of OGT raises global H3K9ac and H3K27me3 level in cells, suggesting that OGT may collaborate with other chromatin remodelers to regulate histone modification (Sakabe and Hart, 2010). Histone H3 is a well-known specific substrate of CARM1 (a co-activator-associated argine methyltrnsferse 1), however the arginine methylation at H3R17 was regulated by *O*-GlcNAc modification of CARM1 (Charoensuksai et al., 2015; Sakabe and Hart, 2010; Schurter et al., 2001). Interestingly, the crosstalk between OGT and chromatin remodelers also exists in DNA damage repair pathway. *O*-GlcNAc-modified H2AX and MDC1 (mediator of DNA damage check point 1) are enriched at DNA damage foci, but suppress the

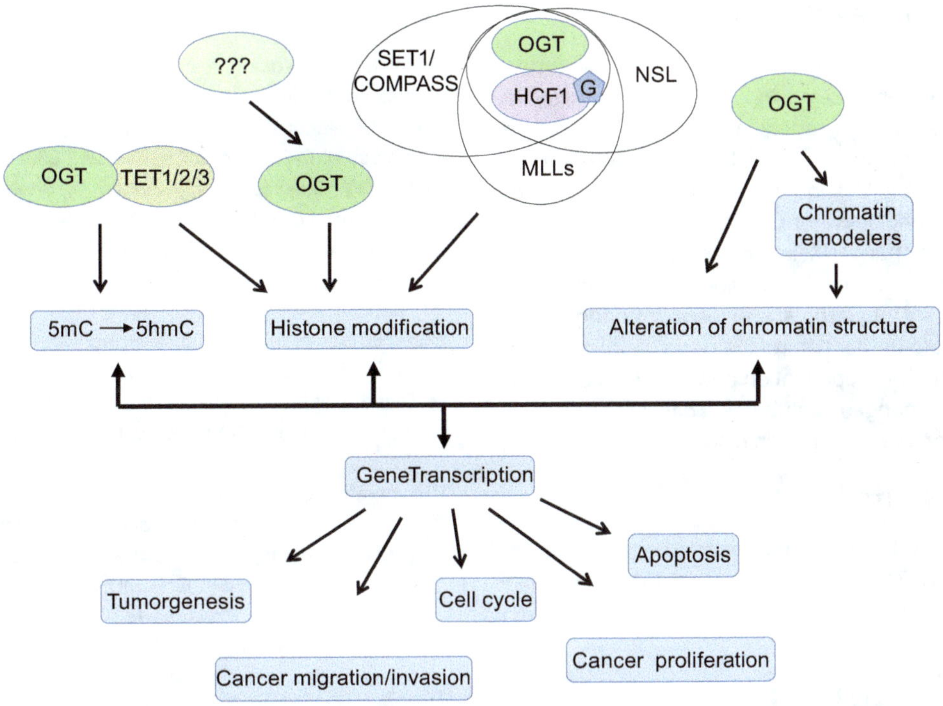

Figure 4. Schematic diagram of coordination between OGT-mediated *O*-GlcNAcylation and chromatin remodelers in intracellular fundamental functions.

expansion of DNA damage-produced phosphorylation at H2AX Ser139 site on the chromatin (Chen and Yu, 2016). What happens on histone variant H3.3 is another example. OGT regulates histone chaperone HIRA complex via *O*-GlcNAc modification, and subsequently deposits histone variant H3.3 to genic regions, then further governs H3.3 nucleosome assembly and cell senescence (Lee and Zhang, 2016; Ricketts and Marmorstein, 2016). Although the mechanism of targeting of histone H3.3 deposition at specific loci in whole genome remains largely unclear, histone H3.3 has been considered as epigenetic regulator that is dispensable for proper expression of genes (Kato et al., 2015). Furthermore, the function of histone variant H3.3 is also regulated by *O*-GlcNAcylated MLL5 which is a specific histone H3K4 trimethylase (Gallo et al., 2015). In cells, *O*-GlcNAcylated MLL5 represses histone H3.3 expression and facilitates the self-renewal of adult glioblastoma (Ding et al., 2015; Sebastian et al., 2009). On the other hand, histone H3.3 substitution and mutation are found in several tumors such as bone tumors and glioblastoma (Kato et al., 2015). However, whether substituted and mutated H3.3 is related to OGT-mediated *O*-GlcNAcylation remains to be further investigated.

Crosstalk of *O*-GlcNAcylation and epigenetics in cancer

Intracellular unbalanced epigenetic changes or abnormal *O*-GlcNAcylation levels can lead to various diseases. However, the evidence of crosslink between epigenetic changes and *O*-GlcNAcylation in tumorigenesis is gradually increased. For example, in addition to modulating the activity of deacetylase SIRT1 in an AMPK-dependent manner in breast cancer cells (Ferrer et al., 2017), OGT-mediated *O*-GlcNAcylation can also affect cancer cell growth and invasion through regulating the oncogenic transcription factor FOXM1 (Caldwell et al., 2010), suggesting the interplay of *O*-GlcNAcylation and SIRT1 in cancer cells. Another deacetylase HDAC1 also can be *O*-GlcNAc modified to promote its phosphorylation level, and simultaneously enhances its enzyme activity. This *O*-GlcNAc modified HDAC1 further functions in regulating the global histone acetylation level, and therefore influences the cell proliferation, invasion, and migration by regulating target genes in HepG2 cells (Zhu et al., 2016), suggesting the potential coordinative role in hepatic carcinoma. Given that abnormal histone acetylation has been implicated in tumorigenesis, the crosslink mechanism of *O*-GlcNAcylation and histone acetylation might provide a new direction for cancer therapy. On the other hand, OGT can modulate the integrity of multi-protein complex through *O*-GlcNAcylation-mediated subunit protein stability. For instance, OGT-mediated *O*-GlcNAcylation of Ezh2 at Ser75 stabilizes the PCR2 complex, which further promotes histone H3K27 tri-methylation in breast cancer MCF-7 cells (Chu et al., 2014). Taken together, the collaborative effects of OGT with histones and other non-histone proteins have been involved in gene transcription and tumorigenesis. Therefore, figuring out the mechanism of the network between OGT and epigenetic changes in tumorigenesis will provide a novel theoretical basis for cancer research.

CONCLUSION AND PERSPECTIVES

In summary, as the only intracellular mono-*O*-GlcNAc transferase, OGT has been implicated in fundamental intracellular biological functions through regulating cellular metabolism, chromatin structure, and gene transcription by cooperating with epigenetic mechanism. Although the precise regulation mechanism of network between OGT-mediated *O*-GlcNAcylation and epigenetic changes still remains largely unclear and to be further investigated, increasing evidence has proven that OGT-mediated *O*-GlcNAcylation and epigenetic changes form a complex network to further control the gene expression and signaling pathways. This coordination between *O*-GlcNAcylation and epigenetic changes contributes a critical role in initiation and progression of many human diseases (Fig. 4).

ACKNOWLEDGEMENTS

This work was supported by the National Natural Science Foundation of China (Grant No. 31571316) and the Project of Jilin Province Science and Technology Development Program (No. 20140414057GH).

ABBREVIATIONS

AMPK, AMP-activated protein kinase; CARM1, a co-activator-associated argine methyltransferse 1; Ezh2, enhancer of Zeste 2; ESC, mouse embryonic stem cells; H2BK120, histone H2B at lysine 120; H2BS112, histone H2B at serine 112; H3K4me3, histone H3K4 tri-methylation; H4K16ac, H4K16 acetylation; HATs, histone acetyltransferases; HBP, hexosamine biosynthetic pathway; HCF1, host cell factor 1; HDACs, histone deacetylases; MLL5, mixed lineage leukemia 5; MOF, males absent on the first; mOGT, the mitochondrial isoform of OGT; MSL, male-specific lethal; ncOGT, nucleocytoplasmic isoform of OGT; Nrf2, nuclear factor erythroid 2-related factor 2; NSL, non-specific lethal; OGA, glycoside hydrolase *O*-GlcNAcase; *O*-GlcNAcylation, *O*-GlcNAc modification; OGT, β-N-acetylglucosamine (*O*-GlcNAc) transferase; sOGT, short isoform of OGT; TET, ten-eleven translocation; TPR, tetratricopeptide-repeat; USP7, ubiquitin specific protease 7.

COMPLIANCE WITH ETHICS GUIDELINES

Donglu Wu, Yong Cai, and Jingji Jin declare that they have no conflict of interest. This article does not contain any studies with human or animal subjects performed by any of the authors.

AUTHORS CONTRIBUTIONS

Donglu Wu, Yong Cai, and Jingji Jin participated in writing, editing and making figures. All authors read and approved the final manuscript.

REFERENCES

Bannister AJ, Kouzarides T (2011) Regulation of chromatin by histone modifications. Cell Res 21:381–395

Bond MR, Hanover JA (2015) A little sugar goes a long way: the cell biology of *O*-GlcNAc. J Cell Biol 208:869–880

Bullen JW, Balsbaugh JL, Chanda D, Shabanowitz J, Hunt DF, Neumann D, Hart GW (2014) Cross-talk between two essential nutrient-sensitive enzymes: *O*-GlcNAc transferase (OGT) and AMP-activated protein kinase (AMPK). J Biol Chem 289:10592–10606

Burén S, Gomes AL, Teijeiro A, Fawal MA, Yilmaz M, Tummala KS, Perez M, Rodriguez-Justo M, Campos-Olivas R, Megías D, Djouder N (2016) Regulation of OGT by URI in response to glucose confers c-MYC-dependent survival mechanisms. Cancer Cell 30:290–307

Butkinaree C, Cheung WD, Park S, Park K, Barber M, Hart GW (2008) Characterization of beta-N-acetylglucosaminidase cleavage by caspase-3 during apoptosis. J Biol Chem 283:23557–23566

Cai Y, Jin J, Swanson SK, Cole MD, Choi SH, Florens L, Washburn MP, Conaway JW, Conaway RC (2010) Subunit composition and substrate specificity of a MOF-containing histone acetyltransferase distinct from the male-specific lethal (MSL) complex. J Biol Chem 285:4268–4672

Caldwell SA, Jackson SR, Shahriari KS, Lynch TP, Sethi G, Walker S, Vosseller K, Reginati MJ (2010) Nutrient sensor *O*-GlcNAc transferase regulates breast cancer tumorigenesis through targeting of the oncogenic transcription factor FoxM1. Oncogene 29:2831–2842

Capotosti F, Guernier S, Lammers F, Waridel P, Cai Y, Jin J, Conaway JW, Conaway RC, Herr W (2011) *O*-GlcNAc transferase catalyzes site-specific proteolysis of HCF-1. Cell 144:376–388

Cassey PJ (1995) Protein lipidation in cell signaling. Science 268:221–225

Charoensuksai P, Kuhn P, Wang L, Sherer N, Xu W (2015) *O*-GlcNAcylation of co-activator-associated arginine methyltransferase 1 regulates its protein substrate specificity. Biochem. J. 466:587–599

Chen Q, Yu X (2016) OGT restrains the expansion of DNA damage signaling. Nucleic Acids Res 44:9266–9278

Chen Q, Chen Y, Bian C, Fujiki R, Yu X (2013) TET2 promotes histone *O*-GlcNAcylation during gene transcription. Nature 493:561–564

Choi HS, Choi BY, Cho YY, Mizuno H, Kang BS, Bode AM, Dong Z (2005) Phosphorylation of Histone H3 at Serine 10 is Indispensable for Neoplastic Cell Transformation. Cancer Res 65:5818–5827

Chou TY, Hart GW, Dang CV (1995) c-Myc is glycosylated at threonine 58, a known phosphorylation site and a mutational hot spot in lymphomas. J Biol Chem 270:18961–18965

Chu CS, Lo PW, Yeh YH, Hsu PH, Peng SH, Teng YC, Kang ML, Wong CH, Juan LJ (2014) *O*-GlcNAcylation regulates EZH2 protein stability and function. Proc Natl Acad Sci USA 111:1355–1360

Comtesse N, Maldener E, Meese E (2001) Identification of a nuclear variant of MGEA5: a cytoplasmic hyaluronidase and a beta-N-acetylglucosaminidase. Biochem Biophys Res Commun 283:634–640

Copeland RJ, Bullen JW, Hart GW (2008) Cross-talk between GlcNAcylation and phosphorylation: roles in insulin resistance and glucose toxicity. Am J Physiol Endocrinol Metab 295:E17–E28

de Queiroz RM, Madan R, Chien J, Dias WB, Slawson C (2016) Changes in O-Linked N-Acetylglucosamine (*O*-GlcNAc) Homeostasis Activate the p53 Pathway in Ovarian Cancer Cells. J Biol Chem 291:18897–18914

Delhommeau F, Dupont S, Della Valle V, James C, Trannoy S, Massé A, Kosmider O, Le Couedic JP, Robert F, Alberdi A, Lécluse Y, Plo I, Dreyfus FJ, Marzac C, Casasevall N, Lacombe C, Romana SP, Dessen P, Soulier J, Viquié F, Fontenay M, Vainchenker W, Bernard OA (2009) Mutation in TET2 in myeloid cancers. N Engl J Med 360:2289–2301

Deplus R, Delatte B, Schwinn MK, Defrance M, Méndez J, Murphy N, Dawson MA, Volkmar M, Putmans P, Calonne E, Shih AH, Levine RL, Bernard O, Mercher T, Solary E, Urh M, Daniels DL, Fuks F (2013) TET2 and TET3 regulate GlcNAcylation and H3K4 methylation through OGT and SET1/COMPASS. EMBO J 32:645–655

Ding X, Jiang W, Zhou P, Liu L, Wan X, Yuan X, Wang X, Chen M, Chen J, Yang J, Kong C, Li B, Peng C, Wong CC, Hou F, Zhang Y (2015) Mixed Lineage Leukemia 5 (MLL5) Protein stability is cooperatively regulated by *O*-GlcNAc transferase (OGT) . and ubiquitin specific protease 7 (USP7). PLoS ONE 10: e0145023

Ferrer CM, Lu TY, Bacigalupa ZA, Katsetos CD, Sinclair DA, Reginato MJ (2017) *O*-GlcNAcylation regulates breast cancer metastasis via SIRT1 modulation of FOXM1 pathway. Oncogene 36:559–569

Fong JJ, Nguyen BL, Bridger R, Medrano EE, Wells L, Pan S, Sifers RN (2005) Beta-N-acetylglucosamine (*O*-GlcNAc) is a novel regulator of mitosis-specific phosphorylations on histone H3. J Biol Chem 287:12195–12203

Fu X, Jin L, Wang X, Luo A, Hu J, Zheng X, Tsark WM, Riggs AD, Ku HT, Huang W (2013) MicroRNA-26a targets ten eleven translocation enzymes and is regulated during pancreatic cell differentiation. Proc Natl Acad Sci USA 110:17892–17897

Fujiki R, Hashiba W, Sekine H, Yokoyama A, Chikanishi T, Ito S, Imai Y, Kim J, He HH, Igarashi K, Kanno J, Ohtake F, Kitagawa H, Roeder RG, Brown M, Kato S (2011) GlcNAcylation of histone H2B facilitates its monoubiquitination. Nature 480:557–560

Gallo M, Coutinho FJ, Vanner RJ, Gayden T, Mack SC, Murison A, Remke M, Li R, Takayama N, Desai K, Lee L, Lan X, Park NI, Barsyte-Lovejoy D, Smil D, Sturm D, Kushida MM, Head R, Cusimano MD, Bernstein M, Clarke ID, Dick JE, Pfister SM, Rich JN, Arrowsmith CH, Taylor MD, Jabado N, Bazett-Jones DP, Lupien M, Dirks PB (2015) MLL5 orchestrates a cancer self-renewal state by repressing the histone variant H3.3 and globally reorganizing chromatin. Cancer Cell 28:715–729

Gambetta MC, Oktaba K, Müller J (2009) Essential role of the glycosyltransferase sxc/Ogt in polycomb repression. Science 325:93–96

Gao Y, Wells L, Comer FI, Parker GJ, Hart GW (2001a) Dynamic *O*-glycosylation of nuclear and cytosolic proteins: cloning and characterization of a neutral, cytosolic beta-N-acetylglucosaminidase from human brain. J Biol Chem 276:9838–9845

Gao Y, Wells L, Comer FI, Parker GJ, Hart GW (2001b) Dynamic *O*-glycosylation of nuclear and cytosolic proteins: cloning and characterization of a neutral, cytosolic beta-N-acetylglucosaminidase from human brain. J Biol Chem 276:9838–9845

Gerhold CB, Gasser SM (2014) INO80 and SWR complexes: relating structure to function in chromatin remodeling. Trends Cell Biol 24:619–631

Gu Y, Mi W, Ge Y, Liu H, Fan Q, Han C, Yang J, Han F, Lu X, Yu W (2010) GlcNAcylation plays an essential role in breast cancer metastasis. Cancer Res 70:6344–6351

Gu Y, Gao J, Han C, Zhang X, Liu H, Ma L, Sun X, Yu W (2014) *O*-GlcNAcylation is increased in prostate cancer tissues and enhances malignancy of prostate cancer cells. Mol. Med. Rep. 10:897–904

Ha C, Lim K (2015) *O*-GlcNAc modification of Sp3 and Sp4 transcription factors negatively regulates their transcriptional activities. Biochem Biophys Res Commun 467:341–347

Hanover JA, Yu S, Lubas WB, Shin SH, Ragano-Caracciola M, Kochran J, Love DC (2003) Mitochondrial and nucleocytoplasmic isoforms of O-linked GlcNAc transferase encoded by a single mammalian gene. Arch Biochem Biophys 409:287–297

Hanover JA, Krause MW, Love DC (2012) Post-translational modifications: bittersweet memories: linking metabolism to epigenetics through *O*-GlcNAcylation. Nat Rev Mol Cell Biol 13:312–321

Hardivillé S, Hart GW (2012) Nutrient regulation of gene expression by *O*-GlcNAcylation of chromatin. Curr Opin Chem Biol 33:88–94

Hirosawa M, Hayakawa K, Yoneda C, Arai D, Shiota H, Suzuki T, Tanaka S, Dohmae N, Shiota K (2016) Novel *O*-GlcNAcylation on Ser(40) of canonical H2A isoforms specific to viviparity. Sci. Rep. 6:31785

Ingham PW (1984) A gene that regulates the bithorax complex differentially in larval and adult cells of Drosophila. Cell 37:815–823

Itkonen HM, Minner S, Guldvik IJ, Sandmann MJ, Tsourlakis MC, Berge V, Svindland A, Schlomm T, Mills IG (2013) *O*-GlcNAc transferase integrates metabolic pathways to regulate the stability of c-MYC in human prostate cancer cells. Cancer Res 73:5277–5287

Ito S, Shen L, Dai Q, Wu SC, Collins LB, Swenberg JA, He C, Zhang Y (2011) Tet proteins can convert 5-methylcytosine to 5-formylcytosine and 5-carboxylcytosine. Science 333:1300–1303

Ito R, Katsura S, Shimada H, Tsuchiya H, Hada M, Okumura T, Sugawara A, Yokoyama A (2014) TET3-OGT interaction increases the stability and the presence of OGT in chromatin. Genes Cells 19:52–65

Jin J, Cai Y, Li B, Conaway RC, Workman JL, Conaway JW, Kusch T (2005) In and out: histone variant exchange in chromatin. Trends Biochem Sci 30:680–687

Jinek M, Rehwinkel J, Lazarus BD, Izaurralde E, Hanover JA, Conti E (2004) The superhelical TPR-repeat domain of O-linked GlcNAc transferase exhibits structural similarities to importin alpha. Nat Struct Mol Biol 11:1001–1007

Kamigaito T, Okaneya T, Kawakubo M, Shimojo H, Nishizawa O, Nakayama J (2014) Overexpression of *O*-GlcNAc by prostate cancer cells is significantly associated with poor prognosis of patients. Prostate Cancer Prostatic Dis 17:18–22

Kang KA, Piao MJ, Ryu YS, Kang HK, Chang WY, Keum YS, Hyun JW (2016) Interaction of DNA demethylase and histone methyltransferase upregulates Nrf2 in 5-fluorouracil-resistant colon cancer cells. Oncotarget 7:40594–40620

Kato S, Ishii T, Kouzmenko A (2015) Point mutations in an epigenetic factor lead to multiple types of bone tumors: role of H3.3 histone variant in bone development and disease. Bonekey. Rep. 4:715

Kreppel LK, Hart GW (1999) Regulation of a cytosolic and nuclear *O*-GlcNAc transferase. Role of the tetratricopeptide repeats. J Biol Chem 274:32015–32022

Kreppel LK, Blomberg MA, Hart GW (1997) Dynamic glycosylation of nuclear and cytosolic proteins. Cloning and characterization of a unique *O*-GlcNAc transferase with multiple tetratricopeptide repeats. J Biol Chem 272:9308–9315

Krzeslak A, Forma E, Bernaciak M, Romanowicz H, Brys M (2012) Gene expression of *O*-GlcNAc cycling enzymes in human breast cancers. Clin Exp Med 12:61–65

Langemeijer SM, Kuiper RP, Berends M, Knops R, Aslanyan MG, Massop M, Stevens-Linders E, van Hoogen P, van Kessel AG, Raymakers RA, Kamping EJ, Verhoef GE, Verburgh E, Hagemeijer A, Vandenberghe P, de Witte T, van der Reijden BA, Jansen JH (2009) Acquired mutations in TET2 are common in myelodysplastic syndromes. Nat. Genet. 41:838–842

Längst G, Manelyte L (2015) Chromatin remodelers: from function to dysfunction. Genes 6:299–324

Lazarus BD, Love DC, Hanover JA (2006) Recombinant *O*-GlcNAc transferase isoforms: identification of *O*-GlcNAcase, yes tyrosine kinase, and tau as isoform-specific substrates. Glycobiology 16:415–421

Lazarus MB, Nam Y, Jiang J, Sliz P, Walker S (2011) Structure of human *O*-GlcNAc transferase and its complex with a peptide substrate. Nature 469:564–567

Lee JS, Zhang Z (2016) O-linked N-acetylglucosamine transferase (OGT) interacts with the histone chaperone HIRA complex and regulates nucleosome assembly and cellular senescence. Proc Natl Acad Sci USA 113:E3213–E3220

Lercher L, Raj R, Patel NA, Price J, Mohammed S, Robinson CV, Schofield CJ, Davis BG (2015) Generation of a synthetic GlcNAcylated nucleosome reveals regulation of stability by H2A-Thr101 GlcNAcylation. Nat. Commun. 6:7978

Liu Y, Li X, Yu Y, Shi J, Liang Z, Run X, Li Y, Dai CL, Grundke-Iqbal I, Iqbal K, Liu F, Gong CX (2012) Developmental regulation of protein *O*-GlcNAcylation, *O*-GlcNAc transferase, and *O*-GlcNAcase in mammalian brain. PLoS ONE 7:43724

Lorsbach RB, Moore J, Mathew S, Raimondi SC, Mukatira ST, Downing JR (2003) TET1, a member of a novel protein family, is fused to MLL in acute myeloid leukemia containing the t (10;11) (q22;q23). Leukemia 17:637–641

Love DC, Kochan J, Cathey RL, Shin SH, Hanover JA (2003) Mitochondrial and nucleocytoplasmic targeting of *O*-linked GlcNAc transferase. J Cell Sci 116:647–654

Lubas WA, Hanover JA (2000) Functional expression of O-linked GlcNAc transferase. Domain structure and substrate specificity. J Biol Chem 275:10983–10988

Lynch TP, Ferrer CM, Jackson SR, Shahriari KS, Vosseller K, Reginato MJ (2012) Critical role of O-linked beta-Nacetylglucosamine transferase in prostate cancer invasion, angiogenesis, and metastasis. J Biol Chem 287:11070–11081

Mailleux F, Gélinas R, Beauloye C, Horman S, Bertrand L (2016) *O*-GlcNAcylation, enemy or ally during cardiac hypertrophy development? Biochim Biophys Acta 1862:2232–2243

Marshall S, Bacote V, Traxinger RR (1991) Discovery of a metabolic pathway mediating glucose-induced desensitization of the glucose transport system. Role of hexosamine biosynthesis in the induction of insulin resistance. J Biol Chem 266:4706–4712

Mazars R, Gonzalez-de-Peredo A, Cayrol C, Lavigne AC, Vogel JL, Ortega N, Lacroix C, Gautier V, Huet G, Ray A, Monsarrat B, Kristie TM, Girard JP (2010) The THAP-Zinc Finger Protein THAP1 Associates with Coactivator HCF-1 and *O*-GlcNAc Transferase, a link between DYT6 and DYT3 dystonias. J Biol Chem 285:13364–13371

Mi W, Gu Y, Han C, Liu H, Fan Q, Zhang X, Cong Q, Yu W (2011) *O*-GlcNAcylation is a novel regulator of lung and colon cancer malignancy. Biochim Biophys Acta 1812:514–519

Nakamura K, Kato A, Kobayashi J, Yanagihara H, Sakamoto S, Oliveira DV, Shimada M, Tauchi H, Suzuki H, Tashiro S, Zou L, Komatsu K (2011) Regulation of homologous recombination by RNF20-dependent H2B ubiquitination. Mol Cell 41:515–528

O'Donnell N, Zachara NE, Hart GW, Marth JD (2004) Ogt-dependent X-chromosome-linked protein glycosylation is a requisite modification in somatic cell function and embryo viability. Mol Cell Biol 224:1680–1690

Ono R, Taki T, Taketani T, Taniwaki M, Kobayashi H, Hayashi Y (2002) LCX, leukemia-associated protein with a CXXC domain, is fused to MLL in acute myeloid leukemia with trilineage dysplasia having t(10;11)(q22;q23). Cancer Res 62:4075–4080

Phoomak C, Vaeteewoottacharn K, Sawanyawisuth K, Seubwai W, Wongkham C, Silsirivanit A, Wongkhamb S (2016) Mechanistic insights of *O*-GlcNAcylation that promote progression of cholangiocarcinoma cells via nuclear translocation of NF-κB. Sci. Rep. 6:27853

Qiao Z, Dang C, Zhou B, Li S, Zhang W, Jiang J, Zhang J, Kong R, Ma Y (2012) O-linked N-acetylglucosamine transferase (OGT) is overexpressed and promotes O-linked protein glycosylation in esophageal squamous cell carcinoma. J. Biomed Res 26:268–273

Ricketts MD, Marmorstein R (2016) A molecular prospective for HIRA complex assembly and H3.3-specific histone chaperone. J Mol Biol. doi:10.1016/j.jmb.2016.11.010

Riu IH, Shin IS, Do SI (2008) Sp1 modulates ncOGT activity to alter target recognition and enhanced thermotolerance in *E. coli*. Biochem Biophys Res Commun 372:203–209

Ronningen T, Shah A, Oldenburg AR, Vekterud K, Delbarre E, Moskaug JO, Collas P (2015) Prepatterning of differentiation-driven nuclear lamin A/C-associated chromatin domains by GlcNAcylated histone H2B. Genome Res 25:1825–1835

Ruan HB, Han X, Li MD, Singh JP, Qian K, Azarhoush S, Zhao L, Bennett AM, Samuel VT, Wu J, Yates JR, Yang X (2012) *O*-GlcNAc transferase/host cell factor C1 complex regulates gluconeogenesis by modulating PGC-1α stability. Cell Metab 16:226–237

Sacoman JL, Dagda RY, Burnham-Marusich AR, Dagda RK, Berninsone PM (2017) Mitochondrial *O*-GlcNAc transferase (mOGT) regulates mitochondrial structure, function and survival in HeLa cells. J Biol Chem 292:4499–4518

Sakabe K, Hart GW (2010) *O*-GlcNAc Transferase Regulates Mitotic Chromatin Dynamics. J Biol Chem 285:34460–34468

Sakabe K, Wang Z, Hart GW (2010) Beta-N-acetylglucosamine (O-GlcNAc) is part of the histone code. Proc Natl Acad Sci USA 107:19915–19920

Schurter BT, Koh SS, Chen D, Bunick GJ, Harp JM, Hanson BL, Henschen-Edman A, Mackay DR, Stallcup MR, Aswad DW (2001) Methylation of histone H3 by coactivator-associated arginine methyltransferase 1. Biochemistry 40:5747–5756

Scourzic L, Mouly E, Bernard OA (2015) TET proteins and the control of cytosine demethylation in cancer. Genome Med. 7:9

Sebastian S, Sreenivas P, Sambasivan R, Cheedipudi S, Kandalla P, Pavlath GK, Dhawan J (2009) MLL5, a trithorax homolog, indirectly regulates H3K4 methylation, represses cyclin A2 expression, and promotes myogenic differentiation. Proc Natl Acad Sci USA 106:4719–4724

Shi FT, Kim H, Lu W, He Q, Liu D, Goodell MA et al (2013) Ten-eleven transloca tion 1 (Tet1) is regulated by O-linked N-acetylglucosamine transferase (Ogt) for target gene repression in mouse embryonic stem cells. J Biol Chem 288:20776–20784

Shin SH, Love DC, Hanover JA (2011) Elevated *O*-GlcNAc-dependent signaling through inducible mOGT expression selectively triggers apoptosis. Amino Acids 40:885–893

Slawson C, Lakshmanan T, Knapp S, Hart GW (2008) A mitotic GlcNAcylation/phosphorylation signaling complex alters the post-translational state of the cytoskeletal protein vimentin. Mol Biol Cell 19:4130–4140

Tefferi A, Pardanani A, Lim KH, Abdel-Wahab O, Lasho TL, Patel J, Gangat N, Finke CM, Schwager S, Mullally A, Li CY, Hanson CA, Mesa R, Bernard O, Delhommeau F, Vainchenker W, Gilliland DG, Levine RL (2009) TET2 mutations and their clinical correlates in polycythemia vera, essential thrombocythemia and myelofibrosis. Leukemia 23:905–911

Toleman C, Paterson AJ, Whisenhunt TR, Kudlow JE (2004) Characterization of the histone acetyltransferase (HAT) domain of a bifunctional protein with activable *O*-GlcNAcase and HAT activities. J Biol Chem 279:53665–53673

Trapannone R, Mariappa D, Ferenbach AT, van Aalten DM (2016) Nucleocytoplasmic human *O*-GlcNAc transferase is sufficient for *O*-GlcNAcylation of mitochondrial proteins. Biochem. J. 473:1693–1702

Van Hooser A, Goodrich DW, Allis CD, Brinkley BR, Mancini MA (1998) Histone H3 phosphorylation is required for the initiation, but not maintenance, of mammalian chromosome condensation. J Cell Sci 111:3497–3506

Vella P, Scelfo A, Jammula S, Chiacchiera F, Williams K, Cuomo A, Roberto A, Christensen J, Bonaldi T, Helin K, Pasini D (2013) Tet proteins connect the O-linked N -acetylglucosamine transferase Ogt to chromatin in embryonic stem cells. Mol Cell 49:645–656

Wells L, Hart GW (2003) *O*-GlcNAc turns twenty: functional implications for post-translational modification of nuclear and cytosolic proteins with a sugar. FEBS Lett 546:154–158

Williams K, Christensen J, Pedersen MT, Johansen JV, Cloos PA, Rappsilber J, Helin K (2011) TET1 and hydroxymethylcytosine in transcription and DNA methylation fidelity. Nature 473:343–348

Williams K, Christensen J, Helin K (2012) DNA methylation: TET proteins-guardians of CpG islands? EMBO Rep 13:28–35

Wu H, D'Alessio AC, Ito S, Xia K, Wang Z, Cui K, Zhao K, Sun YE, Zhang Y (2011) Dual functions of Tet1 in transcriptional regulation in mouse embryonic stem cells. Nature 473:389–393

Yang WH, Kim JE, Nam HW, Ju JW, Kim HS, Kim YS, Cho JW (2006) Modification of p53 with O-linked N-acetylglucosamine regulates p53 activity and stability. Nat Cell Biol 8:1074–1083

Yildirim O, Li R, Hung JH, Chen PB, Dong X, Ee LS, Weng Z, Rando OJ, Fazzio TG (2011) Mbd3/NURD complex regulates expression of 5-hydroxymethylcytosine marked genes in embryonic stem cells. Cell 147:1498–1510

Zhang S, Roche K, Nasheuer HP, Lowndes NF (2011) Modification of histones by sugar β-N-acetylglucosamine (GlcNAc) occurs on multiple residues, including histone H3 serine 10, and is cell cycle-regulated. J Biol Chem 286:37483–37495

Zhu Q, Zhou L, Yang Z, Lai M, Xie H, Wu L, Xing C, Zhang F, Zheng S (2012) *O*-GlcNAcylation plays a role in tumor recurrence of hepatocellular carcinoma following liver transplantation. Med Oncol 29:985–993

Zhu X, Li D, Zhang Z, Zhu W, Li W, Zhao J, Xing X, He Z, Wang S, Wang F, Ma L, Bai Q, Zeng X, Li J, Gao C, Xiao Y, Wang Q, Chen L, Chen W (2016a) Persistent phosphorylation at specific H3 serine residues involved in chemical carcinogen-induced cell transformation. Carcinog, Mol. doi:10.1002/mc.22605

Zhu G, Tao T, Zhang D, Liu X, Qiu H, Han L, Xu Z, Xiao Y, Cheng C, Shen A (2016b) *O*-GlcNAcylation of histone deacetylase-1 in hepatocellular carcinoma promotes cancer progression. Glycobiology 26:820–833

Zippo A, Serafini R, Rocchigiani M, Pennacchini S, Krepelova A, Oliviero S (2009) Histone crosstalk between H3S10ph and H4K16ac generates a histone code that mediates transcription elongation. Cell 138:1122–1136

PERMISSIONS

 Every chapter published in this book has been scrutinized by our experts. Their significance has been extensively debated. The topics covered herein carry significant findings which will fuel the growth of the discipline. They may even be implemented as practical applications or may be referred to as a beginning point for another development.

The contributors of this book come from diverse backgrounds, making this book a truly international effort. This book will bring forth new frontiers with its revolutionizing research information and detailed analysis of the nascent developments around the world.

We would like to thank all the contributing authors for lending their expertise to make the book truly unique. They have played a crucial role in the development of this book. Without their invaluable contributions this book wouldn't have been possible. They have made vital efforts to compile up to date information on the varied aspects of this subject to make this book a valuable addition to the collection of many professionals and students.

This book was conceptualized with the vision of imparting up-to-date information and advanced data in this field. To ensure the same, a matchless editorial board was set up. Every individual on the board went through rigorous rounds of assessment to prove their worth. After which they invested a large part of their time researching and compiling the most relevant data for our readers.

The editorial board has been involved in producing this book since its inception. They have spent rigorous hours researching and exploring the diverse topics which have resulted in the successful publishing of this book. They have passed on their knowledge of decades through this book. To expedite this challenging task, the publisher supported the team at every step. A small team of assistant editors was also appointed to further simplify the editing procedure and attain best results for the readers.

Apart from the editorial board, the designing team has also invested a significant amount of their time in understanding the subject and creating the most relevant covers. They scrutinized every image to scout for the most suitable representation of the subject and create an appropriate cover for the book.

The publishing team has been an ardent support to the editorial, designing and production team. Their endless efforts to recruit the best for this project, has resulted in the accomplishment of this book. They are a veteran in the field of academics and their pool of knowledge is as vast as their experience in printing. Their expertise and guidance has proved useful at every step. Their uncompromising quality standards have made this book an exceptional effort. Their encouragement from time to time has been an inspiration for everyone.

The publisher and the editorial board hope that this book will prove to be a valuable piece of knowledge for researchers, students, practitioners and scholars across the globe.

LIST OF CONTRIBUTORS

Xiaoying Chen, Kunshan Zhang, Liqiang Zhou, Xinpei Gao, Junbang Wang, Yinan Yao, Fei He,Yuping Luo and Liming Cheng
Stem Cell Translational Research Center, Tongji Hospital, Tongji University School of Medicine, Shanghai 200065, China

Yi E. Sun
Stem Cell Translational Research Center, Tongji Hospital, Tongji University School of Medicine, Shanghai 200065, China
Department of Psychiatry and Biobehavioral Sciences, David Geffen School of Medicine, University of California, Los Angeles, CA 90095, USA
Collaborative Innovation Center for Brain Science, Tongji University, Shanghai 200092, China

Siguang Li
Stem Cell Translational Research Center, Tongji Hospital, Tongji University School of Medicine, Shanghai 200065, China
Collaborative Innovation Center for Brain Science, Tongji University, Shanghai 200092, China

Yongchun Yu
Institute of Neurobiology, Institute of Brain Science and State Key Laboratory of Medical Neurobiology, Fudan University, Shanghai 200032, China
Collaborative Innovation Center for Brain Science, Tongji University, Shanghai 200092, China

Ying Gu, Christopher Benner, Fei Yi, Keiichiro Suzuki, Mo Li and Concepcion Rodriguez Esteban
Gene Expression Laboratory, Salk Institute for Biological Studies, 10010 North Torrey Pines Road, La Jolla, CA 92037, USA

Juan Carlos Izpisua Belmonte
Gene Expression Laboratory, Salk Institute for Biological Studies, 10010 North Torrey Pines Road, La Jolla, CA 92037, USA
Center of Regenerative Medicine in Barcelona, Dr. Aiguader 88, 08003 Barcelona, Spain

Jing Qu, Jiping Yang and Weiqi Zhang
National Laboratory of Biomacromolecules, Institute of Biophysics, Chinese Academy of Sciences, Beijing 100101, China

Guang-Hui Liu
National Laboratory of Biomacromolecules, Institute of Biophysics, Chinese Academy of Sciences, Beijing 100101, China
State key Laboratory of Drug Research, Shanghai Institute of Materia Medica, Chinese Academy of Sciences, Shanghai 201203, China

Nongluk Plongthongkum and Kun Zhang
Department of Bioengineering, University of California at San Diego, La Jolla, CA 92093, USA

Nuria Montserrat
Center of Regenerative Medicine in Barcelona, Dr. Aiguader 88, 08003 Barcelona, Spain
Biomedical Research Networking Center in Bioengineering, Biomaterials and Nanomedicine (CIBER-BBN), Poeta Mariano Esquillor s/n, 50018 Sargossa, Spain

Isaac Crespo and Antonio del Sol
Luxembourg Centre for Systems Biomedicine (LCSB), University of Luxembourg, Walferdange, Luxembourg

Zhuoya Gu, Yanyan Huang, Mengyi Hua and Peng Nan
School of Life Sciences, Fudan University, Shanghai 200433, China

Yang Zhong
School of Life Sciences, Fudan University, Shanghai 200433, China
Institute of Biodiversity Science and Institute of High Altitude Medicine, Tibet University, Lhasa 850012, China

Ke Jin
Banting and Best Department of Medical Research, Donnelly Centre, University of Toronto, Toronto, ON M5S 1A1, Canada

Zhaolei Zhang
Banting and Best Department of Medical Research, Donnelly Centre, University of Toronto, Toronto, ON M5S 1A1, Canada
Department of Molecular Genetics, University of Toronto, Toronto, ON M5S 1A1, Canada

M. James C. Crabbe
Department of Zoology, University of Oxford, Tinbergen Building, South Parks Road, Oxford OX1 3PS, UK
Institute of Biomedical and Environmental Science & Technology, Department of Life Sciences, University of Bedfordshire, Park Square, Luton LU1 3JU, UK

Yang Zhang
Department of Bioengineering, University of Illinois at Urbana-Champaign, Champaign, IL 61801, USA

Xiaolin Liu
School of Public Health, University of Michigan, Ann Arbor, MI 48109, USA

Ming Liu, Lingxi Jiang and Xin-Yuan Guan
Department of Clinical Oncology, University of Hong Kong, Hong Kong, China

Peipei Liu, Jinliang Huang, Qian Zheng, Leiming Xie, Xinping Lu, Jie Jin and Geng Wang
MOE Key laboratory of Bioinformatics, Cell Biology and Development Center, School of Life Sciences, Tsinghua University, Beijing 100084, China

Na Qu, Zhao Ma, Mengrao Zhang and Antony K. Chen
Department of Biomedical Engineering, College of Engineering, Peking University, Beijing 100871, China

Muaz N. Rushdi and Christopher J. Krueger
Department of Biomedical Engineering, College of Engineering, Peking University, Beijing 100871, China
Wallace H Coulter Department of Biomedical Engineering, Georgia Institute of Technology, Atlanta, GA 30332, USA

Zhiguo Wang, Ling Hou and Jianfeng Li
Institute of Aging Research, School of Medicine, Hangzhou Normal University, Hangzhou 311121, China

Ruping Chen
Institute of Aging Research, School of Medicine, Hangzhou Normal University, Hangzhou 311121, China
School of Medicine, Shandong University, Jinan 250100, China

Jun-Ping Liu
Institute of Aging Research, School of Medicine, Hangzhou Normal University, Hangzhou 311121, China
School of Medicine, Shandong University, Jinan 250100, China
Department of Immunology, Central Eastern Clinical School, Monash University, Melbourne, VIC 3004, Australia
Center for Cancer Research, Monash Institute of Medical Research, Clayton, VIC 3168, Australia

Ping Wang, Chang Sun and Tingting Zhu
Fudan University Shanghai Cancer Center and Institute of Biomedical Sciences, Shanghai Medical College of Fudan University, Shanghai 200032, China

Yanhui Xu
Fudan University Shanghai Cancer Center and Institute of Biomedical Sciences, Shanghai Medical College of Fudan University, Shanghai 200032, China
State Key Laboratory of Genetic Engineering, School of Life Sciences, Fudan University, Shanghai 200433, China

Donglu Wu
School of Life Sciences, Jilin University, Changchun 130012, China

Yong Cai and Jingji Jin
School of Life Sciences, Jilin University, Changchun 130012, China
National Engineering Laboratory for AIDS Vaccine, Jilin University, Changchun 130012, China
Key Laboratory for Molecular Enzymology and Engineering, The Ministry of Education, Jilin University, Changchun 130012, China

Zeming Wu
State Key Laboratory of Stem Cell and Reproductive Biology, Institute of Zoology, Chinese Academy of Sciences, Beijing 100101, China
National Laboratory of Biomacromolecules, CAS Center for Excellence in Biomacromolecules, Institute of Biophysics, Chinese Academy of Sciences, Beijing 100101, China
University of Chinese Academy of Sciences, Beijing 100049, China
National Clinical Research Center for Geriatric Disorders, Xuanwu Hospital of Capital Medical University, Beijing 100053, China

Jing Qu
State Key Laboratory of Stem Cell and Reproductive Biology, Institute of Zoology, Chinese Academy of Sciences, Beijing 100101, China
University of Chinese Academy of Sciences, Beijing 100049, China

Wei Wang and Chang Chen
National Laboratory of Biomacromolecules, CAS Center for Excellence in Biomacromolecules, Institute of Biophysics, Chinese Academy of Sciences, Beijing 100101, China
University of Chinese Academy of Sciences, Beijing 100049, China

Weiqi Zhang
National Laboratory of Biomacromolecules, CAS Center for Excellence in Biomacromolecules, Institute of Biophysics, Chinese Academy of Sciences, Beijing 100101, China
University of Chinese Academy of Sciences, Beijing 100049, China

National Clinical Research Center for Geriatric Disorders, Xuanwu Hospital of Capital Medical University, Beijing 100053, China

Guang-Hui Liu
National Laboratory of Biomacromolecules, CAS Center for Excellence in Biomacromolecules, Institute of Biophysics, Chinese Academy of Sciences, Beijing 100101, China
University of Chinese Academy of Sciences, Beijing 100049, China
National Clinical Research Center for Geriatric Disorders, Xuanwu Hospital of Capital Medical University, Beijing 100053, China
Key Laboratory of Regenerative Medicine of Ministry of Education, Institute of Aging and Regenerative Medicine, Jinan University, Guangzhou 510632, China

Moshi Song
University of Chinese Academy of Sciences, Beijing 100049, China
State Key Laboratory of Biomembrane and Membrane Biotechnology, Institute of Zoology, Chinese Academy of Sciences, Beijing 100101, China

Wei Li, Jinghui Lei and Piu Chan
National Clinical Research Center for Geriatric Disorders, Xuanwu Hospital of Capital Medical University, Beijing 100053, China

Gang Wei
Key Laboratory of Computational Biology, CAS-MPG Partner Institute for Computational Biology, Shanghai Institutes for Biological Sciences, Chinese Academy of Sciences, Shanghai 200031, China

Yu Huang
Department of Medical genetics, School of Basic Medical Sciences, Peking University Health Science Center, Beijing 100191, China

Yanmei Sang
Department of Pediatric Endocrinology and Genetic Metabolism, Beijing Children's Hospital, Capital Medical University, National Center for Children's Health, Beijing 100045, China

Keiichiro Suzuki
Institute for Advanced Co-Creation Studies, Osaka University, Osaka 560-8531, Japan
Graduate School of Engineering Science, Osaka University, Osaka 560-8531, Japan

Juan Carlos Izpisua Belmonte
Gene Expression Laboratory, Salk Institute for Biological Studies, La Jolla 92037, USA

Hongbin Ji
Key Laboratory of Systems Biology, Shanghai Institutes for Biological Sciences, Chinese Academy of Sciences, Shanghai 200031, China
Innovation Center for Cell Signaling Network, Shanghai Institutes for Biological Sciences, Chinese Academy of Sciences, Shanghai 200031, China
CAS Center for Excellence in Molecular Cell Science, Institute of Biochemistry and Cell Biology, Shanghai Institutes for Biological Sciences, Chinese Academy of Sciences, Shanghai 200031, China
School of Life Science and Technology, Shanghai Tech University, Shanghai 200120, China
Institute of Biochemistry and Cell Biology, Shanghai Institutes for Biological Sciences, Chinese Academy of Sciences, Shanghai 200031, China

Yingjiao Xue, Shenda Hou and Xiangkun Han
Key Laboratory of Systems Biology, Shanghai Institutes for Biological Sciences, Chinese Academy of Sciences, Shanghai 200031, China
Innovation Center for Cell Signaling Network, Shanghai Institutes for Biological Sciences, Chinese Academy of Sciences, Shanghai 200031, China
CAS Center for Excellence in Molecular Cell Science, Institute of Biochemistry and Cell Biology, Shanghai Institutes for Biological Sciences, Chinese Academy of Sciences, Shanghai 200031, China
School of Life Science and Technology, Shanghai Tech University, Shanghai 200120, China

Jianghong Liu and Li Zhu
State Key Laboratory for Brain & Cognitive Science, Institute of Biophysics, Chinese Academy of Sciences, Beijing 100101, China

Yunjia Zhang and Mengmeng Chen
State Key Laboratory for Brain & Cognitive Science, Institute of Biophysics, Chinese Academy of Sciences, Beijing 100101, China
University of Chinese Academy of Sciences, Beijing 100049, China

Jane Y. Wu
State Key Laboratory for Brain & Cognitive Science, Institute of Biophysics, Chinese Academy of Sciences, Beijing 100101, China
Department of Neurology, Center for Genetic Medicine, Lurie Cancer Center, Northwestern University Feinberg School of Medicine, Chicago, IL 60611, USA

Zilong Qiu
Laboratory of Molecular Basis of Neural Plasticity, Institute of Neuroscience, Chinese Academy of Sciences, Shanghai 200031, China

Keping Hu
Institute of Medicinal Plant Development (IMPLAD), Chinese Academy of Medical Sciences, Beijing 100193, China

Warren McGee and Xiaoping Chen
Department of Neurology, Center for Genetic Medicine, Lurie Cancer Center, Northwestern University Feinberg School of Medicine, Chicago, IL 60611, USA

Xuemei Fu, Shengyun Zhu and Xiaocheng
Shenzhen Children's Hospital, Shenzhen 518026, China

Zhili Rong, Wang, Yang Xu and Blue B. Lake
Section of Molecular Biology, Division of Biological Sciences, University of California, San Diego, 9500 Gilman Drive, La Jolla, CA 92093-0322, USA

Qingzhao Feng, Xiaofeng Lu, Yin Zhang and Wenxian Guan
Department of General Surgery, The Affiliated Drum Tower Hospital of Medical School of Nanjing University, Nanjing 210008, China

Fuchao Li
Department of Gerontology, The Affiliated Drum Tower Hospital of Medical School of Nanjing University, Nanjing 210008, China

Ying Pan
Department of Orthopaedics, The Affiliated Drum Tower Hospital of Medical School of Nanjing University, Nanjing 210008, China

Beibei Ning
The Children's Hospital of Zhejiang University School of Medicine, Hangzhou 310003, China

Xionglin Wu
Department of Emergency, The Yiwu Affiliated Hospital of Wenzhou Medical University, Yiwu 322000, China

Shaohong Chen and Guangxia Gao
CAS Key Laboratory of Infection and Immunity, CAS Center for Excellence in Biomacromolecules, Institute of Biophysics, Chinese Academy of Sciences, Beijing 100101, China
University of Chinese Academy of Sciences, Beijing 100101, China

Liang Chen, Zhimin Peng, Qinghang Meng, Maureen Mongan, Jingcai Wang, Maureen Sartor, Jing Chen, Liang Niu and Mario Medvedovic
Department of Environmental Health and Center of Environmental Genetics, University of Cincinnati Medical Center, Cincinnati, OH 45267, USA

Ying Xia
Department of Environmental Health and Center of Environmental Genetics, University of Cincinnati Medical Center, Cincinnati, OH 45267, USA
Department of Ophthalmology, University of Cincinnati Medical Center, Cincinnati, OH 45267, USA

Winston Kao
Department of Ophthalmology, University of Cincinnati Medical Center, Cincinnati, OH 45267, USA

Jiebo Chen, Jiacheng Lin, Lin Xu, Zhiping Zeng, Liqun Chen, Mingfeng Huang, Mengjie Hu, Xiaohong Ye, Yuqi Zhou, Guanghui Wang and Hu Zhou
School of Pharmaceutical Sciences, Xiamen University, Xiamen 361102, China

Ying Su and Xiao-kun Zhang
School of Pharmaceutical Sciences, Xiamen University, Xiamen 361102, China
Cancer Center, Sanford-Burnham Medical Research Institute, 10901 N. Torrey Pines Road, La Jolla, CA 92037, USA

Fan Chen
School of Pharmaceutical Sciences, Xiamen University, Xiamen 361102, China
School of Biological Science and Biotechnology, Minnan Normal University, Zhangzhou 363000, China

Anton V. Cheltsov
Q-MOL LLC, San Diego, CA 92105, USA

Ya Chen
Cancer Center, Sanford-Burnham Medical Research Institute, 10901 N. Torrey Pines Road, La Jolla, CA 92037, USA

Long Zhang
Life Science Institute, Zhejiang University, Hangzhou 310058, China

Fangfang Zhou
Institutes of Biology and Medical Sciences, Soochow University, Suzhou 215123, China

Peijiang Yuan
Robotics Institute, Beihang University, Beijing 100191, China

Feng Su
Robotics Institute, Beihang University, Beijing 100191, China
State Key Laboratory of Membrane Biology, School of Life Sciences, Beijing 100871, China
PKU-IDG/McGovern Institute for Brain Research, Peking University, Beijing 100871, China

Chen Zhang and Yangzhen Wang
State Key Laboratory of Membrane Biology, School of Life Sciences, Beijing 100871, China
PKU-IDG/McGovern Institute for Brain Research, Peking University, Beijing 100871, China

Tian Yu
Intercollege Graduate Degree Program in Molecular, Cellular and Integrative Biosciences, the Pennsylvania State University, University Park, PA 16802, USA

John Bachman
Department of Biology, the Pennsylvania State University, University Park, PA 16802, USA

Zhi-Chun Lai
Intercollege Graduate Degree Program in Molecular, Cellular and Integrative Biosciences, the Pennsylvania State University, University Park, PA 16802, USA
Department of Biology, the Pennsylvania State University, University Park, PA 16802, USA
Department of Biochemistry and Molecular Biology, The Pennsylvania State University, University Park, PA 16802, USA

Chuankai Zhang and Yunda Zhang
Department of Gastrointestinal Surgery, First Affiliated Hospital of Xiamen University, Xiamen 361003, China
State Key Laboratory of Cellular Stress Biology, Innovation Center for Cell Signaling Network, School of Life Sciences, Xiamen University, Xiamen 361102, China

Yancheng Lin
State Key Laboratory of Cellular Stress Biology, Innovation Center for Cell Signaling Network, School of Life Sciences, Xiamen University, Xiamen 361102, China

Weiji Ding, Zhengjie Huang and Qi Luo
Department of Gastrointestinal Surgery, First Affiliated Hospital of Xiamen University, Xiamen 361003, China
Department of Gastrointestinal Surgery, First Clinical Medical College of Fujian Medical University, Fuzhou 350005, China

Donglu Wu
School of Life Sciences, Jilin University, Changchun 130012, China

Jingji Jin and Yong Cai
School of Life Sciences, Jilin University, Changchun 130012, China
National Engineering Laboratory for AIDS Vaccine, Jilin University, Changchun 130012, China
Key Laboratory for Molecular Enzymology and Engineering, the Ministry of Education, Jilin University, Changchun 130012, China

Index

O

P

R

S

T

U

W

www.ingramcontent.com/pod-product-compliance
Lightning Source LLC
LaVergne TN
LVHW081038210926
848485LV00004B/1054
* 9 7 8 1 6 3 9 8 7 3 6 8 5 *